CLONING AND EXPRESSION VECTORS FOR GENE FUNCTION ANALYSIS

CLONING AND EXPRESSION VECTORS FOR GENE FUNCTION ANALYSIS

Edited by

Quinn Lu, PhD
GlaxoSmithKline
King of Prussia, PA, USA

Michael P. Weiner, PhD
GlaxoSmithKline
Research Triangle Park, NC, USA

BioTechniques® Press
Eaton Publishing
Part of the Informa Life Sciences Group

Quinn Lu, PhD
Department of Gene Expression Sciences
Mail Code: UE0548
GlaxoSmithKline
709 Swedeland Road
King of Prussia, PA 19406

Michael P. Weiner, PhD
Department of Genomic Sciences
GlaxoSmithKline
Five Moore Drive
Research Triangle Park, NC 27709

Library of Congress Cataloging-in-Publication Data

Cloning and expression vectors for gene function analysis / edited by Quinn Lu, Michael P. Weiner.
p. cm. -- (BioTechniques bio focus series)
Includes bibliographical references and index.
ISBN 1-881299-21-X
1. Molecular cloning. 2. Genetic vectors. 3. Gene expression. I. Lu, Quinn, 1962- II. Weiner, Michael P., 1958- III. Series.

QH442.2 .C565 2001
572.8'6--dc21
2001018757

ISBN 1-881299-21-X

Printed in the United States of America

9 8 7 6 5 4 3

Eaton Publishing
P.O. Box 1070
Westborough, MA 01581-1070
www.BioTechniques.com

James D. Emerton: *Publisher*
Christine Emerton: *Manager, New Product Development*
Sandra Lamont: *Production Manager*
Robin Davis: *Marketing Projects Coordinator*

Contributors

Zach Adam
The Hebrew University of Jerusalem
Rehovot, Israel
(zach@agri.huji.ac.il)

M. Adams
Clontech Laboratories
Palo Alto, CA, USA

A.P. Alimov
Institute of Protein Research
Russian Academy of Sciences
Pushchino, Moscow Region, Russia

Kensey Amaya
New England Biolabs
Beverly, MA, USA

S.-i. Aota
National Institute of Dental Research
National Institutes of Health
Bethesda, MD, USA

Donna Armentano
Genzyme Corporation
Framingham, MA, USA

K. Benwell
Wyeth-Ayerst Research
Princeton, NJ, USA

Bill Brizzard
Sigma
St. Louis, MO, USA
(bbrizzard@sial.com)

Carsten Carstens
Stratagene
La Jolla, CA, USA

T.-Y. Chang
National Cheng Kung University
Tainan, Taiwan, ROC

A.B. Chetverin
Institute of Protein Research
Russian Academy of Sciences
Pushchino, Moscow Region, Russia
(alexch@vega.protres.ru)

Cheng-Ming Chiang
Case Western University
Cleveland, OH, USA
(c-chiang@biochemistry.cwru.edu)

Te-Hua Tearina Chu
University of Medicine and Dentistry of New Jersey
Piscataway, NJ, USA

Cheng-Ming Chuong
University of Southern California
Keck School of Medicine
Los Angeles, CA, USA

Mark I. Cockett
Bristol-Meyers Squibb
Princeton, NJ, USA
(mark.cockett@bms.com)

Robert E. Cohen
University of Iowa
Iowa City, IA, USA

J. Coote
GlaxoSmithKline
Ware, Herts, and Greenford
Middlesex, UK

Gina L. Costa
Stanford University
Palo Alto, CA, USA

David T. Curiel
University of Alabama
Birmingham, AL, USA

Ralph Dornburg
University of Medicine and Dentistry of New Jersey
Piscataway, NJ, USA

P.-L. Drèze
Université Libre de Bruxelles
Gosselies, Belgium

John J. Dunn
Brookhaven National Laboratory
Upton, NY, USA

Cora-Jean S. Edgell
University of North Carolina
Chapel Hill, NC, USA
(edgellcj@med.unc.edu)

David Engelberg
The Hebrew University of Jerusalem
Jerusalem, Israel
(engelber@vms.huji.ac.il)

Thomas C. Evans
New England Biolabs
Beverly, MA, USA

Katherine A. Felts
Stratagene Cloning Systems
La Jolla, CA, USA

F. Flamant
CNRS-ENS de Lyon
Lyon, France

R. Franco
Wyeth-Ayerst Research
Princeton, NJ, USA

P. Gabant
Université Libre de Bruxelles
Gosselies, Belgium
(pgabant@dbm.ulb.ac.be)

Alfredo Garzino-Demo
University of Maryland
Baltimore, MD, USA
(garzinod@umbi.umd.edu)

Christine Giovane
Ecole Supérieure de Biotechnologie de Strasbourg
Illkirch, France

Erica A. Golemis
Fox Chase Cancer Center
Philadelphia, PA, USA

S. Goodson
GlaxoSmithKline
Ware, Herts, and Greenford
Middlesex, UK

Mark M. Gosink
University of Wisconsin-Madison
Madison, WI, USA

S. Harris
GlaxoSmithKline
Ware, Herts, and Greenford
Middlesex, UK

Ron Hernan
Sigma
St. Louis, MO, USA

Ken Heuermann
Sigma
St. Louis, MO, USA

Rob Hooft van Huijsduijnen
Serono Pharmaceutical Research Institute
Geneva, Switzerland

Tanya Hosfield
Stratagene
La Jolla, CA, USA
(tanya_hosfield@stratagene.com)

Samuel Hou
Case Western University
Cleveland, OH, USA

Ping-Chuan Hu
University of North Carolina
Chapel Hill, NC, USA

Damon Huber
Iowa State University
Ames, IA, USA

Brigitte M. Jockusch
Cell Biology
Zoological Institute
Technical University of Braunschweig
Braunschweig, Germany

Steven R. Kain
Clontech Laboratories
Palo Alto, CA, USA

Niilo Kaldalu
Tartu University
Tartu, Estonia

Peter Kast
Laboratorium für Organische Chemie
Swiss Federal Institute of Technology (ETH)
Zürich, Switzerland
(kast@org.chem.ethz.ch)

Ben-Zion Katz
National Institute of Dental and Craniofacial Research
National Institutes of Health
Bethesda, MD, USA and
The Hematology Institute
Tel-Aviv Medical Center
Tel-Aviv, Israel
(sbkatz@netvision.net.il)

A.Yu. Khmelnitsky
Institute of Protein Research
Russian Academy of Sciences
Pushchino, Moscow Region, Russia

Paul A. Kitts
Clontech Laboratories
Palo Alto, CA, USA

A. Kondepudi
Clontech Laboratories
Palo Alto, CA, USA

Arnold Kristjuhan
Tartu University
Tartu, Estonia

Dmitry Krylov
National Cancer Institute
National Institutes of Health
Bethesda, MD, USA

Reet Kurg
Tartu University and
Estonian Biocentre
Tartu, Estonia

Avraham Laban
Jexys (Jerusalem Expression Systems) Ltd.
Jerusalem, Israel

Heike Launhardt
Hans-Knöll-Institut für Naturstoff-Forschung e.V.
Jena, Germany

Susan F. Law
Fox Chase Cancer Center
Philadelphia, PA, USA

Y. Le Dréan
CNRS
Université de Rennes 1
Rennes, France

C.-H. Lee
National Cheng Kung University
Tainan, Taiwan, ROC

C.-F. Lee
National Cheng Kung University
Tainan, Taiwan, ROC

M.G. Lee
GlaxoSmithKline
Ware, Herts, and Greenford
Middlesex, UK

Staci M. Leffel
University of North Carolina
Greensboro, NC, USA

Dina Lepik
Tartu University
Tartu, Estonia

Siew Pheng Lim
Institute of Molecular and Cell Biology
Singapore

Shi-Lung Lin
University of Southern California
Keck School of Medicine
Los Angeles, CA, USA

Hsai-Sheng Liu
National Cheng Kung University
Tainan, Taiwan, ROC
(A713@mail.ncku.edu.tw)

Quinn Lu
GlaxoSmithKline
King of Prussia, PA, USA
(quinn_2_lu@sbphrd.com)

Hannah M. Lui
University of Southern California
Keck School of Medicine
Los Angeles, CA, USA

Stephen A. Mabon
University of North Carolina
Greensboro, NC, USA

Gavin MacBeath
Center for Genomics Research
Harvard University
Cambridge, MA, USA

Nancy Maizels
Yale University School of Medicine
New Haven, CT, USA

Irit Marbach
The Hebrew University of Jerusalem
Jerusalem, Israel

J. Markwell
University of Nebraska-Lincoln
Lincoln, NE, USA

Henry S. Marr
University of North Carolina
Chapel Hill, NC, USA

Idali Martinez
University of Medicine and Dentistry of New Jersey
Piscataway, NJ, USA

Gerd Mayer
Hans-Knöll-Institut für Naturstoff-Forschung e.V.
Jena, Germany

Marilyn R. McLeod
The John P. Robarts Research Institute and University of Western Ontario
London, ON, Canada

John J. McNulty
Brookhaven National Laboratory
Upton, NY, USA

Denis Michel
CNRS
Université de Rennes 1
Rennes, France
(denis.michel@univ-rennes1.fr)

Avishai Mimran
The Hebrew University of Jerusalem
Jerusalem, Israel

Thomas Munder
Hans-Knöll-Institut für Naturstoff-Forschung e.V.
Jena, Germany
(tmunder@pmail.hki-jena.de)

R. Ochalski
Wyeth-Ayerst Research
Princeton, NJ, USA

M. Olive
National Cancer Institute
National Institutes of Health
Bethesda, MD, USA

Paul Olson
University of Medicine and Dentistry of New Jersey
Piscataway, NJ, USA

Oren Ostersetzer
The Hebrew University of Jerusalem
Rehovot, Israel

Sei-Kyoung Park
Iowa State University
Ames, IA, USA

Ranjan Patnaik
Genentech
South San Francisco, CA, USA

J.C. Pehrson
University of Nebraska-Lincoln
Lincoln, NE, USA

Francine B. Perler
New England Biolabs
Beverly, MA, USA
(perler@neb.com)

Gregory J. Phillips
Iowa State University
Ames, IA, USA
(gregory@iastate.edu)

Cecile M. Pickart
Johns Hopkins University
Baltimore, MD, USA
(cpickart@welchlink.welch.jhu.edu)

Tracey A. Pownder
ZymoGenetics
Seattle, WA, USA

Christopher K. Raymond
University of Washington
Seattle, WA, USA
(craymond@u.washington.edu)

Stephen Rees
GlaxoSmithKline
Stevenage, Herts, England, UK
(esr1353@glaxowellcome.co.uk)

Helene F. Rosenberg
National Institute of Allergy and Infectious Diseases
National Institutes of Health
Bethesda, MD, USA
(hr2k@nih.gov)

Martin Rothkegel
Cell Biology
Zoological Institute
Technical University of Braunschweig
Braunschweig, Germany
(m.rothkegel@tu-bs.de)

Manfred Rüdiger
Cell Biology
Zoological Institute
Technical University of Braunschweig
Braunschweig, Germany

G. Salbert
CNRS
Université de Rennes 1
Rennes, France

G. Sarath
University of Nebraska-Lincoln
Lincoln, NE, USA

Georges Schwalbach
Ecole Supérieure de Biotechnologie de Strasbourg
Illkirch, France

Steven D. Schwartzbach
University of Nebraska-Lincoln
Lincoln, NE, USA
(schwartz@biocomp.unl.edu)

Shannon L. Sexson
ZymoGenetics
Seattle, WA, USA

Pei-Yong Shi
Yale University School of Medicine
New Haven, CT, USA

Timothy G. Simcox
Stratagene
La Jolla, CA, USA

P.N. Simonenko
Institute of Protein Research
Russian Academy of Sciences
Pushchino, Moscow Region, Russia

Parisa Sinai
Clontech Laboratories
Palo Alto, CA, USA

Desmond J. Smith
UCLA School of Medicine
Los Angeles, CA, USA
(dsmith@mednet.ucla.edu)

T. Sourisseau
CNRS
Université de Rennes 1
Rennes, France

Maurice W. Southworth
New England Biolabs
Beverly, MA, USA

David W. Souza
Genzyme Corporation
Framingham, MA, USA
(david.souza@genzyme.com)

A.S. Spirin
Institute of Protein Research
Russian Academy of Sciences
Pushchino, Moscow Region, Russia

J. Stables
GlaxoSmithKline
Ware, Herts, and Greenford
Middlesex, UK

C. Neal Stewart, Jr.
University of North Carolina
Greensboro, NC, USA
(nstewart@uncg.edu)

Craig A. Strathdee
The John P. Robarts Research Institute and University of Western Ontario
London, ON, Canada
(cas@rri.on.ca)

I.-J. Su
National Cheng Kung University
Tainan, Taiwan, ROC

James R. Swartz
Stanford University
Stanford, CA, USA
(swartz@chemeng.stanford.edu)

C. Szpirer
Université Libre de Bruxelles
Gosselies, Belgium

J. Szpirer
Université Libre de Bruxelles
Gosselies, Belgium

Garabet Toby
Fox Chase Cancer Center
Philadelphia, PA, USA
(gg_toby@fccc.edu)

T. Michael Underhill
The John P. Robarts Research Institute and University of Western Ontario
London, ON, Canada

Mart Ustav
Tartu University and
Estonian Biocentre
Tartu, Estonia
(ustav@ebc.ee)

Peter Vaillancourt
Stratagene Cloning Systems
La Jolla, CA, USA
(peter_vaillancourt@stratagene.com)

T. Van Reeth
Université Libre de Bruxelles
Gosselies, Belgium

Richard D. Vierstra
University of Wisconsin-Madison
Madison, WI, USA
(vierstra@facstaff.wisc.edu)

Charles Vinson
National Cancer Institute
National Institutes of Health
Bethesda, MD, USA

W.W. Ward
Cook College
Rutgers University
New Brunswick, NJ, USA

J. Wardwell-Swanson
Wyeth-Ayerst Research
Princeton, NJ, USA

A. Weatherman
University of Nebraska-Lincoln
Lincoln, NE, USA

Alan M. Weiner
Yale University School of Medicine
New Haven, CT, USA
(weiner@biomed.med.yale.edu)

Michael P. Weiner
GlaxoSmithKline
Research Triangle Park, NC, USA
(mpweiner@aol.com)

Étienne Weiss
Ecole Supérieure de Biotechnologie de Strasbourg
Illkirch, France
(eweiss@esbs.u-strasbg.fr)

Shwu-Yuan Wu
Case Western University
Cleveland, OH, USA

Ming-Qun Xu
New England Biolabs
Beverly, MA, USA

Kenneth M. Yamada
National Institute of Dental and Craniofacial Research
National Institutes of Health
Bethesda, MD, USA

Te-Tuan Yang
Clontech Laboratories
Palo Alto, CA, USA

Shao-Yao Ying
University of Southern California
Keck School of Medicine
Los Angeles, CA, USA
(sying@hsc.usc.edu)

Jianxin You
Johns Hopkins University
Baltimore, MD, USA

Lin Zhang
Sigma
St. Louis, MO, USA

Chao-Feng Zheng
Stratagene
La Jolla, CA, USA

INTRODUCTION

The beginning of the new millennium is an exciting time for molecular biology and biomedical researchers. The song lyric "every new beginning comes from some other beginning's end" (Dan Wilson, Semisonic) is appropriate for this new era in our understanding of the human organism. The complete genomes of several model eukaryotes (for examples *Saccharomyces cerevisiae, Caenorhabditis elegans,* and *Drosophila melanogaster*) have just been published, and the first draft of the human genome has been made. A consortium of pharmaceutical and technology companies, along with the UK-based Wellcome Trust, have collaborated on a single nucleotide polymorphism (SNP) map of the human genome that will have major impact on disease gene discovery and pharmacogenomics. All of the above achievements mark the *new* beginning of the genome era in biology.

What is next? While the next steps could be simply defined as functional genomics, the scope of this field remains rather vague. It could vary from bioinformatic data mining and analysis, structural and functional analysis of DNA, RNA, proteins, and their interactions, to high-throughput expression profiling, signaling pathway profiling, and molecular diagnostics. The type of experiments biologists will do include everything they have been doing in the past, though at a larger scale or higher throughput, within a larger context, and most importantly, with a focus on the genome. Nonetheless, cloning and expression vectors are essential tools that are at the heart of almost all of the experimental approaches for functional genomics.

It is within this context that we boldly propose this new volume, which collects and updates many articles that have appeared in the last few years in the journal *BioTechniques*. The selected articles broadly cover both cloning and expression vectors for gene function analysis and have been placed into six sections. However, the decision into which division to place many articles was difficult, and many of the articles could have fit just as well into different sections. The six sections were based on a logical progression of experimental procedures one would take in the laboratory. Hence, the first section includes articles on cloning vectors and cloning strategies; the next section includes articles on protein expression and purification. The following sections become more specialized, and include articles on gene tagging, reporter gene technologies, special purpose vectors, and finally, in vitro protein expression technologies.

In Section I, six articles are selected, each of them representing an important method for DNA cloning and subcloning. While the power of PCR has revolutionized many areas of molecular biology, including cDNA cloning, it inevitably brings problems, such as the generation of mutations and hybrids. Examples of a few pitfalls associated with the use of PCR in cDNA cloning, and possible solutions, are discussed in Chapter 1. We strongly recommend the use of high-fidelity DNA polymerase systems in PCRs and to confirm PCR-amplified DNA by sequencing in cloning and subcloning applications. To secure full-length cloning of novel genes, we recommend the use of mRNA from independent tissue sources, extensive sequence alignment among independently RT-PCR amplified clones, and sequence alignment with orthologues and paralogues.

With the availability of first draft of the human genome, we predict changes will be made in how researchers gain access to genes of interest. Gene analysis programs will be coupled with empirical expressed sequence tag (EST) databases to derive better predictions of gene structure, and potentially, function. In addition, it is likely that repositories of essentially complete genomes will soon be made available (albeit with only some splice variants per gene) for the most important model organisms and man. New vector systems that enable either the in vivo or in vitro transfer of genes and gene fragments between differently functioning subvectors are now commercially available (for example, the Gateway™ and Echo™ cloning systems from Invitrogen). In the future, transfer of genes between these types of "homing" vectors and specialized "transfer" vectors will become easier.

With a defined gene in hand, researchers can start functional studies of the gene and encoded protein in vitro and in vivo. Purified protein is often desired for in vitro applications, such as antibody generation, physical and biochemical studies, or screening for inhibitors. One fundamental issue of protein production is the unpredictability for solubility, yield, and activity of a particular protein. Multiple expression and purification strategies may have to be tested for a given protein before an optimized scheme is obtained.

The second section in this volume concentrates on protein production in various expression systems, and on protein purification. A useful expression strategy is to co-express the target gene with rare codon tRNAs (see Chapter 8 for an example), protein chaperones, subunits, or cofactors. A new set of high-copy number *E. coli* vectors that are compatible with the commonly used expression vectors (i.e., pET, pGEX, and pACYC vectors) are described in Chapter 7, which will facilitate such a co-expression strategy. A set of mammalian expression vectors with different selectable markers, as described in Chapter 14, will facilitate the co-expression strategy in mammalian cells.

Gene fusion technologies are most frequently used for in vivo functional studies. Gene fusion vectors can be divided into gene-epitope tagging vectors (Section III) and reporter gene vectors (Sections IV). Gene-epitope tagging allows convenient detection, analysis, and purification of tagged partners expressed in cells. Of the papers selected in Section III, Chapter 17 describes a genome-wide approach for exon tagging, which is a powerful tool for gene discovery and analysis. Section IV includes six representative papers on the use of reporters for gene-protein functional analysis. The fluorescent proteins (GFP, YFP, and derivatives) and the *E. coli* LacZ are generally used for protein function and localization studies. The firefly luciferase (Luc), the *E. coli* CAT and LacZ, and the secreted alkaline phosphatase (SEAP) are often used in quantitative experiments for an analysis of promoter activities. Although not a part of this BioFocus volume, two of the more exciting developments in the molecular understanding of proteins and genes are the use of display technologies and molecular evolution. Display technologies, including phage display, *lac*I and ribosome display, can be used to generate millions of protein variants to test enzyme activity and substrate specificity. Molecular evolution is often used to look at many millions of enzyme variants and "evolve" out desirable characteristics. These new variants can be selected to have substantially different properties, including temperature stability, substrate specificity, and better enzyme kinetics.

In Section V, we included papers on a variety of specialized vectors for many uses—for example, vectors for inducible gene expression in eukaryotic cells (Chapters 29–31) and vectors for targeting protein domains to specific cellular compartments (Chapters 33 and 34). While viral vectors in gene delivery are not the focus of this book, we would like to bring attention to two articles on the generation of recombinant adenoviruses. In Chapters 38 and 39, the researchers describe two independent systems to produce recombinant adenoviruses by generating recombinant adenoviral DNA as a plasmid in *E. coli*, followed by transfection into a packaging cell line. Compared to the traditional method to select recombinant adenoviruses in mammalian cells, the new methods simplify the procedure significantly. As mentioned above, the use of vectors that facilitate the transfer of cloned genes (e.g., the Gateway and Echo-type cloning vectors) is becoming more popular for gene transfer among vectors. These vectors make it easier to transfer genes to specialized vector systems without having to go through a potentially mutagenic PCR step.

The final section of the book includes three papers that deal with in vitro expression technology. Traditionally, coupled transcription and translation is used to produce a small amount of labeled protein for confirmation of both protein and DNA interactions, protein-protein interactions, protein truncation analysis, and mutation analysis. With the advantage of being admissive to co-expression, in vitro modification of newly synthesized proteins (such as labeling with modified amino acids and incorporation of unnatural amino acids), together with its cell-free nature promises wide applications in gene function analysis and high-throughput drug discoveries. However, the protein yield is generally low, which may limit its application in protein production. Nonetheless, it is certainly a promising field that warrants further investigation, as it is likely that substantial improvements to expression levels will occur in the future.

New technology brings revolution. The genome sequencing projects were made possible by advances in DNA sequencing technology and bioinformatics. At this historic point of the beginning of the genome era, we believe that advances in technologies for gene function analysis and expression profiling analysis will greatly facilitate our understanding of genome function and our achievements in drug discovery and molecular diagnostics. Towards this end, we hope that the topics covered in this book will provide useful information for those both proficient and new in the field. Information about the availability of vectors and strains can be found in the original articles or the updates.

We would like to thank Stephen Weaver, Chance VanWinkle, and Christine McAndrews at Eaton Publishing for editorial management and assistance, Drs. Allan Shatzman and Christine Debouck at the former SmithKline Beecham, and Drs. Dan Burns, Mike Luther, and Allen Roses at the former Glaxo Wellcome, for their encouragement and support. MPW would like to thank Camille Solbrig and his sons, Alan and Evan. QL would like to thank Jenny Xu and his son Louis. We would like to dedicate this book to GlaxoSmithKline, which is the result of the merger between our two heritage companies, and to the researchers who dedicate their lives to making new medicines for unmet medical needs.

Quinn Lu, PhD
Michael P. Weiner, PhD
May 2001

Publisher's note: Figures originally published in color can be viewed at www.biotechniques.com

Contents

Section I

Cloning Vectors and Strategies

CHAPTER 1

PCR-Assisted cDNA Cloning: A Guided Tour of the Minefield

Rob Hooft van Huijsduijnen

Serono Pharmaceutical Research Institute, Geneva, Switzerland

(Reprinted from BioTechniques 24:390-392, 1998)

cDNA cloning based on partial or degenerate sequence information is an essential part of molecular biology. In the past few years, polymerase chain reaction (PCR)-assisted cloning techniques (random amplification of cDNA ends [RACE], reverse transcription [RT]-PCR) have been used extensively to determine the 5′ and 3′ ends of cDNA sequences and also (using degenerate oligonucleotides) to identify new members of emerging gene families. Nevertheless, PCR's unique character also engenders some specific problems (5). The case study described here is a cautionary tale that illustrates the most prominent of these. A detailed description of the techniques used is presented elsewhere (4).

Pitfall 1: Skewed Tissue Expression Information

Degenerate oligonucleotides were designed to amplify from B lymphocytes novel metalloprotease members with a disintegrin and metalloprotease domain (ADAMs) (6). Among the 35 fragments of approximately 200 bp with an open reading frame (ORF) obtained, one was a novel ADAM member (ADAM 20) and was further studied. Only much later (see Pitfall 4) did it become apparent that this novel cDNA is in fact expressed mainly in testis (Figure 2D). This is the only pitfall with a narrow silver lining, in that RT-PCR allows finding novel cDNAs even when starting from the wrong tissue. Unbiased gene expression data can be obtained using quantitative RT-PCR or RNase protection protocols.

Pitfall 2: RACE Tends to Yield Short Products

B-cell or placenta cDNA was linker-ligated, and 5′ and 3′ ends were amplified using a cDNA-specific primer from the novel 200-bp sequence and a linker oligonucleotide using the Marathon™ cDNA Synthesis Kit (Promega, Madison, WI, USA). Newly obtained sequence was then used to repeat the process with more distally 5′- or 3′-located oligonucleotides. Although each time the longest PCR products were selected for subcloning and sequencing, the PCR apparently strongly favors the shortest products the oligonucleotides allow to amplify. As shown in Figure 1, this method yielded fragments of 1 kbp at most per amplification.

One solution for this problem is to ensure that the starting RNA for the cDNA synthesis is full length (size fractionation may help). Another possibility is to serially dilute the linker-ligated cDNA samples for the PCR, so that truncated and full-length cDNA samples do not end up in the same reaction vessel.

Pitfall 3: Mutations

Under the best circumstances, the mutation rate of *Taq* DNA polymerase was reported to be around 10^{-5} (1–3). In this project, however, the mutation rate was at least an order of magnitude higher, at about 2–3 mutations per 400-bp readout (6×10^{-3}). In most cases, comparison of several cloned RACE products identified a clear majority of (probably) original vs. mutated sequences, but occasionally there was an approximately 50% split between two groups of variant sequences, presumably because a mutation had occurred early on in the amplification. These mutations were artifactual because they were not seen in multiple clones obtained from a λ library of testis cDNA (4). After the RACEs, the Titan™ High Fidelity PCR System (Boehringer Mannheim, Indianapolis, IN, USA) was used to amplify full-length cDNA; however, this procedure also failed to yield completely mutation-free products.

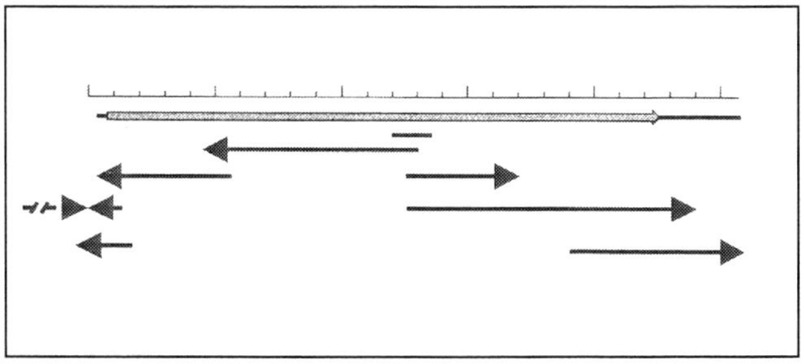

Figure 1. Organization of RACE products obtained (black arrows), starting from a small fragment (black line) of known sequence. The ORF of the corresponding cDNA (gray arrow; total length 2.4 kbp) and untranslated regions are indicated. Double arrow: hybrid extension product (Pitfall 5). The size bar (top) is divided into 100-bp units.

Cloning and Expression Vectors for Gene Function Analysis
Edited by Quinn Lu and Michael P. Weiner

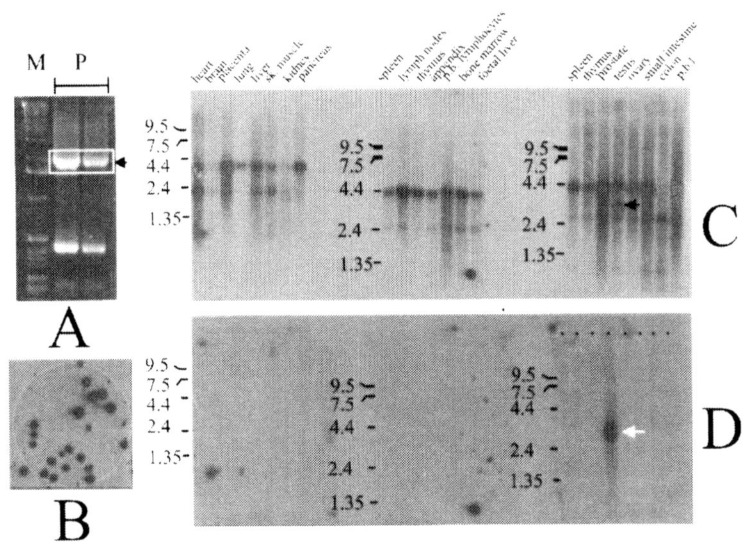

Figure 2. (A) Agarose RT-PCR product (P) of full-length ADAM 20 cDNA. The boxed area was excised for fragment isolation and probe preparation. Lane M: size markers (1-kb ladder; Life Technologies, Gaithersburg, MD, USA). (B) Second-round screening by hybridization of a positive λ colony using this full-length fragment from Panel A as a probe. (C) Multiple-tissue Northern blots probed with the same probe as for Panel B. (D) Same blots as in Panel C, probed with the RT-PCR product after subcloning and repurification.

Pitfall 4: Nonspecific Co-amplification

A full-length RT-PCR product was gel-purified (Figure 2A) and used as probe in multiple-tissue Northern blot hybridizations (Figure 2C) and for the screening of a placenta cDNA λ library (placenta tissue had shown a good Northern signal; Figure 2C). Figure 2B shows a filter from the secondary screening of a positive λ clone with strongly hybridizing plaques. Surprisingly, these positive phages subsequently failed to hybridize with a probe that was prepared from the same fragment after it had been subcloned into a plasmid. When the Northern blots were re-probed with this subcloned fragment, a unique, testis-specific signal was observed (Figure 2D). What apparently had happened was that the RT-PCR had co-amplified a ubiquitously expressed cDNA and that this product had co-migrated with the major ADAM band (Figure 2A). This contaminating fragment must have produced the ubiquitous signal in Figure 2C as well as the positive plaques shown in Figure 2B. The testis-specific band could also be seen in Figure 2C (arrow). Such a probe contamination cannot be eliminated by gel purification and another PCR. The solution here is to use a nested set of oligonucleotides.

Pitfall 5: Hybrid Products

One of the 5′ RACE products was identical to the 5′ sequence of expressed sequence tag (EST) cDNA Clone No. 231995 (Research Genetics, Huntsville, AL, USA). Its 5′ end was re-sequenced. This showed that the similarity to ADAM cDNA abruptly stops at some point and that the RACE product was a hybrid sequence of the EST and ADAM cDNAs. This amplification artifact probably arose because of chance sequence similarity between the two cDNAs.

Many of these pitfalls might have manifested themselves here prominently because the cDNA fragments were amplified from sources (lymphocytes, placenta) in which the ADAM mRNA was extremely rare (Figure 2D). This is probably the origin of the observed high mutation rate and the artifactual amplification of hybrid products (Pitfalls 3 and 5). The correct tissue distribution of the cDNA under study was obscured because of Pitfall 4. A clear lesson from this work is that full-length cDNA cloning must involve library screenings for as much as possible of the coding sequence and that RACE or RT-PCR protocols, if used at all, should be limited to obtain sufficient sequence for hybridizations or for the elucidation of the less critical untranslated regions.

It is well-known that library cDNA clones also often contain extraneous sequences, mostly in the form of un-spliced exons, but these are nearly always derived from the same gene. Usually, only a single clone is found per aberrant construct.

Subcloned, gel-purified DNA fragments are usually contaminated with the plasmid DNA from which they had been excised. When used as a probe, these contaminations often manifest themselves by their strong hybridization to the size markers. These contaminations are usually harmless in Northern blot hybridizations or library screenings (except when plasmid libraries are screened). However, RT-PCR-generated contaminations such as those described in Pitfall 4 are far more insidious and should be avoided at all times.

REFERENCES

1. **Cline, J., J.C. Braman and H.H. Hogrefe.** 1996. PCR fidelity of *Pfu* DNA polymerase and other thermostable DNA polymerases. Nucleic Acids Res. *24*:3546-3551.
2. **Eckert, K.A. and T.A. Kunkel.** 1990. High fidelity DNA synthesis by the *Thermus aquaticus* DNA polymerase. Nucleic Acids Res. *18*:3739-3744.
3. **Erlich, H.A., D. Gelfand and J.J. Sninsky.** 1991. Recent advances in the polymerase chain reaction. Science 252:1643-1651.
4. **Hooft van Huijsduijnen, R.** 1998. ADAM 20 and 21; two novel human testis-specific membrane metalloproteases with similarity to fertilin-α. Gene *206*:273-282.
5. **Kenzelmann, M. and K. Mühlemann.** 1997. Pitfalls of PCR: cross-reactivity with joyride *E. coli* nucleic acid. BioTechniques *23*:204-206.
6. **Wolfsberg, T.G., P.D. Straight, R.L. Gerena, A.P. Huovila, P. Primakoff, D.G. Myles and J.M. White.** 1995. ADAM, a widely distributed and developmentally regulated gene family encoding membrane proteins with a disintegrin and metalloprotease domain. Dev. Biol. *169*:378-383.

CHAPTER 2

Positive Selection Vectors to Generate Fused Genes for the Expression of His-Tagged Proteins

T. Van Reeth, P.-L. Drèze, J. Szpirer, C. Szpirer and P. Gabant

Université Libre de Bruxelles, Rhode-Saint-Genèse, Belgium

(Reprinted from BioTechniques 25:898-904, 1998)

ABSTRACT

Epitope tagging simplifies detection, characterization and purification of proteins. Gene fusion to combine the coding region of a well-characterized epitope with the coding region for a protein of interest generally requires several subcloning steps. Alternatively, a PCR strategy can be used to generate such a chimeric gene. In addition to its simplicity, this approach allows one to limit the size of the multiple cloning sites present in conventional expression vectors, thus reducing the introduction of artifactual amino-acid sequences into the fused protein. In this communication, we describe new vectors that allow PCR cloning and selection of chimeric genes coding for N- or C-terminal His-tagged proteins. These vectors are based on the control of cell death CcdB direct selection technology and are well adapted to the cloning of blunt-ended PCR products that were generated by using thermostable polymerases that provide proofreading activity.

INTRODUCTION

In vitro analysis of protein function (i.e., enzymatic assays, protein-protein and protein-DNA interactions) requires large amounts of purified polypeptide. Often, this is hindered by difficulties in purifying the expressed protein and/or the lack of specific antibodies. To circumvent these problems, a technique known as epitope tagging has been developed. This approach involves the construction of a recombinant fusion protein containing a Tag epitope attached to either the N or the C terminus of the protein of interest. The chimeric protein can then be easily identified in whole cell extracts and purified by immunoprecipitation using antibodies raised against the Tag epitope. Epitope tagging has been widely used for the study of newly isolated gene products in biological systems as diverse as *Escherichia coli*, yeast, insect cells and mammalian cells.

Several Tag epitopes have been used including: β-galactosidase, glutathione-S-transferase (GST) and peptides like c-myc or poly-histidine peptide (His-Tag). Some chimeric expression systems include a specific endopeptidase cleavage site (e.g., Enterokinase [EK], Thrombin or Xa factor) that allows one to clear the purified protein from its Tag epitope. Different genetic systems are now available to express tagged proteins (8,12). The His-Tag is a short region consisting generally of six histidine residues [6×(His)-Tag] with a high affinity for Cu_2^+, Ni_2^+, Co_2^+ or Zn_2^+, allowing the fusion product to be separated from a complex protein mixture by metal chelate affinity chromatography (6). The fusion proteins that bind to the affinity matrix are eluted with a solution containing either ammonium chloride, glycine, histidine, imidazol and EDTA or with high concentrations of competing metal ions. In contrast with long, fused Tag sequences that can interfere with the activity of the fused protein of interest, it can be expected that in most cases the short His-Tag epitope will not affect protein activities.

Complex strategies have to be used to fuse the Tag epitope coding sequence in-frame with the gene of interest. To simplify the generation of such fusions, expression systems containing three vectors presenting the His-Tag coding sequence in the different frames were developed. However, this approach is limited by the possibility that compatible restriction sites, both in the studied gene and/or in the multiple cloning site (MCS) of the vector, are present.

The fused proteins produced by these vectors generally contain additional amino acids (aa) because of the translation of vector sequences that might introduce artifacts in the final protein to be analyzed.

Alternatively, polymerase chain reaction (PCR) could be used to generate the fusion between the Tag coding region and the sequence of interest. However, current expression vectors are not well adapted to the cloning of PCR products. We have recently developed new direct-selection vectors adapted to the cloning of PCR-generated inserts (4). They are based on the insertional inactivation of the control of cell death *ccdB* gene (3) and were shown to be particularly efficient for the cloning of blunt-ended PCR products generated by thermostable DNA polymerases, that also provide proofreading activity (e.g., *Pfu* and Vent®). Commercial versions of these vectors (e.g., pCR®-Blunt) are now available in the Zero Blunt™ PCR Cloning Kit (Invitrogen, Carlsbad, CA, USA). We generated new vectors to use the *ccdB* technology for the expression of His-tagged pro-

Cloning and Expression Vectors for Gene Function Analysis
Edited by Quinn Lu and Michael P. Weiner

teins in *E. coli*. Their advantages are presented and illustrated in this communication.

Construction of pKILHIS-1

The fragment containing pT7-RBS-ATG-6×(His)-EK from pRSET-B (Invitrogen) was amplified by PCR and cloned (*Hind*III-*Eco*RV) into the *Hind*III-*Eco*RV fragment of pZERO-2 (Invitrogen). The terminator of pRSET-B (fragment *Xho*I-*Bgl*I) was cloned into the *Xho*I-*Bgl*I sites of this first construction. The fragment containing *lacZα-ccdB* (fragment *Eco*RV-*Eco*RI of pZERO-2) was then cloned in the *Eco*RV-*Eco*RI sites of the construction to generate the pKILHIS-1 plasmid. The sequence of pKILHIS-1 appears in the GenBank® database (Accession No. AF028837).

Construction of pKILHIS-2

The *Nhe*I-*Hind*III fragment from pKILHIS-1, containing EK*-*lacZα-ccdB* was cloned into the *Nhe*I-*Hind*III sites of

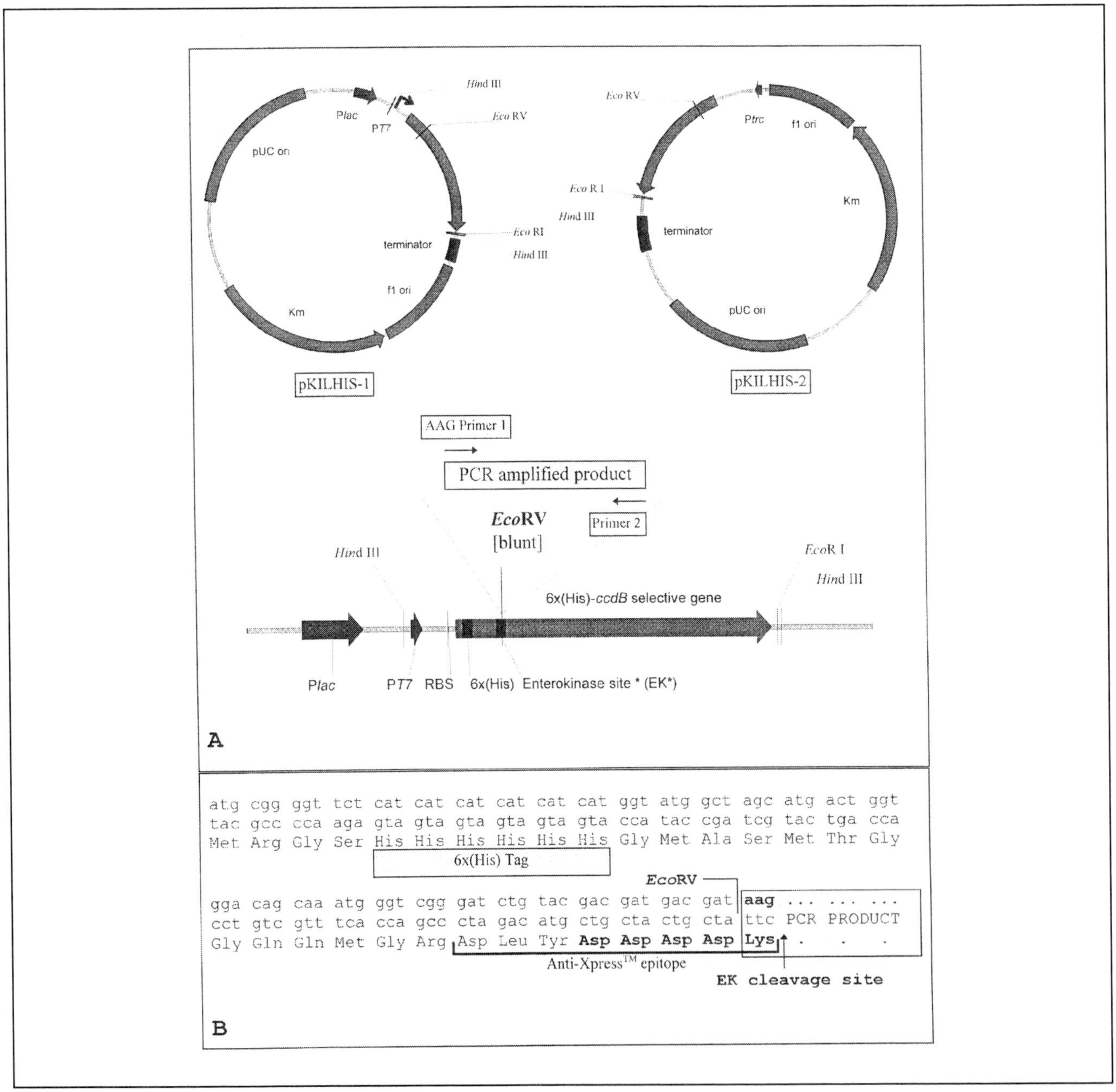

Figure 1. Structure of pKILHIS-1 and pKILHIS-2. The two plasmids contain a 6×(His)-*ccdB* fusion under the control of strong promoters (P*lac* and P*T7* for pKILHIS-1 and P*trc* for pKILHIS-2). (A) General map of pKILHIS-1 and pKILHIS-2, details of the different elements of the fused gene of pKILHIS-1. The *Eco*RV (blunt) site that can be used to clone the PCR-generated insert, the partial Enterokinase site (EK*) overlapping the *Eco*RV restriction site and the ribosome binding site (RBS) are indicated. Primers (primer 1 and 2) represent the oligonucleotides designed to produce the insert. A lysine (AAG) codon is added at the 5′ end of the primer 1 to generate a complete EK recognition coding sequence (see text). (B) Schematic representation of the insertion of a PCR-generated insert in the *Eco*RV cloning site of pKILHIS-1 vector, showing the EK cleavage site and the Anti-Xpress epitope generated by this insertion.

pTrcHisB (Invitrogen), and the fragment *Ssp*I-*Ssp*I of this new construction (pKILHIS-B) was cloned in the *Afl*III(blunted)-*Stu*I sites of pZERO-2 to generate the pKILHIS-2 plasmid. The sequence of pKILHIS-2 appears in the GenBank database (Accession No. AF028838).

N-Terminal His Tagging

Direct selection vectors pKILHIS-1 and pKILHIS-2 were genetically engineered to generate and express N-terminal His-tagged proteins.

Figure 1 shows the structure of the pKILHIS-1 and -2 vectors. These vectors differ from the original pKILPCR-2 plasmid (4) in two ways: *(i)* they contain a 6×(His)-Tag coding sequence introduced in the *ccdB* gene, and *(ii)* they contain strong prokaryotic promoters for the expression of the fused protein. No difference was observed between the killing activity of these derivatives and that of the parental pKILPCR-2 plasmid (5), showing that the 6×(His)-CcdB poison produced by these vectors is not less toxic than the selective marker of the parental construct. Amplified fragments can be cloned into the blunt-end *Eco*RV site of the plasmids. The two vectors have different promoters (Figure 1A; P*lac* and P*T7* for pKILHIS-1; P*trc* for pKILHIS-2) that are used to regulate the expression of the fused protein in *E. coli*. The 5′ primer (primer 1) is chosen in the gene of interest and designed in such a way that the amplified fragment is inserted in-frame with the 6×(His)-Tag coding sequence of both vectors. Because the reading frame is imposed by the 5′ primer, a single vector can be used to generate any N-terminal fusion with the 6×(His)-Tag epitope of these vectors. In its C-terminal end, the fusion is limited by either the natural stop codon present in the amplified gene or by a stop codon inserted in the 3′ primer.

It may be important for certain applications to remove the Tag epitope from the fused protein. Most systems are designed in such a way that an endopeptidase cleavage site allows clearance of the epitope from the fused protein, generally leaving artifactual aa at the N terminal of the cleared protein that, consequently, do not correspond to the natural one. This drawback can be overcome by using the vectors described and a PCR strategy that generates a peptidase site at the junction between the epitope and the polypeptide of interest. To limit the length of sequence added in the 5′ primer, we searched for restriction sites that generate blunt-ends and contain at least part of the coding sequence for a peptidase recognition site. EK and Xa factor are the most useful peptidases because they cleave at the carboxy-terminal side of their respective recognition sequence. The *Eco*RV site of pKILHIS-1 or -2 is part of the coding sequence of a partial EK cleavage site (Figure 1A, EK*), and a complete site is generated if a lysine codon (AAG) is included at the 5′ end of the 5′ primer (Figure 1B, primer 1).

To analyze the potential of fusion strategies using these vectors, the *E. coli* chloramphenicol acetyltransferase (*cat*) gene present in the pBLCAT2 plasmid (9) was amplified and cloned into pKILHIS-2. This amplification was performed using the *Pfu* DNA Polymerase (Catalog No. 600153; Stratagene, La Jolla, CA, USA) and the following two primers: 5′-AAGATGGAGAAAAAAATCACTGG-3′ (primer 1), in which the 5′ nucleotides (AAG) are added to generate the EK cleavage site after ligation, and 5′-TTACGCCCCGCCCTGCCACTC-3′ (primer 2), in which the three 5′ nucleotides (TTA) constitute the *cat* stop codon. The resulting 663-bp amplification product was directly cloned into *Eco*RV-digested pKILHIS-2. Recombinants were selected on kanamycin plates after transformation of the CcdB sensitive TOP10F′ Competent Cells (Invitrogen). One recombinant containing the *cat* gene fused to the 6×(His)-Tag of the pKILHIS-2 vector was chosen to over-express the fusion protein (see Figure 2).

The selected *E. coli* TOP10F′ strain (Invitrogen) harboring pKILHIS-2 containing the *cat* coding sequence was grown in 30 mL LB medium at 37°C to an optical density (OD) of 0.6. Expression was then induced by adding isopropyl-β-D-thiogalactopyranoside (IPTG) to a final concentration of 1 mM. The culture was grown for 2 h under induced conditions. After centrifugation at 4500× *g*, the bacterial pellet was suspended in 50 mM Tris-HCl, pH 7.5. The protein lysate was obtained by two freeze-thaw cycles. This crude protein extract was cleared by centrifugation and divided in two aliquots. One aliquot was treated with 3 µg of EK (Boehringer Mannheim, Mannheim, Germany) overnight at 37°C, and the other aliquot was used as a control (placed overnight at 37°C without treatment). For analysis, 5 µL of each sample were electrophoresed on a 12% sodium dodecyl sulfate (SDS) polyacrylamide gel, and the gel was stained with Coomassie® Blue. The protein samples were also blotted on a nitrocellulose membrane and revealed with the Xpress™ Antibody (Invitrogen), raised against the epitope containing the EK recognition motif.

The 30-kDa 6×(His)-CAT fusion protein we obtained was treated with an excess of EK to generate the wild type CAT protein (28 kDa). A Western blotting experiment using Anti Xpress Antibody (Invitrogen) raised against the Tag polypeptide was performed to confirm that the epitope had been cleaved from the fused CAT protein. To assess that the N-terminal end of CAT was restored in the cleaved polypeptide, the product obtained after EK digestion was sequenced. The correct N terminus was found (Figure 2B).

C-Terminal His Tagging

Direct selection expression vectors aimed at generating and expressing C-terminal, His-tagged proteins were constructed because the insertion of an epitope at the N terminus of the protein of interest could alter its function. Another vector, pKILHIS-3, has been designed and allows the generation of a fusion protein with an epitope at its C terminus. Insertion of the sequence of interest in the blunt-ended *Nru*I site of pKILHIS-3 generates a fusion protein with the 6×(His) epitope. As described in Figure 3, the cloning site of pKILHIS-3 is designed so that the insertion of a PCR product introducing a +2 frame shift not only disrupts CcdB but also imposes the translation of *(i)* an EK cleavage site, *(ii)* a 6×(His)-Tag and *(iii)* two stop codons limiting the tagged protein at its C terminus. Because EK has a recognition site in the C terminus, cleavage by EK of any fused protein expressed by this vector will leave 6 additional aa at the C terminus of the expressed protein.

We observed that the pKILHIS-3 plasmid is toxic for the CcdB-sensitive TOP10F′ strain even without IPTG induction of the P*trc* promoter (which controls the *ccdB* gene). This unexpected observation opens the possibility of selecting recombinants in the absence of P*trc* induction. This is a real advantage, especially if high-level expression of the studied protein is anticipated to be toxic.

To test the usefulness of pKILHIS-3 as a 6×(His)-Tag fusion expression vector, new primers to the *E. coli cat* gene were designed as follows and then used for amplifcation: 5′-ATGGAGAAAAAAATCACTGG-3′ (primer 1) and 5′-TCC-GCCCCGCCCTGCCACTC-3′ (primer 2). The 659-bp PCR product was inserted into *Nru*I-digested pKILHIS-3. The ligation reaction was transformed in TOP10F′. Selection in the presence or absence of IPTG led to more than 90% of recombinants. However, in recombinants selected without induction, inserts were found in both orientations; whereas in recombinants selected in the presence of the inducer (IPTG), only *cat* genes inserted in the opposite orientation to the P*trc* promoter were recovered. This observation suggests that high-level production of C-terminal-tagged CAT protein is toxic to the bacterial host. Moreover, recombinant plasmids selected without induction and containing the fused *cat* gene under the control of P*trc* were killed when grown in the presence of IPTG. The toxic effect of the C-terminal 6×(His)-CAT protein was surprising because no toxic effect was observed by induction of the N-terminal-tagged *cat* gene in pKILHIS-2. Because the toxicity could potentially be due to residual activity of *ccdB* in the pKILHIS-3-derived construct, the CcdB resistant strain B462 (1) was transformed either with pKILHIS-3 or with a recombinant containing the C-terminal 6×(His)-tagged *cat* coding sequence. In presence of IPTG, pKILHIS-3 carrying the fused *cat* gene was lethal, while no killing was observed with the vector. This strongly suggests that the C-terminal 6×(His)-CAT protein is toxic for *E. coli* when over-expressed.

To definitively show that the killing was caused by the 6×(His) C-terminal tagged CAT, the corresponding coding sequence was cloned in another expression vector, namely Xpress Vector pTrcHis B (Invitrogen). Bacteria harboring recombinant plasmids obtained from pTrcHis B were also killed by induction conditions. These observations demonstrate that high-level expression of the C-terminal tagged CAT protein is toxic for *E. coli* and illustrate the potential value of selecting for recombinants in non-induced conditions.

DISCUSSION

The *ccdB* gene of the F plasmid *ccd* operon has been shown to be an efficient marker for the selection of recombinant clones. The original vectors pKIL18–19 were designed to clone inserts generated by conventional restriction enzyme digestion (3). The positive selection system could be seen as a technology that allows one to avoid the vector dephosphorylation step and/or the cost of the 5-bromo-4-chloro-3-indolyl-β-D-galactopyranoside (X-gal) reagent used in *lac*Z inactivation screening. To our knowledge, the major advantage of the CcdB technology over other selection methods developed so far, is that this approach is based on a small gene whose function is known (*ccdB* encodes a natural gyrase poison) and that a resistant strain has been selected (2). The function of the CcdB gyrase poison and the conservation of its target in different bacterial species opens the door to a broader use of the *ccdB* gene in bacterial genetics (5).

Cloning of PCR-generated fragments is currently very popular; however, this approach precludes vector dephosphorylation and is rather inefficient. The possibility of selecting vectors with insertion singularly improves the power of PCR cloning, especially with the new generation of thermostable polymerases that generate blunt-ended products (4). Unlike the original *Taq* DNA polymerases, these proofreading polymerases generate blunt-ended products, thereby preventing the TA cloning used to limit vector self-ligation and to increase cloning efficiency of *Taq*-generated inserts (7,10,11).

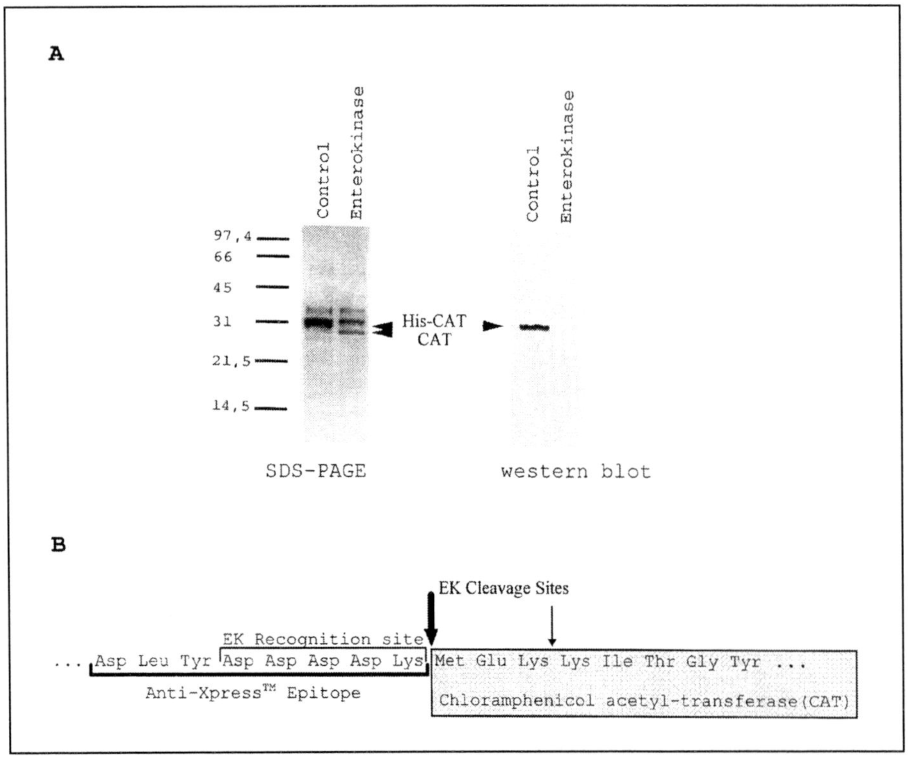

Figure 2. Analysis of the removal by EK digestion of the 6×(His)-Tag from the CAT protein expressed by pKILHIS-2. (A) SDS-polyacrylamide gel electrophoresis (PAGE) (left) and Western blot (right) of the fused CAT protein, before (control) and after EK treatment of the protein extract. As shown on the left, a 30-kDa polypeptide corresponding to the 6×(His)- CAT fused protein was detected in the control extract, and a smaller band (28 kDa) corresponding to the natural CAT protein was detected after EK treatment. As shown on the right, Anti-Xpress antibody gave a single band in the control and no signal after EK treatment of the extract, indicating that the epitope has been cleaved from the fusion protein. (B) To determine whether the cleavage had occurred at the EK site, the CAT protein obtained after epitope cleavage was extracted from the gel, and the N-terminal end of the protein was sequenced. As indicated by a bold arrow, 80% of the product was shown to correspond to a cleavage at the expected site, with 20% corresponding to a CAT internal secondary cleavage site (indicated by an arrow).

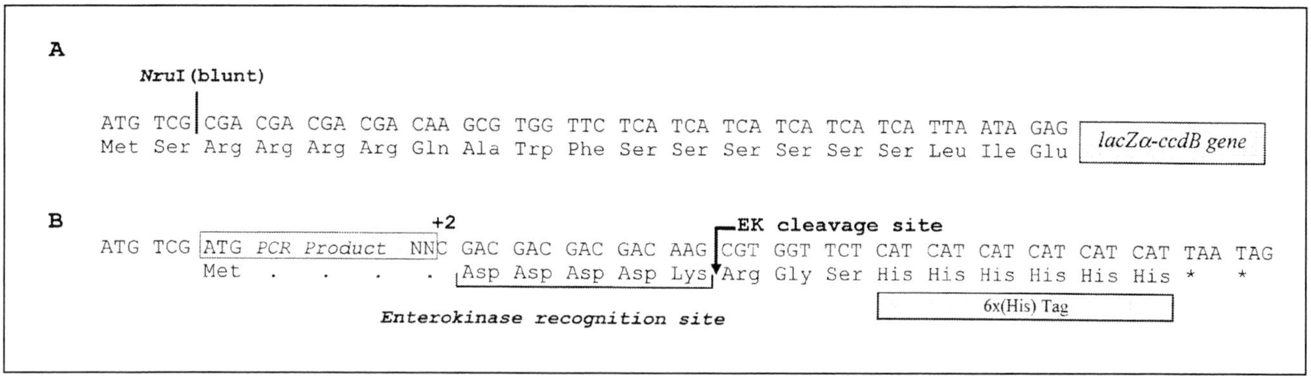

Figure 3. Structure of the pKILHIS-3 cloning site. pKILHIS-3 was generated by insertion in the *Nco*I-*Bam*HI sites of pKILHIS-B (for description see text) of the following sequence: 5′-CATGTCGCGACGACGACGACAAGCGTGGTTCTCATCATCATCATCATCATTAATAGAGG-3′. The sequence of pKILHIS-3 appears in the GenBank database (Accession No. AF0500464). (A) Sequence of the pKILHIS-3 cloning site. The 5′ end of the *lacZα-ccdB* sequence is indicated with its corresponding translation. The *Nru*I restriction site that can be used to clone the PCR-generated insert is indicated. (B) Schematic representation of the structure obtained after insertion of a PCR product in the *Nru*I site. The translation of the C-terminal peptide obtained after introduction of a +2 frameshift is indicated by the two additional bases added at the 5′ end of primer 2 are represented (NN). The EK recognition site, the EK cleavage site, the 6×(His)-Tag and the stop codons (*) generated by the insertion are also shown.

In this communication, we describe new expression vectors, combining the CcdB selection systems and His tagging. The major advantage of these new vectors is that they allow, in one cloning step, the generation of a chimeric gene generated by PCR. In addition, His tagging either at the N or C terminus is possible. The combination of PCR flexibility and selection of recombinants allowed us to propose the possibility of removing the N-terminus epitope from the chimeric protein to release the "authentic" polypeptide.

This new application demonstrates the versatility of the positive selection using *ccdB* gene derivatives.

The potential of this selection technique to generate fusion proteins opens the door to new eukaryotic expression vectors containing coding sequences for signal peptides to target the expressed protein to the extracellular medium or to specific cell compartments.

Positive selection relying on insertional inactivation of *ccdB* genes is filed for patent (2) protection and is licensed by the Université Libre de Bruxelles to Invitrogen Corporation.

ACKNOWLEDGMENTS

We are grateful to Michel Faelen and Laurence Van Melderen for critical review of the manuscript. This work was supported by CGER-Assurances, the Association contre le Cancer, the Communauté française de Belgique (ARC) and the Fund for Medical Scientific Research (FRSM). T.V.R. is a fellow of the Fonds pour la Recherche et l'Industrie Agroalimentaire (FRIA). C.S. is a Research Director of the Fonds National pour la Recherche Scientifique (FNRS).

REFERENCES

1. **Bernard, P.** 1996. Positive selection of recombinant DNA by CcdB. BioTechniques *21*:320-323.
2. **Bernard, P. and P. Gabant,** inventors. Université Libre de Bruxelles, assignee. Cloning and/or sequencing vector. PCT patent PCT/BE93/00051 and Belgian patent 09200696.
3. **Bernard, P., P. Gabant, El.M. Bahassi and M. Couturier.** 1994. Positive-selection vectors using the F plasmid CcdB killer gene. Gene *148*:71-74.
4. **Gabant, P., P.-L. Drèze, T. Van Reeth, J. Szpirer and C. Szpirer.** 1997. Bifunctional *lacZα-ccdB* genes for selective cloning of PCR products. BioTechniques *23*:938-941.
5. **Gabant, P., C.Y. Szpirer, M. Couturier and M. Faelen.** 1998. Direct selection cloning vectors adapted to the genetic analysis of Gram-negative bacteria and their plasmids. Gene *207*:87-92.
6. **Hochuli, E., W. Bannwarth, H. Döbeli, R. Gentz and D. Stüber.** 1988. Genetic approach to facilitate purification of recombinant proteins with a novel metal chelate adsorbent. Bio/Technology *6*:1321-1325.
7. **Invitrogen Corporation,** inventor. T/A cloning. US patent 5,487,993 and European patent 0550693.
8. **LaVallie, E.R., D.B. Smith and P. Riggs.** 1994. Introduction to expression by fusion protein vectors, p. 16.4.1-16.4.17. *In* F.M. Ausubel, R. Brent, R.E. Kingston, D.D. Moore, J.G. Seidman, J.A. Smith and K. Struhl (Eds.), Current Protocols in Molecular Biology. John Wiley & Sons, New York.
9. **Luckow, B. and G. Schütz.** 1987. CAT constructions with unique restriction sites for the functional analysis of eukaryotics promoters and regulatory elements. Nucleic Acids Res. *15*:5490.
10. **Marchuck, D., M. Drumm, A. Saulino and F.S. Collins.** 1991. Construction of T-vectors, a rapid and general system for direct cloning of unmodified PCR products. Nucleic Acids Res. *19*:1154.
11. **Mead, D.A., N.K. Pey, C. Herrnstadt, R.A. Marcil and L.M. Smith.** 1991. Universal method for the direct cloning of PCR amplified nucleic acid. Bio/Technology 9:657-663.
12. **Zhang, N. and J.-L. Chen.** 1998. Purification of recombinant proteins and study of protein interaction by epitope tagging, p. 10.15.1-10.15.9. *In* F.M. Ausubel, R. Brent, R.E. Kingston, D.D. Moore, J.G. Seidman, J.A. Smith and K. Struhl (Eds.), Current Protocols in Molecular Biology. John Wiley & Sons, New York.

Update to: Positive Selection Vectors to Generate Fused Genes for the Expression of His-Tagged Proteins

T. Van Reeth, P.-L. Drèze, J. Szpirer, C. Szpirer and P. Gabant

Université Libre de Bruxelles, Gosselies, Belgium

Plasmid poison-antidote loci (also termed postsegregational killing systems) contribute to low copy number plasmid stability in bacterial populations by killing daughter cells that have not inherited a plasmid copy. The *ccd* (control of cell death) operon of the *Escherichia coli* F plasmid is one of these poison-antidote systems. It contains two genes: *ccdB*, encoding a natural gyrase poison, and *ccdA*, encoding its antidote. *ccdB* is used as a selective marker in different cloning vectors, and its inactivation by insertion leads to a rapid and efficient selection for recombinants (1). Commercial versions of such vectors allowing a "Zero Background™" selection have been available since 1995 (pZerO®; Invitrogen, Carlsbad, CA, USA). As the insertion rate of blunt-ended fragments generated by polymerase chain reaction (PCR) using proofreading thermostable polymerases is rather low, new positive selection tools were derived for the cloning of such blunt-ended PCR fragments (pCR-Blunt®; Invitrogen). The combination of PCR flexibility and *ccdB* positive selection is illustrated by the generation of expression vectors presented in this paper. On the other hand, *ccdB*-based selection was also combined with TOPO®-cloning (pCR-blunt TOPO, pCR-XL-TOPO; Invitrogen), a ligase-independent cloning technology using topoisomerase I (5). Selection for recombinants is especially important when a shotgun approach is used for a sequencing project. It is essential to eliminate nonrecombinant vectors from the clones to be sequenced; a specific kit has been developed for this purpose (TOPO shotgun subcloning kit; Invitrogen). The main advantage of the small *ccdB* gene for the selection is that plasmids containing the toxic gene can be amplified in a CcdB-resistant host (*E. coli* strain containing a gyrase mutation, gyrA462). This strain allows the propagation of *ccdB* cloning vectors without any harmful effect on its growth rate. Consequently, no selective advantage is given to possible mutations in the *ccdB* gene that would generate a significant background at the step of selection.

More recently, the unique properties of *ccdB* have been applied to select recombinants in the Gateway® cloning system (Life Technologies, Rockville, MD, USA). In this new cloning approach, the different steps of restriction enzymes digestion, ligation and gel electrophoresis are replaced by an in vitro site-specific recombination step. This cloning "operating system" allows the parallel transfer of genes into or out of multiple expression vectors by simply mixing a donor plasmid (vector with the insert) with a destination plasmid (vector containing *ccdB*) in the presence of a recombinase mixture (clonase mixture). After transformation, only destination vectors containing the insert (plasmids without CcdB activity) will give rise to viable colonies. This recombinase-based high-throughput cloning system uses the unique properties of CcdB to obtain only recombinants after reaction.

Until very recently, CcdB was the only natural poison used to select recombinants. However, other poison-antidote systems exist in numerous natural plasmids (for reviews see References 2 and 4). To enrich and widen the range of selective approaches, we derived new vectors based on the postsegregational killing *parD* operon of the R1 plasmid (3). *parD*, like *ccd*, contains two major players (two small proteins), one of which being the toxin (Kid) and the other being its antidote (Kis). As mentioned above, the existence of a toxin-resistant strain is a prerequisite for the development of a robust selection system. We anticipated that bacterial strains expressing Kis would be the ideal host for *kid*-based vector (pKID) amplification. The pKID vectors were shown to be as efficient as the *ccdB*-based vectors for the selection of recombinants (3). Moreover, the observed independence of both selections and the engineering of a resistant strain to both systems will allow the development of cloning tools adapted to technologies requiring several cloning steps. This is the case for the design of targeting vectors (insertion of the two recombination arms) used to generate knock-out mice for mammalian functional genomic approaches. The development of such vectors is in progress in our laboratory.

Note: Since publication of the original article, patents have been issued and are broader in scope: Positive selection relying on inactivation of poison genes is patented (Bernard, P., P. Gabant, inventors, Université de Bruxelles, assignee, Cloning and/or sequencing vector. US patent 5,910,438 and European patent 0652963) and is currently licensed by the Université Libre de Bruxelles to Invitrogen Corporation.

REFERENCES

1. **Bernard, P., P. Gabant, El.-M. Bahassi and M. Couturier.** 1994. Positive-selection vectors using the F plasmid CcdB killer gene. Gene *148*:71-74.
2. **Couturier, M., El.-M. Bahassi and L.Van Melderen.** 1998. Bacterial death by DNA gyrase poisoning. Trends Microbiol. *6*:269-275.
3. **Gabant, P., T. Van Reeth, P.L. Dreze, M. Faelen, C. Szpirer and J. Szpirer.** 2000. New positive selection system based on the *parD* (*kis/kid*) system of the R1 plasmid. BioTechniques *28*:784-788.
4. **Jensen, R.B. and K. Gerdes.** 1995. Programmed cell death in bacteria: proteic plasmid stabilization systems. Mol. Microbiol. *17*:205-210.
5. **Shuman, S.** 1994. Novel approach to molecular cloning and polynucleotide synthesis using vaccinia DNA topoisomerase. J. Biol. Chem. *269*:32678-32684.

General Method for Plasmid Construction Using Homologous Recombination

Christopher K. Raymond, Tracey A. Pownder and Shannon L. Sexson

ZymoGenetics, Seattle, WA, USA

(Reprinted from BioTechniques 26:134-141, 1999)

ABSTRACT

We describe a general method for plasmid assembly that uses yeast and extends beyond yeast-specific research applications. This technology exploits the homologous recombination, double-stranded break repair pathway in Saccharomyces cerevisiae *to join DNA fragments. Synthetic, double-stranded "recombination linkers" were used to "subclone" a DNA fragment into a plasmid with >80% efficiency. Quantitative data on the influence of DNA concentration and overlap length on the efficiency of recombination are presented. Using a simple procedure, plasmids were shuttled from yeast into* Escherichia coli *for subsequent screening and large-scale plasmid preps. This simple method for plasmid construction has several advantages.* (i) *It bypasses the need for extensive PCR amplification and for purification, modification and/or ligation techniques routinely used for plasmid constructions.* (ii) *The method does not rely on available restriction sites, thus fragment and vector DNA can be joined within any DNA sequence. This enables the use of multifunctional cloning vectors for protein expression in mammalian cells, other yeast species,* E. coli *and other expression systems as discussed.* (iii) *Finally, the technology exploits yeast strains, plasmids and microbial techniques that are inexpensive and readily available.*

INTRODUCTION

The subcloning and manipulation of DNA fragments from one plasmid context to another is a central activity in molecular biology. In many cases, experimental direction dictates the precise fusion of one sequence to another, and cumbersome, multistep cloning procedures are required to obtain the desired constructs. The well-documented features of double-stranded (ds) break repair by homologous recombination in the yeast *Saccharomyces cerevisiae* offer unique opportunities for simplified plasmid construction and the precise joining of DNA fragments.

Plasmids routinely used in *S. cerevisiae* possess a selectable marker and sequences for replication and mitotic segregation. These regions of such plasmids are comprised of either an autonomously replicating sequence (ARS) and a centromere (CEN) or replication and partitioning sequences from the endogenous 2-μm plasmid. They are maintained with a circular topology, and linearized plasmids lacking telomeres fail to replicate. Moreover, when linear fragments that possess overlapping DNA sequences are co-transformed into yeast, they can undergo homologous mitotic recombination through the dominant ds break repair pathway (15). Hence, a linearized vector and fragment that shares DNA sequence overlap on either side of the plasmid break point (Figure 1A) can undergo recombination that restores the circular topology of the plasmid (7,12). Plasmid repair is manifest as a "positive transformant", and this provides a means to detect plasmid recombination events.

Several investigators have presented methods for plasmid assembly in yeast by co-transformation of overlapping DNA fragments and subsequent plasmid repair (6,9,13,14). Recent reports indicate that plasmid repair can be an extemely efficient process (12) and that recombination between short overlapping segments can be readily detected (6,10,13,14). In this report, we explore the influence of DNA concentration and overlap length on plasmid repair efficiency. We find that short, synthetic recombination linkers are efficient substrates for homologous recombination, and we demonstrate their use in plasmid assembly. Applications of this plasmid assembly technology are discussed.

MATERIALS AND METHODS

Strains and Methods

The yeast strain used in this study was SF838-9D (17), and the *Escherichia coli* strain was MC1061 (2). Yeast was transformed by electroporation (11). The numbers of uracil-prototrophic (Ura^+) transformants reported in Figures 1, 3 and 4 are the average of three separate transformation experiments.

Cloning and Expression Vectors for Gene Function Analysis
Edited by Quinn Lu and Michael P. Weiner

For routine plasmid constructions, an entire plate of yeast transformants was pooled by resuspension in water, and plasmid DNA was shuttled from yeast into *E. coli* as described (5).

Recombination Linkers and Plasmids

Figure 2 shows diagrams of recombination linkers that were synthesized from four oligonucleotides. Sense and antisense central oligonucleotides of approximately 60 nucleotides were designed to synthesize an initial segment of dsDNA by primer extension. This segment was then both extended and amplified by using sense and antisense primers. The final lengths of the linker segments were determined by the site of primer annealing and the length of the primer oligonucleotides. Overlaps between oligonucleotides were generally 20–25 bp in length with calculated melting temperatures in the range of 60°–70°C. For linker synthesis, 1 pmol of central oligonucleotides and 100 pmol of primers were amplified with *Pwo* DNA Polymerase (Boehringer Mannheim, Indianapolis, IN, USA) in a 100-µL reaction containing the manufacturer's supplied buffer and 0.25 mM dNTPs. Synthesis conditions were 10 cycles of amplification at 94°C for 30 s, 50°C for 60 s and 72°C for 60 s. The yield of linker was generally 2–4 µg/100 µL reaction, and linkers were ethanol-precipitated and resuspended at 1 µg/µL for use in transformations.

The donor fragment was a 450-bp *Eco*RI-*Xba*I fragment encoding human leptin (20); it was digested from a Bluescript® vector (Stratagene, La Jolla, CA, USA) and used in transformations (without purification) at a final concentration of 1 µg/µL of donor fragment. The acceptor vectors were all derived from the yeast CEN-ARS, *URA3* and Bluescript-derived vector pRS316 (19), and they were used in transformations at a concentration of 1 µg/µL or 100 ng/µL. The acceptor vector for Figure 1, pCZR185, carries a unique *Bam*HI site separating two 50-bp blocks of sequence identical to the 5′ and 3′ ends of the human leptin fragment, respectively. The acceptor vector for Figure 3, pCZR182, carries a subclone of the *S. cerevisiae* α-factor preprotein (prepro) coding region (8), and it was digested at *Hin*dIII and *Xho*I sites at the 3′ end of the coding sequence. The linkers used in Figure 3 span from the α-factor prepro coding region into pRS316 sequences. The acceptor vector used in Figure 4, pCZR213, is identical to pCZR182, except it carries a unique, blunt-end *Sna*BI site in place of the *Hin*dIII and *Xho*I sites. The linkers for Figure 4 overlap from α-factor prepro sequences into the 5′ end of the leptin donor fragment on one end and from the 3′ leptin coding sequence into the pRS316 sequences on the other. DNA fragments and linkers were quantified by gel electrophoresis and compared with standards of known concentrations.

RESULTS

Two-Part Plasmid Recombination

To quantify the influence of DNA concentration on plasmid repair efficiency, a linearized acceptor vector and donor fragment were co-transformed into yeast (Figure 1). The acceptor vector harbored sequences for selection and replication in *S. cerevisiae* (*URA3* and CEN-ARS) and *E. coli* [ampicillin-resistance (Amp^R) and colE1 ori]; it was derived from the Bluescript-based yeast vector pRS316 (19). It carried two 50-bp blocks of DNA sequence separated by a unique *Bam*HI site that, when digested, overlaps with the two ends of the donor fragment (Figure 1A). Plasmid repair was monitored by scoring the frequency of Ura$^+$ transformants (Figure 1B). DNA concentration was tested over a range from 0.1–10 µg. Tenfold increases in uncut, circular plasmid led to modest increases in transformation efficiency. Transformations with *Bam*HI-linearized vector alone yielded transformation efficiencies comparable with uncut DNA, particularly at higher concentrations of DNA. Plasmid repair by illegitimate end-joining of staggered ds breaks has been characterized previously (1,12). In all cases, the addition of donor fragment stimulated the efficiency of plasmid repair significantly. At higher concentrations of DNA, the transformation efficiencies with co-transformed acceptor vector and donor fragment exceeded the transformation frequency with equivalent amounts of vector DNA by as much as 20-fold. Characterization of recombinant plasmids by transfer into *E. coli* and polymerase chain reaction (PCR) screening/restriction digestion indicate that >90% of the recovered plasmids possess the anticipated structure shown in Figure 1A (data not shown).

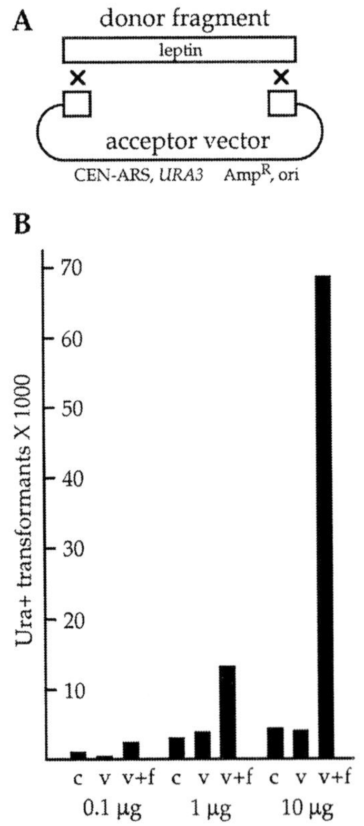

Figure 1. Two-part plasmid recombination. (A) The *Bam*HI linearized acceptor vector has sequences for selection and replication in yeast (*URA3* and *CEN-ARS*) and *E. coli* (Amp^R and *ori*), and it carries two 50-bp blocks of sequences that overlap with the 5′ and 3′ ends of the coding region of human leptin, respectively. The donor fragment includes the coding region of human leptin. (B) Transformation efficiencies with plasmid DNA that is circular (c), linearized vector (v) and linear acceptor vector rescued with an overlapping donor fragment (v+f). Equal quantities of plasmid and fragment DNAs were used at the indicated amounts.

The data in Figure 1 indicated that homologous recombination between short regions of DNA overlap can occur with high frequency. A second set of reactions was used to test whether short segments of dsDNA (60–140 bp; Figure 2) could serve as donor fragments for plasmid repair (Figure 3A). These reactions also examined the influence of acceptor/donor DNA overlap length on repair efficiency. Short, ds recombination linkers were synthesized by PCR from component oligonucleotides. Figure 3B shows donor fragments as short as 60 bp, with sequence overlap of 30 bp on either side of the ds break in the acceptor vector were capable of stimulating plasmid repair. Recombination linkers of 120 bp (60-bp overlaps) and 140 bp (70-bp overlaps) stimulated plasmid repair with identical eightfold efficiency over the control reaction.

Multi-Fragment Plasmid Recombination

The observation that short, synthetic linker segments stimulated plasmid repair suggested that they could be used to recombine donor fragments with acceptor vectors. In the simplest scheme, two synthetic recombination linkers that possess sequence overlap on one end with the acceptor vector and on the other end with the donor fragment are used to "subclone" DNA by recombination (Figure 4A). To test this, two synthetic 140-bp linkers, donor fragment and blunt-end linearized acceptor vector were co-transformed into yeast. Blunt-ended ds breaks are poor substrates for illegitimate end-joining repair (1), and vectors linearized in this manner were found to minimize the background of plasmids repaired by illegitimate end-joining reactions. This point is illustrated by the observation that 0.1 µg of circular pRS316 (19) yielded 1500 Ura$^+$ transformants, 0.1 µg of *Bam*HI-linearized pRS316 yielded 680 transformants and 0.1 µg of *Sma*I-linearized plasmid gave 160 Ura$^+$ colonies. Plasmid repair efficiency of a fixed amount of *Sna*BI (blunt end) linearized acceptor vector (0.1 µg) was measured for various concentrations of linkers and donor fragment (0.1 and 1.0 µg; Figure 4B). At the highest concentration of linkers and donor fragment, the recovery of Ura$^+$ yeast transformants was stimulated 40-fold relative to the cut plasmid control. Plasmids with the predicted linker-mediated subclone of the donor fragment into acceptor plasmid were recovered at high frequency from these yeast transformants. Total yeast DNA was prepared from a pool of approximately 500 Ura$^+$ transformants and used to transform *E. coli* that were then grown and selected for Amp^R. PCR analysis indicated that at least 34 of 40 (85%) colonies harbored the desired construct (data not shown).

The influence of DNA overlap length of the recombination linkers on the subcloning efficiency of the donor fragment into the acceptor vector was measured by monitoring recovery of Ura$^+$ transformants for linkers of various lengths (Figure 4C). Linkers as short as 60 bp, with 30 bp of overlap on either end with the donor fragment and acceptor vector, respectively, stimulated plasmid repair more than tenfold. Repair efficiencies increased with increasing linker/DNA overlap length as expected.

DISCUSSION

We have described a general method for plasmid assembly that exploits the homologous recombination-mediated DNA repair pathway in the yeast *S. cerevisiae*. The technology is

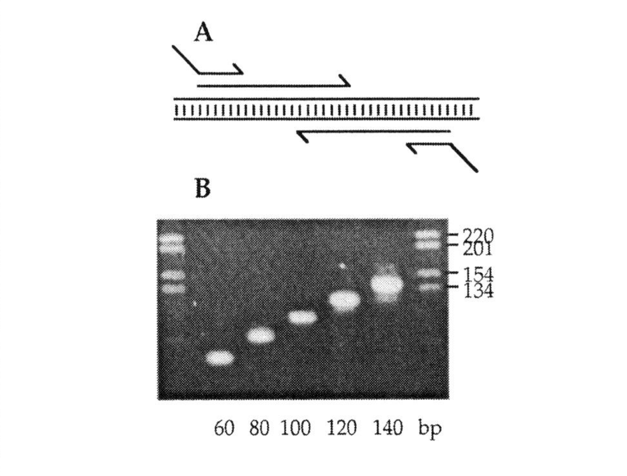

Figure 2. Synthesis of recombination linkers. (A) dsDNA linkers were synthesized from constituent oligonucleotides. Sense and antisense central oligonucleotides (straight lines) were used to create a duplex by primer extension. Sense and antisense primers (bent lines) then amplify the core duplex; the length of the primers and their site of annealing determined the final linker length. (B) Synthetic linkers of various sizes resolved on a 2.5% agarose gel.

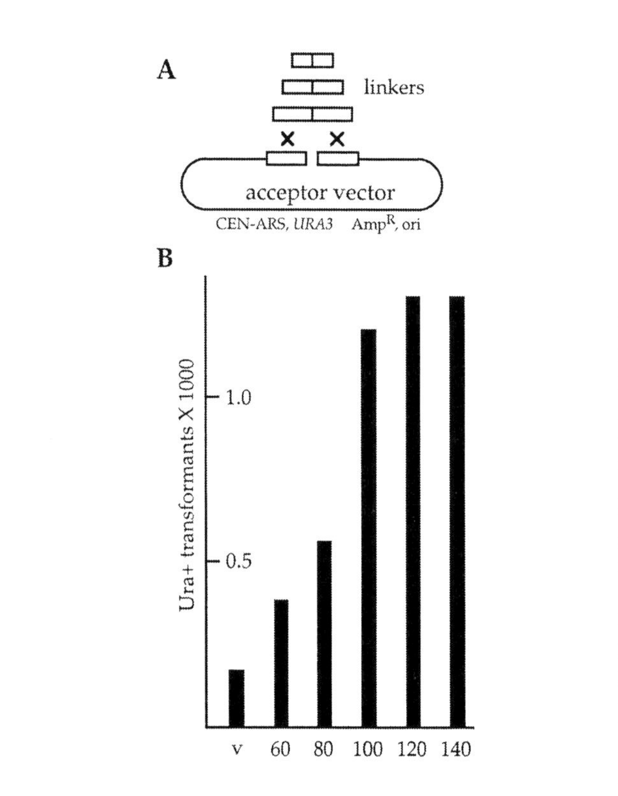

Figure 3. Recombination linker-mediated plasmid repair. (A) Linker-mediated plasmid repair was examined for linkers that overlap the acceptor vector on either side of the ds break. The length of linker overlap with the plasmid was varied, as shown. (B) Transformation efficiencies with 0.1 µg of linear vector DNA (v) and 0.1 µg of linear vector co-transformed with 1 µg of synthetic linker of the indicated lengths.

simple to use—DNA fragments to be joined are simply co-transformed into yeast. Therefore, as discussed below, the technology has generalized applications for plasmid assembly that extend beyond yeast-specific research. Importantly, the inherent flexibility in linker design permits the precise joining of two sequences without the constraint of available restriction sites. Three parameters were found to influence successful plasmid assembly: *(i)* The appropriate concentrations and ratios of acceptor vector (100 ng) to linkers and donor fragment (≥1 μg) were critical for the efficient assembly of the desired plasmid construct. In general, higher concentrations of linkers and donor fragments were found to favor recombination-mediated plasmid repair. *(ii)* The length of DNA overlap influenced repair efficiency, with overlaps ≥50 bp yielding the greatest number of recombinants. *(iii)* Background plasmid repair by illegitimate end-joining of the acceptor vector is minimized when the vector is linearized with a restriction enzyme that generates blunt ends. While electroporation was used as the method for yeast transformation in this study, plasmid recombination using the popular lithium acetate transformation method (18) has been described previously (10,12–14), and the technology described here seems likely to work with this transformation procedure as well.

Several methods have been described that exploit the homologous recombination pathway of *S. cerevisiae* for plasmid assembly (6,9,14). The technology most similar to that reported here involves the use of "tailed" PCR primers that extend an amplified segment of DNA with sequences that overlap a yeast cloning vector (6,14). The PCR product is then ligated into the vector by homologous recombination. The current study complements and extends this concept in several important respects. Synthetic recombination linkers are simple to design and synthesize, and we have rarely encountered difficulty synthesizing linker segments from constituent oligonucleotides (C. Raymond and T. Pownder, unpublished). In contrast, PCR amplification of target sequences with tailed PCR primers has been unsuccessful in some cases and has been impractical in other cases, particularly with long target sequences. Linker-mediated subcloning has the additional advantage that the target sequence is not PCR-amplified, thus DNA sequencing to confirm that the

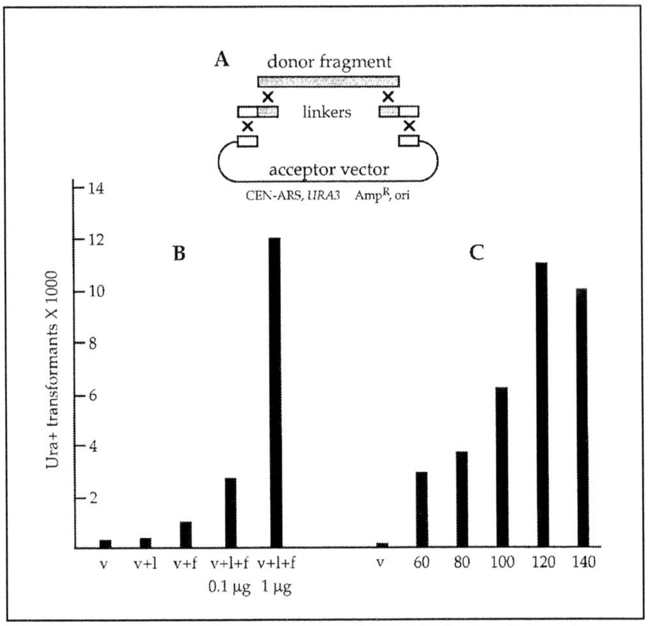

Figure 4. Linker-mediated subcloning. (A) Recombination linkers that overlap the acceptor vector on one end and the donor fragment on the other were used to subclone the fragment into the vector. (B) The quantities of DNA required for efficient linker-mediated plasmid assembly were determined. Transformation efficiencies were determined for controls that included 0.1 μg of linear acceptor vector (v), 0.1 μg of vector plus 1 μg of each linker (v+l) and 0.1 μg of vector plus 1 μg of donor fragment (v+f). The efficiencies of linker-mediated subcloning were measured with 0.1 μg of linear acceptor vector co-transformed with 0.1 or 1 μg of the two linkers and donor fragment (v+l+f), as indicated. (C) Transformation efficiencies of co-transformations that included 0.1 μg of acceptor vector alone (v) and 0.1 μg of acceptor vector, 1 μg of two linkers of the indicated lengths and 1 μg of donor fragment.

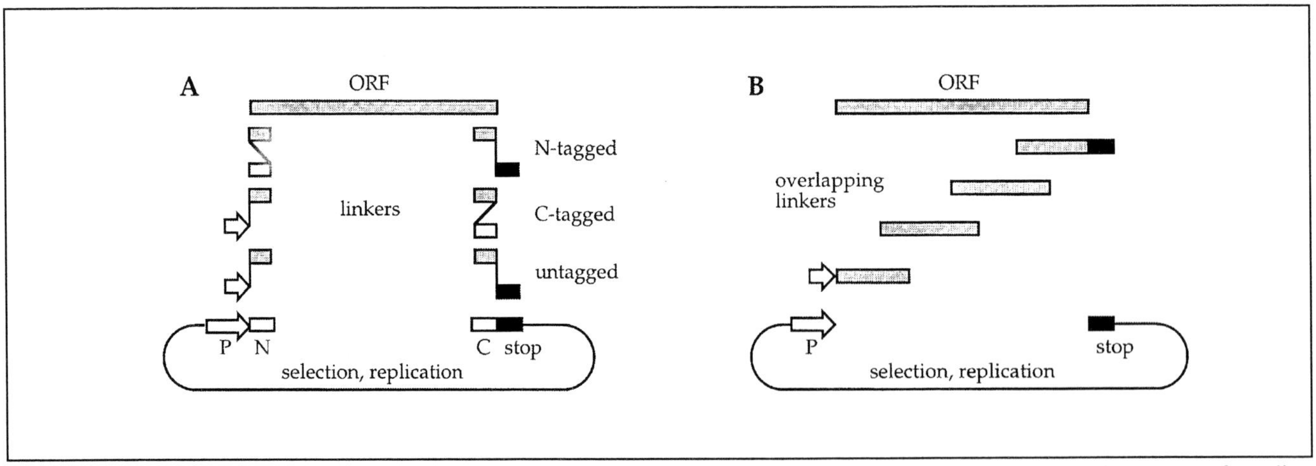

Figure 5. Applications of recombination linker-mediated plasmid assembly. (A) A multifunctional protein expression vector carries sequences for replication and selection in yeast, *E. coli* and the host expression organism. The vector carries a strong promoter (P), a translation-intiating site (ATG), a coding region for an amino-terminal epitope tag (N), a blunt-end restriction site, a coding region for a carboxy-terminal epitope tag (C) and appropriate vector stop signals for translation and transcription (stop). Linkers used to subclone the indicated ORF are represented by blocks of homology to the vector and fragment that are connected by a thin line. By appropriate linker design, the target ORF can be subcloned such that an amino-terminal tagged (N-tagged), a carboxy-terminal tagged (C-tagged) or untagged protein is expressed. (B) A series of linkers can be used to assemble a synthetic DNA sequence. In the example shown, overlapping linkers that encode the target ORF and that overlap the acceptor vector promoter region (P) and stops are used to assemble a synthetic (e.g., codon-optimized) protein coding region.

desired construct has been obtained can be confined to the junction regions between the vector and insert. Finally, synthetic linker segments can be used to introduce relatively long stretches of nucleotide sequence that do not exist in either the donor fragment or acceptor vector.

The use of recombination linkers has facilitated the design of multifunctional protein expression vectors for use in *E. coli*, the methylotrophic yeast *Pichia methanolica* (16) or mammalian cells (Figure 5A). Each vector is derived from pRS316 (19). We have found that the yeast-specific sequences do not interefere with protein expression levels in mammalian cells (M. Moore and C. Raymond, personal communication, unpublished). In addition, these plasmids harbor a marker for selection in the organism used for expression, a strong promoter, coding sequences for amino-terminal and carboxy-terminal epitope tags (e.g., poly-histidine or *myc;* Reference 4) and a blunt-end restriction site between the tags for linearization of the vector. By designing linkers that recombine within different sites in the vector, it is possible to generate expression constructs for a given open reading frame (ORF) that are N-terminally tagged, C-terminally tagged or untagged. We have also used a multiple set of recombination linkers to generate synthetic, codon-optimized genes (Figure 5B). As an example, a mutant form of green fluorescent protein (GFP) (3) was assembled into a pRS316-derived vector (19) from four 250-bp linkers that overlapped one another and/or the vector by 40 bp (T. Pownder and C. Raymond, unpublished). Fifty eight Ura^+ colonies were recovered, compared to 13 colonies observed on a control plate of vector alone. Yeast DNA prepared from the pooled 58 transformants was transferred into *E. coli*, and subsequent analysis indicated that 14 of 24 colonies harbored a plasmid of the correct structure. The DNA sequence of at least three of these clones yielded a GFP gene of the anticipated sequence. In summary, the method described is a simple and powerful tool to generate plasmids for a wide range of research applications.

ACKNOWLEDGMENTS

The authors would like to thank Andrzej Sledzieski, Betty Haldeman and Margaret Moore for constructive critical discourse, Virginia Mugford for critical reading of the manuscript and Will Lint and Nels Hamacher for synthesis of the oligonucleotides used in this study.

REFERENCES

1. **Boulton, S. and S. Jackson.** 1996. Identification of a *Saccharomyces cerevisiae* Ku80 homologue: roles in DNA double strand break joining and in telomeric maintenance. Nucleic Acids Res. *24*:4639-4648.
2. **Casadaban, M. and S. Cohen.** 1980. Analysis of gene control signals by DNA fusion and cloning in *Escherichia coli*. J. Mol. Biol. *138*:179-207.
3. **Cormack, B., R. Valdivia and S. Falkow.** 1996. FACS-optimized mutants of the green fluorescent protein (GFP). Gene *173*:33-38.
4. **Evan, G., G. Lewis, G. Ramsay and J. Bishop.** 1985. Isolation of monoclonal antibodies specific for human c-myc proto-oncogene product. Mol. Cell. Biol. *5*:3610-3616.
5. **Hoffman, C. and F. Winston.** 1987. A ten-minute DNA preparation from yeast efficiently releases autonomous plasmids for transformation of *Escherichia coli*. Gene *57*:267-272.
6. **Hudson, J., Jr., E. Dawson, K. Rushing, C. Jackson, D. Lockshon, D. Conover, C. Lanciault, J. Harris, S. Simmons, R. Rothstein and S. Fields.** 1997. The complete set of predicted genes from *Saccharomyces cerevisiae* in a readily usable form. Genome Res. *7*:1169-1173.
7. **Kunes, S., H. Ma, K. Overbye, M. Fox and D. Botstein.** 1987. Fine structure recombinational analysis of cloned genes using yeast transformation. Genetics *115*:73-81.
8. **Kurjan, J. and I. Herskowitz.** 1982. Structure of a yeast pheromone gene (*Mf*α): a putative α-factor precursor contains four tandem copies of mature α-factor. Cell *39*:933-943.
9. **Ma, H., S. Kunes, P. Schatz and D. Botstein.** 1987. Plasmid construction by homologous recombination in yeast. Gene *58*:201-216.
10. **Marykwas, D. and S. Passmore.** 1995. Mapping by multifragment cloning in vivo. Proc. Natl. Acad. Sci. USA *92*:11701-11705.
11. **Meilhoc, E., J.-M. Masson and J. Teisse.** 1990. High efficiency transformation of intact yeast cells by electric field pulses. Bio/Technology *8*:223-227.
12. **Mezard, C. and A. Nicolas.** 1994. Homologous, homeologous, and illegitimate repair of double-strand breaks during transformation of a wild-type strain and a *rad*52 mutant strain of *Saccharomyces cerevisiae*. Mol. Cel. Biol. *14*:1278-1292.
13. **Mezard, C., D. Pompon and A. Nicolas.** 1992. Recombination between similar but not identical DNA sequences during yeast transformation occurs within short stretches of identity. Cell *70*:659-670.
14. **Oldenburg, K., K. Vo, S. Michaelis and C. Paddon.** 1997. Recombination-mediated PCR-directed plasmid construction in vivo in yeast. Nucleic Acids Res. *25*:451-452.
15. **Orr-Weaver, T. and R. Szostak.** 1985. Fungal recombination. Microbiol. Rev. *49*:33-58.
16. **Raymond, C., T. Bukowski, S. Holderman, A. Ching, E. Vanaja and M. Stamm.** 1998. Development of the methylotrophic yeast *Pichia methanolica* for the expression of the 65 kilodalton isoform of human glutamate decarboxylase. Yeast *14*:11-23.
17. **Rothman, J., I. Howald and T. Stevens.** 1989. Characterization of genes required for protein sorting and vacuolar function in the yeast *Saccharomyces cerevisiae*. EMBO J. *8*:2057-2065.
18. **Schiestl, R. and R. Gietz.** 1989. High efficiency transformation of intact yeast cells using single stranded nucleic acids as a carrier. Curr. Genet. *16*:339-346.
19. **Sikorski, R. and P. Hieter.** 1989. A system of shuttle vectors and yeast host strains designed for efficient manipulation of DNA in *Saccharomyces cerevisae*. Genetics *122*:19-27.
20. **Zhang, Y., R. Proenca, M. Maffei, M. Barone, L. Leopold and J. Friedman.** 1994. Positional cloning of the mouse obese gene and its human homologue. Nature *372*:425-427.

Update to: General Method for Plasmid Construction Using Homologous Recombination

Christopher K. Raymond

University of Washington, Seattle, WA, USA

SIMPLE SEQUENCE-SPECIFIC CLONING USING YEAST

The unique strength of plasmid assembly using the yeast *Saccharomyces cerevisiae* is that DNA fragments can be joined at precise user-defined sequence junctions without the constraint of available restriction enzyme sites. The technolo-

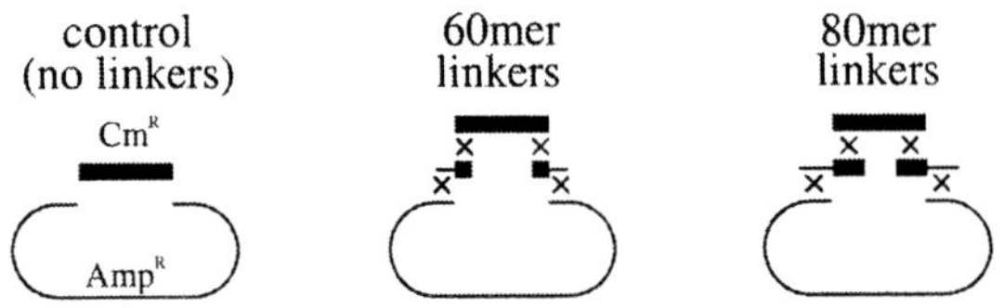

	control (no linkers)	60mer linkers	80mer linkers
	Ura⁺ yeast transformants		
Exp1	300	2300	16,500
Exp2	740	3900	56,000
	% Cm^R (Cm^R/Amp^R)		
Exp1	1.6% (3/185)	72% (131/181)	97% (176/181)
Exp2	1.4% (4/282)	66% (146/222)	98% (99/101)

Figure 1. Linker-mediated plasmid assembly. The data are from two independent transformation experiments, Exp1 and Exp2.

gy is simple. Plasmids are created by cotransforming overlapping DNA fragments into yeast. It is robust. Plasmid assemblies involving five or more overlapping DNA fragments are routine. Finally, plasmid assembly in yeast is efficient. Desired plasmid constructs are generally recovered over background with greater than 90% efficiency.

In this update, I describe a simple method to generate recombination linkers. Linker-mediated recombination refers to a technique where two nonoverlapping DNA fragments are joined by short, synthetic, double-stranded DNA "linkers" (Figure 1 and in original paper see Figure 4). This technology is useful when the generation of overlaps using polymerase chain reaction (PCR) (4) is either impractical or unsuccessful. In the experiments described here, the popular lithium acetate yeast transformation method (1) was used to assemble plasmids. A simple assay was used to determine that cloning efficiency using this technology was greater than 95%.

Previous work (see original paper) suggested that recombination linkers as short as 60–80 bp supported plasmid assembly. In principle, linkers of this length should be simple to assemble by annealing complementary oligonucleotides. To test this, a set of four 60-mer or 80-mer oligonucleotides were synthesized (custom primers; Life Technologies, Rockville, MD, USA). These four oligonucleotides encoded the sense and antisense strands for the linker to the left and the sense and antisense strands for the linker to the right of the fragment to be cloned. In this 'proof of principle' experiment, the 'acceptor vector' was the URA3-marked ampicillin resistant (Amp^R) yeast/*E. coli* shuttle plasmid pRS316 (5) linearized at the unique *Sma*I restriction site. The 'donor fragment' was the chloramphenicol resistance gene (Cm^R) amplified as a PCR fragment from pBACe3.6 (2). The oligonucleotides were designed to provide sequence overlaps of 30 or 40 bp with the vector on one end and the fragment on the other for the annealed 60 or 80 bp linkers, respectively.

The four 60-mer or four 80-mer oligonucleotides encoding the left and right recombination linkers were mixed in TE (10 mM Tris, pH 8.0, 1 mM EDTA) at room temperature and examined by gel electrophoresis using a 4% MetaPhor® agarose gel (BMA Bioproducts, Rockland, ME, USA). Surprisingly, the samples migrated as discrete 60 or 80 bp bands, suggesting that the double-stranded linkers spontaneously assembled under these conditions. Control samples of single or mismatched oligonucleotides migrated as smaller, more diffuse entities. As described below, these self-assembled linkers promoted recombination of the donor fragment into the acceptor vector, indicating that linker assembly can be accomplished by simple mixing of the constituent oligonucleotides.

To test the efficiency of linker-mediated plasmid assembly, 100 ng of pRS316 (*Sma*I) acceptor vector, 1000 ng of Cm^R PCR donor fragment and 1000 ng of each double-stranded linker (25 pmol of 60 mers or 20 pmol of 80 mers) were added to competent yeast cells that were then transformed by the lithium acetate transformation method (Reference 1 and Figure 1). Transformants were quantified as Ura+ colonies. The cloning efficiency of the Cm^R marker into the Amp^R vector was measured directly, by transferring plasmid DNA from pooled yeast transformants into *E. coli* (3), selecting for Amp^R colonies. These were replica-plated to chloramphenicol-containing media, and the frequency of Amp^R colonies that were Cm^R was quantified. These experiments demonstrated that 80-mer recombination linkers were very efficient in promoting plasmid assembly, while 60-mer linkers were somewhat less effective.

The technology described here presents an extremely simple method to subclone DNA at precisely defined sequence junctions. It involves the simple mixing and cotransformation of vector, fragment and an oligonucleotide mixture that specifies the sequence junctions between the two. Given that the assembly of DNA constructs is a core activity in many different avenues of molecular biology, it is anticipated that this technology could find utility in a variety of diverse applications.

The author would like to thank Maynard Olson, in whose laboratory this work was conducted. This work was supported by National Institutes of Health Grant No. HG02043.

REFERENCES

1. **Agatep, R., R.D. Kirkpatrick, D.L. Parchaliuk, R.A. Woods and R.D. Gietz.** 1998. Transformation of *Saccharomyces cerevisiae* by the lithium acetate/single-stranded carrier DNA/polyethylene glycol (LiAc/ss-DNA/PEG) protocol. Technical Tips Online (http://tto.trends.com).
2. **Frengen, E., D. Weichenhan, B. Zhao, K. Osoegawa, M. van Geel and P. de Jong.** 1999. A modular, positive selection bacterial artificial chromosome vector with multiple cloning sites. Genomics *58*:250-253.
3. **Hoffman, C. and F. Winston.** (1987) A ten-minute DNA preparation from yeast efficiently releases autonomous plasmids for transformation of *Escherichia coli*. Gene *57*:267-272.
4. **Oldenburg, K., K. Vo, S. Michaelis and C. Paddon.** 1997. Recombination-mediated PCR-directed plasmid construction in vivo in yeast. Nucleic Acids Res. *25*:451-452.
5. **Sikorski, R. and P. Hieter.** 1989. A system of shuttle vectors and yeast host strains designed for efficient manipulation of DNA in *Saccharomyces cerevisisae*. Genetics *122*:19-27.

CHAPTER 4

Generation of Full-Length cDNA Library from Single Human Prostate Cancer Cells

Shao-Yao Ying, Hannah M. Lui, Shi-Lung Lin and Cheng-Ming Chuong

University of Southern California School of Medicine, Los Angeles, CA, USA

(Reprinted from BioTechniques 27:410-414, 1999)

Several existing methods to construct cDNA libraries require several thousand cells and involve the lengthy procedure of reverse transcription (RT), restriction, adaptor ligation and vector cloning (5,6), which usually fail to maintain the completeness of a cDNA library, resulting in a loss of rare c-DNAs. However, gene expression analysis of specific cell populations within a heterogeneous tissue is essential for research in vivo, requiring a method to generate cDNA libraries from a very small number of specific cells. The generation of amplified antisense RNA (aRNA) by incorporating an oligo(dT) primer coupled to a T7 RNA polymerase promoter sequence during RT has been developed to increase transcriptional copies of mRNAs from a limited amount of promoter-linked cDNAs (2,4). However, the aRNAs prepared from a single live neuron has been reported to cover 50%–75% of the total mRNA population (1,2), indicating that rare mRNAs were not protected during the amplification procedure.

Here, we present a novel method for generating cDNA libraries that combines in situ RT (3) with amplified aRNAs followed by polymerase chain reaction (PCR) amplification. To protect mRNAs from degradation, the initial steps of in situ RT, 3′ oligonucleotide tailing and aRNA amplification are performed on fixed and permeabilized cells. Subsequent RT-PCR generates full-length cDNA libraries in a quantity easily manipulatable for gene expression analysis; the resultant cDNAs are amplified to total more than 10^9 the amount of original mRNAs.

The method can be broken down into six steps (Figure 1, a–f): (a) cell fixation and permeabilization, (b) first RT, (c) terminal transferase tailing and cDNA double-stranding, (d) in-cell transcription, (e) second RT and (f) PCR amplification.

Cell fixation and permeabilization (3). We have used LNCaP cells, cultured in RPMI 1640 medium (Irvine Scientific, Santa Ana, CA, USA) supplemented with 2% fetal calf serum, which were trypsinized and suspended in 1 mL of ice-cold 10% formaldehyde solution in 0.15 M NaCl. After a 1-h incubation on ice with occasional agitation, the cells were centrifuged at 11 000× *g* for 2 min and washed three times in ice-cold phosphate-buffered saline (PBS) with vigorous pipetting. The collected cells were permeabilized in 0.5% Nonidet® P-40 (NP40; BDH Ltd., Poole, Dorset, England, UK) for 1 h with frequent agitation. After three washes in ice-cold PBS containing 0.1 M glycine, the cells were resuspended in 1 mL of the same buffer, aliquotted and stored at -70°C for up to a month.

First RT. For RT of mRNAs in cells, twenty fixed cells were picked up under microscopy and suspended in 20 µL of ice-cooled diethyl pyrocarbonate (DEPC)-treated double-distilled (dd)H_2O. An RT reaction (50 µL) was prepared, containing the above cold cells, 5 µL of 10× RT buffer [1.2 M KCl, 0.5 M Tris-HCl, 80 mM $MgCl_2$, 50 mM dithiothreitol (DTT), pH 8.1 at 42°C], 25 pmol oligo(dT)-T7 primer (2), 0.2 mM of each dNTP and 80 U RNase inhibitor. After 6 U *C. therm.* Polymerase with Reverse Transcriptase Activity (Roche Molecular Biochemicals, Indianapolis, IN, USA) were added, the RT reaction was vortex mixed and incubated at 55°C for 2 min and then 68°C for 1 h. The cells were washed once with PBS and collected with an Amicon® Microcon-50 Microconcentrator (Millipore, Bedford, MA, USA).

Terminal transferase tailing and cDNA double-stranding. To increase the intracellular copies of whole mRNAs, the T7 promoter region of the poly(dA)-tailed cDNAs served as an initial template for transcriptional amplification by T7 RNA polymerase (Roche Molecular Biochemicals) (2). The cells collected from the first RT were suspended in a 50-µL tailing reaction solution, containing 0.2 mM dATP and 10 µL of 5× tailing buffer (Roche Molecular Biochemicals). The mixture was denatured at 94°C for 3 min and then chilled on ice immediately for mixing with 20 U Terminal Transferase (Roche Molecular Biochemicals), followed by further incubation at 37°C for 20 min. The reaction was stopped by warming to 94°C for 3 min, mixed with poly(dT) 26-mer primer and chilled on ice immediately; this formed poly(dA)-tailed cDNAs. The double-stranding of the poly(dA)-tailed cDNAs was completed by adding 0.2 mM of each dNTP, 1.5 mM

Cloning and Expression Vectors for Gene Function Analysis
Edited by Quinn Lu and Michael P. Weiner

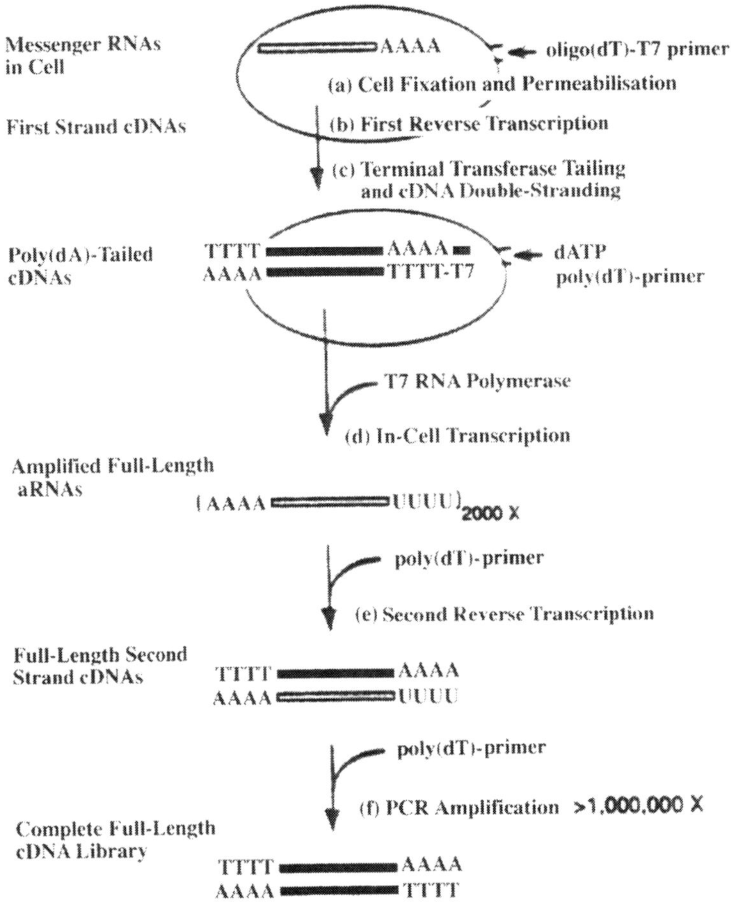

Figure 1. Flowchart of current method for generating a full-length cDNA library from single cells. Cell fixation and permeabilization protect mRNA from degradation and allow enzymes to penetrate cells. Upon RT using an oligo(dT)-T7 promoter and tailing with poly(dA) oligonucleotide, the cDNA library was amplifiable by in-cell transcription and PCR. The final PCR products yield one billion times the amount of original mRNAs.

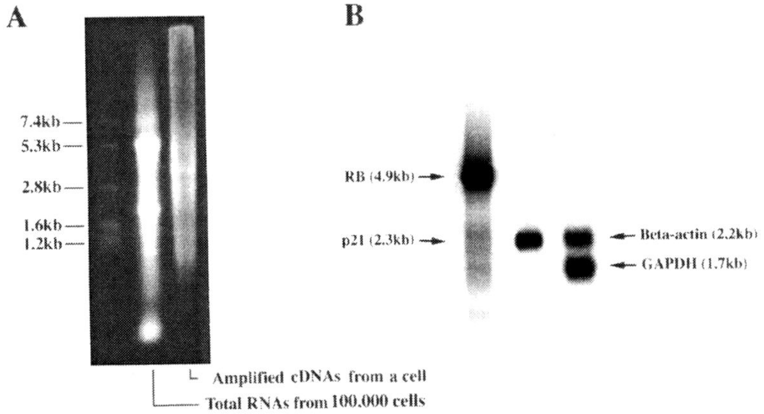

Figure 2. (A) 1% formaldehyde-agarose gel analysis of cDNA library from different numbers of cells. Lane 1, RNA markers from 1.2–7.4 kb; lane 2, total RNA repertoire from 100 000 cells; lane 3, cDNA library from a few single cells. cDNA library obtained from single cells was sufficient to produce a smear visualizable upon staining with ethidium bromide. (B) Southern blot analysis of Rb, p21, β-actin and GAPDH in cDNA library generated from LNCaP cells. After electrophoresis, the cDNA library was denatured and transferred onto a nylon membrane and UV cross-linked. Specific cDNA probes were radiolabeled by Prime-It® II Kit (Stratagene, La Jolla, CA, USA) with [^{32}P]dATP (>3000 Ci/mM; Amersham Phamacia Biotech, Piscataway, NJ, USA). Hybridization was carried out in QuikHyb® (Stratagene) for 4 h at 68°C according to the manufacturer's protocol and washed in 2% standard saline citrate (SSC)/0.1% sodium dodecyl sulfate (SDS) at 20°C for 15 min, then in 0.1% SSC/0.1% SDS at 55°C for 30 min before autoradiography.

$MgCl_2$ and 6 U *C. therm.* polymerase at 55°C for 2 min and then at 70°C for 5 min. As few as one cell in 5 µL of the tailing reaction can be used to accomplish aRNA amplification.

In-cell transcription. The in-cell transcription reaction mixture (50 µL) contained 10 µL of 5× transcription buffer (Roche Molecular Biochemicals), 80 U RNase inhibitors, 0.5 mM of each rNTP and 2000 U T7 RNA Polymerase (Roche Molecular Biochemicals). After a 3-h incubation at 37°C, the cDNA transcripts were isolated from both cells and supernatant and used for another round of RT.

Second RT. The second RT reaction consisted of 10 µL of 5× RT buffer (Roche Molecular Biochemicals), 0.5 mM of each dNTP, 25 pmol poly(dT) 26-mer, 80 U RNase inhibitor, DEPC-treated ddH_2O and 5 µL of the above aRNA-containing supernatant. After 6 U *C. therm.* polymerase were added, the RT reaction was vortex mixed and incubated at 55°C for 1 min and then 68°C for 1 h.

PCR amplification. The resulting products of the second RT were directly used for the PCR (50 µL), which contained 5 µL of 10× PCR buffer (Roche Molecular Biochemicals), 0.2 mM of each dNTP, 25 pmol poly(dT) 26-mer primer, 2 µL of above RT product and 3.5 U of Taq/Pwo DNA Polymerase Mixture (Expand™ Long Template PCR System; Roche Molecular Biochemicals). The PCR underwent thirty cycles of denaturation at 94°C for 1 min, annealing at 55°C for 1 min and extension at 72°C for 4 min.

The quality of the final amplified cDNA library (20 µL) was assessed on a 1% denaturing-formaldehyde-agarose gel. It ranged from 300 bp to above 7.4 kb, demonstrating the completeness of cDNA libraries generated by this method (Figure 2A). We observed that mRNA was well preserved by cell fixation and RNase inhibitors during the amplification process. With an oligo(dT)-promoter sequence as primer, the intracellular mRNAs were reverse transcribed into cDNAs, and consequently the cDNAs were used to transcribe aRNAs. A terminal transferase reaction was performed so that a poly(dA) tail was added to the furthermost 3′ end of the first-strand cDNAs for full-length generation by complementary poly(dT) primers. An RNA polymerase reaction from the promoter region of the double-stranded cDNAs generated full-length aRNAs, which were amplified approximately 1500- to 2000-fold (2). Subsequently, the full-length aRNAs were reverse transcribed and further amplified with PCR extension of poly(dT) primers, by which another one millionfold amplification was achieved. The resulting product represents a complete library enriched in full-length cDNAs from the individual fixed cells. Furthermore, with this method, as few as a single cell can be used to produce a complete cDNA library. We have routinely cloned specific gene products from this cDNA library using PCR amplification with related primers (not shown); we have also detected Rb (4.9 kb), p21 (2.3 kb), glyceraldehyde-3-phosphate dehydrogenase (GAPDH) (1.7 kb) and β-actin (2.2 kb) cDNAs by Southern blot analysis, shown to be relatively the same size as their original mRNAs (Figure 2B).

Although the above procedure was shown to synthesize a cDNA library from cultured human prostate cancer cells, the method can also be used to analyze gene expression patterns of specific cells in vivo. A challenge in studying the organogenesis and pathogenesis of complex organs such as the prostate is that these organs are composed of heterogeneous cell populations (i.e., epithelial vs. stromal, hormone-dependent vs. hormone-independent). The analysis of gene expression at the single cellular level would aid in our understanding of molecular mechanisms involved in a variety of developmental and pathological processes.

REFERENCES

1. **Crino, P.B., J.Q. Trajanowski, M.A. Dichter and J. Eberwine.** 1996. Embryonic neuronal markers in tumerous sclerosis: single-cell molecular pathology. Proc. Natl. Acad. Sci. USA *93*:14152-14157.
2. **Eberwine, J., H. Yeh, K. Miyashiro, Y. Cao, S. Nair, R. Finnell, M. Zettel and P. Coleman.** 1992. Analysis of gene expression in single live neurons. Proc. Natl. Acad. Sci. USA *89*:3010-3014.
3. **Embleton, M.J., G. Gorochov, P.T. Jones and G. Winter.** 1992. In-cell PCR from mRNA: amplifying and linking the rearranged immunoglobulin heavy and light chain V-genes within single cells. Nucleic Acids Res. *20*:3831-3837.
4. **O'Dell, D.M., P. Raghupathi, P.B. Crino, B. Morrison III, J.H. Eberwine and T.K. McIntosh.** 1998. Amplification of mRNAs from single, fixed, TUNEL-positive cells. BioTechniques *25*:566-570.
5. **Patanjali, S.R., S. Parimoo and S.M. Weissman.** 1991. Construction of a uniform abundance (normalized) cDNA library. Proc. Natl. Acad. Sci. USA *88*:1943-1947.
6. **Sambrook, J., E.F. Fritsch and T. Maniatis.** 1989. Construction and analysis of cDNA libraries, p. 8.11-8.35. *In* Molecular Cloning: A Laboratory Manual, 2nd ed. CSH Laboratory Press, Cold Spring Harbor, NY.

Update to: Generation of Full-Length cDNA Library from Single Human Prostate Cancer Cells

Shi-Lung Lin and Shao-Yao Ying

University of Southern California, Keck School of Medicine, Los Angeles, CA, USA

Since publication of the original article, we have further improved the technique of generating full-length cDNA library from single human cells. To repeat the transcriptional amplification of a full-length cDNA library without polymerase chain reaction (PCR) disadvantages, an improved cycling procedure was devised to generate high quality and quantity messenger RNAs from as few as 20 cells (2 pg mRNAs). The principle relies upon the cycling steps of promoter-linked double-stranded cDNA synthesis and promoter-driven transcription to amplify mRNAs up to 250-fold/cycle with good representation of high and low copy transcript profiles. The amplified mRNA libraries were tested to possess high fidelity, purity, specificity and reproducibility for in vivo analyses of cancerous gene expressions in human prostate cancers (1).

Promoter-Linked Double-Stranded cDNA Synthesis

Few fixed cells (>20 but <200) were applied to a reverse transcription reaction (20 μL) on ice, comprising 2 μL of 10× RT&T buffer (400 mM Tris-HCl, pH 8.3 at 25°C, 400 mM KCl, 80 mM $MgCl_2$, 5 M betaine, 100 mM dithiothreitol [DTT] and 20 mM spermidine), 1 μM oligo(dT)26-mer primer, dNTPs (2 mM each) and RNase inhibitors (10 U). After avian myeloblastosis virus (AMV) reverse transcriptase (6 U) was added, the reaction was incubated at 42°C for 30 min. The first-strand cDNAs so obtained were collected by a 100-bp cutoff filter column in a tailing reaction (20 μL), comprising 4 μL of 5× tailing buffer (250 mM KCl, 100 mM Tris-HCl, 4 mM $CoCl_2$, 10 mM $MgCl_2$, pH 8.3 at 20°C) and 0.5 mM dGTP. After terminal transferase (75 U) was added, the reaction was incubated at 37°C for 15 min, stopped by denaturation at 94°C for 3 min and mixed immediately with 1 μM oligo(dC)12N-T7 promoter primers (N = dA, dT or dG in equal amount). *Taq*/*Pwo* DNA polymerase mixture (3 U each) and dNTPs (2 mM each) were then added to double-stranding the poly(dG)-tailed cDNAs at 42°C for 3 min and then 68°C for 10 min.

Promoter-Driven Transcriptional Cycling Amplification

A transcription reaction (40 μL) was prepared, containing 4 μL of 10× RT&T buffer, 1 μM oligo(dT)26-mer primer, 1 μM oligo(dC)12N-T7 promoter primers, 2 mM dNTPs, 2 mM rNTPs, T7 RNA polymerase (>100 U), AMV reverse transcriptase (>10 U), *Taq*/*Pwo* DNA polymerases (3 U each) and above promoter-linked double-stranded cDNAs. The incubation of this reaction was based on previous procedure without tailing reaction as follows: at 37°C for 1 h (transcription), at 42°C for 30 min (reverse transcription), at 94°C for 2 min (denaturation), at 42°C for 3 min (promoter-primer annealing), at 68°C for 10 min (cDNA double-standing). After one cycle of procedure, new T7 RNA polymerase and AMV reverse transcriptase need to be added for another round of transcriptional amplification due to the denaturation step. Such reaction can be repeated until a required cDNA amount is reached. The quality of final products should be assessed on a 1% formaldehyde-agarose gel, ranging up to 12 kb.

REFERENCES

1. **Lin, S.L., C.M. Chuong, R.B. Widelitz and S.Y. Ying.** 1999. In vivo analysis of cancerous gene expression by RNA-polymerase chain reaction. Nucleic Acids Res. 27:4585-4589.

CHAPTER 5

High-Throughput Transformation and Plating Using Petristrips

John J. McNulty and John J. Dunn

Brookhaven National Laboratory, Upton, NY, USA

(Reprinted from BioTechniques 26:390-392, 1999)

We describe a method and simple device that allows the rapid transformation (2) and plating of *Escherichia coli*. Cells are transformed in 96-well cycle plates and plated onto the novel petristrips. Petristrips are simply 12 wells of agar in a plate format compatible with a standard 12-channel pipet. This method, which allows the rapid cloning of a large number of DNAs in a relatively short time, is compatible with a variety of tasks. Our primary focus has been to use it for the production of nested deletion clones (1) as part of a high-throughput sequencing strategy (Figure 1). One of the difficulties we encountered when trying to adapt standard nested deletion protocols to large-scale sequencing was the large number of transformation and plating events that were required. Using the method described here, a single researcher can easily perform 96 transformations and platings in approximately 2 h.

This method was designed to be compatible with a 12-channel pipet. To perform transformations, we used a chilled 0.2-mL, 96-well CyclePlate (Robbins Scientific, Sunnyvale, CA, USA). Two microliters of DNA (ca. 50 ng) are transferred from a 96-well source plate to each well of the CyclePlate using a 12-channel pipet. Ten microliters of Epicurian Coli® XL1-Blue MRF′ Supercompetent Cells (Stratagene, La Jolla, CA, USA) are added to each well of the plate and incubated on ice for 30 min. The cells are then heat-shocked by incubating the plate at 42°C for 50 s. It is then returned to ice for 3 min, and 90 µL of 37°C 2× YT media (per liter; 16 g tryptone, 10 g yeast extract, 5 g NaCl) are added to each well. The plate is sealed using an adhesive sealing film and placed at 37°C shaking for 1 h at 300 rpm in a Model 4628 Shaker (Lab-Line Instruments, Melrose Park, IL, USA).

For plating cells, we used petristrips rather than standard petri dishes. These are made by using a 12-channel reservoir liner plate (Corning Costar, Cambridge, MA, USA). Each lane is filled with 2 mL of MacConkey agar (Difco Laboratories, Detroit, MI, USA) supplemented with appropriate antibiotics. After the plates set, they are placed at 37°C for approximately 1.5 h to allow slight drying. Then, they can be used directly or stored at 4°C for several weeks. To plate cells, 25 µL of transformed cells are transferred to each lane of a 12-lane petristrip using a 12-channel pipet. For each plate of 96 transformants, 8 petristrips are used. Cells are spread along the lanes by simply tilting the petristrips. The unused portion of the transformation mixture can be saved by adding 25 µL of 40% glycerol to each well and storing at -70°C. Petristrips are incubated at 37°C overnight. Colonies are picked individually into 96-well, deep-well plates containing 1 mL of 2× YT media containing antibiotics.

We developed this protocol for simplifying the use of nested deletions in a sequencing strategy. Cycle plates were chosen for their compatibility with polymerase chain reaction

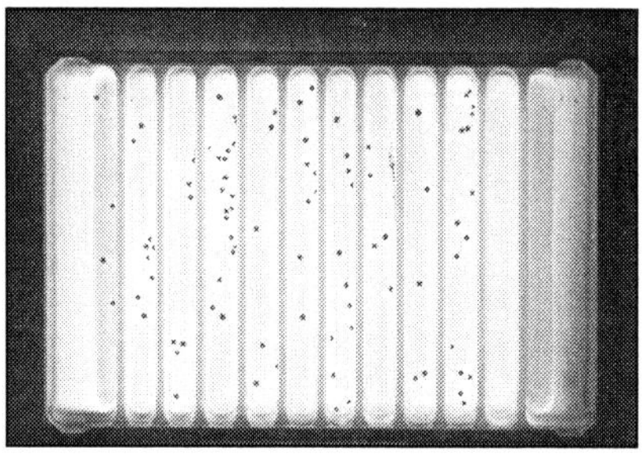

Figure 1. Nested deletion subclones on petristrip. Photograph of a 12-lane petristrip. Each petristrip contains 12 lanes of equal width, with an area on either end that does not contain agar. This area is labeled with the numbers 1 and 12 to designate orientation. MacConkey agar containing ampicillin (40 µg/mL) and kanamycin (25 µg/mL) was used. Cells used in this experiment contained plasmids generated from a nested deletion time course of a pHIND2.1 plasmid (unpublished) containing a 2344-bp insert. This plasmid contains ampicillin- and kanamycin-resistance genes flanking the insert. In this example, the deletions, which progress beyond the insert, disrupt the kanamycin-resistance gene. The disruption of the kanamycin resistance is the reason no colonies are seen on the last two lanes.

Cloning and Expression Vectors for Gene Function Analysis
Edited by Quinn Lu and Michael P. Weiner

(PCR) and other commercially available heating and cooling blocks. We chose MacConkey agar (3) because the dark red colonies are much easier to visualize on petristrips containing this light-pink agar. We also used 8- and 12-lane petristrips for both streaking and plating pGEM clones using color selection on TB agar containing ampicillin (40 µg/mL), 5-bromo-4-chloro-3 indolyl-β-D-galactopyranoside (X-gal) (40 µg/mL) and 0.5 mM isopropyl β-D-thiogalactopyranoside (IPTG) and found that they work well for this. We chose to use the 12-lane format due to simplified sample tracking. The ability to transform and plate cells in a format compatible with a 12-channel pipet is useful for high-throughput plating and could prove useful to a variety of other areas.

ACKNOWLEDGMENTS

We would like to thank Neal R. Tempel for help with creating the figure and Laura Praissman and Barbara Lade for critical reading of the manuscript. This research was supported by the Office of Biological and Environmental Research of the U.S. Department of Energy under Prime Contract No. DE-AC02-98CH10886 with Brookhaven National Laboratory. Address correspondence to John J. McNulty, Biology Department, Bldg. 463, Brookhaven National Laboratory, Upton, NY 11973-5000, USA. Internet: jack@bnl.gov

REFERENCES

1. **Guo, L., R.C.A. Yang and R. Wu.** 1983. An improved strategy for the rapid direct sequencing of both strands of long DNA molecules cloned in a plasmid. Nucleic Acids Res. *11*:5521-5540.
2. **Hanahan, D.** 1983. Studies on the transformation of *Escherichia coli* with plasmids. J. Mol. Biol. *166*:557-580.
3. **Jennings, M.P. and I.R. Beacham.** 1989. MacConkey agar as an alternative to Xgal in the detection of recombinant plasmids. BioTechniques *7*:1082.

CHAPTER 6

Directional Cloning of Blunt-Ended PCR Products

Michael P. Weiner

Stratagene Cloning Systems, La Jolla, CA, USA

(Reprinted from BioTechniques 15:502-505, 1993)

ABSTRACT

*A method that allows the directional cloning of blunt-ended polymerase chain reaction (PCR) fragments is described. One PCR primer must be 5′ phosphorylated. Extra bases are not required on either PCR primer. A linearized vector is enzymatically processed to contain a single 5′-terminal phosphate. The monophosphorylated vector is amenable to recombinant-insertion during ligation when the fragment is in the correct orientation. Increased recombinant yield results from incubating the monophosphorylated vector with a restriction enzyme (*SrfI*) that relinearizes nonrecombinant plasmids during the ligation reaction.*

INTRODUCTION

Several methods exist for the cloning of PCR-generated fragments. These include *(i)* T/A cloning methods that rely on the limited terminal deoxynucleotidyl transferase activity of some exonuclease-deficient DNA polymerases (4,5,10,11); *(ii)* blunt-ended fragment cloning (2,8); *(iii)* restriction digestion of sites incorporated at the ends of PCR primers (7,9,14); *(iv)* 3′→5′ exonuclease treatment of fragments and ligation by overlap extension (1,15); and *(v)* incorporation of uracil into PCR primers followed by treatment with uracil-*N*-glycosylase (12).

Up to now, all of the methods that could be used to clone directionally PCR products (*iii*, *iv* and *v* above) required the addition of extra bases to the PCR primers. We have developed a method to alleviate this drawback to directional cloning.

MATERIAL AND METHODS

Enzymes

*Srf*I, *Sma*I, *Sca*I, *Sac*I, *Kpn*I, T4 DNA ligase, calf intestinal alkaline phosphatase (CIAP) and 10× Universal buffer were obtained from Stratagene (La Jolla, CA). All enzymatic reactions (*Srf*I, *Sma*I, *Sca*I, *Sac*I and *Kpn*I, CIAP and T4 DNA ligase) were performed in 1× Universal buffer; T4 DNA ligase reactions contained 0.5 mM ATP (final).

Media

Luria-Bertani (LB) medium contained tryptone (10g/L), yeast extract (5 g/L), and NaCl (10 g/L). LB plates were supplemented with ampicillin (Amp) and methicillin (Meth) at 20 mg and 8 mg/L, respectively, for β-lactamase resistance and at 30 mg/L for chloramphenicol resistance.

Plasmid Purification

Large-scale (3 liters) plasmid DNA was purified by the alkaline lysis method (3). Miniprep plasmid DNA was purified by the boiling method (6).

Creation of pMW239

Plasmid pBlueScript® II SK(+) (Stratagene) was restricted with *Sac*I and *Kpn*I and ligated to annealed primers MW846 (5′-CCGCGGCCGCCCGGGCTGCAGGATCCCGGGCAT-ATGTATATCTCCTTGGCGCGCCGGTAC-3′) and MW847 (5′-CGGCGCGCCAAGGAGATATACATATGCCCGGGAT-CCTGCAGCCCGGGCGGCCGCGGAGCT-3′) at a vector concentration of 100 ng/μL and annealed oligonucleotide concentration of 0.1 ng/μL. The ligation reaction was transformed into XL1-Blue (Stratagene) and plated onto LB + Meth/Amp. Blue colonies on 5-bromo-4-chloro-3-indolyl-β-D-galactopyranoside (X-gal) + isopropyl-β-D-thio-galactopyranoside (IPTG)-containing LB + Meth/Amp plates after overnight incubation at 37°C. were isolated and several sequenced. A colony with the correct cloning site DNA sequence was labeled pMW239 and chosen for all further experimentation.

Preparation of Monophosphorylated Vector DNA

Twenty micrograms of plasmid pMW239 at 0.5 μg/μL were digested with 40 units of *Srf*I for 1 h at 37°C, then CIAP (0.2 units) was added and the plasmid incubated for an additional 30 min at 37°C. The plasmid DNA was extracted twice with Tris-buffered phenol and twice with $CHCl_3$. The plasmid DNA was incubated for 20 min at 70°C, then cooled on ice, diluted to 0.1 μg/μL with 1× Universal buffer and then 40 units of *Sma*I were added. The restriction digestion was

Cloning and Expression Vectors for Gene Function Analysis
Edited by Quinn Lu and Michael P. Weiner

allowed to proceed for 1 h at 34°C, and the restriction endonuclease was heat-killed for 20 min at 70°C. Plasmid DNA that was treated with CIAP and restriction endonuclease was precipitated by the addition of 0.1 vol of 10 M LiCl and 2.5 vol of ice-cold ethanol. The plasmid was centrifuged at 15 000 rpm (Eppendorf; Brinkmann, Westbury, NY) and dried in vacuo. The monophosphorylated vector was resuspended to 50 ng/µL in TE (10 mM Tris-HCl, pH 7.5, 1 mM EDTA).

PCR Protocol

The chloramphenicol gene from pBC SK- (Stratagene) was amplified by PCR using the oligonucleotides MW227 (5′-CCTGTGACGGAAGATCACTTCGC-3′) and MW228 (5′-pCCTCCACGGGGAGAGCCTGAGC-3′). MW228 was synthesized with a 5′-terminal phosphate. The PCR was performed with, per 100 µL, 1 µg of each primer, 50 nM deoxyribonucleoside triphosphate (dNTP), 100 ng template DNA (restriction endonuclease *Pst*I-digested pBC SK+ DNA) and 10 units of *Taq* DNA polymerase (Perkin-Elmer, Norwalk, CT) in 1× *Taq* DNA Polymerase buffer. PCR cycle conditions were 1 min at 54°C, 1 min at 72°C, 1 min at 90°C for 40 cycles, followed by 10 min at 72°C.

Ligation Procedure and Transformation

The ligation reactions were performed in 10 µL of 1× Universal buffer containing 50 ng plasmid DNA, 200 ng PCR product, 500 nM ATP, 4 units T4 DNA ligase and 15 units of *Srf*I. The reaction was incubated for 1 h at room temperature and then heated for 10 min at 70°C. *E. coli* strain XL1-Blue (100 µL) was transformed with 2 µL of the ligation reaction and plated onto LB + Meth/Amp + IPTG + X-gal. The plates were incubated at 37°C overnight and phenotypically β-galactosidase- (β-gal-) colonies were patched onto LB + chloramphenicol plates. Colonies that were chloramphenicol-resistant were inoculated into 20 mL of LB + Amp, incubated overnight and plasmid mini-preparations purified from them.

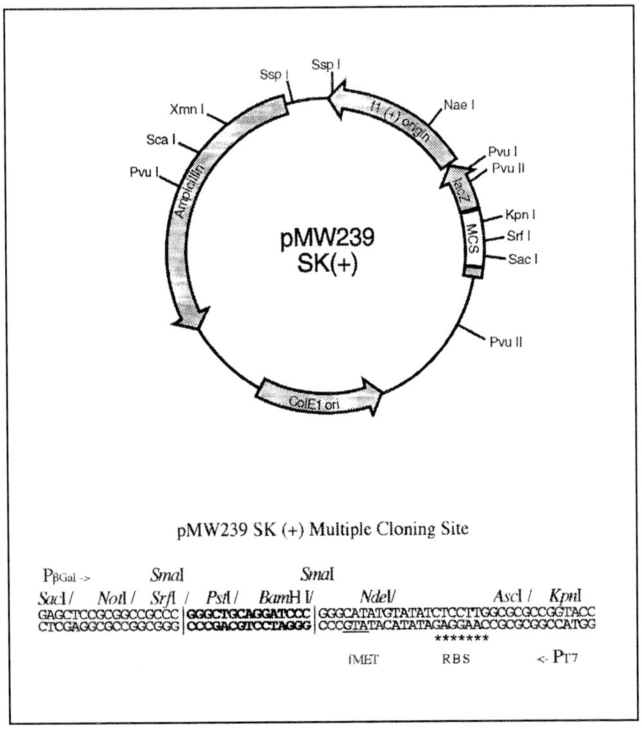

Figure 1. Plasmid pMW239. Shown is the plasmid pMW239 and the multiple cloning site; including the RBS and fMet initiation codon for expression of genes from the T7 RNA polymerase binding site in the orientation opposite to that of the β-galactosidase promoter. The region removed by digestion with both *Srf*I and *Sma*I is shown in bold. Promoter direction is indicated by the arrow (←) for $P_{\beta gal}$ and PT7; RBS, ribosome binding site; fMet, methionine start codon.

RESULTS AND DISCUSSION

Design of the pMW239 Multiple Cloning Site

Liu and Schwartz (8) and Bauer et al. (2) have shown that incubation of a ligation reaction in the presence of an excess amount of restriction enzyme dramatically increases the yield and efficiency of recombinant plasmid formation. Presumably, this is due to a high steady-state concentration of linearized plasmid DNA in solution that is able to interact with an insert. Religation of vector DNA without insert recreates a restriction target site. In preliminary experiments we were able to optimize the restriction/ligation reaction for approximately 80% completion in 1 h at room temperature (unpublished results).

The plasmid pMW239 (Figure 1) was created in order to clone directionally monophosphorylated insert fragments by a variation of the Liu and Schwartz procedure (8). The restriction enzyme *Srf*I contains within its DNA target site (5′-GCCC/GGGC-3′) the site for *Sma*I (5′-CCC/GGG-3′) (13). A multiple cloning site of the sequence 5′-GCCCGGGC-//-CCCGGGC-3′ contains a single *Srf*I site and two *Sma*I sites. Digestion with *Srf*I, followed by treatment with alkaline phosphatase yields a dephosphorylated vector that when digested with *Sma*I recreates an *Srf*I site because of a C deoxynucleotide 3′ to the *Sma*I site (see Figures 1 and 2). The new *Srf*I site is uniterminally monophosphorylated at one side. Ligation of this vector creates a nicked circle that is amenable to restriction by the *Srf*I endonuclease.

Additional Features of pMW239

The cloning site in *Srf*I, CIAP, *Sma*I-treated pMW239 is bracketed by two octanucleotide-recognizing restriction enzymes, *Not*I (5′-GC/GGCCGC-3′) and *Asc*I (5′-GG/CGCGCC-3′) and three hexanucleotide-recognizing restriction enzymes, *Sac*I (5′-GAGCTC-3′), *Kpn*I (5′-GGTACC-3′) and *Nde*I (5′-CATATG-3′).

For protein synthesis, the pMW239 vector contains a β-galactosidase promoter 5′ to the insertion site. A T7 promoter, *E.coli* ribosome binding site (RBS) and fMet start site 3′ to the insertion site in the opposite direction. The β-galactosidase gene is functional, and the reading frame of ligated *Srf*I, *Sma*I-digested vector is conserved, allowing phenotypic color selection for insert-containing clones on X-gal plates. Protein fusions using the T7 RNA polymerase site will have an N-terminal methionine, proline fusion. The *Nde*I restriction site can be used to delete the two amino acids and fuse coding sequences directly to the fMet start site downstream from the RBS.

The pMW239 plasmid also contains an F1 origin for single-stranded DNA rescue and the ampicillin gene for transformant selection.

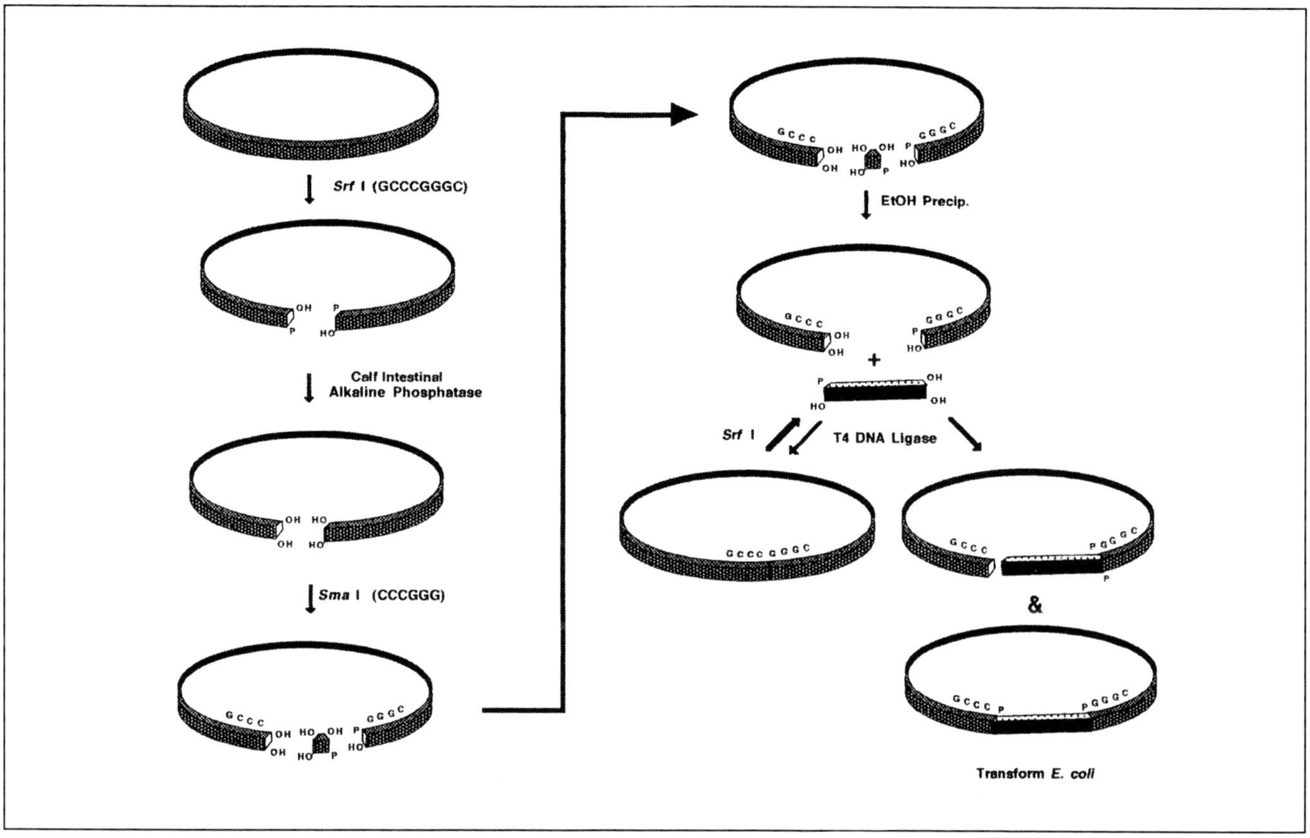

Figure 2. Schematic of method for directional cloning. Plasmid pMW239 is digested with the restriction enzyme *Srf*I, treated with CIAP and alkaline phosphatase removed by phenol-chloroform extraction. The non-phosphorylated plasmid is then restricted with *Sma*I and ethanol precipitated to remove the 15-bp spacer region. The uniterminally phosphorylated vector is then added to PCR fragment containing a single 5′ phosphate. The ligation reaction is described in the text.

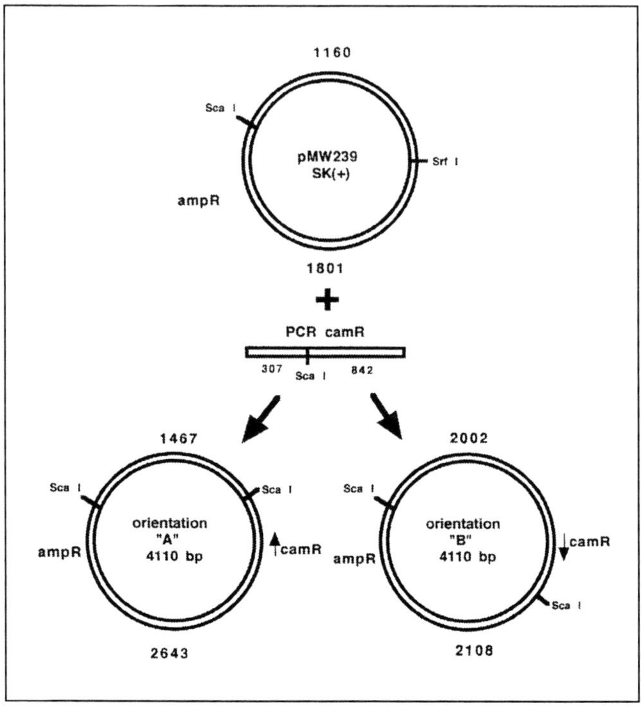

Figure 3. Orientation of cloned inserts. Confirmation of the method was performed on a PCR fragment of the cam^R gene that contained a single *Sca*I site. Orientations A and B would be expected to produce two different sets of fragments when digested with *Sca*I.

Directional Cloning

*Srf*I only and *Srf*I, CIAP, *Sma*I-treated pMW239 DNA were used to determine if monophosphorylated PCR insert could be cloned bi- and unidirectionally. A PCR product of the chloramphenicol-resistant (cam^R) gene from restriction endonuclease-digested pBC was made with a primer set containing one phosphorylated primer (MW228). The PCR fragment was cloned into the two differently treated vectors, and colonies containing the cam^R gene were isolated. Twenty ampicillin, chloramphenicol-resistant colonies were chosen for analysis of clonal-insert orientation. *Sca*I restriction endonuclease was used to determine orientation (see Figure 3). In one orientation, "A," *Sca*I digestion will yield fragments of 1467 and 2643 bp. In the "B" orientation the fragments will be 2108 and 2002 bp. The vector treated with *Srf*I alone contained 40% (8 of 20) of the inserts in the "A" and 60% in the "B" orientation. The *Srf*I, CIAP, *Sma*I-treated pMW239 vector produced 5% (1 of 20) in the "A" orientation and 95% (19 of 20) in the "B" orientation.

CONCLUSIONS

A method is presented to clone directionally blunt-ended PCR fragments containing a single phosphorylated terminus. The PCR cloning vector is modified enzymatically using *Srf*I, CIAP and *Sma*I. The ligation reaction is performed in the presence of both T4 DNA ligase and *Srf*I.

It should be possible to use this procedure with other restriction enzymes in order to enzymatically create monophosphorylated vector DNA. Alternatively, it should be possible to ligate appropriate linkers containing modified 5′ sites onto a suitable vector for either blunt, sticky or T/A-type cloning. It should also be possible to use T4 DNA kinase to phosphorylate a PCR primer before the start of the reaction.

With this method it is now possible to clone directionally PCR fragments without the need to modify the primers with uracil or the addition of extra bases. In future studies we will examine the transcriptional control of protein synthesis using this new vector.

REFERENCES

1. **Aslandis, C. and P. deJong.** 1990. Ligation-independent cloning of PCR products (LIC-PCR). Nucleic Acids Res. *18*:6-69-6074.
2. **Bauer, J., D. Deely, J. Viola, J. Braman and M. Weiner.** 1992. PCR-Script SK(+) cloning system: a simple and fast method for PCR cloning. Strategies Mol. Biol. *5*:56-58.
3. **Birnboim, H.C. and J. Doly.** 1979. A rapid alkaline extraction procedure for screening recombinant plasmid DNA. Nucleic Acids Res. *7*:1513-1523.
4. **Clark, J.M.** 1988. Novel non-template nucleotide addition catalyzed by prokaryotic and eukaryotic DNA polymerases. Nucleic Acids Res. *16*:9677.
5. **Helmley, D.S., N. Arnheim, M.D.Toney, G. Cortopossi and D.J. Galas.** 1989. A simple method for site-directed mutagenesis using the polymerase chain reaction. Nucleic Acids Res. *17*:6545.
6. **Holmes, D.S. and M. Quigley.** 1981. A rapid boiling method for the preparation of bacterial plasmids. Anal. Biochem. *114*:193-197.
7. **Kaufman, D.L. and G.A. Evans.** 1990. Restriction endonuclease cleavage at the termini of PCR products. BioTechniques 9:304-306.
8. **Liu, Z.-G. and L.M. Schwartz.** 1992. An efficient method for blunt-end ligation of PCR products. BioTechniques *12*:28-30.
9. **Lorens, J.** 1991. Rapid and reliable cloning of PCR products. PCR Methods Appl. *1*:140-141.
10. **Marchuk, D., M. Drumm, A. Saulino and F. Collins.** 1991. Construction of T-vectors, a rapid and general system for direct cloning of unmodified PCR products. Nucleic Acids Res. *19*:1154.
11. **Mead, D., N. Kristy, C. Herrnstadt, R. Marcil and L. Smith.** 1991. A universal method for direct cloning of PCR amplified nucleic acids. Bio/Technology 9:657-663.
12. **Nisson, P., A. Rashtchian and P. Watkins.** 1991. Rapid and efficient cloning of Alu-PCR products using uracil-DNA glycosylase. PCR Methods Appl. *1*:120-123.
13. **Simcox, T., S. Marsh, E. Gross, W. Lernhardt, S. Davis and M. Simcox.** 1991. *Srf*I, a new type-II restriction endonuclease that recognizes the octanucleotide sequence, 5′-GCCC/CCCG-3′. GENE *109*:121-123.
14. **Scharf, S.** 1990. Cloning with PCR, p. 84-91. *In* M. Innis, D. Gelfand, J. Sninsky and T. White (Eds.), PCR Protocols: A Guide to Methods and Applications. Academic Press, San Diego.
15. **Shuldiner, A., K. Tanner, L. Scott, G. Moore and J. Roth.** 1991. Ligase-free subcloning: a versatile method to subclone polymerase chain reaction (PCR) products in a single day. Anal. Biochem *194*:9-15.
16. **Stoker, A.** 1990. Cloning of PCR products after defined cohesive termini are created with T4 DNA polymerase. Nucleic Acids Res. *18*:4290.

Update to: Directional Cloning of Blunt-Ended PCR Products

Gina L. Costa and Michael P. Weiner[1]

Stanford University, Palo Alto, CA and [1]Glaxo Wellcome Research, Research Triangle Park, NC, USA

ALTERNATIVES TO SINGLE-SITE RESTRICTION ENZYMES AND THE DEVELOPMENT OF RETROVIRAL VECTORS FOR THE CLONING OF BLUNT-ENDED PCR PRODUCTS

Several methods have been developed for cloning polymerase chain reaction (PCR) products that obviate the need for added bases (and associated additional expense) onto the 5′ ends of the PCR primers. These methods can generally be grouped into T/A and blunt-ended cloning. T/A-type cloning methods rely on the base- and polymerase-specific extension of a single nucleotide onto the 3′ end of PCR products (2,3,8). Initial studies of this phenomenon first showed that several polymerases extended predominately a single A nucleotide onto the 3′ end of double-stranded DNA in a nontemplate-directed fashion (2). However, other studies have since shown that this extension depends upon the polymerase used (not all polymerases contain a terminal extendase activity) and the 3′ base on the end of the double-stranded DNA (4,8). Polymerases with extendase activity can be used for T/A cloning only when the base extended is complementary to the 3′ base extension on the cloning vector (Table 1).

Blunt-ended cloning methods have been developed to efficiently clone PCR fragments generated with blunt ends. Initial methods generally involved the addition of the insert, vector and T4 DNA ligase to a reaction mixture. Because blunt-ended cloning is inefficient, either dephosphorylation of the vector or a phenotypic selection is typically used to select recombinant clones. A modification of the blunt-ended cloning procedure to increase its efficiency was the addition of a restriction enzyme to the ligation reaction to increase the amount of linear vector available for recombinant insertion (4,9). We had previously described a method whereby the restriction enzyme *Srf*I (target sequence 5′-GCCC|GGGC-3′) was used to regenerate the linear vector. *Srf*I has the advantage of a relatively rare octameric sequence that would occur, on average, once every 64 thousand bases. Having a rare target site in the DNA to be cloned is critical to the success of this procedure.

Alternatives to Single-Site Restriction Enzymes

With the advent of PCR methods for generating very large DNA molecules (on the order of 10–40 kb), it has become apparent that there is an increased chance that the *Srf*I target site may be present on the PCR fragments. An alternative to *Srf*I is the use of the methylation-requiring restriction enzyme *Dpn*I, and a vector-methylase pretreatment method whereby the restriction enzyme target site will never be present in the PCR fragment. Because *Dpn*I has a requirement for home- or

Table 1. 3′ End Modification by DNA Polymerase

Polymerase	5′ Nucleotide	3′ End Modification[a]
Taq Polymerase	A	-T, +A
	C	+G>+A>+C
	G	+A>+C
	T	(+A)
Klenow Polymerase	A	+A
	C	+C>>+G, +A
	G	+A>>+C
	T	(+A)
T4 Polymerase	A	Blunt end
	C	Blunt end
	G	Blunt end
	T	Blunt end
Pfu Polymerase	A	Blunt end
	C	Blunt end
	G	Blunt end
	T	Blunt end

[a]3′ end nucleotide extension (+) or nonaddition (-) of DNA fragments by DNA polymerase in conditions as specified by Hu (8).

hemimethylated adenosines in its target sequence, PCR fragments generated with nonmodified adenosine nucleotides will not contain any target sites. A plasmid can be either systematically modified to contain a single *Dpn*I (5′-G^{m6} A|TC-3′) target site, or, in a second method, pretreated with overlapping methylation sites which can also be used to protect it against the restriction enzyme used in the ligation reaction. As a specific example, Nelson and McClelland have shown that methylase-target sites can be overlapped to create larger *Dpn*I restriction target sites (11). For example, *Cla*I methylase (M. *Cla*I) modifies the second A nucleotide in the recognition sequence 5′-ATCGAT-3′ to 5′-ATCG^{m6}AT-3′. By use of a 2-base overlap of two M. *Cla*I target sites, (5′-ATC<u>GATC</u>GAT-3′), it is possible to synthesize a single *Dpn*I site. The ligation reaction can then be performed in the presence of both T4 DNA ligase and *Dpn*I.

Development of Retroviral Vectors for Cloning of Blunt-Ended PCR Products

Gene therapy involves the insertion and expression of foreign DNA in a host cell (6). To date, utilization of viral vectors is the most efficient way to introduce genes into cells by the process of transduction. Many vectors exist, each with advantages and disadvantages. The murine oncoretroviruses, which include the Moloney murine leukemia virus (MMLV), are widely studied and utilized for gene transfer in murine models, as well as in human studies. The MMLV genome consists of two long terminal repeats (LTRs) that contain promoter and enhancer elements important for the synthesis and transcription of viral genes. In a conventional retrovirus, LTRs flank the *gag*, *pol* and *env* genes and promote expression of essential viral structural proteins, including envelope glycoproteins that confer the "host range" or tropism of the virus. The packaging signal, psi (ψ), lies 3′ to the 5′ LTR and is necessary for the packaging of viral genomes.

The application of gene therapy using retroviruses involves the use of cell lines that have been designed for stable expression, in trans, of the *gag-pol* and *env* genes to allow production of intact virions (5,6). Retroviral vectors are engineered to contain LTRs that flank the internal expression cassette(s) and encode a psi signal sequence. The gene(s) of interest can

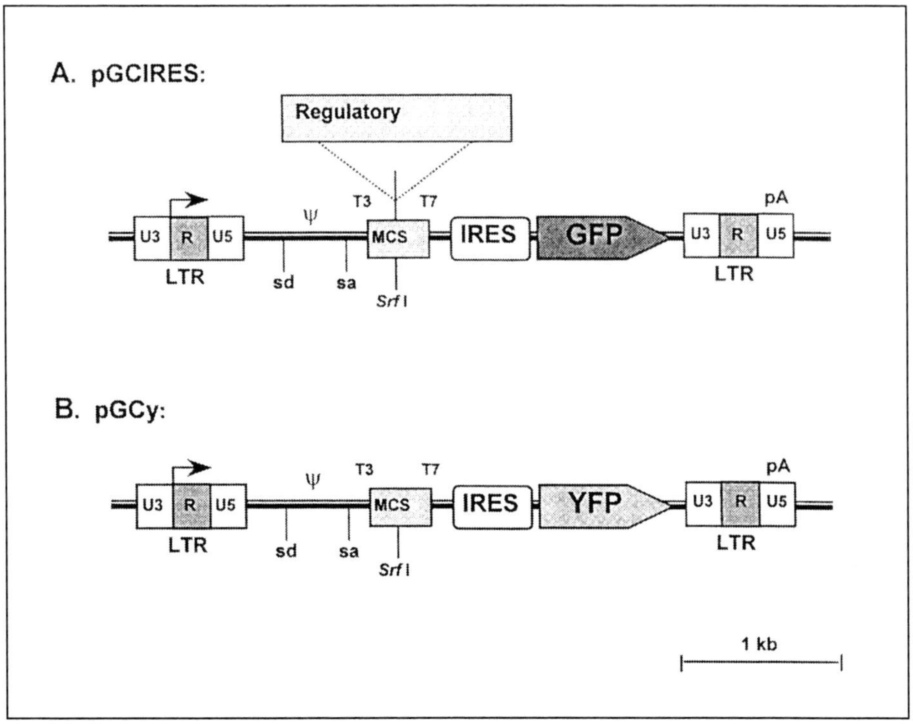

Figure 1. Schematic diagram of MMLV-based pGC bicistronic retroviral vectors. (A) pGCIRES (6691 bp), retroviral vector containing the MMLV-MFG retroviral packaging signal, psi (ψ) and 5′ and 3′ LTRs; a *Srf*I-containing multiple cloning site (MCS) flanked with the T3 and T7 oligonucleotide primer sites for the cloning and sequencing of regulatory genes; the encephalomyocarditis virus IRES; and a mammalian codon-enriched GFP variant; or (B) YFP variant, termed pGCy. Arrow indicates transcriptional start site and direction of transcription.

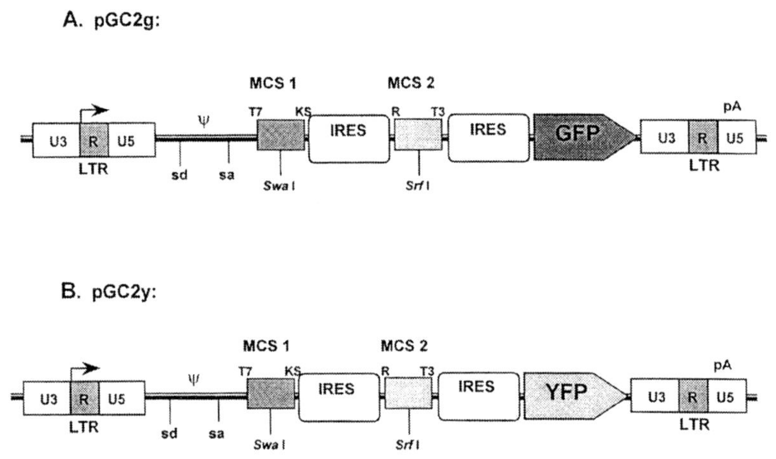

Figure 2. Schematic diagram of second generation MMLV-based pGC2 tricistronic retroviral vectors. (A) pGC2g (7344-bp) retroviral vector containing the MMLV-MFG retroviral packaging signal, psi (ψ) and 5′ and 3′ LTRs; a *Swa*I-containing multiple cloning site (MCS 1) flanked with the T3 and KS oligonucleotide primer sites; the Harvey sarcoma virus IRES [IRES (HaSV)], a *Srf*I-containing multiple cloning site (MCS 2) flanked with the reverse (R) and T7 oligonucleotide primer sites; the encephalomyocarditis virus IRES [IRES (emcv)]; and a mammalian codon-enriched GFP variant; or (B) YFP variant, termed pGC2y. Arrows indicate transcriptional start site and direction of transcription.

be subcloned into the pGC retroviral vectors by restriction ligation using either *Srf*I (5′-GCCC|GGGC-3′) or *Swa*I restriction endonuclease (5′-ATTT|AAAT-3′). A second open reading frame has been incorporated for the expression of additional genes by including an internal ribosome entry site (IRES) (12). IRES elements allow cap-independent translation of an open reading frame from a single polycistronic mRNA transcript. Because translation of each protein is from the same transcript, upstream and downstream gene products are made in approximately equivalent amounts. Reporter proteins such as green fluorescent protein (GFP) and/or yellow fluorescent protein (YFP) can be used to identify cells and provide an indirect measurement of the upstream gene expression.

Construction of retroviral vectors engineered to express fluorescent proteins to serve as markers for transduction has become common practice (1). Fluorescent proteins, such as GFP, allow direct visualization of proteins in living cells without the need for substrate addition or cellular manipulation. Although blue forms of fluorescent proteins such as the blue-shifted variant of GFP (BFP) and the red fluorescent protein (RFP) are available for use as reporter markers, these blue variants require excitation in the UV range and can therefore cause DNA damage to living cells (13). Recently GFP has been mutagenized to give rise to a "yellow" form, YFP, which has the intrinsic brightness and stability of GFP, and excites with visible light (7). Although the excitation of YFP is the same as GFP (488 nm), the emission of YFP is "red-shifted" (600 nm) with respect to GFP (509 nm) and therefore allows separation of the two spectra (10).

Schematic representation of the pGCIRES vectors are outlined in Figures 1 and 2. The pGCIRES vectors are MMLV-based IRES-containing retroviral vectors with different reporter proteins (Figure 1). Second generation retroviral vectors, termed pGC2IRES, contain two unique IRES elements and allow for tricistronic gene expression with either GFP or YFP reporter expression (Figure 2). Following transfection of the retroviral vector DNA into a packaging cell line, viral RNA transcripts containing the psi (ψ) packaging signal will be preferentially incorporated into the virions. Because retroviral vectors lack the *gag-pol* and *env* viral genes, the recombinant retroviral particles produced are replication-defective and capable of infecting only one target cell without further propagation of virus. Therefore, once the viral vector is inserted into the packaging cell line, the system can be used to retrieve replication-defective, infection-competent virus particles for transduction of a target cell population (5).

REFERENCES

1. **Chalfie, M., Y. Tu, G. Euskirchen, W.W. Ward and D.C. Prasher.** 1994. Green fluorescent protein as a marker for gene expression. Science *263*:802-805.
2. **Clark, J.M.** 1988. Novel non-templated nucleotide addition reactions catalyzed by procaryotic and eucaryotic DNA polymerases. Nucleic Acids Res. *16*:9677-9686.
3. **Costa, G. and M. Weiner.** 1994. Polishing of PCR fragments with either T4 or *Pfu* polymerase improves the efficiency of blunt-ended cloning. Nucleic Acids Res. *22*:2423.
4. **Costa, G.L., A. Grafsky and M. Weiner.** 1994. Cloning and analysis of PCR-generated DNA fragments. PCR Methods Appl. *3*:338-345.
5. **Costa, G.L., J.M. Benson, C.M. Seroogy, P. Achacoso, C.G. Fathman and G.P. Nolan.** 2000. Targeting rare populations of murine antigen-specific T lymphocytes by retroviral transduction for potential application in gene therapy for autoimmune disease. J. Immunol. *164*:3581-3590.
6. **Fathman, C.G., G.L. Costa and C.M. Seroogy.** 2000. Gene therapy for autoimmune disease. Clin. Immunol. *95*:S39-S43.
7. **Heim, R. and R.Y. Tsien.** 1996. Engineering green fluorescent protein for improved brightness, longer wavelengths and fluorescence resonance energy transfer. Curr. Biol. *6*:178-182.
8. **Hu, G.** 1993. DNA-polymerase-catalyzed addition of nontemplated extra nucleotides to the 3′ end of a DNA fragment. DNA Cell Biol *12*:763-770.
9. **Liu, Z. and L. Schwartz.** 1992. An efficient method for blunt-end ligation of PCR products. BioTechniques *12*:28-30.
10. **Lybarger, L., D. Dempsey, G.H. Patterson, D.W. Piston, S.R. Kain and R. Chervenak.** 1998. Dual-color flow cytometric detection of fluorescent proteins using single-laser (488-nm) excitation. Cytometry *31*:147-152.
11. **Nelson, M. and M. McClelland.** 1987. The effect of site-specific methylation on restriction-modification enzymes. Nucleic Acids Res. *15*:r219-r230.
12. **Sugimoto, Y., I. Aksentijevich, M.M. Gottesman and I. Pastan.** 1994. Efficient expression of drug-selectable genes in retroviral vectors under control of an internal ribosome entry site. BioTechnology *12*:694-698.
13. **Wildt, S. and U. Deuschle.** 1999. cobA, a red fluorescent transcriptional reporter for *Escherichia coli*, yeast, and mammalian cells. Nat. Biotechnol. *17*:1175-1178.

Section II

Protein Expression and Purification

CHAPTER 7

High Copy Number Plasmids Compatible with Commonly Used Cloning Vectors

Gregory J. Phillips, Sei-Kyoung Park and Damon Huber

Iowa State University, Ames, IA, USA

(Reprinted from BioTechniques 28:400-408, 2000)

A number of genetic and biotechnological applications are facilitated by the ability to transform *Escherichia coli* with more than one recombinant plasmid. For example, a plasmid that carries a gene of interest may require that a *trans*-acting regulatory protein be expressed from a second plasmid in the cell to effectively regulate its expression (9,16,29). In addition, purification of multisubunit complexes, or characterization of the interaction between gene products, can also be facilitated by elevated expression of the interactive components. In *E. coli*, for example, the Ffh protein interacts with a 4.5S RNA species to form a ribonucleoprotein complex known as the signal recognition particle (SRP) (17,18). To better understand the interactions between Ffh and 4.5S RNA we sought to jointly over-express both of the SRP components in *E. coli*. One strategy to do this was to individually clone *ffh* and *ffs*, which encode 4.5S RNA, into cloning vectors that are members of different incompatibility groups and transform both plasmids into the same *E. coli* strain.

Some of the most commonly used plasmid vectors are derivatives of the ColE1-like plasmid pMB1, including pBR322 (2), and high copy number versions comprising the pUC (27,32) and pBluescript® (Stratagene, La Jolla, CA, USA) family of vectors. Multicopy cloning vectors that are compatible with pMB1 plasmids include those that use the p15A origin of replication (*ori*) such as pACYC177, PACYC184 (4), as well as other derivatives (1). In general, these plasmids exist at a medium copy number of 15–20 plasmids/chromosome. Also described are cloning vectors derived from pSC101 (13,28) that are maintained at low copy number (6–8 copies/chromosome). These vectors permit construction of bacterial strains that harbor compatible plasmids. While medium to low copy number vectors may be suitable for many applications, their use can be limiting when high-level expression of a gene product is required. Therefore, to maximize expression of two different gene products we sought to construct new cloning vectors that are compatible with many commonly used plasmids and that replicate at a high copy number. Although replication of ColE1-like plasmids is dependent on DNA polymerase I and their replication is regulated by the interaction of two complementary RNA transcripts, distinct incompatibility groups have been identified (20,22). The ColE1-like plasmid RSF1030 (10), for example, is able to co-reside with both pMB1 and p15A-derivative plasmids, as well as with non-ColE1 vectors such as pSC101. The *ori* from RSF1030 was therefore selected for construction of a series of new cloning vectors.

Wild-type RSF1030 is maintained at a copy number of 15–20 per chromosome equivalent, a copy number that is comparable to pBR322 (6), pACYC177 and pACYC184 (4). To construct a cloning vector that replicates at a high copy number yet maintains its compatibility with commonly used cloning vectors, a high copy number mutant of RSF1030 was isolated. An RSF1030 derivative, pST19 (22), was converted to a high copy number mutant by selecting transformants able to form colonies in the presence of elevated levels of ampicillin and methicillin. An overnight culture of DH5α transformed with pST19 was plated onto LB agar plates (21) containing 400 µg/mL ampicillin and 400 µg/mL methicillin (31). Several colonies that arose under these selective conditions were screened using a rapid cell lysis procedure (19) to identify transformants that contained elevated quantities of plasmid DNA. A single mutant that yielded significantly more plasmid DNA than cells transformed with pST19 was selected for further analysis.

DNA sequence analysis of a 1.5 kb segment containing the *ori* from the high copy number mutant revealed a single base pair alteration located in a region that encodes the RNAII replication primer and the antisense RNAI transcript (Figure 1A). Although this mutation does not appear to have been previously reported in selections for high copy number mutants of ColE1 plasmids (7), the location of this mismatch within the REF1030 *ori* suggests a possible mechanism for increased copy number. It is well established that the copy number of ColE1-like plasmids is regulated by interaction between folded RNAI and RNAII molecules in a way that prevents RNAII from functioning effectively as a primer for replication (14,15,25). Initial interaction between RNAI and RNAII occurs by contact between the loop regions of the folded RNA molecules (the regions I′, II′ and III′ are shown for RNAI in Figure 1) (24). If this initial interaction is weak-

Cloning and Expression Vectors for Gene Function Analysis
Edited by Quinn Lu and Michael P. Weiner

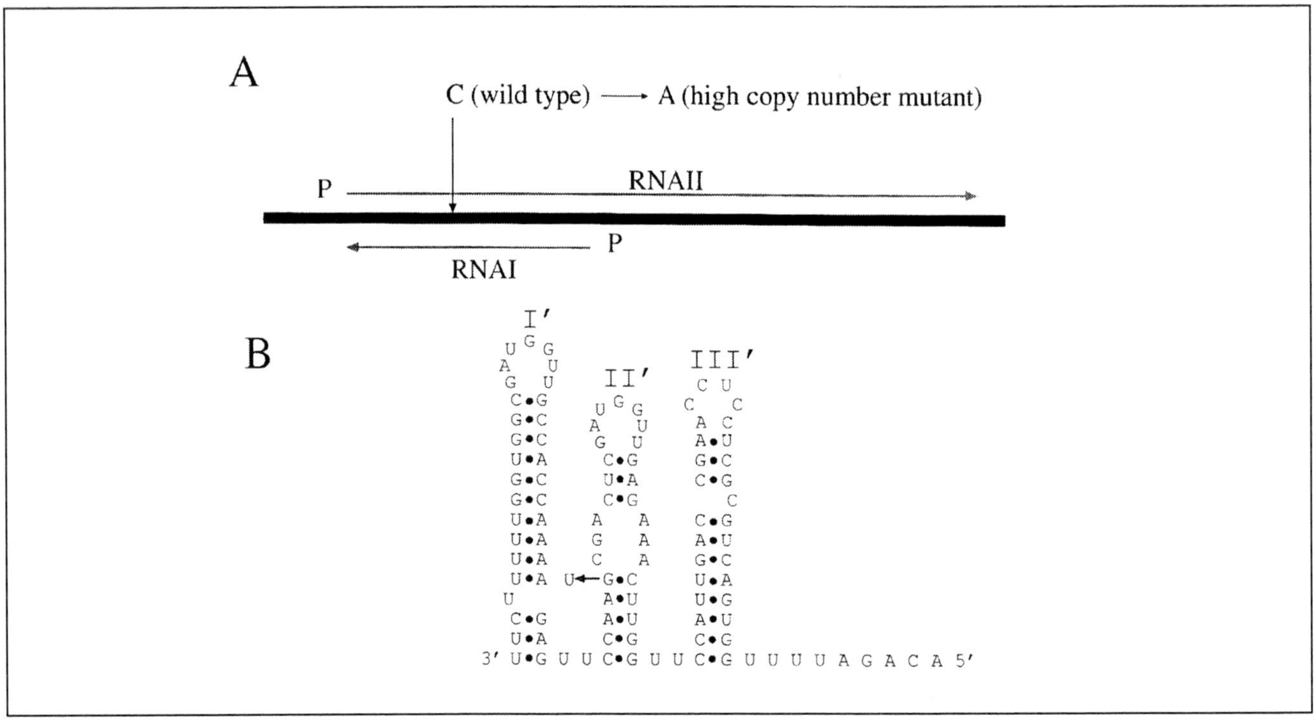

Figure 1. The *ori* region of RSF1030. (A) The RSF1030 *ori* region, including the RNAI (antisense) and RNAII (replication primer) transcripts, is schematically shown. The location of the C to A transversion responsible for the elevated copy number is also indicated as it would appear on the sense strand. (B) The predicted secondary structure of the antisense RNAI molecule (20,26). The arrow indicates the result of the high copy number mutation. The stem loop structures I′, II′ and III′ are also labeled.

ened then RNAII can more frequently assume a confirmation required for it to prime the initiation of replication. Figure 1B shows that the RSF1030 high copy number mutation disrupts a G-C intramolecular base pair in the stem region comprising the II′ structure of the RNAI molecule (20). This mismatch is positioned so that it would extend a stretch of unpaired bases within the II′ stem to a total of four. It is reasonable to suggest that this alteration could also disrupt the II′ loop structure and, as a consequence, weaken the ability of RNAI to interact with the RNAII primer, leading to an increased frequency of initiation of plasmid replication. It is also worthy to note that, although significant homology exists between the RNAI molecules of several ColE1-like plasmids, the high copy number mutation corresponds to a region of RNAI that is unique to RSF1030 (20).

The high copy number RSF1030 mutant was modified to

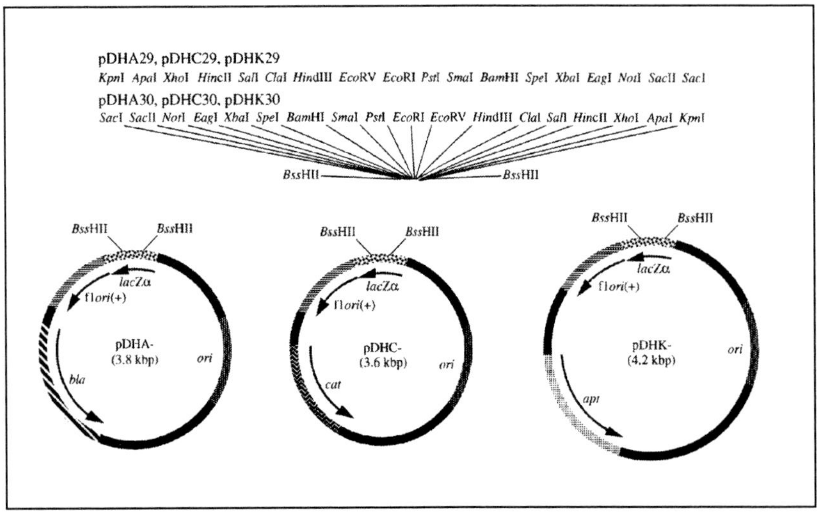

Figure 2. High copy number RSF1030-derivative cloning vectors. Features of the plasmids include a number of unique restriction enzyme recognition sites within the *lacZα* region; antibiotic resistance markers imparting resistance to Amp, Cam and Kan; f1*ori*(+); and the *ori* region of the RSF1030 high copy number mutant.

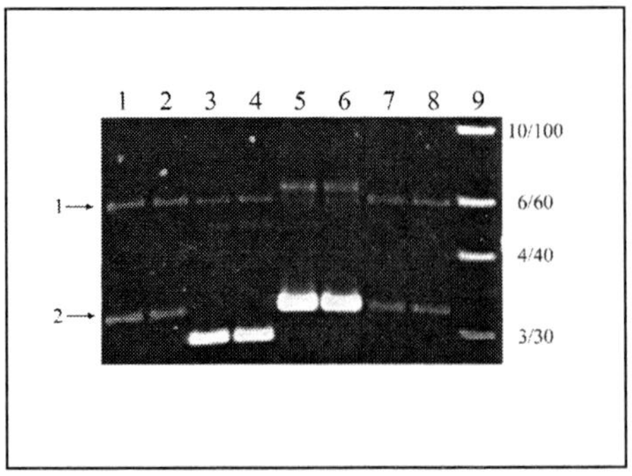

Figure 3. Copy number determination of high copy number cloning vectors. Plasmid DNA samples were resolved in duplicate and includes pBR322 (lanes 1 and 2); pUC18 (lanes 3 and 4); pDHA29 (lanes 5 and 6); pACYC184 (lanes 7 and 8); DNA mass ladder (Life Technologies) showing the ratio of the size (kb) and quantity of DNA (ng) in each band (lane 9). Arrow 1 indicates the location of pWSK129 and arrow 2 shows the location of the co-resident plasmids. Isolation of plasmid DNA and copy number determinations were performed as described in the text.

introduce a variety of features to convert it to a series of versatile cloning vectors. Using standard techniques of recombinant DNA (19), a 1.5 kb *Bst*YI fragment containing the *ori* from the high copy pST19 mutant was ligated to a 2.3-kb *Bgl*II fragment from pWSK29 and pWSK30 (28), creating pDHA29 and pDHA30, respectively (Figure 2). These plasmids impart resistance to high levels of ampicillin (Amp^R, 100 μg/mL). Derivatives of pDHA29 and pDHA30 were also constructed that impart resistance to different antibiotics. The chloramphenicol (Cam^R, 20 μg/mL) derivatives pDHC29 and pDHC30 were constructed by replacing 0.7 kb *Dra*I fragment containing the β-lactamase gene (*bla*) from pDHA29 and pDHA30 with a similar sized *Bsa*AI-*Bst*UI fragment encoding *cat* from pSU23 (1), respectively. pDHK29 and pDHK30 are kanamycin resistant derivatives (Kan^R, 30 μg/mL) and were constructed by replacing *bla* on pDHA29 and pDHA30 with a 1.3 kb *Stu*I fragment encoding the Kan^R determinant (*apt*) from pBSK8 (23), respectively. As shown in Figure 2, all vectors carry the multiple cloning site and *lacZ*α from pBluescriptII to facilitate screening of recombinant plasmids by blue/white colony color on X-gal. The vectors also carry f1*ori*(+) for isolation of single-stranded DNA.

To characterize the cloning vectors, plasmid copy number determinations were made by a modification of the technique described by Wegrzyn et al. (30). DH5α (Life Technologies, Gaithersburg, MD, USA) was transformed with pDHA29, pBR322 or pBluescriptIIKS(+) as described (11) and then co-transformed with the pSC101-derivative plasmid pWSK129 (28). Cultures were grown at 30°C to saturation in the presence of antibiotics, and the cells were lysed by treatment with alkaline lysis solution (19). Plasmid DNA was resolved directly from the cell lysates by electrophoresis through a 0.8% agarose gel. Following staining with ethidium bromide, the gel image (Figure 3) was captured and analyzed by a Gel-Doc™ 2000 imaging system (Bio-Rad Laboratories, Hercules, CA, USA) using pWSK129 DNA as an internal standard. Copy number determinations were made by comparing the relative fluorescence of pDHA29 with that of pBR322 and pBluescriptIIKS(+). The copy number of pDHA29 was observed to be 150–200 copies/chromosome, a value somewhat elevated over pUC- and pBluescript-derivative plasmids (5). Similar copy numbers were also observed for the Cam^R and Kan^R derivative plasmids.

The new cloning vectors were shown to be compatible with pBR322, pBluescriptII, pACYC184 and pWSK129. Different pDH- vectors were transformed into *E. coli* DH5α that had previously been transformed with other cloning vectors encoding different antibiotic resistances. After culturing the transformants in the absence of antibiotic selection overnight (representing approximately 20 generations), cells were plated onto media to select for the RSF1030-derivative transformants. Colonies were subsequently replica plated onto media to screen for the antibiotic resistance encoded by the co-resident plasmid. A typical result indicated that 98% of the population retained both the high copy number plasmid and the second cloning vector. In contrast, plasmids with known incompatibilities showed a nearly complete loss of one of the plasmids. In a typical experiment, only 12% of a population retained two incompatible plasmids when cultured overnight without antibiotics. Two RSF1030-derivatives with different antibiotic resistance markers were also shown to be incompatible with each other (data not shown).

The newly constructed cloning vectors were used to express both components of the *E. coli* SRP in the same bacterial stain. To accomplish this, a derivative of pDHC29 that carried *ffs*+ was constructed. This plasmid, pDHC29-*ffs*, was made by introducing a 0.6 kb *Hin*dIII-*Bam*HI fragment encoding 4.5S RNA from pSB832 (3) into the multiple cloning site of pDHC29. pDHC29-*ffs* was transformed into a derivative of *E. coli* MC4100 (21) that also carried pFfh18, a pUC18-based plasmid previously constructed in our laboratory that expresses Ffh. Control plasmids were also assembled

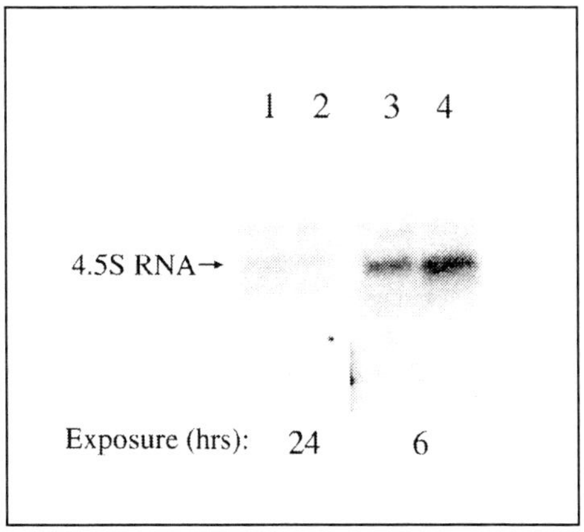

Figure 4. Northern blot analysis of 4.5S RNA expression. RNA samples were prepared in duplicate from *E. coli* co-transformed with pFfh18 and pDHC29-Δ*ffs* (lanes 1 and 2); or pFfh18Δ*ffh* and pDHC29-*ffs*+ (lanes 3 and 4). The location of 4.5S RNA is indicated along with the time of exposure of the blots.

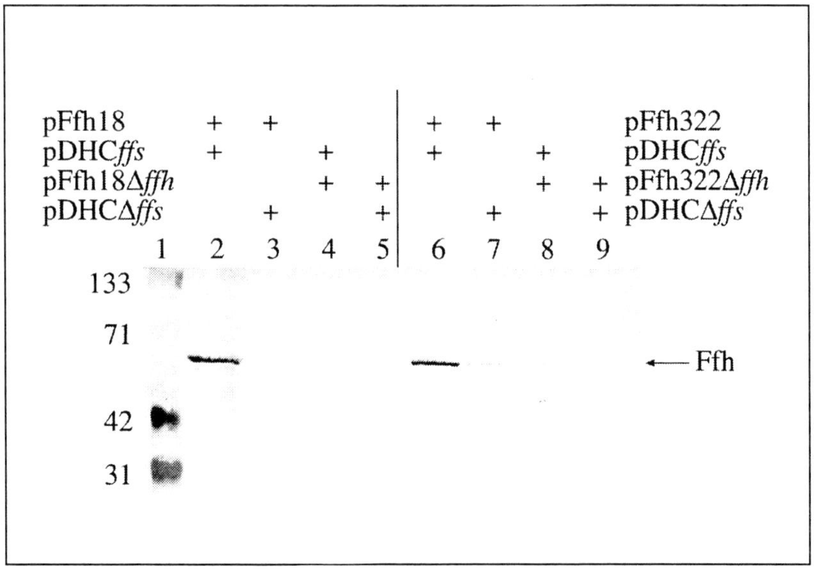

Figure 5. Western blot analysis of Ffh expression. Ffh protein was detected from cells transformed with different combinations of plasmids, as indicated above the blot. Lane 1, molecular weight markers with the corresponding values given in kd. Lanes 2–5 represent expression of Ffh from pUC18 derivative plasmids while lanes 6–9 are from pBR322 derivatives, described in the text. The location of the Ffh protein is also shown.

by deleting the coding region for either *ffs* or *ffh*. pDHC29-*ffs* was digested with *Mlu*I and religated to yield pDHC29-Δ*ffs*. pFfh18Δ*ffh* was constructed by removing an *Eco*RV fragment to delete a significant portion of the *ffh* gene. Two additional plasmids were constructed by inserting *ffh*+, and the *Eco*RV deletion derivative into pBR322, yielding pFfh322 and pFfh322Δ*ffh*, respectively.

The levels of 4.5S RNA expressed from pDHC29-*ffs* were determined by Northern hybridization. Cultures of MC4100 transformed with pFfh18 along with either pDHC29-*ffs* or pDHC29-Δ*ffs* were grown overnight in the presence of both Amp and Cam. Total cellular RNA was isolated from the cultures (Tri-Pure® Isolation reagent, Roche Molecular Biochemicals, Indianapolis, IN, USA) and resolved by electrophoresis on a 1.3% agarose gel under denaturing conditions (19). RNA was electrophoretically transferred to a nylon membrane (Hybond®; Amersham Pharmacia Biotech, Piscataway, NJ, USA) and probed with a 0.6 kb *Hin*dIII-*Bam*HI *ffs* fragment from pSB832 that had been radioactively labeled with [^{32}P]-dATP (HexaLabel™ DNA labeling kit; MBI-Fermentas, Amherst, NY, USA). Blots were hybridized at 68°C and washed twice at room temperature before exposure by phosphorimager analysis (GS-363 Molecular Imager System™; Bio-Rad Laboratories). As shown in Figure 4 (lanes 3 and 4), the strain transformed with the pDHC29-*ffs*+ plasmid yielded significantly more 4.5S RNA than the control strain where the RNA was expressed only from a single chromosomal copy of *ffs* (lanes 1 and 2). As noted in Figure 4, the exposure times of the filters had to be adjusted to ensure adequate resolution of both samples.

Western immunoblot analysis was then used to determine the amount of Ffh expressed from various MC4100 transformants. Cultures of different transformants were grown overnight, as described above. Samples were prepared and total cellular proteins were resolved by SDS-PAGE on a 10% gel, using a standard protocol (8). Proteins were electrophoretically transferred to nitrocellulose, and the filters were reacted with polyclonal antibodies against Ffh as described (8). The decorated proteins were visualized by addition of horseradish peroxidase-conjugated secondary antibodies and substrate (Opti-4CN™; Bio-Rad Laboratories). Although Figure 5 shows that Ffh could be detected in all samples, we observed that significant overproduction of the protein occurred only when 4.5S RNA was also expressed at elevated levels in the cell (compare Figure 5, lanes 2 and 3). Strikingly, the levels of Ffh expressed from pFfh18 alone were not significantly higher than that seen when Ffh was expressed from the chromosome (Figure 5, lanes 3 and 5). Similar results were also seen when Ffh was expressed from a pBR322-derivative plasmid (Figure 5, lanes 6–9). This result is consistent with a previous report that the stability of Ffh is dependent on the levels of 4.5S RNA (12). Apparently, Ffh is unstable when not assembled into the SRP and so significant overproduction of the protein actually requires increased expression of 4.5S RNA.

In conclusion, the pDH-vectors have been constructed that replicate at high copy numbers are compatible with commonly used cloning vectors. Derivatives of these vectors have been used to produce both components of the *E. coli* SRP and likewise should be useful to increase expression of a number of gene products in combination with products expressed from a number of other cloning vectors.

ACKNOWLEDGMENTS

The authors thank Sidney Kushner and Tapan Som for plasmids and P.C. Tai for antibodies. This is journal paper No. J-18212 of the Iowa Agriculture and Home Economics Experiment Station, Ames, IA, Project No. IOW3220, and was supported by the National Institutes of Health (GM50836-04), Hatch Act and State of Iowa funds. Address correspondence to Dr. Gregory J. Phillips, Department of Microbiology, 207 Science I Building, Iowa State University, Ames, IA 50011, USA. Internet: gregory@iastate.edu

REFERENCES

1. **Bartolomé, B., Y. Jubete, E. Martínez and F. de la Cruz.** 1991. Construction and properties of a family of pACYC184-derived cloning vectors compatible with pBR322 and its derivatives. Gene *102*:75-78.
2. **Bolivar, F., R.L. Rodriguez, P.J. Greene, M.C. Betlach, H.L. Heyneker, H.W. Boyer, J.H. Crosa and S. Falkow.** 1977. Construction and characterization of new cloning vehicles. II. A multipurpose cloning system. Gene *2*:95-113.
3. **Brown, S. and M.J. Fournier.** 1984. The 4.5S RNA gene of *Escherichia coli* is essential for cell growth. J. Mol. Biol. *178*:533-550.
4. **Chang, A.C.Y. and S.N. Cohen.** 1978. Construction and characterization of amplifiable multicopy DNA cloning vehicles derived from the P15A cryptic miniplasmid. J. Bacteriol. *134*:1141-1156.
5. **Chao, S.L.-C., W.-T. Chen and T.-T. Wong.** 1992. High copy number of the pUC plasmid results from a rom/rop-suppressible point mutation in RNAII. Mol. Microbiol. *6*:3385-3393.
6. **Covarrubias, L., L. Cervantes, A. Covarrubias, X. Soberón, I. Vichido, A. Blanco, Y.M. Kuperstoch-Portnoy and F. Boliver.** 1981. Construction and characterization of new cloning vehicles. V. Mobilization and coding properties of pBR322 and several deletion derivatives including pBR327 and pBR328. Gene *13*:25-35.
7. **Davison, J.** 1984. Mechanism of control of DNA replication and incompatibility in ColE1-type plasmids: a review. Gene *28*:1-15.
8. **Harlow and Lane.** 1994. Antibodies, A Laboratory Manual. 2nd ed., CSH Laboratory Press, Cold Spring Harbor, NY.
9. **Hasan, N. and W. Szybalski.** 1995. Construction of *lacI*^{ts} and *lacI*^{qts} expression plasmids and evaluation of the thermosensitive *lac* repressor. Gene *163*:35-40.
10. **Heffron, G., R. Sublett, R.W. Hedges, A. Jacob and S. Falkow.** 1975. Origin of the TEM beta-lactamase gene found on plasmids. J. Bacteriol. *122*:250-256.
11. **Inoue, H., H. Nojima and H. Okayama.** 1990. High efficiency transformation of *Escherichia coli* with plasmids. Gene *96*:23-28.
12. **Jensen, C.G., S. Brown and S. Pedersen.** 1994. Effect of 4.5S RNA depletion on *Escherichia coli* protein synthesis and secretion. J. Bacteriol. *176*:2502-2506.
13. **Lerner, C.G. and M. Inouye.** 1990. Low copy number plasmids for regulated low-level expression of cloned genes in *Escherichia coli* with blue/white insert screening capability. Nucleic Acids Res. *18*:4631.
14. **Masukata, H. and J. Tomizawa.** 1984. Effects of point mutations on formation and structure of the RNA primer for ColE1 DNA replication. Cell *36*:513-522.
15. **Masukata, H. and J. Tomizawa.** 1986. Control of primer formation for ColE1 plasmid replication: conformational change of the primer transcript. Cell *44*:125-136.
16. **Munson, M., P.F. Predki and L. Regan.** 1994. ColE1-compatible vectors for high-level expression of cloned DNAs from the T7 promoter. Gene *144*:59-62.
17. **Poritz, M.A., H.D. Bernstein, K. Strub, D. Zopf, H. Wilhelm and P. Walter.** 1990. An *E. coli* ribonucleoprotein containing 4.5S RNA resembles mammalian signal recognition particle. Science *250*:1111-1117.
18. **Ribes, V., K. Romisch, A. Giner, B. Dobberstein and D. Tollevey.** 1990. *E. coli* 4.5S RNA is part of a ribonucleoprotein particle that has properties related to signal recognition particle. Cell *63*:591-600.
19. **Sambrook, J., E.F. Fritsch and T. Maniatis.** 1989. Molecular Cloning: A Laboratory Manual. CSH Laboratory Press, Cold Spring Harbor, NY.
20. **Selzer, G., T. Som, T. Itoh and J. Tomizawa.** 1983. The origin of repli-

cation of plasmid p15A and comparative studies on the nucleotide sequences around the origin of related plasmids. Cell *32*:119-129.

21. **Silhavy, T.J., M.L. Berman and L.W. Enquist.** 1984. Experiments with Gene Fusions. CSH Laboratory Press, Cold Spring Harbor, NY.
22. **Som, T. and J. Tomizawa.** 1982. Origin of replication of *Escherichia coli* plasmid RSF1030. Mol. Gen. Genet. *187*:375-383.
23. **Spratt, B.G., P.J. Hedge, S. te Heesen, A. Edelman and J.K. Broome-Smith.** 1986. Kanamycin-resistant vectors that are analogues of plasmids pUC8, pUC9, EMBL8 and EMBL9. Gene *41*:337-342.
24. **Tomizawa, J.** 1984. Control of ColE1 plasmid replication: the process of binding of RNA I to the primer transcript. Cell *38*:861-870.
25. **Tomizawa, J.** 1986. Control of ColE1 plasmid replication: binding of RNA I to RNA II and inhibition of primer formation. Cell *47*:89-97.
26. **Tomizawa, J.** 1990. Control of ColE1 plasmid replication. Intermediates in the binding of RNA I and RNA II. J. Mol. Biol. *212*:683-694.
27. **Vieira, J. and J. Messing.** 1982. The pUC plasmids, an M13mp7-derived system for insertion mutagenesis and sequencing with synthetic universal primers. Gene *19*:259-68.
28. **Wang, R.F. and S.R. Kushner.** 1991. Construction of versatile low-copy-number vectors for cloning, sequencing and gene expression in *Escherichia coli*. Gene *100*:195-199.
29. **Warne, S.R., C.M. Thomas, M.E. Nugent and W.C.A. Tacon.** 1986. Use of a modified *Escherichia coli trpR* gene to obtain tight regulation of high-copy-number expression vectors. Gene *46*:103-112.
30. **Wegrzyn, G., P. Neubauer, S. Krueger, M. Hecker and K. Taylor.** 1991. Stringent control of replication of plasmids derived from coliphage λ. Mol. Gen. Genet. *225*:94-98.
31. **Wong, E.M., M.A. Muesing and B. Polisky.** 1982. Temperature-sensitive copy number mutants of *Col*E1 are located in an untranslated region of the plasmid genome. Proc. Natl. Acad. Sci. USA *79*:3570-3574.
32. **Yanisch-Perron, C., J. Viera and J. Messing.** 1985. Improved M13 phage cloning vectors and host strain: nucleotide sequence of the M13mp18 and pUC19 vectors. Gene *33*:103-119.

Update to: High Copy Number Plasmids Compatible with Commonly Used Cloning Vectors

Gregory J. Phillips

Iowa State University, Ames, IA, USA

The high copy number cloning vectors described here are available upon request from the corresponding author. In addition, files containing the DNA sequences of the vectors can be found at: (**http://www.micro.iastate.edu/faculty/Phillipslab/Frame2.html#plasmids**).

CHAPTER 8

Construct for High-Level Expression and Low Misincorporation of Lysine for Arginine During Expression of pET-Encoded Eukaryotic Proteins in *Escherichia coli*

Jianxin You, Robert E. Cohen[1] and Cecile M. Pickart

Johns Hopkins University, Baltimore, MD and [1]University of Iowa, Iowa City, IA, USA

(Reprinted from BioTechniques 27:950-954, 1999)

ABSTRACT

The arginine codon AGA is rarely used in E. coli *but is common in eukaryotic genes. Prior studies have shown that the low level of* $tRNA_{UCU}^{Arg}$ *can lead to low expression and misincorporation of lysine for arginine, during expression of genes containing AGA codons in* E. coli. *The chloramphenicol-selectable plasmid pJY2 is designed to facilitate the expression of such genes cloned into pET vectors: it encodes T7 lysozyme (to depress constitutive expression of the cloned gene) and* $tRNA_{UCU}^{Arg}$ *(to suppress lysine misincorporation at AGA codons). Using pJY2, we observed robust and translationally faithful expression of mutant ubiquitin genes in which 14% (11 out of 76) of the total codons were AGA. Competent BL21(DE3)-pJY2 cells can be used to suppress lysine misincorporation and achieve high-level expression of pET-encoded target genes without modification of AGA codons in the target gene sequence.*

INTRODUCTION

The arginine codon AGA occurs commonly in eukaryotic genes but is extremely rare in *Escherichia coli* (7). When eukaryotic genes harboring AGA codons are expressed in *E. coli*, this discordance in codon usage can lead to (*i*) low expression level and (*ii*) misincorporation of lysine for arginine (3–5, 7,9,15,17). That these problems arise from a low level of $tRNA_{UCU}^{Arg}$ is indicated by the restoration of translational fidelity and robust expression on the substitution of AGA with a different arginine codon (5,7,9). In some cases, it has been possible to achieve the same effect by using a helper plasmid carrying the argU gene (also known as dnaY), which encodes $tRNA_{UCU}^{Arg}$ (3–5,7,9,16,17).

We are interested in ubiquitin conjugation reactions. Conjugation involves the formation of an isopeptide bond between the carboxyl terminus of ubiquitin and a lysine residue of a target protein (12,14). A favored in vitro strategy for identifying ubiquitin conjugation sites involves the mutation of candidate lysines to arginines. This strategy will be defeated if there is misincorporation of lysine for arginine during target-protein expression.

Among the targets of ubiquitin conjugation is ubiquitin itself (14). The unmutated ubiquitin gene has four AGA-encoded arginine residues (R42, R54, R72 and R74) and seven lysine residues. Here, we show that these endogenous AGA codons, and other AGA codons introduced by mutagenesis, are subject to extensive lysine misincorporation during ubiquitin expression in *E. coli* strain BL21(DE3)pLysS (18). Attempts to overcome this problem by replacing pLysS with an argU-encoding plasmid caused a severe reduction in the expression of ubiquitin. We have developed a new plasmid, pJY2, which carries both the argU and T7 lysozyme (LysS) genes. The use of pJY2 (in place of pLysS) allowed for robust expression of the above-described ubiquitins while reducing lysine misincorporation to a nearly undetectable level.

MATERIALS AND METHODS

Plasmids and Strains

pLysS was obtained from F. Studier (Brookhaven National Laboratory, Upton, NY, USA) (18). pDC952 (5,17) was obtained from J. Walker (University of Texas, Austin, TX, USA) and is identical to pACYC184 (6) except that pDC952 carries a 650-bp *Bam*HI/ *Hin*dIII fragment containing the *E. coli* argU gene (15). A set of pET3a vectors encoding "single-lysine" ubiquitins were provided by M. Ellison (University of Alberta, Edmonton, Alberta, Canada) (1,13). K0 is a ubiquitin gene in which all seven lysine codons are mutated to AGA, while K6-, K11-, K27-, K29-, K33-, K48- and K63-Ub are genes in which every lysine codon, except the one indicated, has been changed to AGA. DNA sequence analysis showed that relative to the yeast ubiquitin sequence, pET3a-K63-Ub specified P19 (vs. S19) and E24 (vs. D24). The calculated mass of K63-ubiquitin is 8732 Da. K48R-ubiquitin, which has a calculated mass of 8593 Da, was expressed from

Cloning and Expression Vectors for Gene Function Analysis
Edited by Quinn Lu and Michael P. Weiner

pET3d-K48R-Ub (2).

A plasmid carrying both the argU and LysS genes was constructed by cloning the *Bam*HI fragment of pLysS (carrying the T7 lysozyme gene) into the *Bam*HI site of pDC952. We obtained clones with the lysozyme gene in both orientations relative to the *tet* promoter upstream of the argU gene. Clone pJY2, carrying the lysozyme gene in the antisense orientation, was selected for this work (Figure 1). As in pLysS (18), a weak T7 polymerase promoter (Ø3.8) immediately following the lysozyme gene allows for weak constitutive expression of lysozyme by transcription around the plasmid.

Protein Expression and Purification

The pET vectors encoding mutant ubiquitins were transformed into mineral-competent BL21(DE3) cells harboring one of the following helper plasmids: pLysS, pDC952 or pJY2 (each helper plasmid carried a chloramphenicol marker). Ubiquitin expression was induced with 0.4 mM isopropyl-1-thio-β-D-galactoside (IPTG) (2). Ubiquitins were purified at neutral pH by sequential filtration through Q-Sepharose® and S-Sepharose, followed by gel filtration on Sephacryl®-200 (all resins were from Amersham Pharmacia Biotech, Piscataway, NJ, USA).

Polyubiquitin Chain Assembly

The assembly of ubiquitin into K48-linked chains, which was catalyzed by the ubiquitin-conjugating enzyme $E2_{25K}$, was assayed in the presence of 10 μM $E2_{25K}$ and 0.3 mg/mL purified ubiquitin (pH 7.3, 37°C) (11). Incubations were sampled at time zero and 2 h. Chains were resolved by sodium dodecyl sulfate polyacrylamide gel electrophoresis (SDS-PAGE), transferred to Immobilon™-P (Millipore, Bedford, MA, USA) for 3 h at 68 V (5°C) and probed with affinity-purified anti-ubiquitin antibodies (10). Immunocomplexes were detected by enhanced chemiluminescence (Amersham Pharmacia Biotech).

Electrospray Ionization Mass Spectrometry

Mass spectra were obtained at the University of Iowa Mass Spectrometry Facility using an Autospec Mass Spectrometer (Micromass, Manchester, England, UK) equipped with an electrospray interface. Samples in water/acetonitrile/formic acid (50:50:1) were introduced by direct infusion at a flow rate of 10 μL/min. A mass range (m/z) of 700–2200 Da was scanned, and for each sample, approximately 30 scans were acquired, averaged and transformed using OPUS™ software (Micromass). The instrument was calibrated with tetraethyl ammonium iodide to within 0.2 Da.

RESULTS AND DISCUSSION

Extensive Lysine Misincorporation During Expression of Ubiquitin Genes with Multiple AGA Codons in Strain BL21(DE3)pLysS

Eight pET3a plasmids encoding mutant ubiquitins were expressed in BL21(DE3)pLysS cells. Each ubiquitin gene specified four natural AGA-encoded arginine residues (R42, R54, R72 and R74). In seven of the plasmids, six (of seven) natural lysine residues were mutated to arginine using the AGA codon (K6-, K11-, K27-, K29-, K33-, K48- and K63-ubiquitin); the eighth plasmid encoded a lysine-free protein, K0-ubiquitin (1,13). Each mutant protein was expressed at a high level, as shown for K63-ubiquitin in Figure 2. Coomassie® Blue staining revealed a prominent band of approximately 8 kDa (Figure 2A, lane 2); the identity of this protein as ubiquitin was confirmed by western blotting (Figure 2B, lane 2). Thus, the presence of a high density of AGA codons (14%–16% of total codons) did not inhibit ubiquitin expression.

The presence of AGA codons can also cause misincorporation of lysine for arginine (5,9). As an initial test of whether such misincorporation occurred with the mutant ubiquitins, we evaluated the purified proteins as substrates for ubiquitin-ubiquitin conjugation (i.e., polyubiquitin chain assembly) catalyzed by ubiquitin-conjugating enzyme $E2_{25K}$. This enzyme exclusively used K48 in chain assembly, as determined by direct tryptic peptide mapping (8) and its inability to polymerize K48R-ubiquitin (2,11). If there was no lysine misincorporation, then only K48-ubiquitin would be a substrate for $E2_{25K}$. Instead, all of the mutant ubiquitins, including the "lysine-free" protein K0-ubiquitin, were active as substrates (Figure 3A and data not shown). Thus, in spite of robust expression, extensive misincorporation of lysine at the AGA_{48} codon occurred during the expression of each mutant ubiquitin gene.

The extent of misincorporation was quantitatively analyzed by electrospray ionization mass spectrometry (ESIMS) of purified K63-ubiquitin. The spectrum displayed seven peaks: one of approximately the expected mass (8732 Da) and six peaks of lower mass (Figure 4A). Each peak differed in mass from its neighbor(s) by an increment of 28 Da, corresponding to the mass difference between lysine and arginine. Thus, successively lower-mass peaks represented proteins that contain an increasing number of misincorporated lysine residues. Only a tiny fraction (ca. 3%) of the molecules corresponded to the translation product actually specified by the K63-ubiquitin gene. Calderone et al. reported 36%–42% misincorporation at AGA codons during recombinant protein expression in *E. coli* (5). The distribution of translation products seen in Figure 4A fits well to the binomial distribution predicted by a 35%–40% probability of lysine misincorporation at each of the 11 AGA codons in the K63-ubiquitin gene.

Poor Expression in BL21(DE3)pDC952 Cells

Co-expression of $tRNA_{UCU}^{Arg}$, encoded by the argU gene in pDC952, has been reported to suppress lysine misincorporation at AGA codons (5). However, when BL21(DE3)pDC952 cells were used for mutant ubiquitin expression, the level of expressed ubiquitin was reduced to less than 1% (Figure 2, A and B, lane 3). Although it is likely that lysine misincorporation was significantly suppressed (5,9), the low expression of K63-ubiquitin prevented us from carrying out ESIMS analysis to address this issue. We suspected that the low expression was because of the absence of T7 lysozyme, which was expressed from pLysS in the studies described above. T7 lysozyme suppresses constitutive expression of pET-encoded genes while allowing a high level of expression upon induction of T7 RNA polymerase; the reduction in constitutive expression can reduce toxicity correlated with recombinant protein expression (18).

Restoration of Translational Fidelity and Robust Expression in BL21(DE3) pJY2 Cells

To reduce constitutive ubiquitin expression and simultaneously reduce lysine misincorporation, we sought to co-express lysozyme and $tRNA_{UCU}^{Arg}$. Because pLysS and pDC952 are both derived from pACYC184 (17,18) and thus cannot be stably maintained in the same cell (6), we generated a new plasmid, pJY2, that carried the lysozyme and argU genes (see Materials and Methods and Figure 1). pJY2 was transformed into BL21(DE3) cells. Competent cells were then transformed with the pET-ubiquitin plasmids.

The presence of pJY2 restored high expression of the mutant proteins (Figure 2, A and B, lane 4), suggesting that the low expression in BL21(DE3)pDC952 cells was indeed due to toxicity associated with premature ubiquitin expression. Moreover, the presence of the argU gene in pJY2 strongly enhanced translational fidelity, as shown by the finding that only K48-ubiquitin was well-utilized by $E2_{25K}$ (Figure 3B, the band labeled E2-Ub corresponds to an auto-ubiquitinated form of $E2_{25K}$, i.e., it is not a polyubiquitin chain). Traces of diubiquitin seen in several reactions in Figure 3B may be due to a low level of lysine misincorporation at AGA_{48} or to a weak ability of $E2_{25K}$ to utilize other lysine residues.

A strong reduction in lysine misincorporation was rigorously confirmed by ESIMS analysis of purified K63-ubiquitin. Figure 4B shows a single major translation product, whose mass (8732 Da) agreed with that calculated from the deduced amino acid sequence. The several minor peaks do not show the differences in mass (±28 Da) that are expected if these peaks arise from lysine misincorporation (see Figure 4B legend). ESIMS analysis of K0-ubiquitin purified from BL21(DE3)pJY2 cells indicated that this protein also exhibited negligible lysine misincorporation (not shown).

The presence of AGG arginine codons can also be problematic for gene expression in *E. coli*. The presence of tandem AGG codons can lead to frameshifting and thereby decrease the expression of recombinant proteins, while the presence of multiple dispersed AGG codons can reduce expression level (3,4,7,16,17). Prior studies of $tRNA_{UCU}^{Arg}$-dependent suppression of frameshifting at tandem AGG and tandem AGA codons showed that this tRNA is AGA-specific when it is expressed from the endogenous argU gene, but is able to decode AGG when overexpressed (17). This divergence in properties has been suggested to reflect incomplete

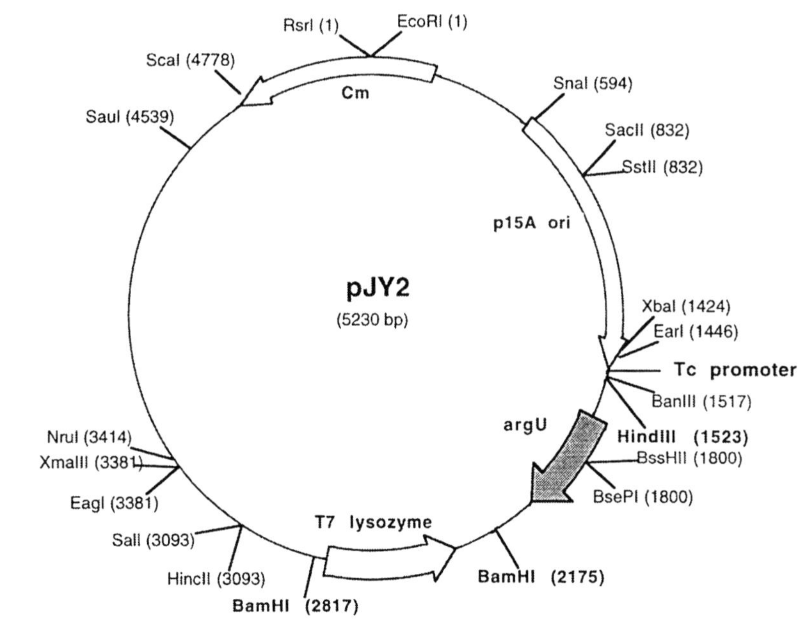

5'-AAGCTTGGCTGCAGGTCACTTTGATACATGAAAATACGGGTTTTCTTG
ATTCAGACGCGCAGCGGTGTGCGTTTGTTTGCCGCTATAGCGAAATAA
ATCAGAAAATCAGACGCGGTCGTTCACTTGTTCAGCAACC[**AGA----**
ATC]AGCCAGCTTGCCCAGGTGTGTCTGAGGTCATGGAAACGGAAATC
TTCAATTCCTGCACGACGACAAGCTGATAGCCATGATGTCTTGCTGTC
GATGCGCATCTTCCTGACCGCAGGCGTTGATGTTCCATCTGCTCGCTT
AGCCGCCTTGGTATGTGACGGATCC-3'

Figure 1. Map of pJY2. Top: diagram of vector; except for the LysS and argU genes, the sequence of the vector is identical to the sequence of pACYC184. Selected unique restriction sites are shown with their map positions. Bottom: sequences flanking argU gene; the *Bam*HI (5′) and *Hind*III sites (3′) used in cloning are underlined; the start and stop codons of the argU gene are bracketed and shown in bold.

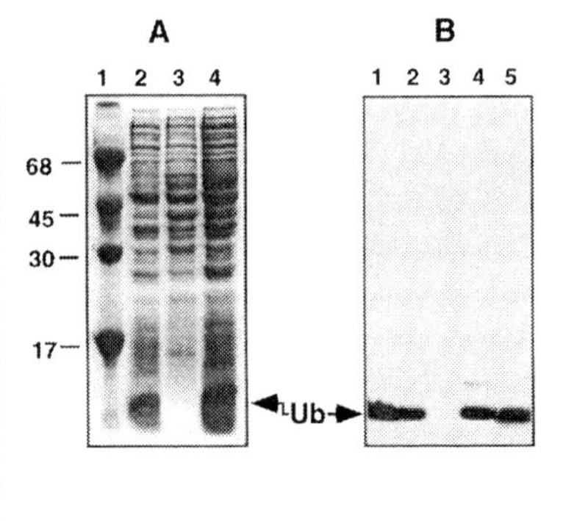

Figure 2. K63-ubiquitin expression in different strains. (A) Coomassie Blue-stained gel. Lane 1, molecular mass standards (in kDa); lanes 2-4, 1 µL of soluble extract from 4-h induced BL21 (DE3) *E. coli* cells transformed with pET3a-K63Ub and the following helper plasmid: pLysS (lane 2); pDC952 (lane 3); pJY2 (lane 4). (B) Western blot developed with anti-ubiquitin antibodies. Lane 1, standard (20 ng); lanes 2–4, 1 µL of 1000-fold dilutions of samples in lanes 2–4 of panel A; lane 5, 2 µL of undiluted extract from lane 3 of panel A. All lanes in panel B derived from one blot; intervening lanes were excised between lanes 4 and 5.

modification of bases in the overexpressed tRNA (17). Because pJY2 is a high copy plasmid (10), most likely pJY2 will ameliorate problems associated with the presence of multiple AGG codons. However, we did not address this question experimentally.

In contrast to frameshifting or pausing, lysine misincorporation at AGG codons has not been reported. In the case of ubiquitin, we did not observe misincorporation at a single AGG codon in the gene encoding a different mutant protein, K48R-ubiquitin. This gene has four natural AGA codons, plus an AGG-encoded arginine at residue 48 (2). As in the case of K63-ubiquitin, K48R-ubiquitin was highly expressed in BL21(DE3)pLysS cells and was subject to extensive lysine misincorporation (Figure 4C), indicating that $tRNA_{UCU}^{Arg}$ was limiting. However, this misincorporation occurred exclusively at the AGA codons, because K48R-ubiquitin expressed in BL21(DE3)pLysS cells cannot be polymerized by $E2_{25K}$ (2,11). This finding is consistent with the conclusion that the product of the endogenous argU gene specifically decodes AGA vs. AGG (17). Lysine misincorporation at the AGA codons of K48R-ubiquitin was suppressed when this protein was expressed in BL21(DE3)pJY2 cells (Figure 4D).

Our results demonstrate that co-expression of $tRNA_{UCU}^{Arg}$ and T7 lysozyme provides great advantages: $tRNA_{UCU}^{Arg}$ expression from the plasmid-borne argU gene virtually eliminated lysine misincorporation at AGA codons, even when these codons were present at a high density, while lysozyme expression reduced constitutive expression of the target gene and allowed for robust expression upon induction. Thus, eukaryotic genes cloned into pET vectors (or other T7 polymerase-based vectors) can be efficiently and accurately expressed in BL21(DE3)pJY2 cells even when these genes harbor multiple AGA codons.

These benefits are documented here for the case of ubiqui-

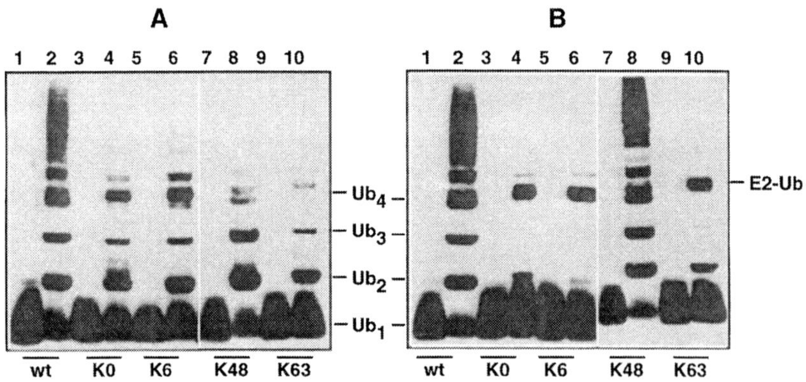

Figure 3. $E2_{25K}$-catalyzed assembly of K48-linked polyubiquitin chains from wild-type and mutant ubiquitins. Chain assembly reactions were carried out as described in Materials and Methods, with wild-type ubiquitin (Sigma, St. Louis, MO, USA) or purified mutant ubiquitins (recombinant, as indicated at the bottom). Odd-numbered lanes: time zero; even-numbered lanes, after 2 h of incubation. E2-Ub, auto-monoubiquitinated $E2_{25K}$ (text). (A) Ubiquitins purified from BL21(DE3)pLysS cells. (B) Ubiquitins purified from BL21(DE3)pJY2 cells. Intervening lanes were excised between lanes 6 and 7 each panel.

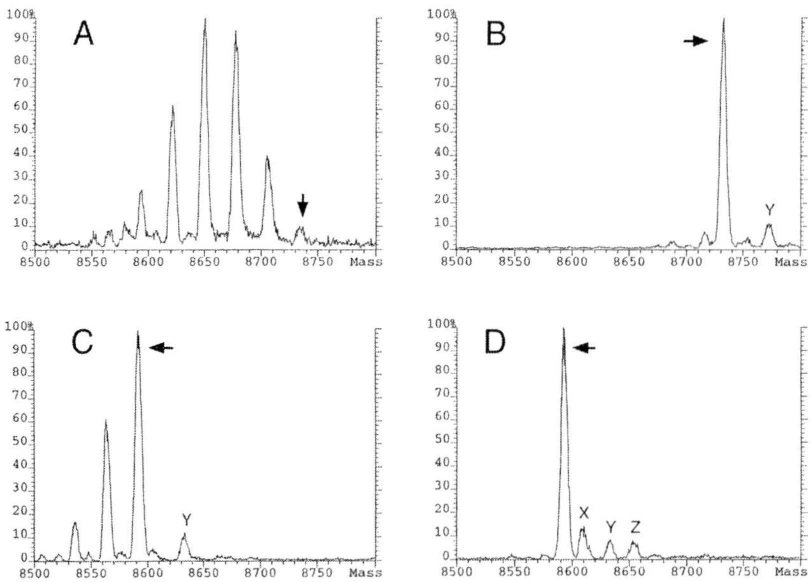

Figure 4. ESIMS analysis of purified recombinant ubiquitins. (A) K63-ubiquitin expressed in BL21(DE3)pLysS cells. (B) K63-ubiquitin expressed in BL21(DE3)pJY2 cells. (C) K48R-ubiquitin expressed in BL21(DE3)pLysS cells. (D) K48R-ubiquitin expressed in BL21(DE3)pJY2 cells. The arrow in each panel indicates the peak for the molecular ion (MH^+) of the ubiquitin species with the expected composition; in panels A and C, the major peaks with lower mass than the expected MH^+ species are smaller by increments of 28 Da and correspond to protein with arginine(s) replaced by lysine(s). The minor peaks X, Y and Z correspond to oxidized (as MetSO) ubiquitin, a potassium adduct and ubiquitin containing both sodium and potassium.

tin; we have also observed that pJY2 affords enhanced expression of non-ubiquitin proteins whose genes harbor AGA codons (data not shown). The use of pJY2 is more convenient than expressing $tRNA_{UCU}^{Arg}$ and T7 lysozyme from separate plasmids (e.g., 4). Moreover, pJY2 is more versatile than the previously described argU-encoding plasmids pUBS520 (3) and pSBET (16), which carry kanamycin markers and therefore cannot be used with kanamycin-selectable pET vectors; the chloramphenicol marker in pJY2 is compatible with all currently available pET vectors. pJY2 can also be used as a source of $tRNA_{UCU}^{Arg}$ with expression systems, such as the pGEX series of vectors, that do not rely on T7 polymerase.

ACKNOWLEDGMENTS

We thank Dr. Lynn Teesch for her help with the mass spectrometry and members of our laboratories for helpful comments on the manuscript. Funded by National Institutes of Health Grant Nos. DK46984 (to C.M.P.) and GM37666 (to R.E.C.). Address correspondence to Cecile M. Pickart, Department of Biochemistry, School of Public Health, Johns Hopkins University, 615 North Wolfe Street, Baltimore, MD 21205, USA. Internet: cpickart@welchlink.welch. jhu.edu

REFERENCES

1. **Arnason, T. and M.J. Ellison.** 1994. Stress resistance in *Saccharomyces cerevisiae* is strongly correlated with assembly of a novel type of multiubiquitin chain. Mol. Cell. Biol. *14*:7876-7883.
2. **Beal, R., Q. Deveraux, G. Xia, M. Rechsteiner and C.M. Pickart.** 1996. Surface hydrophobic residues of multiubiquitin chains essential for proteolytic targeting. Proc. Natl. Acad. Sci. USA *93*:861-866.
3. **Brinkmann, U., R.E. Mattes and P. Buckel.** 1989. High-level expression of recombinant genes in *Escherichia coli* is dependent on the availability of the *dnaY* gene product. Gene *85*:109-114.
4. **Buhler, C., D. Gadelle, P. Forterre, J.C. Wang and A. Bergerat.** 1998. Reconstitution of DNA topoiosomerase VI of the thermophilic archaeon *Sulfolobus shibatae* from subunits separately overexpressed in *Escherichia coli*. Nucleic Acids Res. *26*:5157-5162.
5. **Calderone, T.L., R.D. Stevens and T.G. Oas.** 1996. High-level misincorporation of lysine for arginine at AGA codons in a fusion protein expressed in *Escherichia coli*. J. Mol. Biol. *262*:407-412.
6. **Chang, A.C. and S.N. Cohen.** 1978. Construction and characterization of amplifiable multicopy DNA cloning vehicles derived from the P15A cryptic miniplasmid. J. Bacteriol. *134*:1141-1156.
7. **Chen, G.T. and M. Inouye.** 1994. Role of the AGA/AGG codons, the rarest codons in global gene expression in *Escherichia coli*. Genes Dev. *8*:2641-2652.
8. **Chen, Z. and C.M. Pickart.** 1990. A 25-kilodalton ubiquitin carrier protein (E2) catalyzes multi-ubiquitin chain synthesis via lysine 48 of ubiquitin. J. Biol. Chem. *265*:21835-21842.
9. **Forman, M.D., R.F. Stack, P.S. Masters, C.R. Hauer and S.M. Baxter.** 1998. High level, context dependent misincorporation of lysine for arginine in *Saccharomyces cerevisiae* a1 homeodomain expressed in *Escherichia coli*. Protein Sci. *7*:500-503.
10. **Haas, A.L. and P.M. Bright.** 1985. The immunochemical detection and quantitation of intracellular ubiquitin-protein conjugates. J. Biol. Chem. *260*:12464-12473.
11. **Haldeman, M.T., G. Xia, E.M. Kasperek and C.M. Pickart.** 1997. Structure and function of ubiquitin conjugating enzyme E2-25K: the tail is a core-dependent activity element. Biochemistry *36*:10526-10537.
12. **Hochstrasser, M.** 1996. Ubiquitin-dependent protein degradation. Annu. Rev. Genet. *30*:405-439.
13. **Hodgins, R., C. Gwozd, T. Arnason, M. Cummings and M.J. Ellison.** 1996. The tail of a ubiquitin-conjugating enzyme redirects multi-ubiquitin chain synthesis from the lysine 48-linked configuration to a novel nonlysine-linked form. J. Biol. Chem. *271*:28766-28771.
14. **Pickart, C.M.** 1997. Targeting of substrates to the 26S proteasome. FASEB J. *11*:1055-1066.
15. **Saxena, P. and J.R. Walker.** 1992. Expression of argU, the *Escherichia coli* gene coding for a rare arginine tRNA. J. Bacteriol. *174*:1956-1964.
16. **Schenk, P.M., S. Baumann, R. Mattes and H.-H. Steinbiss.** 1995. Improved high-level expression system for eukaryotic genes in *Escherichia coli* using T7 RNA polymerase and rare ^{Arg}tRNAs. BioTechniques *19*:196-200.
17. **Spanjaard, R.A., K. Chen, J.R. Walker and J. van Duin.** 1990. Frameshift suppression at tandem AGA and AGG codons by cloned tRNA genes: assigning a codon to argU tRNA and T4 tRNA(Arg). Nucleic Acids Res. *18*:5031-5036.
18. **Studier, F.W.** 1991. Use of bacteriophage T7 lysozyme to improve an inducible T7 expression system. J. Mol. Biol. *219*:37-44.

Update to: Construct for High-Level Expression and Low Misincorporation of Lysine for Arginine During Expression of pET-Encoded Eukaryotic Proteins in *Escherichia coli*

Jianxin You, Robert E. Cohen[1] and Cecile M. Pickart

Johns Hopkins University, Baltimore, MD and [1]University of Iowa, Iowa City, IA, USA

Numerous requests for pJY2 have been received; the vector is now commercially available from Affiniti Research Products (product number DW8720; www.proteasome.com). Although we routinely use pJY2 as a helper plasmid when expressing proteins encoded by AGA-containing open reading frames, we have not directly compared protein expression in the presence and absence of pJY2 except in our original report. Shortly before this volume went to press, scientists who had requested pJY2 were queried concerning the plasmid's efficacy in facilitating the expression of AGA-containing open reading frames. Results obtained to date suggest that pJY2 is efficacious in facilitating the expression of many, but not all, open reading frames harboring AGA codons. In some cases, the failure of pJY2 to improve expression may be attributed to the presence of other rare codons.

CHAPTER 9

UGA Read-Through Artifacts—When Popular Gene Expression Systems Need a pATCH

Gavin MacBeath and Peter Kast

The Scripps Research Institute, La Jolla, CA, USA

(Reprinted from BioTechniques 24:789-794, 1998)

ABSTRACT

pET and similar vectors are widely used for efficient gene expression in Escherichia coli *and subsequent protein purification, often by means of a C-terminal histidine (His) tag. We found that the TGA translation termination signal following the His-tag sequence in pET constructs gives rise to a significant fraction of read-through protein extended by 21 amino acids. Mass spectrometry indicated that tryptophan is inserted at the UGA (opal) stop codon in the examined non-opal suppressor strains; no evidence for translational frameshifting was detected. We have shown that the problem of obtaining heterogeneous protein preparations can easily be corrected. Plasmid pATCH1 provides a replacement sequence for the inefficient stop signal and can be used to repair both pET vectors and existing pET-based expression constructs. Our observation illustrates the largely ignored fact that a UGA codon is the worst choice for proper translation termination in efficient overexpression vectors.*

INTRODUCTION

T7 promoter-based gene expression (18,21,22) combined with affinity chromatography (23) is the current strategy of choice for highly efficient production and purification of recombinant proteins using *Escherichia coli*. Suitable expression vectors, such as the frequently utilized and commercially available pET series (Novagen, Madison, WI, USA), often encode a hexapeptide tag consisting of six consecutive histidine (His) residues fused to the C terminus of the protein of interest. $(His)_6$-tagged proteins can usually be purified in a single step by affinity chromatography on a matrix containing chelated Ni^{2+} ions (23).

We have successfully overexpressed a variety of genes in our laboratory using the pET system and purified the corresponding proteins with the aid of a C-terminal His tag. Among these were genes for two structurally unrelated chorismate mutases (CMs) from *E. coli* and *Bacillus subtilis* (12). Unexpectedly, we found that both preparations of the enzymes contained a significant fraction of a contaminating protein with a larger molecular weight. We then set out to identify the contaminants using the method of matrix-assisted laser desorption/ionization mass spectrometry (MALDI-MS) (14,20). Our discovery of a systematic heterogeneity in the protein preparations caused by some improperly designed expression systems has important general implications on gene expression strategies.

MATERIALS AND METHODS

Bacterial Strains, Plasmids and Oligonucleotides

Routine cloning was done in *E. coli* strain XL1-Blue (Stratagene, La Jolla, CA, USA). All other *E. coli* strains used carry a chromosomally integrated, isopropyl β-D-thiogalactopyranoside (IPTG)-inducible gene for T7 RNA polymerase (21). Strains HMS174(DE3) and BL21(DE3) were obtained from Novagen. KA13 was constructed by site-specific integration of λDE3 prophage into the chromosome of strain KA12 (9,10) using the λDE3 Lysogenization Kit (Novagen). Vector pET-22b(+) was also from Novagen. pKSS was constructed in this laboratory (8). The detailed structure of plasmids pET-EcCM and pKET4-W is irrelevant to this work and will be described elsewhere. Importantly, they contain, respectively, a 3′-truncated *pheA* gene from *E. coli* and the entire *aroH* from *B. subtilis* inserted between the *Nde*I and *Xho*I sites and *Nde*I and *Eag*I sites of vector pET-22b(+). The initial Met codons of the CM genes overlap with the *Nde*I site, and the protein's C termini are fused in-frame to the $(His)_6$ tag. pKET3-W is identical to pKET4-W except for the presence of the natural *aroH* stop codon and 6 bp additional nucleotides (ACTAGT) downstream of the *aroH* coding sequence. Oligonucleotides PATX (5′-CGATAACTCGAGCACCACCACCACCACCACTAA-TAATGACTAGTCAGCTGATCCGGCTGCTAACAAAG), 22S-B (5′-AAACCGCTATCAGGGCGAT) and M13-Rev (5′-GGAAACAGCTATGACCATG) were synthesized by Life Technologies (Gaithersburg, MD, USA).

Cloning and Expression Vectors for Gene Function Analysis
Edited by Quinn Lu and Michael P. Weiner

Nucleotide Sequence Analysis

DNA sequences were determined by the Protein and Nucleic Acid Core Facility of The Scripps Research Institute (La Jolla, CA, USA) using a Model 373 DNA Sequencer (PE Applied Biosystems, Foster City, CA, USA). The RNA sequence surrounding the critical stop codon was examined for secondary structure with the MFold program (Genetics Computer Group, Madison, WI, USA).

Construction of pATCH1

A polymerase chain reaction (PCR) fragment was generated on template pET-22b(+) with 0.5 µM of oligonucleotides PATX and 22S-B using *pfu* DNA Polymerase (Stratagene). The PCR consisted of initial denaturation (4 min at 94°C), 25 amplification cycles (1 min at 94°C, 1 min at 50°C and 2 min at 72°C) and a final extension (5 min at 72°C). The 457-bp PCR product was purified with the QIAquick™ PCR Kit (Qiagen GmbH, Hilden, Germany), digested with *Xho*I-*Dra*III, and the gel-purified 420-bp *Xho*I-*Dra*III fragment was ligated to the gel-purified 2443-bp *Xho*I-*Dra*III fragment of pKSS (*caveat*: *Dra*III is prone to star activity). The DNA sequence of the 420-bp *Xho*I-*Dra*III fragment in pATCH1 was confirmed using primers 22S-B and M13-Rev to verify the absence of possible PCR-introduced mutations. All DNA manipulations were carried out according to standard procedures (19) and using enzymes from New England Biolabs (Beverly, MA, USA).

Cell Culturing and Isolation of Chorismate Mutase Protein

For each sample, 40 mL LB medium containing 150 µg/mL sodium ampicillin were inoculated with a single colony of the relevant transformed strain, incubated under good aeration at 37°C to an optical density $(OD)_{600}$ of 0.8 (±0.1) and then induced with 0.4 mM IPTG. After 5 h at 37°C, the cells were harvested, lysed by sonication, and 25% of the soluble protein fraction was mixed with 100 µL Ni-NTA™ agarose (Qiagen GmbH) in a total volume of 1 mL for 5 min at room temperature. The resin was washed 4 times with 1 mL 40 mM imidazole (in 10 mM sodium phosphate buffer, pH 7.5, 160 mM NaCl) followed by elution of the purified proteins with 100 µL of 500 mM imidazole in the same buffer.

Protein Gel Electrophoresis

Sodium dodecyl sulfate polyacrylamide gel electrophoresis (SDS-PAGE) was performed with the PhastSystem™ and on 20% homogeneous gels (Amersham Pharmacia Biotech, Uppsala, Sweden). Gels were stained with a Coomassie® Brilliant Blue R 350 dye. The broad-range protein standard from Bio-Rad (Hercules, CA, USA) was used as the molecular weight marker.

Sample Preparation for Mass Spectrometry

Proteins to be submitted for MALDI-MS were prepared as described above and subsequently desalted on a NAP™-5 column (Amersham Pharmacia Biotech), which was equilibrated and eluted with 1% aqueous acetic acid. One MALDI-MS measurement required less than 0.5 µg of protein (at 0.1 mg/mL). MALDI-MS was performed by The Scripps Mass Spectrometry Laboratory on a Voyager™-Elite Mass Spectrometer with delayed extraction (PerSeptive Biosystems, Framingham, MA, USA).

RESULTS AND DISCUSSION

Heterogeneous Protein Preparations from the pET System

To characterize mechanistic and structural aspects of CMs (9,10,13), we have overexpressed a truncated *pheA* gene from *E. coli* and the *aroH* gene from *B. subtilis*, each encoding a monofunctional CM [termed EcCM (13) and BsCM (10), respectively]. A His tag was fused to the C termini of both CMs by inserting their genes into the appropriate sites of vector pET-22b(+) yielding plasmids pET-EcCM (for EcCM) and pKET4-W (for BsCM). When analyzing purified protein from the T7 RNA polymerase-producing *E. coli* strain KA13 by SDS-PAGE, we noticed that both preparations of the two structurally distinct CMs contained a contaminating protein whose molecular mass was 2–4 kDa higher (Figure 1). This contaminant could not be separated (data not shown) from the expected CM bands by anion exchange (9) or His-tag affinity chromatography and constituted 2% (BsCM) to 5% (EcCM) of the purified protein sample. As shown in Figure 1, the additional protein band was not only present in preparations from

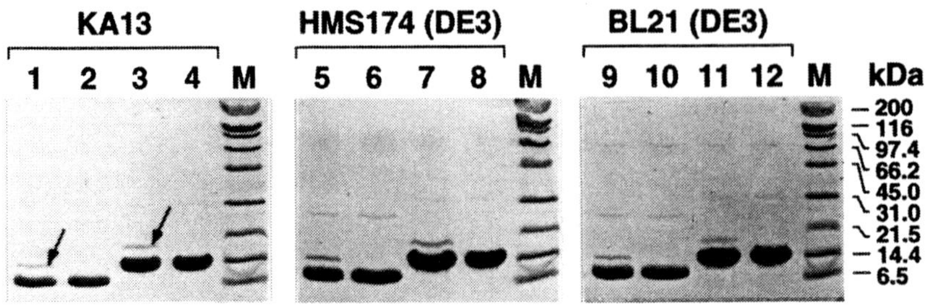

Figure 1. SDS-PAGE analysis of EcCM (12.0 kDa) and BsCM (14.5 kDa) overproduced by three different *E. coli* strains. The respective CM genes were expressed from either pET-EcCM (lanes 1, 5 and 9) and pKET4-W (lanes 3, 7 and 11) or from the corresponding plasmids with corrected translation stop signals pET-EcCM-pATCH (lanes 2, 6 and 10) and pET-BsCM-pATCH (lanes 4, 8 and 12). Arrows in lanes 1 and 3 point to the read-through products. Some additional faint bands corresponding to multiples of the subunit mass are probably artifacts of the electrophoresis/sample preparation leading to covalent bond formation. Lanes marked "M" contain the protein molecular weight marker.

KA13 but also from the commonly used *E. coli* K-12 and B strains HMS174(DE3) and BL21(DE3), respectively. No such contamination was observed when *aroH* was expressed in pET-22b(+) using its native TAA (ochre) stop codon (i.e., using plasmid pKET3-W without translational fusion to the sequence encoding the His tag; data not shown).

This observation prompted us to examine translational read-through at the stop codon following the His-tag sequence in the pET vectors (Figure 2A). Most pET vectors encoding a C-terminal His-tag fusion possess a single TGA (opal) termination codon at this location (Novagen). None of the strains that we tested carry opal suppressor tRNAs (4), but two natural *E. coli* tRNAs are known to decode opal stop codons (16). By using a dedicated translation apparatus, the special $tRNA^{Sec}$ efficiently and co-translationally inserts the "21st amino acid" selenocysteine if the UGA codon is followed by a distinct mRNA stem-loop structure (2). Computer-assisted examination of the sequence in the pET vectors following the TGA codon revealed the possibility of a stable (-15.3 kcal/mol) RNA stem-loop structure (Figure 2A), but it lacks essential features of the selenocysteine incorporation signal (5). The natural $tRNA^{Trp}$ (normal codon UGG) can decode UGA with tryptophan with efficiencies of up to 3% (4,6); in fact, various biological processes rely on such "leaky" termination at opal codons, as first reported for *E. coli* phage Q_{β} (7,15,24). The presence of the stem-loop structure and a purine base (G; Figure 2A) 3′ to the TGA may even enhance $tRNA^{Trp}$-mediated translational read-through (3,11).

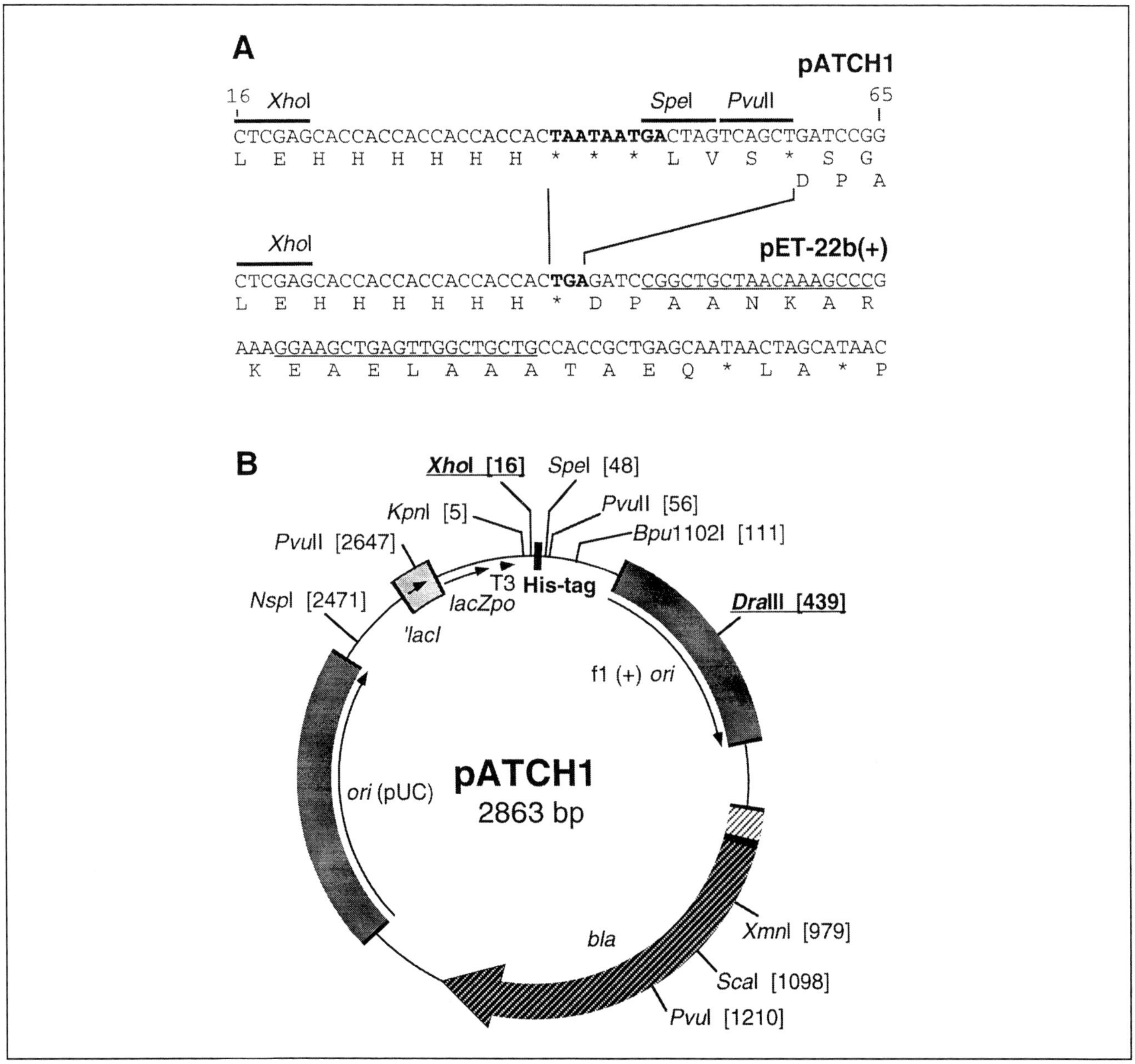

Figure 2. Structural features of the pATCH1 plasmid. (A) Relevant nucleotide and protein sequences around the His tag of pATCH1 and pET-22b(+). Underlined bases are involved in a predicted stem-loop structure. (B) Circular representation of pATCH1 with several restriction sites (and coordinate of 3′ base after cleavage). The His-tag/stop-codon sequence can easily be moved as an *Xho*I-*Dra*III fragment. All other features of pATCH1 are as described in detail for its parent plasmid pKSS (8). The GenBank® Accession Number for pATCH1 is AF013597 (1).

Mass Spectrometric Analysis Indicates Translational Read-Through after Trp Insertion

Analysis of an EcCM protein preparation from strain KA13 by MALDI-MS (14,20) yielded a mass for the extended polypeptide consistent only with the insertion of a tryptophan residue at the stop codon (Figure 3A). As reported earlier for BsCM (9) and shown here for EcCM, a variable fraction of the protein produced under our high-level gene expression conditions has its amino-terminal methionine removed. For each of these two expected EcCM species, an additional mass peak with a 2221 Da higher molecular mass (±0.1% error of MALDI-MS) was observed (Figure 3A). The additional mass can be explained with a C-terminal extension of EcCM by the peptide sequence WDPAANKARKEAELAAATAEQ (Figure 2A) if the TGA triplet following the His-tag sequence is decoded as tryptophan (calculated mass increase: +2223.5 Da).

A statistical analysis of about half of the *E. coli* genome has previously revealed that the use of stop codons and the base on their 3′ side is strongly biased in highly expressed genes (17). The tetranucleotide TGAG present in the pET vector examined here is particularly underrepresented in that set of genes. After discovering this bias, Poole et al. measured termination efficiencies of all possible combinations of stop codons and 3′ nucleotides in competition with +1 frameshifting (17). In the rather specific sequence context of their in vivo assay, TGAG was found to permit significant translational frameshifting, but read-through was not apparent. In contrast to those results, no signals compatible with frameshifting before the stop codon were observed in our highly resolving MALDI-MS experiments.

Correction of Inefficient Translation Termination

We then set out to correct the problem of heterogeneous protein production due to partial read-through at the TGA codon of the pET vectors. Plasmid pATCH1 (Figure 2) was designed to serve as the donor of a DNA fragment to replace the leaky opal codon in the relevant pET constructs. pATCH1 is based on the cloning vector pKSS (8) and features a 420-bp *Xho*I-*Dra*III fragment encoding the $(His)_6$ tag followed by three stop codons (two TAA in tandem and a TGA) to ensure efficient translation termination. The tetranucleotide TAAT introduced at the position of the desired termination site corresponds to the statistically most-favored translational stop signal of highly expressed genes (17). Furthermore, it was shown to minimize frameshifting and allow for the fastest release factor selection rates of all stop signals in vivo (17). An *Spe*I and a *Pvu*II site were added to allow for easy identification of correct clones and, if desired, for subsequent transfer of any target gene together with its His-tag sequence (Figure 2).

To prove the effectiveness of the pATCH1 substitution sequence, we replaced the 404-bp *Xho*I-*Dra*III fragments of the pET-22b(+)-based overproduction plasmids pET-EcCM and pKET4-W with the corresponding 420-bp fragment from pATCH1. The resulting plasmids pET-EcCM-pATCH and pET-BsCM-pATCH were introduced into host strains KA13, HMS174(DE3) and BL21(DE3) and the isolated CM proteins analyzed by SDS-PAGE (Figure 1). In each case, the protein bands corresponding to the read-through polypeptides were fully eliminated by the introduction of the pATCH1 fragments. This is further confirmed by a MALDI-MS experiment with EcCM isolated from KA13/pET-EcCM-pATCH, which shows the absence of the additional mass peaks corresponding to the read-through products (Figure 3B). These stop-codon substitution experiments provide clear functional evidence for the proposed protein elongation by UGA read-through.

CONCLUSIONS

Read-through at a UGA codon by insertion of a Trp residue, rather than translational frameshifting, was found to be responsible for the generation of improperly terminated polypeptides in the gene expression system studied. By

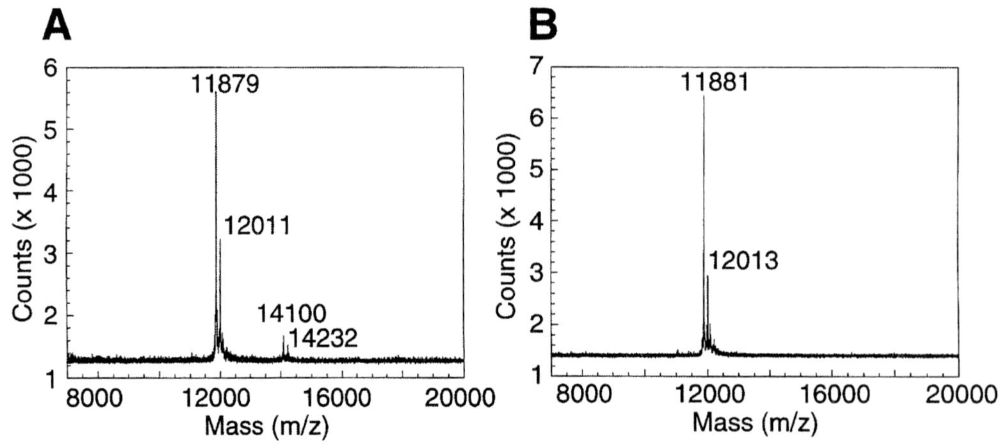

Figure 3. MALDI-MS analysis of EcCM prepared from *E. coli* strain KA13. The theoretical molecular masses per subunit are 12016.9 and 11885.7 Da for EcCM with and without the initial methionine, respectively. The corresponding calculated masses for the extended proteins with Trp insertion at the stop codon are 14240.4 and 14109.2 Da. The host strain harbored either the original expression plasmid pET-EcCM (A) or the stop codon-corrected version pET-EcCM-pATCH (B).

replacing the pET sequences responsible for this read-through with the newly designed fragment from pATCH1, we present a convenient, efficient and error-proof strategy to repair individual pET (or similar) vectors or the many already existing pET-based expression constructs. Alternatively, the inefficient stop signal identified in this work may be directly replaced using custom-made oligonucleotides and PCR with subsequent sequence verification, as detailed in Materials and Methods. The avoidance of inherent heterogeneity in protein preparations is crucial to a variety of biochemical and biophysical studies, including protein crystallization and fluorescence experiments in which the spurious presence of even a single additional tryptophan residue can have dramatic consequences. Everyone involved in protein production in *E. coli* should be aware of our observation and, if necessary, use constructs with tight stop codons or "upgrade" affected pET and similar gene expression systems using the strategies discussed in this report.

ACKNOWLEDGMENTS

We are very grateful to Donald Hilvert and the Skaggs Institute for Chemical Biology for supporting this work. G.M. is the recipient of a Natural Sciences and Engineering Research Council of Canada 1967 Centennial Postgraduate Scholarship and an Eli Lilly Graduate Student Fellowship.

REFERENCES

1. **Benson, D.A., M.S. Boguski, D.J. Lipman and J. Ostell.** 1997. GenBank®. Nucleic Acids Res. *25*:1-6.
2. **Böck, A., K. Forchhammer, J. Heider, W. Leinfelder, G. Sawers, B. Veprek and F. Zinoni.** 1991. Selenocysteine: the 21st amino acid. Mol. Microbiol. *5*:515-520.
3. **Chen, G.-F.T., L. Fang and M. Inouye.** 1993. Effect of the relative position of the UGA codon to the unique secondary structure in the *fdhF* mRNA on its decoding by selenocysteinyl tRNA in *Escherichia coli*. J. Biol. Chem. *268*:23128-23131.
4. **Eggertsson, G. and D. Söll.** 1988. Transfer ribonucleic acid-mediated suppression of termination codons in *Escherichia coli*. Microbiol. Rev. *52*:354-374.
5. **Heider, J., C. Baron and A. Böck.** 1992. Coding from a distance: dissection of the mRNA determinants required for the incorporation of selenocysteine into protein. EMBO J. *11*:3759-3766.
6. **Hirsh, D. and L. Gold.** 1971. Translation of the UGA triplet *in vitro* by tryptophan transfer RNA's. J. Mol. Biol. *58*:459-468.
7. **Horiuchi, K., R.E. Webster and S. Matsuhashi.** 1971. Gene products of bacteriophage Q_β. Virology *45*:429-439.
8. **Kast, P.** 1994. pKSS—A second-generation general purpose cloning vector for efficient positive selection of recombinant clones. Gene *138*:109-114.
9. **Kast, P., M. Asif-Ullah and D. Hilvert.** 1996. Is chorismate mutase a prototypic entropy trap?—Activation parameters for the *Bacillus subtilis* enzyme. Tetrahedron Lett. *37*:2691-2694.
10. **Kast, P., M. Asif-Ullah, N. Jiang and D. Hilvert.** 1996. Exploring the active site of chorismate mutase by combinatorial mutagenesis and selection: the importance of electrostatic catalysis. Proc. Natl. Acad. Sci. USA *93*:5043-5048.
11. **Kopelowitz, J., C. Hampe, R. Goldman, M. Reches and H. Engelberg-Kulka.** 1992. Influence of codon context on UGA suppression and readthrough. J. Mol. Biol. *225*:261-269.
12. **Lee, A.Y., J.D. Stewart, J. Clardy and B. Ganem.** 1995. New insight into the catalytic mechanism of chorismate mutases from structural studies. Chem. Biol. *2*:195-203.
13. **MacBeath, G., P. Kast and D. Hilvert.** 1998. Exploring sequence constraints on an interhelical turn using *in vivo* selection for catalytic activity. Protein Sci. *7*:325-335.
14. **Mann, M. and G. Talbo.** 1996. Developments in matrix-assisted laser desorption/ionization peptide mass spectrometry. Curr. Opin. Biotechnol. *7*:11-19.
15. **Moore, C.H., F. Farron, D. Bohnert and C. Weissmann.** 1971. Possible origin of a minor virus specific protein (A_1) in Q_β particles. Nature New Biol. *234*:204-206.
16. **Parker, J.** 1989. Errors and alternatives in reading the universal genetic code. Microbiol. Rev. *53*:273-298.
17. **Poole, E.S., C.M. Brown and W.P. Tate.** 1995. The identity of the base following the stop codon determines the efficiency of *in vivo* translational termination in *Escherichia coli*. EMBO J. *14*:151-158.
18. **Rosenberg, A.H., B.N. Lade, D.-S. Chui, S.-W. Lin, J.J. Dunn and F.W. Studier.** 1987. Vectors for selective expression of cloned DNAs by T7 RNA polymerase. Gene *56*:125-135.
19. **Sambrook, J., E.F. Fritsch and T. Maniatis.** 1989. Molecular Cloning: A Laboratory Manual, 2nd ed. CSH Laboratory Press, Cold Spring Harbor, NY.
20. **Siuzdak, G.** 1996. Mass Spectrometry for Biotechnology. Academic Press, San Diego.
21. **Studier, F.W. and B.A. Moffatt.** 1986. Use of bacteriophage T7 RNA polymerase to direct selective high-level expression of cloned genes. J. Mol. Biol. *189*:113-130.
22. **Studier, F.W., A.H. Rosenberg, J.J. Dunn and J.W. Dubendorff.** 1990. Use of T7 RNA polymerase to direct expression of cloned genes. Methods Enzymol. *185*:60-89.
23. **Van Dyke, M.W., M. Sirito and M. Sawadogo.** 1992. Single-step purification of bacterially expressed polypeptides containing an oligo-histidine domain. Gene *111*:99-104.
24. **Weiner, A.M. and K. Weber.** 1971. Natural read-through at the UGA termination signal of Q_β coat protein cistron. Nature New Biol. *234*:206-209.

Update to: UGA Read-Through Artifacts—When Popular Gene Expression Systems Need a pATCH

Gavin MacBeath[1] and Peter Kast[2]

[1]Center for Genomics Research, Harvard University, Cambridge, MA, USA; [2]Laboratorium für Organische Chemie, Swiss Federal Institute of Technology (ETH), Zürich, Switzerland

Since the article was published, we have received a number of comments (and pATCH1 plasmid requests) by other researchers. They can now rationalize their unexpected, approximately 2 kDa larger contaminants in preparations of purified proteins by the suppression of the leaky UGA stop codon in the pET system. Based on our findings, read-through at the weak UGA stop was also invoked to explain co-purification of a considerably extended version of a GTP-binding protein produced in *Escherichia coli* (6) and to interpret affinity selection results in phage display experiments (1).

We now routinely use "pATCHed" versions of the pET vectors in all of our work (2–5) and so far have had no further problems with heterogeneity due to read-through artifacts. It should also be noted that the newest versions of the pET vectors distributed by Novagen (Madison, WI, USA) employ the UAA (ochre) stop codon recommended here.

Corrections and updates to the manuscript: please note that the position of the two horizontal bars indicating the *Spe*I and *Pvu*II restriction sites in Figure 2A of the original article must be moved by one nucleotide to the right. The detailed descriptions of plasmids pET-EcCM (3) and pKET3-W (5) have now been published.

REFERENCES

1. **Cárcamo, J., M.W. Ravera, R. Brissette, O. Dedova, J.R. Beasley, A. Alam-Moghé, C. Wan, A. Blume and W. Mandecki.** 1998. Unexpected frameshifts from gene to expressed protein in a phage-displayed peptide library. Proc. Natl. Acad. Sci. USA *95*:11146-11151.
2. **Gamper, M., D. Hilvert and P. Kast**. 2000. Probing the role of the C-terminus of *Bacillus subtilis* chorismate mutase by a novel random protein-termination strategy. Biochemistry *39*:14087-14094.
3. **MacBeath, G., P. Kast and D. Hilvert.** 1998. A small, thermostable, and monofunctional chorismate mutase from the archeon *Methanococcus jannaschii*. Biochemistry *37*:10062-10073.
4. **MacBeath, G., P. Kast and D. Hilvert.** 1998. Probing enzyme quaternary structure by combinatorial mutagenesis and selection. Prot. Sci. *7*:1757-1767.
5. **Mattei, P., P. Kast and D. Hilvert.** 1999. *Bacillus subtilis* chorismate mutase is partially diffusion-controlled. Eur. J. Biochem. *261*:25-32.
6. **Sturm, N.R., H. Van Valkenburgh, R.A. Kahn and D.A. Campbell.** 1998. Characterization of a GTP-binding protein in the ADP-ribosylation factor subfamily from *Leishmania tarentolae*. Biochim. Biophys. Acta *1442*:347-352.

CHAPTER 10

S. pombe Expression Vector with 6×(His) Tag for Protein Purification and Potential for Ligation-Independent Cloning

Tanya Hosfield and Quinn Lu[1]

Stratagene Cloning Systems, La Jolla, CA and [1]SmithKline Beecham Pharmaceuticals, King of Prussia, PA, USA

(Reprinted from BioTechniques 27:58-60, 1999)

In the area of protein production and purification, researchers have turned to eukaryotic organisms, such as *Schizosaccharomyces pombe*, as hosts for expression of many eukaryotic proteins instead of bacterial organisms. The reasons are many: some bacterial hosts cannot manage eukaryotic genes due to their toxicity, insolubility or their need for posttranslational modifications, such as phosphorylation and acetylation. It is in this area that *S. pombe* is greatly utilized due to its unique traits that closely resemble higher eukaryotic organisms. This includes chromosomal structure and function, cell-cycle control and RNA splicing (15). We have constructed and tested a new plasmid vector, pESP-5 (Figure 1), for gene expression and protein production in *S. pombe*. pESP-5 is derived from pESP-2 (9). This vector contains the *S. pombe* inducible *nmt*1 promoter for high-level gene expression (3,11,12) and the 6×(His) affinity tag for protein purification. Similar vectors that contain the 6×(HIS) and GST purification tags and the c-myc, HA and Pk epitope tags have also been constructed for use in fission yeast (2,4). A novel aspect of pESP5 includes the potential of ligation-independent cloning (LIC) (1). The gene of interest can be cloned in the multiple cloning sites by using the conventional restriction fragment ligation method or the LIC method (1,5,7). The LIC strategy allows efficient and directional cloning of the target gene. When specific polymerase chain reaction (PCR) primers are used, the target gene can be placed immediately downstream of the DNA sequences encoding for the epitope tag, FLAG® (Scientific Imaging Systems [Eastman Kodak], New Haven, CT, USA) (6), and sequences for recognition by enterokinase. Because enterokinase cleaves the peptide bond C terminal to its recognition site (8), polypeptides with native amino acid sequences can be obtained after removal of the

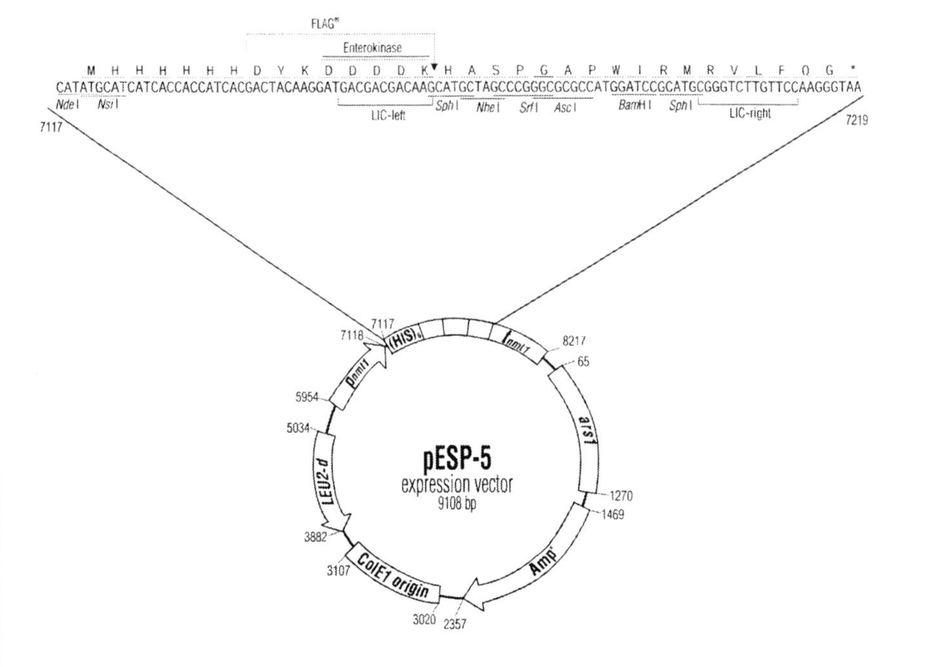

Figure 1. The pESP-5 vector. Shown is the nucleotide sequence with the corresponding amino acids around the multiple cloning sites. The LIC site and the unique restriction sites in the region are also shown. The stop codon is indicated by an asterisk. Protease recognition sites are indicated on top of the sequences, and the corresponding cleavage sites are indicated with arrows.

Cloning and Expression Vectors for Gene Function Analysis
Edited by Quinn Lu and Michael P. Weiner

6×(His) tag by enterokinase.

To prepare pESP-5 for LIC cloning, the vector was linearized with *Sph*I and treated with *Pfu* DNA Polymerase (Stratagene, La Jolla, CA, USA) in the presence of 1 mM dTTP. *Pfu* DNA polymerase has both 5′ to 3′ polymerase activity and 3′ to 5′ exonuclease activity (10). The 3′ to 5′ exonuclease activity of *Pfu* DNA polymerase removes nucleotides at the 3′ ends of the vector until a dTMP residue is reached, resulting in a vector with 5′ extended tails of defined sequence and length at both ends. Because the two tails are noncomplementary to each other, the "LIC-ready" vector ensures directional cloning of the insert with low background, avoiding self-ligation of the vector.

The chicken calmodulin gene was used as a test gene for cloning into pESP-5 using the LIC cloning strategy. Two specifically designed primers with defined "LIC-specific" tails were synthesized and used to PCR-amplify the chicken calmodulin gene. After 25 cycles of amplification with *Pfu* DNA polymerase in a RoboCycler® 40 (Stratagene), the PCR mixture was electrophoresed on a 1% agarose gel, the gel slice containing the PCR product was excised, and the DNA was recovered using a StrataPrep™ Gel Extraction Kit (Stratagene). The purified PCR fragment was further treated with *Pfu* DNA polymerase in the presence of dATP to generate the LIC-specific tails. *Pfu* DNA polymerase was subsequently removed by extraction with StrataClean® Resin (Stratagene). The LIC-ready pESP-5 vector and the LIC-ready insert DNA fragments can then be annealed at room temperature and directly used for transformation. The length of complementary sequences between the vector and the insert (12 and 13 bases, respectively) are long and stable enough to eliminate the ligation step. The annealed mixture was transformed into *Escherichia coli* cells and selected on LB+amp plates. The cloning efficiency of the calmodulin gene into pESP-5 using LIC strategy was determined by *E. coli* colony PCR, with 92% containing an insertion at the size expected of the calmodulin gene (data not shown).

pESP-5-calmodulin plasmid DNA was then transformed into the *S. pombe* strain SP-Q01. One microgram of the expression construct was mixed with 100 µL of cells (ca. 2 × 10^8 cells) of the host strain SP-Q01 (*leu1-$^{32}h^-$*) made competent by using a standard protocol (13). The mixture was incubated at 30°C for 30 min. After the addition of 300 µL of transformation mixture [40% (wt/vol) polyethylene glycol (PEG) 3350, 0.1 M LiCl, 10 mM Tris-HCl (pH 7.5), 1 mM EDTA], the cells were further incubated at 30°C for another 30 min. Following a 20-min heat shock at 42°C, the cells were plated on EMM plates (BIO 101, Vista, CA, USA) supplemented with 25 µM thiamine (repressed conditions) and incubated at 30°–32°C for 4–5 days. For storage, single colonies were streaked out on an EMM+25 µM thiamine plate, and the strains were stored at -70°C as a cell suspension in 20% glycerol.

For protein induction, the expression strain was inoculated into 5 mL yeast extract plus supplements (YES) media and incubated at 30°C overnight. A portion of the overnight culture was used to inoculate another 10 mL YES (optical density $[OD]_{600} = 0.1$). After 4–5 h of further incubation at 30°C for the culture to reach mid-log phase ($OD_{600} = 0.5$), the cells were collected by centrifugation at 1000× *g* in a benchtop centrifuge and washed once with 50 mL sterilized water. The cells were then resuspended into 10 mL of EMM media with 25 µM thiamine (for repressed condition) or without thiamine (for induced condition) and grown at 30°C for 18–20 h. The cells were collected and washed once with lysis buffer (50 mM NaH_2PO_4, pH 8.0, 300 mM NaCl, 10 mM imidazole), and the cell pellet was further resuspended in 500 µL lysis buffer. Five grams of acid-washed glass beads (0.4–0.6 mm in diameter; Sigma, St. Louis, MO, USA) were added, and the cells were broken by vortex mixing at 4°C for 5 min. After centrifugation in a microcentrifuge for 5 min at 12000× *g*, the supernatant (crude lysate) was saved and stored at -70°C for further purification and analysis.

For purification of 6×(His) fusion proteins, cleared lysate was subjected to a one-step purification using the Ni-NTA resin (14) provided in the QIA*express*™ Ni-NTA Spin Kit (Qiagen, Valencia, CA, USA). Briefly, 200 µL of cleared lysate were loaded onto a pre-equilibrated Ni-NTA spin column. After centrifugation at 700× *g*, the column was washed three times with the wash buffer (50 mM NaH_2PO_4, pH 8.0,

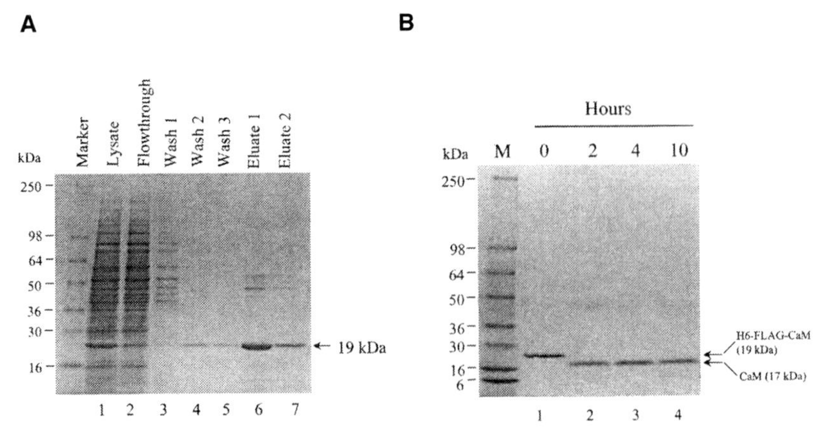

Figure 2. Expression and purification of chicken calmodulin protein. A chicken calmodulin gene was inserted at the LIC site of pESP-5 vector and expressed in *S. pombe*. (A) Shown is a Coomassie Blue stained 4%–20% Tris-glycine-sodium dodecyl sulfate (SDS) gel (Novex) with samples derived from various stages of the purification procedure. Lane 1 is crude lysate; lane 2 is a flow-through sample of the lysate after subjection to the one-step purification for 6×(His) fusion proteins, using the Ni-NTA resin; lanes 3, 4 and 5 are washes with 20, 20 and 40 mM imidazole, respectively; lanes 5 and 6 are products eluted with 250 mM imidazole. The lane containing marker proteins (Novex) is indicated. (B) Shown is a Coomassie Blue stained 4%–20% Tris-glycine-SDS gel with 6×(His)-calmodulin before and after enterokinase digestion. Lane 1, 6×(His)-FLAG-calmodulin; lanes 2, 3 and 4 are samples of the fusion protein incubated with enterokinase for 2, 4 and 10 h, respectively. The cleavage reaction was carried out in a vol of 80 µL with 20 µg of 6×(His)-calmodulin fusion protein and 1 U of enterokinase at 37°C.

300 mM NaCl) containing 20 and 40 mM imidazole, respectively. The fusion protein was then eluted with 50 µL elution buffer (50 mM NaH_2PO_4, pH 8.0, 300 mM NaCl, 250 mM imidazole) twice and combined. Twenty micrograms of purified HIS6-calmodulin fusion protein were incubated with 1 U of enterokinase (Stratagene) at room temperature for 0, 2, 4 and 10 h, respectively. A fraction of the sample was electrophoresed on a 4%–20% Tris-glycine-acrylamide gel (Novex, San Diego, CA, USA) and stained with Coomassie® Blue. Figure 2A shows the expression and purification of 6×(His)-calmodulin fusion protein. The level of induced expression of 6×(His)-calmodulin is estimated to be 5%–10% of the total soluble protein. Note the extra bands appearing on the gel are probably due to the endogenous His proteins contained within the cell that are co-purified with the target. As shown in Figure 2B, calmodulin protein without any extraneously added amino acid residues was obtained by removing the 6×(His) purification tag with enterokinase. Before enterokinase cleavage, the fusion protein was dialyzed against the enterokinase cleavage buffer (20 mM Tris-HCl, pH 7.4, 50 mM NaCl, 2 mM $CaCl_2$) using Amicon® Microcon® Microconcentrators (Millipore, Bedford, MA, USA). The FLAG epitope can be tagged to the protein of interest and used in immunoprecipitation experiments. The 6×(His)-FLAG-calmodulin fusion protein was detected through a western blot using the M2 anti-FLAG antibody (Scientific Imaging [Eastman Kodak]) (data not shown).

In conclusion, the new pESP-5 *S. pombe* expression vector allows efficient and directional cloning of PCR products immediately downstream of the enterokinase recognition/cleavage site. Thus, polypeptides without extraneously added amino acid residues can be obtained by removing the purification tag with enterokinase after purification of the recombinant 6×(His) fusion protein.

ACKNOWLEDGMENTS

FLAG and anti-FLAG are registered trademarks of Immunex Corporation. We would like to thank Cathy Chang, Peter Vaillancourt, Alan Greener, John C. Bauer and members of the Research and Development Department at Stratagene for discussions and suggestions. Address correspondence to Tanya Hosfield, Stratagene Cloning Systems, Inc., Research and Development, 11011 North Torrey Pines Road, La Jolla, CA 92037, USA. Internet: tanya_hosfield@ stratagene.com

REFERENCES

1. **Aslanidis, C. and P.J. de Jong.** 1990. Ligation-independent cloning of PCR product (LIC-PCR). Nucleic Acids Res. *18*:6069-6074.
2. **Craven, R.A., D.J.F. Griffiths, K.S. Sheldrick, R.E. Randall, I.M. Hagan and A.M. Carr.** 1998. Vectors for the expression of tagged proteins in *Schizosaccharomyces pombe*. Gene *221*:59-68.
3. **Forsburg, S.** 1993. Comparison of *Schizosaccharomyces pombe* expression systems. Nucleic Acids Res. *21*:2955-2956.
4. **Forsburg, S.L. and D.A. Sherman.** 1997. General purpose tagging vectors for fission yeast. Gene *191*:191-195.
5. **Haun, R. and J. Moss.** 1992. Ligation-independent cloning of glutathione S-transferase fusion genes for expression in *Escherichia coli*. Gene *112*:37-43.
6. **Hopp, H.P., K.S. Prickett, V.L. Price, R.T. Libby, C.J. March, D.P. Cerretti, D.L. Urdal and P.J. Conlon.** 1988. A short polypeptide marker sequence useful for recombinant protein identification and purification. BioTechnology *6*:1204-1210.
7. **Huan, R.S., I.M. Serventi and J. Moss.** 1992. Rapid reliable ligation-independent cloning of PCR products using modified plasmid vectors. BioTechniques *13*:515-518.
8. **La Vallie, E.R., A. Rehemtulla, L.A. Racie, E.A. DiBlasio, C. Ferenz, K.L. Grank, A. Light and J.M. McCoy.** 1993. Cloning and functional expression of a cDNA encoding the catalytic subunit of bovine enterokinase. J. Biol. Chem. *268*:23311-23317.
9. **Lu, Q., J.C. Bauer and A. Greener.** 1997. Using *Schizosaccharomyces pombe* as a host for expression and purification of eukaryotic proteins. Gene *200*:135-144.
10. **Lundberg, K.S., D.D. Shoemaker, M.W.W. Adams, J.M. Short, J.A. Sorge and E.J. Mather.** 1991. High-fidelity amplification using a thermostable DNA polymerase isolated from *Pyrococcus furiosus*. Gene *108*:1-6.
11. **Maundrell, K.** 1989. *Nmt1* of fission yeast. A highly transcribed gene completely repressed by thiamine. J. Biol. Chem. *265*:10857-10864.
12. **Maundrell, K.** 1993. Thiamine-repressible expression vectors pREP and pRIP for fission yeast. Gene *123*:127-130.
13. **Moreno, S., A. Klar and P. Nurse.** 1991. Molecular genetic analysis of fission yeast *Schizosaccharomyces pombe*. Methods Enzymol. *194*:795-823.
14. **Schmitt, J., H. Hess and H.G. Stunnenberg.** 1993. Affinity purification of histidine-tagged proteins. Mol. Biol. Rep. *18*:223-230.
15. **Sipiczki, M.** 1989. Taxonomy and phylogenesis, p. 431-452. *In* A. Nasim, P. Young and B.F. Johnson (Eds.), Molecular Biology of the Fission Yeast. Academic Press, San Diego.

Update to: *S. pombe* Expression Vector with 6×(His) Tag for Protein Purification and Potential for Ligation-Independent Cloning

Tanya Hosfield and Quinn Lu[1]

Stratagene, La Jolla, CA and [1]SmithKline Beecham Pharmaceuticals, King of Prussia, PA, USA

ADDITIONAL EXPRESSION VECTORS FOR HETEROLOGOUS PROTEIN EXPRESSION IN *S. POMBE*

The vectors described in this article are new additions to the ESP (Expression *Schizosaccaromyces pombe*) yeast protein expression and purification system (1). This system was developed to provide an easy alternative to protein expression in *Escherichia coli*. Unlike *E. coli* expression hosts, *S. pombe* allows for eukaryotic posttranslational modification of the expressed proteins, which is often critical for the biological activity of eukaryotic proteins.

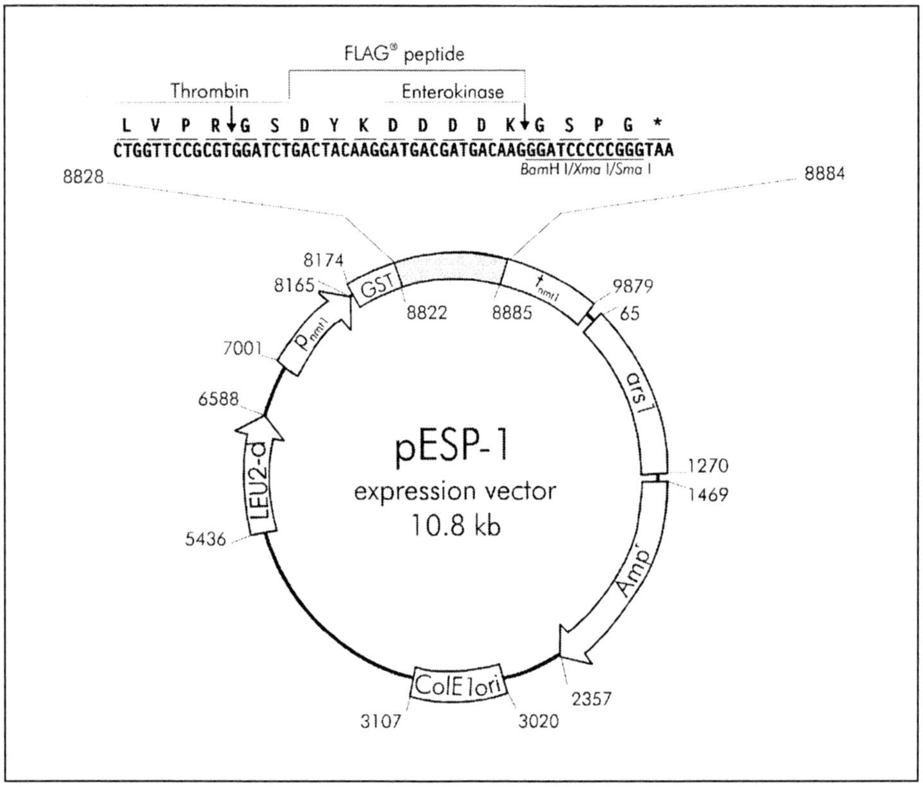

Figure 1. Circular map of the pESP-1 plasmid. Nucleotide sequences around the MCS with corresponding translation products are shown. Protease recognition sites are indicated on top of the respective sequences, and the cleavage sites are indicated with arrows. The FLAG peptide is indicated with a bracket.

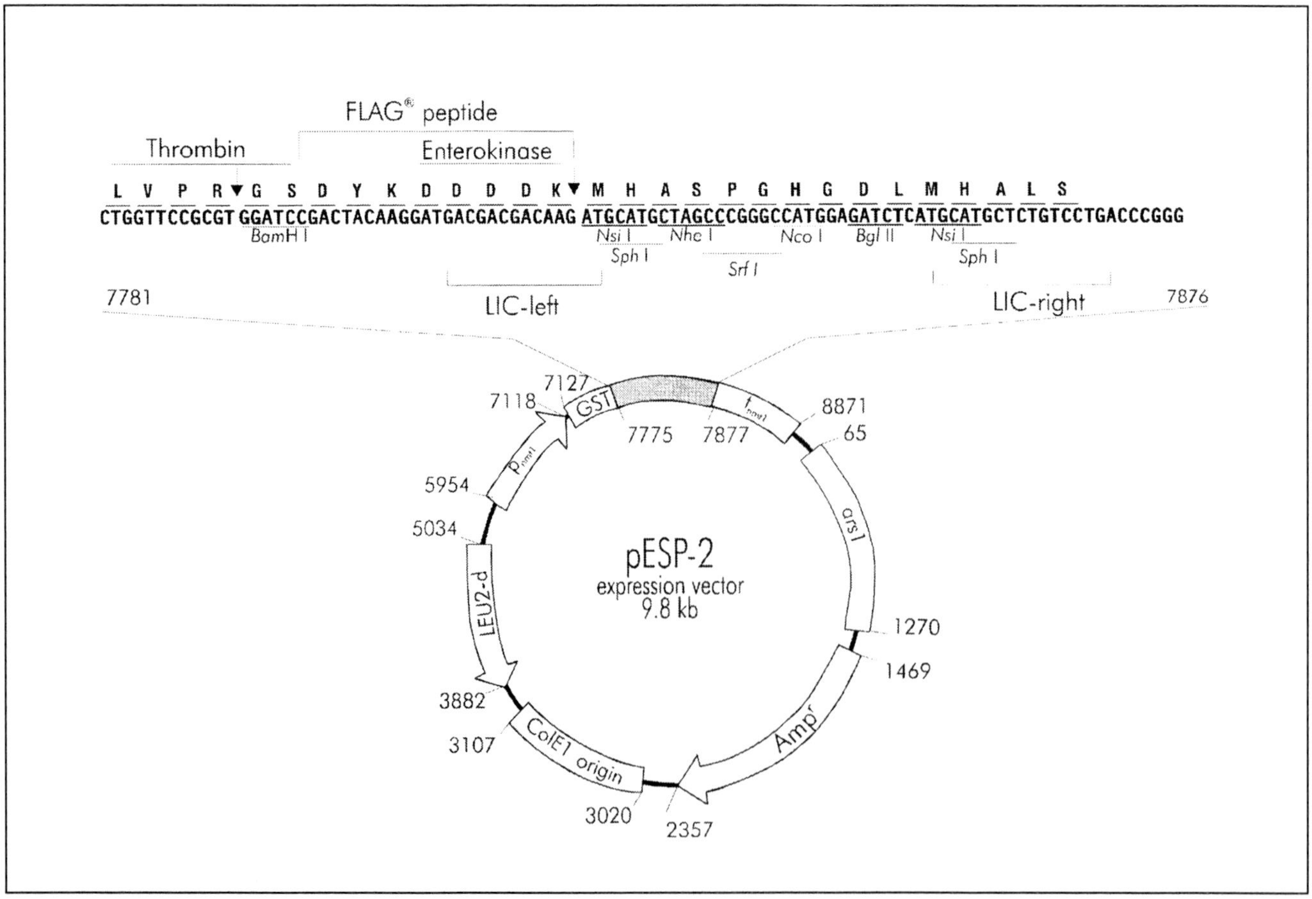

Figure 2. The pESP-2 plasmid contains a sequence encoding a recognition target for EK positioned between the GST fusion tag and the MCS. The positions of the ligation-independent cloning (LIC) regions are indicated.

Since the publication of the original system (1), additional vectors were constructed for heterologous protein expression in *S. pombe*, including pESP-1, pESP-2, pESP-3 and pESP-4. These vectors all include the *nmt1* promoter for constitutive or induced protein expression. The protein of interest is expressed as a fusion protein with the GST polypeptide. Protein expression is also tightly repressed under conditions in which expression is undesirable and can be induced to a high level.

The pESP-1, pESP-2, pESP-3, and pESP-4 plasmids (Figures 1, 2, 3 and 4) are derived from pBluescript® II phagemid (Stratagene, La Jolla, CA, USA) and are composed of four parts: *(i)* a *ColE1* origin of replication and an ampicillin resistance (Amp^r) gene, which allows antibiotic selection and replication of the vector in *E. coli* for cloning; *(ii)* the *ars1* sequence, which provides the origin of replication to allow the vector to replicate autonomously in *S. pombe*; *(iii)* a *LEU2-d* gene from *Saccharomyces cerevisiea*, which serves as a selection marker for transformation of the expression cassette, which contains the *S. pombe nmt1* promoter (P_{nmt1})-terminator(t_{nmt1}) flanking the *GST* gene and a multiple cloning site (MCS) for generating N-terminal fusion genes with the GST peptide. The pESP-2 vector, originally designed for ligation independent cloning, has an expanded MCS for greater ease of sequence insertion. pESP-1 and pESP-2 only allow N-terminal tagging with GST, and this restrictive tagging site may limit utilization of the system. We have also constructed pESP-3 for C-terminal tagging of the protein of interest with GST for expression and purification. pESP-3 can also be used to express proteins without the GST tag for in vivo gene function analysis. The pESP4 *S. pombe* expression vector is an improved version of the pESP2 vector, which eliminates the requirement to start the protein of interest with methionine (2).

Plasmid vectors with the P_{nmt1} or its mutant variants have been used widely to study gene function *in vivo*. The P_{nmt1} of *S. pombe* is strictly regulated by the concentration of thiamine in the media. The P_{nmt1} is tightly repressed in the presence of >0.5 μM thiamine, hence the name nmt (no message in thiamine), and is highly activated on deprivation of thiamine in the media (3). The induction ration of the promoter in *S. pombe* is estimated to be approximately 300-fold (4).

The pESP plasmids also contain DNA sequences coding for amino acid residues that can be specifically recognized and cleaved by enterokinase and thrombin. These protease recognition and cleavage sites are located between the *GST* gene and the MCS. In many cases, removal of the GST affinity tag is nonessential because the enzymatic activities of the GST fusion proteins are similar or identical to their native counterparts. However, in cases in which the presence of the GST affinity tag interferes with the structure and function of the fusion proteins, removal of the tag is necessary. The GST affin-

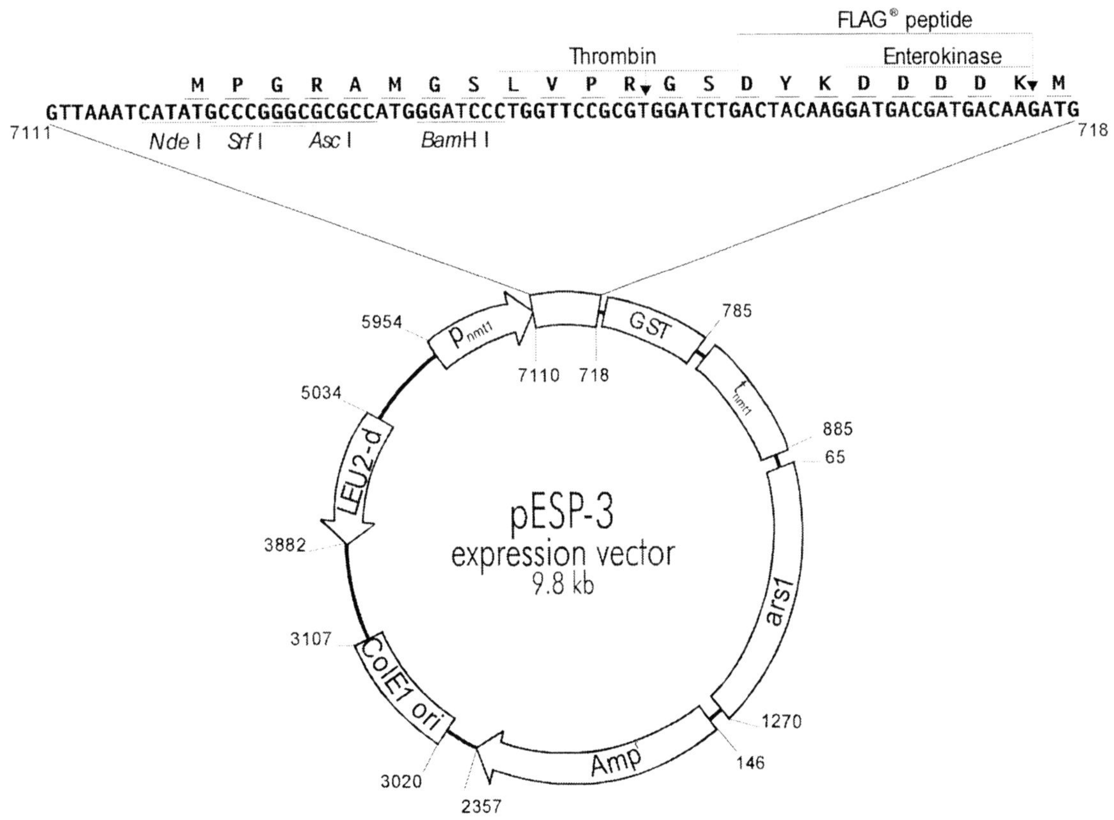

Figure 3. Circular map of the pESP-3 plasmid. The nucleotide sequences around the multiple cloning sites (MCS) with the corresponding amino acid sequences are shown. The unique restriction sites in the MCS region are shown. Protease recognition sites are indicated on top of the sequences and the corresponding cleavage site are indicated with arrows.

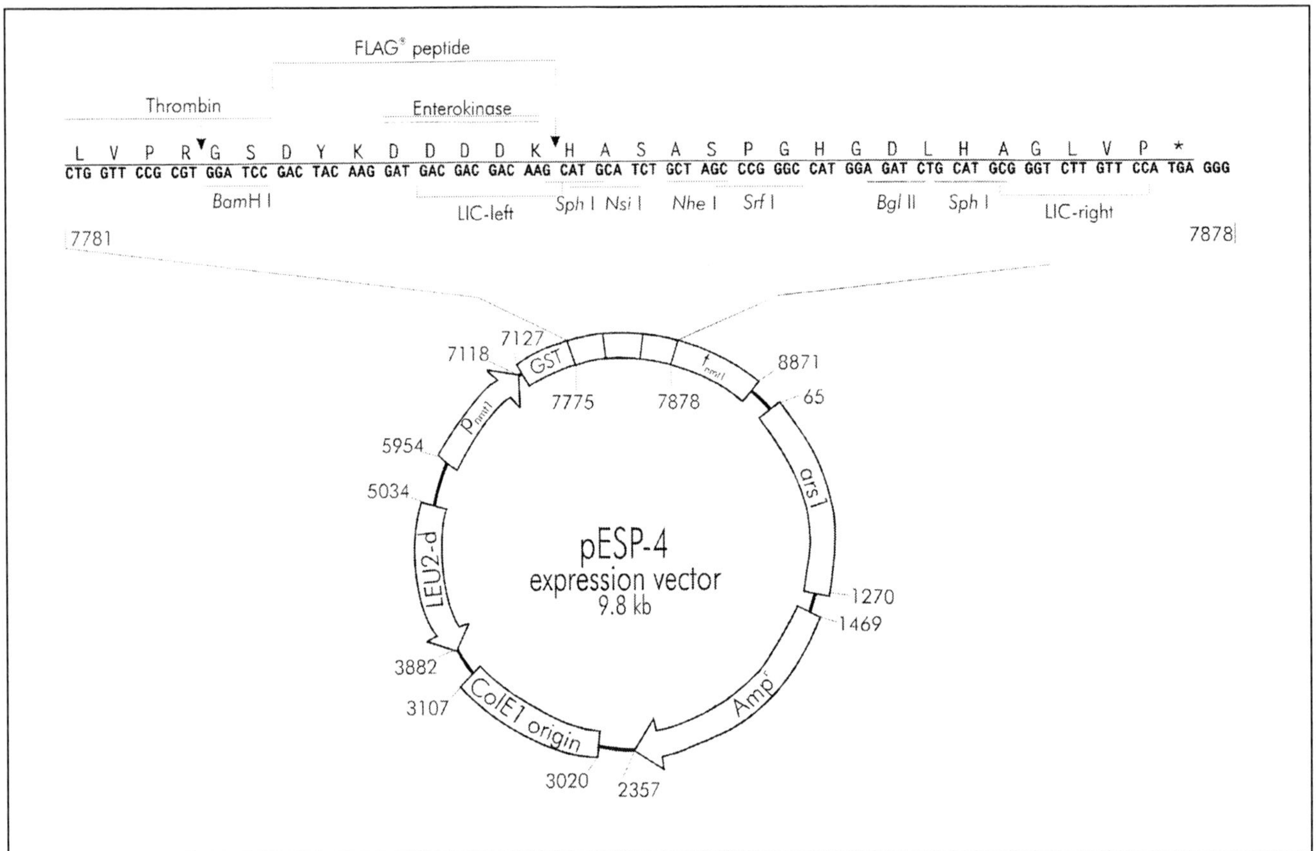

Figure 4. The pESP4 *S. pombe* expression vector map. Nucleotide sequences around the multiple cloning sites with corresponding translation products are shown. Protease recognition sites and the corresponding cleavage sites are indicated. The FLAG peptide is bracketed. The unique multiple cloning sites are shown.

ity tag can be removed by specific proteolytic cleavage with either enterokinase or thrombin. Cleavage with thrombin allows researchers to obtain the protein of interest with a FLAG® peptide (DYKDDDDK) (5) attached to its N terminus, which facilitates further study of the protein.

pESP vectors are commercially available from Stratagene Cloning Systems.

REFERENCES

1. **Lu, Q., J.C. Bauer, and A. Greener.** 1997. Using Schizosaccharomyces pombe as a host for expression and purification of eukaryotic proteins. Gene *200*:135-140.
2. **Hosfield, T. and Q. Lu.** 1999. Influence of the amino acid residue downstream of (Asp)4Lys on enterokinase cleavage of a fusion protein. Anal. Biochem. *269*:10-16.
3. **Maundrell, K.** 1990. nmt1 of fission yeast. A highly transcribed gene completely repressed by thiamine. J. Biol. Chem. *265*:10857-10864.
4. **Forsburg, S.** 1993. Comparison of Schizosaccharomyces pombe expression systems. Nucleic Acids Res. *21*:2955-2956.
5. **Hopp, H.P., K.S. Prickett, V.L. Price, R.T. Libby, C.J. March, D.P. Cerretti, D.L. Urdal, and P.J. Conlon.** 1998. A short polypeptide marker sequence useful for recombinant protein identification and purification. Bio/Technology *6*:1204-1210.

CHAPTER 11

Purification of Proteins Fused to Either the Amino or Carboxy Terminus of the *Mycobacterium xenopi* Gyrase A Intein

Maurice W. Southworth, Kensey Amaya, Thomas C. Evans, Ming-Qun Xu and Francine B. Perler
New England Biolabs, Beverly, MA, USA

(Reprinted from BioTechniques 27:110-120, 1999)

ABSTRACT

The Mycobacterium xenopi *gyrase A mini-intein has been engineered to yield a controllable N-terminal or C-terminal, single-splice-junction autocleavage element. When combined with an affinity tag, these modified mini-inteins can be used to purify target proteins after a single combined chromatography/cleavage step. Cleavage at the intein N terminus was induced with thiol reagents, while cleavage at the intein C terminus was induced by a temperature shift to 16°–25°C. Different preferences for the residue immediately preceding the intein were observed during thiol-induced, N-terminal splice-junction cleavage of the* M. xenopi *gyrase A mini-intein vs. the* Saccharomyces cerevisiae *vacuolar ATPase, subunit A (VMA) intein present in the IMPACT™ purification system. Furthermore, the* M. xenopi *gyrase A mini-intein C-terminal autocleavage vector allows isolation of polypeptides with N-terminal cysteine residues that are active in the Intein Mediated Protein Ligation method of protein semisynthesis.*

INTRODUCTION

Many single-step systems are available to over-express and purify proteins fused to affinity tags, such as the *Escherichia coli* maltose binding protein (MBP), the *Schistosoma japonicum* glutathione S-transferase (GST), His-tag, chitin binding domain (CBD), antibody epitope tags etc. (7). The subsequent proteolytic cleavage of the fusion protein to separate the target protein from the affinity tag can be problematic in these systems (7). Proteases are often inefficient, especially when the cleavage site is buried within the fusion protein. Frequently, one has to optimize the protease reaction to balance separation from the fusion partner with cleavage at internal sites in the target protein. Many vectors also leave additional amino acids (aa) linked to the target protein after protease digestion. After successful cleavage of the target from the affinity tag, a second purification step is required to separate the protease and the fusion protein affinity tag away from the target protein.

Several of these problems have been solved in the IMPACT™ modified *Saccharomyces cerevisiae* (Sce) vacuolar ATPase, subunit A (VMA) intein-based purification systems (1,2). Inteins are protein-splicing elements that mediate cleavage of the peptide bonds linking them to N-terminal and C-terminal flanking regions (termed exteins). Inteins excise themselves from precursor proteins and ligate the flanking exteins. The mechanism of protein splicing has been extensively reviewed (6,10). Most inteins begin with Ser or Cys and end with Asn or Gln (9). Splicing is initiated by an N-S or N-O acyl shift of Cys1 or Ser1 to form a thioester or ester linkage between the N-extein and the intein. If subsequent steps in the splicing pathway are blocked, the precursor is stalled as a (thio)ester intermediate. The (thio)ester bond at the N-terminal splice junction is more reactive than a normal peptide bond and can be easily cleaved with sulfur or nitrogen nucleophiles, such as dithiothreitol (DTT) (1) or hydroxylamine (14). The C-terminal splice junction is cleaved when the intein C-terminal Asn cyclizes to form a succinimide (13). If Cys1 or Ser1 are mutated to a residue other than Cys, Ser or Thr, then splicing and N-terminal splice-junction cleavage are blocked, while C-terminal splice-junction cleavage can still occur (2,14). The challenge in using inteins for protein engineering is in learning how to control these reactions so that a precursor can be purified and autocleavage initiated at will. Previously, the Sce VMA intein was modified for use as a cleavage vector for both N-terminal (1,4) and C-terminal (2) splice-junction cleavage. In both instances, the modified Sce VMA intein was fused to the *Bacillus circulans* CBD, and the autocleavage reaction was initiated by treatment with a reducing agent, such as DTT. Some of the advantages of intein-based vectors are: *(i)* no exogenous proteases are required, *(ii)* autocleavage is specific to the intein-target protein junction, *(iii)* cloning sites are available so that the target protein can be purified with no additional vector-derived aa and *(iv)* the fusion partner remains bound to the resin after autocleavage, so no further steps are necessary to purify the target protein. However, a small peptide is released along with the target protein when using the pIMC vectors (with the target protein fused to the Sce VMA intein C terminus), and this peptide can

Cloning and Expression Vectors for Gene Function Analysis
Edited by Quinn Lu and Michael P. Weiner

be removed by dialysis along with any remaining thiol reagent (2). Intein-based purification systems are limited by the fact that each intein has preferred proximal sequences that promote or inhibit splicing and single-splice junction cleavage (4,12). It would therefore be useful to have several intein-based purification vectors, since each intein has different preferred flanking residues. It might also be advantageous to have an intein C-terminal splice-junction cleavage vector that does not require thiol treatment and that does not generate a small fusion partner peptide.

In this study, we describe N-terminal and C-terminal splice-junction cleavage vectors based on the *Mycobacterium xenopi* gyrase A (Mxe GyrA) intein. The Mxe GyrA intein is a natural mini-intein (198 aa), which lacks a central intein endonuclease domain. As a bacterial protein, it can be very highly expressed in *E. coli* (12). Previous studies have also demonstrated that splicing and single-splice-junction cleavage of the Mxe GyrA intein can be controlled by temperature; in the absence of native GyrA extein sequences, the Mxe GyrA intein remains as a precursor at 37°C and can then be induced to splice at 16°–23°C (12).

MATERIALS AND METHODS

Construction of Mxe GyrA Intein Clones

The Mxe GyrA intein (I or X), with Ala preceding the intein (at the -1 position), was previously cloned between the *E. coli* MBP (M) and a fragment of *Dirofilaria immitis* paramyosin (P) to form pMXP(Ala), which was previously called pMIP(Ala) (12). Ala(-1) in pMXP(Ala) was mutated to 19 other amino acids by cassette replacement. Each cassette also contained an Ile2Val mutation, which did not affect splicing of Mxe GyrA MXP(Ala) when compared with the native Mxe GyrA intein sequence. pMBXP was constructed by cloning the *B. circulans* CBD (B) (1) between MBP and the intein in pMXP(Ser), with Ser at the -1 position. pMBXt was generated from pMBXP by replacing paramyosin with *E. coli* thioredoxin (3). A C-terminal splice-junction cleavage vector, pMBX1 (Figure 1), including a multiple cloning site (MCS) polylinker at the C terminus of the Mxe GyrA mini-intein and Ser preceding the N terminus of the mini-intein, was constructed by removing a *Sap*I site from pMBXP and replacing the paramyosin fragment with a MCS cassette containing unique *Sap*I GCTCTTC(N1/4), *Nco*I (CCATGG), *Not*I (GCGGCCGC) and *Age*I (ACCGGT) cloning sites. All cloning enzymes and oligonucleotides were from New England Biolabs (Beverly, MA, USA) and used according to the manufacturer's instructions. All polymerase chain reaction (PCR) fragments were sequenced in both directions by the New England Biolabs DNA Sequencing Core Facility. For details of vector construction, contact the authors.

Expression and Purification

All plasmids were transformed into *E. coli* strain ER2267 and incubated at 37°C in LB medium supplemented with 100 µg/mL ampicillin as previously described (1). The cultures were grown to mid log (optical density $[OD]_{600}$ of 0.6) and induced with 0.4 mM isopropy-B-D-thiogalactoside (IPTG), after which they were further incubated at 37°C for 2 h or 19°C overnight. All samples were electrophoresed in 10%–20% sodium dodecyl sulfate polyacrylamide gel electrophoresis (SDS-PAGE) gels (Novex, San Diego, CA, USA). All gels were stained with Coomassie® Blue, digitized with a Microtek Scanmaker III (Microtek International, Taiwan, R.O.C.) and analyzed with NIH Image 1.51 software (**http://rsb.info.nih. gov/nih-image**) for quantitation as described (11). Protein concentrations were determined by the Bradford assay (Bio-Rad, Hercules, CA, USA).

In vitro DTT cleavage studies were performed with precursor protein purified over amylose resin (New England Biolabs) or with lysates from cultures induced at 37°C. Samples were incubated in vitro at room temperature (23°C) in 0.5 M NaCl, 20 mM phosphate buffer, pH 7.2, in the presence or absence of 50 mM DTT. Cleavage reactions were sampled at 0, 1, 2, 5, 10, 30, 60, 120 and 240 min and overnight. The rate of DTT cleavage was determined from the slope of the line obtained by plotting the percent N-terminal splice-junction cleavage

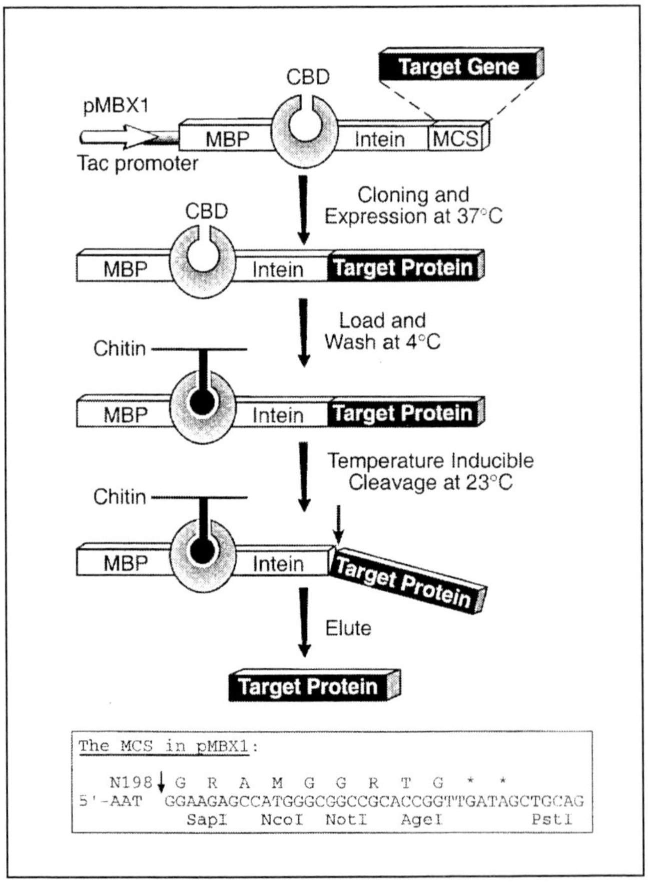

Figure 1. Purification scheme for the pMBX1 C-terminal splice-junction cleavage vector. A target protein is cloned into the pMBX1 MCS such that its N terminus is in-frame with the Mxe GyrA mini-intein C terminus. After expression at 37°C, the cleared cell lysate is loaded onto a chitin resin at 4°C, and unbound proteins are washed away. Autocleavage at the intein C-terminal splice junction is induced by shifting the temperature of the entire column to room temperature overnight. The target protein is then eluted in a suitable buffer for the target protein (pH 6.0–8.0). The pMBX1 MCS is shown with the single letter aa code above the DNA sequence and restriction enzyme sites below. N198 is the C terminus of the intein, and cleavage occurs after this aa. When a target protein is cloned into the *Sap*I site, the native sequence is obtained after autocleavage. Intein, Mxe GyrA intein with Ser at the -1 position; *, stop codon; vertical arrow, site of autocleavage.

Table 1. The Effect of the -1 Residue on Splicing and Cleavage of the Mxe GyrA Intein

-1 Amino Acid	% MXP Precursor and Products Produced In Vivo: Spliced Products 19°C	N-Terminal Cleavage 37°C	C-Terminal Cleavage 37°C	DTT Cleavage In Vitro: MXP Precursor 37°C	Rate Product Formation 23°C	% Cleaved Overnight 23°C
Ser	0	0	37	63	0.04	24
Gly	5	0	23	76	0.48	99
His	7	5	12	81	1.91	86
Thr	0	6	17	77	0.24	91
Asp	0	100	12	0	ND[a]	ND[a]
Ile	3	2	0	98	0.12	64
Trp	54	0	0	98	2.48	83
Leu	16	24	1	74	2.73	95
Lys	41	4	5	77	2.33	86
Ala	10	0	11	88	0.43	93
Arg	36	11	2	84	3.51	77
Met	30	15	3	74	2.88	81
Tyr[b]	83	0	0	99	2.17	73
Phe	86	0	0	98	2.91	72
Pro	1	0	4	96	0.08	39
Val	0	5	1	94	0.18	81
Asn	0	0	2	98	0.40	96
Gln	15	0	0	100	0.63	97
Cys	13	21	3	72	0.48	97
Glu	0	1	0	98	0.10	45

The other 19 naturally occurring aa were substituted for the wild type Tyr preceding the Mxe GyrA intein at the -1 position. The molar percent of precursor and products present after in vivo induction is listed for proteins expressed at the temperatures indicated.
Samples induced at 37°C were treated with 50 mM DTT at 23°C, and cleavage at the intein N terminus was quantitated during overnight time courses.
The rate of product formation (M + XP) per minute and the percent substrate cleaved after the overnight incubation in DTT are listed in the last two columns.
[a]ND, not determined; [b]wild type -1 aa.

products generated vs. time while in the linear range of the curve. The percentage of precursor (MXP), spliced product (MP), N-terminal splice-junction cleavage products (M+XP) and C-terminal splice-junction cleavage products (MX+P) was determined from at least two separate experiments.

For purification of target proteins fused to the intein C terminus (Figure 1), the cell pellet from 500 mL of culture was resuspended in 25 mL Column Buffer (20 mM Tris-HCl, pH 7.4, 0.5 M NaCl, 0.1% Triton® X-100, 1 mM EDTA), sonicated and centrifuged to yield a cleared lysate. Five milliliters of soluble extract were diluted to a final volume of 25 mL with Column Buffer (30 mg total protein), loaded onto 10 mL chitin resin (New England Biolabs) at 4°C and washed with 200 mL of Column Buffer. To induce C-terminal splice-junction cleavage, the entire column containing the bound precursor (MBXP or MBXt) was removed from the refrigerator, incubated for 18 h at room temperature (23°C) to induce the cleavage reaction and then transferred back to 4°C for 2 h. Free paramyosin or thioredoxin was eluted with Column Buffer.

RESULTS AND DISCUSSION

The pTXB1 and pTXB2 Mxe GyrA intein N-terminal splice-junction cleavage vectors for purification of proteins whose C terminus is fused to the intein N terminus have been previously described (5). In this system, mutation of the Mxe GyrA intein C-terminal Asn to Ala blocked C-terminal splice-junction cleavage, while allowing thiol-induced cleavage at the Mxe GyrA intein N terminus at 4°C overnight. The pTXB1 and pTXB2 vectors have a T7 promoter (T), followed by different MCSs, the modified Mxe GyrA intein (X), a CBD affinity tag (B) and a stop codon (5). These vectors were successfully used to generate proteins with C-terminal thioesters for use in Intein Mediated Protein Ligation (IPL), a protein semisynthesis technique (5). In the present study, 20 naturally occurring aa were cloned immediately preceding a splicing-competent Mxe GyrA intein (at the -1 position) in the MXP context (with the MBP preceding the intein and paramyosin following the intein) (12) to determine preferred residues for thiol-induced cleavage. In all cases, except with MXP(Asp), expression at 37°C resulted in predominantly precursor protein, while expression at 19°C yielded differing amounts of spliced product (Table 1). MXP precursors synthesized at 37°C could be cleaved by DTT at their N-terminal splice junction with differing rates, although most -1 aa resulted in $>70\%$ N-terminal splice-junction cleavage product after overnight incubation with 50 mM DTT (Table 1). The exceptions were those precursors whose rate of product formation was <0.13 per min. Some percentage of each precursor might have failed to cleave because of misfolding or aggregation. Note that DTT cleavage can also be affected by intein mutations or extein residues beyond the -1 position (1,4).

In studying the effect of the -1 aa on N-terminal splice-junction cleavage, we observed that both Ser and Gly precursors yielded mostly C-terminal splice-junction cleavage products (>80%) in vivo when induced at 19°C, but not at 37°C (data not shown and Table 1). Moreover, these Ser and Gly precursors synthesized at 37°C could be induced to efficiently cleave the C-terminal splice junction when incubated at 19°C. These properties suggested that constructs with Ser and Gly at the -1 position would be good candidates for temperature-controllable, C-terminal splice-junction cleavage vectors. All other precursors yielded much less C-terminal splice-junction cleavage product after the temperature shift (data not shown). The MXP(Ser) precursor was modified to generate an Mxe GyrA intein-based, C-terminal splice-junction cleavage vector for protein purification when the N terminus of the target protein is fused to the C terminus of the intein (Figure 1). A CBD (B) was inserted between MBP (M) and the Mxe GyrA intein (X), with Ser at the -1 position, to generate pMBXP. Paramyosin was later replaced with a MCS to generate pMBX1. If target genes are cloned into the *Sap*I site in pMBX1, the purified target protein will have no additional vector-derived aa at its N terminus (Figure 1). To purify the target protein, soluble lysates from cultures induced at 37°C for 2 h are loaded onto either amylose or chitin resin at 4°C. The columns are then washed to remove unbound protein. Autocleavage at the intein C terminus is induced by overnight incubation at room temperature, followed by elution of the target protein. Chitin columns are preferred because of the high affinity of the CBD. However, amylose can be used to easily purify the precursor after eluting with maltose. Proteins bound to chitin can be released by treatment with 1% SDS or other denaturants (1).

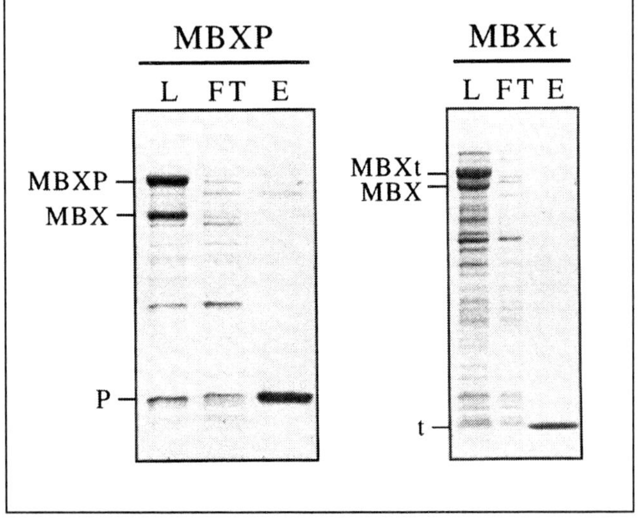

Figure 2. Purification of paramyosin and thioredoxin using the Mxe GyrA intein C-terminal fusion system. Paramyosin (P) and thioredoxin (t) were cloned as fusions in the pMBX system (see Figure 1). Soluble crude MBXP and MBXt extracts from cells induced at 37°C for 2 h (lane L) were loaded onto a chitin column at 4°C. After collecting the flow through (lane FT), the column was washed to remove unbound protein. The column was then incubated overnight at 23°C to activate Asn cyclization, which results in C-terminal splice-junction cleavage and release of free paramyosin or thioredoxin. Free paramyosin or thioredoxin were then eluted from the column after autocleavage (lane E). Samples were visualized on Coomassie® blue-stained SDS-PAGE. MBXP, paramyosin-containing precursor; MBXt, thioredoxin-containing precursor; MBX, the fusion partner after C-terminal splice-junction cleavage; P, paramyosin; t, thioredoxin.

Two target proteins (paramyosin and thioredoxin) were purified using the pMBX system. When induced at 37°C for 2 h, pMBXP yielded 37 ± 5 mg precursor (99 kDa) per L culture upon elution from amylose resin (data not shown) and 10 ± 0.5 mg of purified paramyosin (29 kDa) after elution from chitin resin (Figure 2). Thioredoxin (t) was cloned in place of paramyosin to yield MBXt. Purification on chitin resin yielded 2 mg of thioredoxin (11 kDa) per L culture (Figure 2). Although some in vivo C-terminal splice-junction cleavage was observed, levels of expression were so high that 2–10 mg of purified target proteins could still be obtained. Varying induction times and minimizing preparation and storage times all reduced the amount of cleavage observed before loading on the column (data not shown). It might be possible in the future to reduce in vivo C-terminal splice-junction cleavage without compromising in vitro C-terminal splice-junction cleavage by mutating residues within the intein.

The chitin resin (1) and the pMBX1 autocleavage system are amenable to many buffer systems. Similar results were obtained with buffers at pH 6.0–8.0, while autocleavage of MBXt was reduced by 60% at pH 9.0–9.5 and by 35% at pH 5.0. The optimal temperature for in vitro autocleavage at the Mxe GyrA intein C terminus was 23°–30°C. Autocleavage was 20% less at 16°C, compared to 23°C, and was reduced by >75% at 37°–42°C. Autocleavage at 4°C was very slow; however, if the column was incubated at 4°C for 3 days before elution, similar amounts of purified target protein were obtained with pMBXt.

The Mxe GyrA intein C-terminal purification method is simple and straightforward (Figure 1). The pMBX1 Mxe GyrA intein C-terminal vector yields a purified target protein that is eluted into the desired buffer free of any other enzyme, peptide or chemical. pMBX1 combines the advantages of previous intein purification systems with the high expression levels of soluble protein observed with MBP (7,11). The newly developed IPL method provides a means of joining polypeptides that has great potential for protein engineering and selective protein modification (5,6,8). Unlike the Sce VMA intein-based, C-terminal splice-junction cleavage system (2), the pMBX1 vector works efficiently when Cys is at the N terminus of the target protein (data not shown) and can thus be used to generate the ligation partner with an N-terminal Cys required for IPL.

In summary, the Mxe GyrA intein has been modified to yield vectors for purification of proteins fused to the intein N or C terminus. pMBX1 provides a second method for inducing C-terminal splice-junction cleavage to complement the existing method (2), especially when thiol treatment is incompatible with the target protein. The pTXB1 and pTXB2 Mxe GyrA intein N-terminal splice-junction cleavage system provides an alternative to the Sce VMA intein IMPACT system for purifying proteins fused to the intein N terminus. The system has different preferences for acceptable sequences at the intein-target protein junction, expanding the repertoire of allowable target protein C-terminal residues compatible with intein N-terminal autocleavage vectors. For example, precursors with Asn, Cys and Pro at the -1 position were not cleaved by DTT in the Sce VMA intein system (1,4), but were in the Mxe GyrA intein system. Little in vivo N-terminal splice-junction cleavage was observed in the Mxe GyrA intein system when His, Thr, Arg or Glu were present at the -1 position, although significant amounts of in vivo cleavage were observed with similar Sce VMA intein precursors (1,4). In both cases, Asp preceding the intein resulted in nearly complete in vivo cleavage at the intein N terminus.

ACKNOWLEDGMENTS

We thank Eric Adam and Lixin Chen for helpful discussions and Donald G. Comb for support and encouragement. Address correspondence to Dr. Francine B. Perler, New England Biolabs, Inc., 32 Tozer Road, Beverly, MA 01915, USA. Internet: perler@neb.com

REFERENCES

1. **Chong, S., F.B. Mersha, D.G. Comb, M.E. Scott, D. Landry, L.M. Vence, F.B. Perler, J. Benner, R.B. Kucera, C.A. Hirvonen, J.J. Pelletier, H. Paulus and M.Q. Xu.** 1997. Single-column purification of free recombinant proteins using a self-cleavable affinity tag derived from a protein splicing element. Gene *192*:271-281.
2. **Chong, S., G.E. Montello, A. Zhang, E.J. Cantor, W. Liao, M.Q. Xu and J. Benner.** 1998. Utilizing the C-terminal cleavage activity of a protein splicing element to purify recombinant proteins in a single chromatographic step. Nucleic Acids Res. *26*:5109-5115.
3. **Chong, S., Y. Shao, H. Paulus, J. Benner, F.B. Perler and M.Q. Xu.** 1996. Protein splicing involving the *Saccharomyces cerevisiae* VMA intein. The steps in the splicing pathway, side reactions leading to protein cleavage, and establishment of an in vitro splicing system. J. Biol. Chem. *271*:22159-22168.
4. **Chong, S., K.S. Williams, C. Wotkowicz and M.Q. Xu.** 1998. Modulation of protein splicing of the *Saccharomyces cerevisiae* vacuolar membrane ATPase intein. J. Biol. Chem. *273*:10567-10577.
5. **Evans, T.C., J. Benner and M.-Q. Xu.** 1998. Semisynthesis of cytotoxic proteins using a modified protein splicing element. Protein Sci. *7*:2256-2264.
6. **Gimble, F.S.** 1998. Putting protein splicing to work. Chem. Biol. *5*:R251-R256.
7. **LaVallie, E.R. and J.M. McCoy.** 1995. Gene fusion expression systems in *Escherichia coli*. Curr. Opin. Biotechnol. *6*:501-506.
8. **Muir, T.W., D. Sondhi and P.A. Cole.** 1998. Expressed protein ligation: a general method for protein engineering. Proc. Natl. Acad. Sci. USA *95*:6705-6710.
9. **Perler, F.B.** 1999. InBase: the New England Biolabs intein database. Nucleic Acids Res. *27*:346-347.
10. **Perler, F.B., M.-Q. Xu and H. Paulus.** 1997. Protein splicing and autoproteolysis mechanisms. Curr. Opin. Chem. Biol. *1*:292-299.
11. **Southworth, M.W., E. Adam, D. Panne, R. Byer, R. Kautz and F.B. Perler.** 1998. Control of protein splicing by intein fragment reassembly. EMBO J. *17*:918-926.
12. **Telenti, A., M. Southworth, F. Alcaide, S. Daugelat, W.R. Jacobs, Jr. and F.B. Perler.** 1997. The *Mycobacterium xenopi* GyrA protein splicing element: characterization of a minimal intein. J. Bacteriol. *179*:6378-6382.
13. **Xu, M., D.G. Comb, H. Paulus, C.J. Noren, Y. Shao and F.B. Perler.** 1994. Protein splicing: an analysis of the branched intermediate and its resolution by succinimide formation. EMBO J. *13*:5517-5522.
14. **Xu, M. and F.B. Perler.** 1996. The mechanism of protein splicing and its modulation by mutation. EMBO J. *15*:5146-5153.

Update to: Purification of Proteins Fused to Either the Amino or Carboxy Terminus of the *Mycobacterium xenopi* Gyrase A Intein

Maurice W. Southworth, Kensey Amaya, Thomas C. Evans, Ming-Qun Xu and Francine B. Perler

New England Biolabs, Beverly, MA, USA

See Noren et al. (2) for an updated review of protein splicing and Perler and Adam (3) for an updated review of intein vectors and applications. Vectors for purification of proteins fused to the Mxe GyrA intein N-terminal splice junction are now commercially available (New England BioLabs, Beverly, MA, USA). The pTXB1 and pTXB3 vectors have different polylinkers for target protein insertion. The Mxe GyrA intein has also been used in the pTwin1 system to cyclize the backbone of expressed polypeptides by the intein-mediated protein ligation technique (1). The pTwin1 vector is also commercially available (New England BioLabs).

REFERENCES

1. **Evans, T.C., Jr., J. Benner and M.Q. Xu.** 1999. The cyclization and polymerization of bacterially expressed proteins using modified self-splicing inteins. J. Biol. Chem. *274*:18359-18363.
2. **Noren, C.J., J. Wang and F.B. Perler.** 2000. Dissecting the chemistry of protein splicing and its applications. Angew. Chem. Int. Ed. Engl. *39*:450-466.
3. **Perler, F.B. and E. Adam.** 2000. Protein splicing and its applications. Curr. Opin. Biotechnol. *11*:377-383.

CHAPTER 12

In Vivo Labeling of Over-Expressed Recombinant Proteins in *E. coli*

Christine Giovane, Georges Schwalbach and Étienne Weiss

Ecole Supérieure de Biotechnologie de Strasbourg, Illkirch, France

(Reprinted from BioTechniques 22:796-798, 1997)

Recombinant proteins produced in *Escherichia coli* are widely used for protein-protein or protein-DNA interaction studies and generally have to be labeled for most experiments. For carrying out very sensitive assays, radioactively labeled proteins are preferred. These are prepared either by growing the appropriate strain in M9 minimal medium (6) complemented with radioactive amino acids, such as L-[^{35}S]methionine, or by phosphorylating in vitro the purified over-expressed protein with [γ-^{32}P]ATP (1,4). Growing bacteria in minimal medium leads to a poor yield of the over-expressed protein, and in vitro labeling requires the presence of an appropiate tag that can be phosphorylated. We found that these two drawbacks can be overcome by growing the *E. coli* cells in a eukaryotic cell culture medium recommended for the metabolic labeling of the cellular proteins, such as modified Eagle's medium (MEM) with Earle's salts and without L-methionine. We have indeed been able to produce in vivo highly ^{35}S-labeled recombinant protein with high yield by feeding the *E. coli* cells with this medium, which contains inorganic salts, glucose, essential amino acids and vitamins, complemented with L-[^{35}S]methionine. We report here the in vivo labeling of two recombinant fusion proteins consisting of the viral oncoprotein E6 (2) of the human papillomavirus 16 (HPV16) fused to the carboxyl end of either the glutathione S-transferase (GST) or the maltose binding protein (MBP). These fusion systems are widely used for rapidly preparing recombinant proteins since the over-expressed fusion proteins can be easily purified by a one-step affinity purification on glutathione Sepharose® 4B (Pharmacia Biotech, Uppsala, Sweden) or amylose resin (New England Biolabs, Beverly, MA, USA), respectively (5,7).

The coding region of the E6 protein was amplified by the polymerase chain reaction (PCR) with two flanking primers containing appropriate restriction sites (Giovane et al., unpublished) and cloned into the pGEX-2T (Pharmacia Biotech) and the pMAL-c2 (New England Biolabs) expression vectors. Each vector contains, in addition to an ampicillin resistance gene, a strong P_{tac} promoter and the translation initiation signal of GST or MBP, respectively. The resulting plasmids, pGEX-E6 and pMAL-E6, were transfected into the XL1-blue *E. coli* strain (Stratagene, La Jolla, CA, USA), and individual colonies of transformed cells were grown at 37°C in LB medium (6) containing 100 µg/mL ampicillin and 15 µg/mL tetracycline. At an optical density (OD_{600}) of 0.5, the cells were pelleted by centrifugation for 10 min at 4500 rpm in a low-speed centrifuge (Jouan, Saint-Nazaire, France) and resuspended in either LB medium or MEM (Life Technologies, Paisley, Scotland, UK) with the antibiotics. The MEM was also supplemented with 1 mM glutamine and 25 mM HEPES buffer, pH 7.5. After addition of 1 mM isopropyl-1-thio-β-D-galactoside (IPTG) and L-[^{35}S]methionine (1000 Ci/mmol; 10 mCi/mL; Amersham International, Little Chalfont, Bucks, England, UK), cells were further incubated at 37°C for 2.5 h in the rotating shaker at 300 rpm and finally stored on ice. The expression of the fusion proteins was analyzed by lysing the recovered cells in sodium dodecyl sulfate (SDS)-gel loading buffer (6) and electrophoresis on a 15% SDS-polyacrylamide gel. After migration, the gel was stained with Coomassie Brilliant Blue R®-250 Protein Stain (Life Technologies) and exposed to X-ray film (Eastman Kodak, Rochester, NY, USA) after drying. Alternatively, the IPTG-treated cells were lysed in 20 mM Tris-HCl, pH 8.0, 200 mM NaCl, 1 mM dithiothreitol, 0.2 mM phenylmethylsulfonyl fluoride (buffer A) by sonication and subjected to centrifugation for 20 min in a JA-20 Rotor (Beckman Instruments, Fullerton, CA, USA) at 12000 rpm. The supernatant containing the soluble proteins was then incubated in the presence of either glutathione or amylose resin (0.1 mL beads/20 mL of culture) for 1 h at room temperature. After extensive washing with buffer A, bound proteins were eluted with SDS-gel loading buffer and analyzed on gel as described above.

The results of the over-expression of the GST-E6 and MBP-E6 fusions after induction in either MEM or LB medium are presented in Figure 1. Figure 1A shows a typical gel after staining. As expected, both fusion proteins, which have an apparent molecular weight of about 45 and 60 kDa, respectively, are the most represented proteins in the whole cell lysate of the induced cells in LB medium (lanes 3 and 7). Iden-

Cloning and Expression Vectors for Gene Function Analysis
Edited by Quinn Lu and Michael P. Weiner

tical results are obtained when the bacteria are incubated with IPTG in MEM (lanes 2 and 6). Since almost no fusion protein can be observed before induction (lanes 1 and 5), it demonstrates that the eukaryotic cell culture MEM can be successfully used for expressing recombinant proteins in *E. coli*. Figure 1B shows the autoradiogram of the same gel. No radioactivity is detected before induction (lanes 1 and 5) or after incubation in LB medium with IPTG (lanes 3 and 7). LB medium is a rich culture medium that contains all amino acids in excess. In contrast, most of the proteins are labeled when the bacteria are induced in MEM (lanes 2 and 6) and, in particular, a strong radioactive band is observed at the migration level of each fusion protein (lane 2 for GST-E6 and lane 6 for MBP-E6). This suggests that the simple induction of the expression of recombinant proteins in cells growing in MEM leads to a high incorporation of L-[^{35}S]methionine within the bacterial proteins.

To demonstrate definitively that the two fusion proteins were metabolically highly labeled by culturing the pGEX-E6 and pMAL-E6-transformed cells in MEM, soluble proteins of each clone were absorbed on glutathione or amylose beads after induction, and bound material was analyzed on the same gel. As shown in Figure 1A, essentially one species of protein is retained on either the glutathione or the amylose beads (lanes 4 and 8, respectively). When probed by Western blotting, these products react with anti-E6 antibodies, and no additional band is observed (not shown). The purified GST-E6 and MBP-E6 proteins co-migrate with the over-expressed proteins present in the whole cell extracts (Figure 1A) and are ^{35}S-labeled (Figure 1B). The radioactivity incorporated in these purified fusions was estimated by counting precise amounts of purified protein in a liquid scintillation counter (Packard Instruments, Rungis, France). By adding 0.1 mL of L-[^{35}S]methionine per 100 mL of bacterial culture, we routinely obtained 2–3 × 10^5 cpm/μg of expressed fusion protein.

We show here for the first time that the eukaryotic culture MEM can be advantageously used to prepare in vivo-labeled recombinant proteins, compared to LB medium. Expression of GST-E6 and MBP-E6 fusions was also performed with transformed cells incubated in M9 minimal medium, as a control (not shown). In these conditions, the expression level of the fusions was far below that obtained in MEM, and no significant improvement was observed when saturating amounts of the amino acids present in the MEM solution were added to the M9 minimal medium, suggesting that the salts and vitamins of the MEM are beneficial for the over-expression of plasmid-encoded proteins. The MEM is relatively inexpensive and can be purchased ready-to-use from many companies. Radioactive amino acids other than methionine may also be used in a similar manner since MEM can be requested "à la carte", and labeled leucine, glycine, cysteine or glutamine, for instance, are commercially available. It is possible that other media available for the culture of eukaryotic cells could be more adequate than MEM, but these have not yet been tested.

The GST-E6 and MBP-E6 proteins prepared as described have been successfully used for screening bacterial colonies expressing antibody fragments to E6 (Giovane et al., unpublished). This simple method of labeling in vivo could be applicable not only to the over-expression of fusion proteins, but also to the production of any recombinant polypeptides that are synthesized in reasonable amounts within *E. coli* cells. Compared to other methods (i.e., in vitro phosphorylation), the main limitation of this procedure may be linked to the specific activity of the recovered proteins when very sensitive assays are needed. This could probably be overcome, at least in part, by adding a higher proportion of L-[^{35}S]methionine to the culture. Finally, whether this method could also be exploited for nuclear magnetic resonance (NMR) studies by over-expressing polypeptides with *E. coli* cells incubated in MEM containing specific ^{15}N- or ^{13}C-labeled amino acids (Reference 3 and references therein) remains to be investigated.

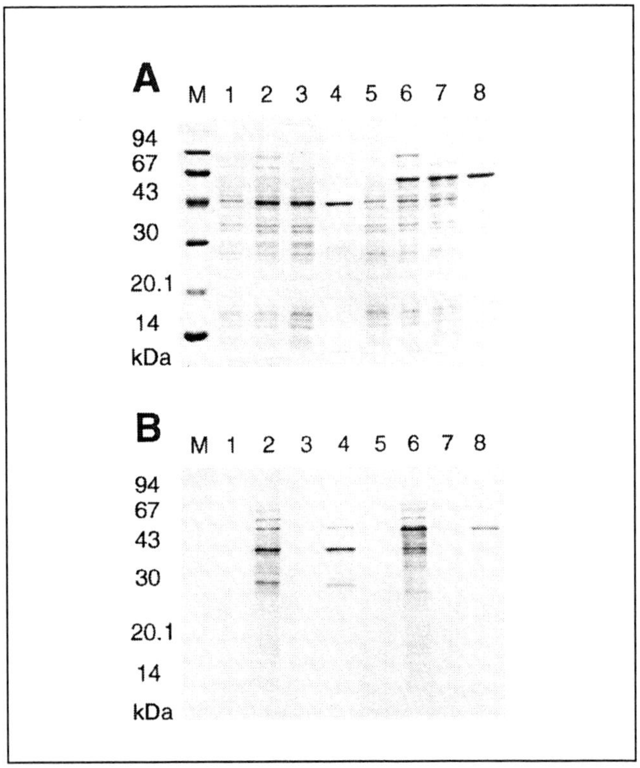

Figure 1. Analysis of the GST-E6 and MBP-E6 proteins over-expressed and labeled in vivo. Bacterial pellets over-expressing either GST-E6 (lanes 1–3) or MBP-E6 (lanes 5–7) and corresponding to 0.1 mL of culture before (lanes 1 and 5) and after incubation in MEM (lanes 2 and 6) or LB medium (lanes 3 and 7) were analyzed on an SDS gel. Purified GST-E6 and MBP-E6 fusions (lanes 4 and 8, respectively) isolated from lysates of cells incubated in MEM were analyzed in parallel. After electrophoresis, the gel was first stained with Coomassie Brilliant Blue (A) and then autoradiographed (B). M indicates low molecular weight markers (Pharmacia Biotech).

ACKNOWLEDGMENTS

We thank Dr. Bohdan Wasylyk for providing the pGOMT plasmid bearing the 16E6 cDNA. We are grateful to Drs. Gilles Travé and Annick Dejaegere for the critical reading of the note. This work was supported in part by the Association pour la Recherche sur le Cancer (No. 2016).

REFERENCES

1. **Chatton, B., A. Bahr, J. Acker and C. Kedinger.** 1995. Eukaryotic GST fusion vector for the study of protein-protein association *in vivo*: application to interaction of ATFa with Jun and Fos. BioTechniques *18*:142-145.

2. **Gissmann, L.** 1992. Human papillomaviruses and genital cancer. Semin. Cancer Biol. *3*:253-261.
3. **Muchmore, D.C., L.P. Macintosh, C.B. Russel, D.E. Anderson and F.W. Dahlquist.** 1989. Expression and nitrogen-15 labelling of proteins for proton and nitrogen-15 nuclear magnetic resonance. Methods Enzymol. *3*:44-73.
4. **Neri, D., H. Petrul, G. Winter, Y. Light, R. Marais, K.E. Britton and A.M. Creighton.** 1996. Radioactive labeling of recombinant antibody fragments by phosphorylation using human casein kinase II and [γ-^{32}P]ATP. Nat. Biotechnol. *14*:485-489.
5. **Riggs, P.** 1996. Expression and purification of maltose-binding protein fusions, p. 16.6.1-16.6.14. *In* F.M. Ausubel et al. (Eds). Current Protocols in Molecular Biology, Vol. 2. John Wiley & Sons, New York.
6. **Sambrook, J., E.F. Fritsch and T. Maniatis.** 1989. Molecular Cloning: A Laboratory Manual, 2nd ed. CSH Laboratory Press, Cold Spring Harbor, NY.
7. **Smith, D.B. and K.S. Johnson.** 1988. Single-step purification of polypeptides expressed in Escherichia coli as fusions with glutathione S-transferase. Gene *67*:31-40.

Update to: In Vivo Labeling of Over-Expressed Recombinant Proteins in *E. coli*

Christine Giovane, Georges Schwalbach and Étienne Weiss

Ecole Supérieure de Biotechnologie de Strasbourg, Illkirch, France

The described technique allowed us to reproducibly prepare in vivo labeled GST-E6 or MBP-E6 proteins. The purified fusions were successfully used to screen bacterial colonies expressing low amounts of antibody fragments to the E6 domain as described previously (1). By incubating the nitrocellulose lifts with 5×10^6 cpm/filter, we were able to identify the clones that produced functional single chain Fvs by autoradiography after about 3 h of exposure (2). Recently, we found that the yield of labeled over-expressed protein in these conditions can be improved by a factor of two when the modified Eagle's medium (MEM) was modified by doubling the concentration of the essential amino acids (R, C, H, I, L, K, F, T, W, Y, V) initially present in this medium. Furthermore, we also observed that the MEM supplemented with 1 mM glutamine, 25 mM HEPES buffer, 1 mM isopropyl-β-D-thiogalactoside (IPTG), L-[^{35}S]methionine and the indicated amino acids can be reused for a second round of recombinant protein production.Thus, when these modifications of the protocol are taken into account, about four times more labeled protein can be prepared with the same quantity of labeling medium. Because of its simplicity, we are currently using this method for labeling several E6 mutant proteins with increased solubility, as compared to the wild-type E6 polypeptide, to analyze their binding capacity to a panel of peptides derived from relevant human proteins.

REFERENCES

1. **Chatellier, J., N. Rauffer-Bruyere, M.H.V. Van Regenmortel, D. Altschuh and E. Weiss.** 1996. Comparative interaction kinetics of two recombinant Fabs and the corresponding antibodies directed to the coat protein oftobacco mosaic virus. J. Mol. Recognit. *9*:39-51.
2. **Giovane, C., G. Travé, A. Briones, Y. Lutz, B. Wasylyk and E. Weiss.** 1999. Targetting of the N-terminal domain of the human papilloma virus type16 E6 oncoprotein with monomeric scFvs blocks the E6-mediated degradation of cellular p53. J. Mol. Recognit. *12*:141-152.

CHAPTER 13

Recovery of Soluble, Active Recombinant Protein from Inclusion Bodies

Pei-Yong Shi, Nancy Maizels and Alan M. Weiner
Yale University School of Medicine, New Haven, CT, USA

(Reprinted from BioTechniques 23:1036-1038, 1997)

Overexpression of recombinant proteins in bacteria frequently produces inactive, insoluble inclusion bodies instead of active, soluble proteins. It is not entirely clear why high intracellular concentrations of nascent or newly synthesized recombinant protein tend to misfold and aggregate, although the yield of correctly folded, active protein can often be increased by reduced expression levels, slower growth, alternative host strains, redesign of the recombinant construct, fusion to a more soluble protein or coexpression of chaperones to facilitate correct folding (1,7,10). Nonetheless, even when these experimental parameters are varied systematically, the outcome is not always satisfactory, and the search for conditions that maximize correct folding and prevent formation of inclusion bodies can consume a great deal of precious research time.

We have developed a reliable method for recovering soluble, active recombinant protein from inclusion bodies. This method does not attempt to avoid formation of inclusion bodies. Instead, protein is recovered from the inclusion bodies, renatured and solubilized on an affinity matrix and finally eluted as active enzyme from the matrix. The method allows recombinant protein bearing a His_6 tag to be recovered in fully active form at a yield of 10 mg/L culture. This method builds on several earlier protocols developed to refold denatured proteins by immobilization on a solid matrix, including an ion-exchange resin (2), a hydrophobic interaction resin (5), a nickel ion affinity resin (8) and other affinity columns (9). It is a particularly useful protocol because it allows for gradual renaturation of protein on the affinity matrix and results in elution of soluble, active protein.

The CCA-adding enzyme [ATP(CTP):tRNA nucleotidyltransferase] catalyzes the addition and regeneration of the 3′-terminal CCA sequence of all tRNAs (4). By cloning the *E. coli cca* gene (3) into the pET-22b(+) vector (Novagen, Madison, WI, USA), we generated a construct, pCCA, in which the CCA-adding enzyme, fused with six His codons at its C terminus, is expressed upon isopropyl-β-D-thiogalactopyranoside (IPTG) induction. For production of recombinant protein, a single colony of *E. coli* strain BL21 carrying the pCCA plasmid was inoculated into liquid YT medium containing 50 μg/mL of ampicillin and grown at 37°C until optical density $(OD)_{600}$ = 0.6, when 0.6 mM IPTG was added to induce *cca* gene expression and the culture incubated for an additional 2 h at 37°C. Cells were then harvested by centrifugation and broken by sonication, and insoluble material was pelleted. Inclusion bodies were dissolved in binding buffer (20 mM Tris base, 0.5 M NaCl, 5 mM imidazole, adjusted to pH 7.8 with HCl) containing 8 M urea at 4°C, and this material was loaded directly onto a nickel ion chelate affinity column pre-equilibrated with the same buffer containing 8 M urea. The column was washed with 5 vol binding buffer containing 8 M urea followed by 5 vol wash buffer (20 mM Tris base, 0.5 M NaCl, 20 mM imidazole, adjusted to pH 7.8 with HCl) containing 8 M urea. The increased imidazole concentration in the wash buffer compared to the loading buffer is suggested by the manufacturer (Novagen) to diminish nonspecific binding to the nickel column. CCA-adding enzyme bound to the nickel column was renatured with a gradient of 8–0 M urea, in 80 column vol binding buffer. Finally, the column was washed with 3 vol binding buffer without denaturant before eluting the renatured CCA-adding enzyme with 3 vol elution buffer (20 mM Tris base, 0.5 M NaCl, 0.3 M imidazole, adjusted to pH 7.8 with HCl). Imidazole was removed by dialysis against 20 mM Tris-HCl (pH 8.0) containing 50 mM NaCl.

Figure 1 compares the recovery of protein purified from inclusion bodies (as described above) or from the soluble fraction, following the standard Novagen protocol for nickel chelate chromatography. IPTG induction resulted in a dramatic increase in the levels of the 53-kDa tagged CCA-adding enzyme (lanes 2 and 3), but more than 90% of the overexpressed *E. coli* CCA-adding enzyme was in the form of inclusion bodies and found not in the supernatant (lane 4) but in the cell pellet (lane 5). Treatment of the cell pellet according to the above protocol permitted recovery of most of this material as soluble protein that migrates identically with protein purified from the supernatant fraction (lanes 6 and 7). Because 90% of the recombinant protein was in inclusions bodies, immobilization-assisted renaturation increased the yield of protein tenfold.

Activity of protein renatured from inclusion bodies was compared with that of protein purified from the supernatant

Cloning and Expression Vectors for Gene Function Analysis
Edited by Quinn Lu and Michael P. Weiner

by assaying CCA addition to a wheat germ tRNA substrate (11). Figure 2 shows that specific activities of the two preparations are essentially identical. In addition, both protein preparations displayed identical kinetics of CTP incorporation in the presence of ATP (data not shown). Kinetic analysis of the renatured, recombinant enzyme determined the K_m to be approximately 9 μM and the V_{max} approximately 17 mmol/h/mg, comparable to the activity (K_m = 1 - 11 μM, V_{max} = 19 - 24 mmol/h/mg) of the CCA-adding enzyme purified from *E. coli* using classical techniques (4).

These data show that high yields of active *E. coli* CCA-adding enzyme can be readily solubilized from inclusion bodies, renatured on the solid affinity resin and eluted as soluble protein. We have successfully carried out this procedure with both urea (8 M) and guanidine HCl (6 M), and we have used continuous gradients of either denaturant for renaturation (8–0 M urea, 6–0 M guanidine HCl). Guanidine HCl is generally preferable to urea for initial solubilization because it is the stronger denaturant, and isocyanate formed by decomposition of urea might carbamylate free amino groups (6). We have also found that inclusion bodies that have been solubilized in 6 M guanidine HCl can be bound directly to the nickel column and then renatured with a reverse 8–0 M urea gradient in a protocol that combines superior solubilization by ionic guanidine HCl and the possibly superior renaturation by nonionic urea.

We note that initial solubilization of some proteins may require denaturation at higher temperatures than 4°C or addition of reducing agents. If reducing agents are added, these must be removed by gel filtration or dialysis to prevent reaction with the nickel ion. In many cases, a step gradient can be as effective as a continuous gradient for protein renaturation. We suspect that immobilization of denatured protein on the solid support favors renaturation by minimizing protein-protein interactions that can cause aggregation when the concentration of denaturant is reduced.

In this purification, formation of inclusion bodies is actually advantageous because: *(i)* the recombinant protein is greatly enriched in the inclusion bodies; *(ii)* inclusion bodies protect the recombinant protein from proteolysis by intracellular proteases; and *(iii)* inclusion bodies may also facilitate overexpression of recombinant proteins that would be lethal to the host cell in an active, soluble form. Because many overexpressed recombinant proteins form inclusion bodies, we believe that immobilization-assisted renaturation and purification should be applicable to other recombinant proteins with inclusion body problems.

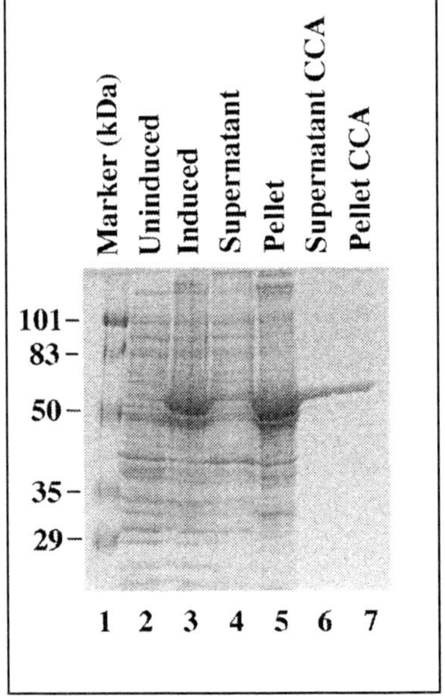

Figure 1. Recovery of soluble, recombinant CCA-adding enzyme from inclusion bodies. Samples were analyzed by 10% sodium dodecyl sulfate polyacrylamide gel electrophoresis (SDS-PAGE). Lane 1: prestained low-molecular-weight protein standards (Bio-Rad, Hercules, CA, USA). Lanes 2 and 3: *E. coli* BL21/pCCA culture before and after IPTG induction. Lanes 4 and 5: supernatant and pellet of IPTG-induced culture after sonication and centrifugation. Lanes 6 and 7: CCA-adding enzyme purified from supernatant or renatured from inclusion bodies; 2 μg of protein were loaded on both lanes.

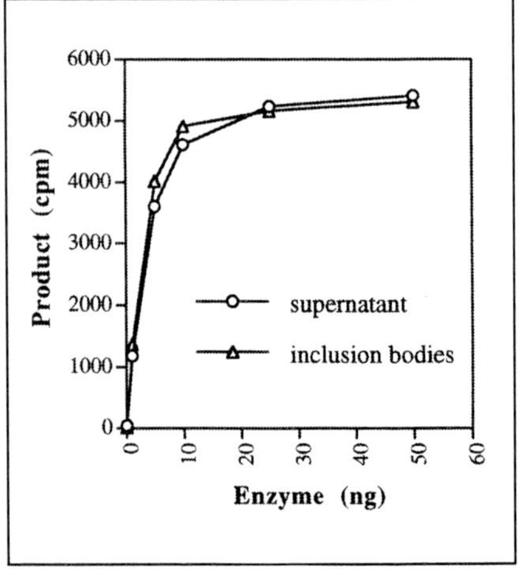

Figure 2. Specific activity of CCA-adding enzyme purified directly from the supernatant or renatured from inclusion bodies. CCA-adding activity was assayed as previously described (11). Data shown represent average of the three experiments.

ACKNOWLEDGMENTS

We thank Tom Griffin for timely advice. This research was supported by the US National Science Foundation Award No. MCB-9507046 to A.M.W. and N.M.

REFERENCES

1. **Bunchner, J., M. Schmidt, M. Fuchs, R. Jaenicke, R. Ridolph, F.X. Schmid and T. Kiefhaber.** 1991. GroE facilitates refolding of citrate synthase by suppressing aggregation. Biochemistry *30*:1586-1591.
2. **Creighton, T.E.** 1985. Folding of proteins adsorbed reversibly to ion-exchange resins. UCLA Symp. Mol. Cell. Biol. *39*:249-258.
3. **Cudny, H., J.R. Lupski, G.N. Godson and M.P. Deutscher.** 1986. Cloning, sequencing, and species relatedness of the *Escherichia coli cca* gene encoding the enzyme tRNA nucleotidyltransferase. J. Biol. Chem. *261*:6450-6453.
4. **Deutscher, M.P.** 1982. tRNA nucleotidyltransferase. The Enzymes *15*:183-215.
5. **Geng, X. and X. Chang.** 1992. High-performance hydrophobic interaction chromatography as a tool for protein refolding. J. Chromatogr. *599*:185-194.
6. **Hagel, P., J.J.T. Gerding, W. Fieggen and H. Bloemendal.** 1971. Cyanate formation in solutions with urea. Biochim. Biophys. Acta *243*:366-372.
7. **King, J., C. Haase-Pettingell, A.S. Robinson, M. Speed and A. Mitraki.** 1996. Thermolabile folding intermediates: inclusion body precursors and chaperonin substrates. FASEB J. *10*:57-66.
8. **Sinha, D., M. Bakhshi and R. Vora.** 1994. Ligand binding assays with recombinant proteins refolded on an affinity matrix. BioTechniques *17*:509-514.
9. **Stempfer, G., B. Höll-Neugebauer and R. Rudolph.** 1996. Improved refolding of an immobilized fusion protein. Nat. Biotechnol. *14*:329-334.
10. **Todd, M.J., P.V. Viitanen and G.H. Lorimer.** 1994. Dynamics of the chaperonin ATPase cycle: implications for facilitated protein folding. Science *269*:659-665.
11. **Yue, D., N. Maizels and A.M. Weiner.** 1996. CCA-adding enzymes and poly(A) polymerases are all members of the same nucleotidyltransferase superfamily: characterization of the CCA-adding enzyme from the archaea hyperthermophile *Sulfolobus shibatae*. RNA 2:895-908.

CHAPTER 14

Simultaneous Expression of Multi-Subunit Proteins in Mammalian Cells Using a Convenient Set of Mammalian Cell Expression Vectors

M.I. Cockett, R. Ochalski, K. Benwell, R. Franco and J. Wardwell-Swanson

Wyeth-Ayerst Research, Princeton, NJ, USA

(Reprinted from BioTechniques 23:402-407, 1997)

To achieve fully functional heterologous expression of multi-subunit recombinant proteins, it is necessary to express several polypeptides within the cell at once. To address this requirement, we have developed a panel of four mammalian expression vectors in which all the components are identical except for the selectable markers. We present use of three of these vectors to express the α1, α2 and β3 subunits of the human N-type calcium channel simultaneously in HEK293 cells (Catalog No. CRL 1573; ATCC, Rockville, MD, USA). Optimum concentrations of each selection were determined, and cells were transfected by either calcium phosphate with supercoiled or linear plasmid DNA or by electroporation with linear plasmid DNA. After triple selection, transfectant colonies were screened for expression of each cDNA by reverse transcription polymerase chain reaction (RT-PCR). Transfection of linear plasmid DNA by electroporation yielded the highest percentage of transfectant cells expressing all three subunits.

Normally, many proteins exist as multiple-subunit complexes. Within the cell membrane, many receptors and ion channels comprise multiple subunits, different combinations of which give them their precise biological activity. This is well-illustrated by the voltage-gated calcium channels, which exist as a complex of pore-forming α1 subunits combined with α2 and β and, in some cases, γ subunits. In the case of the N-type calcium channel, the α1, α2 and β subunits are required for full-channel activity (1). In addition, each of the subunits exists in multiple forms, potentially giving rise to a whole array of different channels, depending on the expression and assembly of the various subunits (1).

We have constructed a set of mammalian cell expression vectors (Figure 1), in which all of the features of each vector are identical except for the four selectable markers. The vectors, named pWE 1 through 4, have a common backbone for propagation in bacteria, a human cytomegalovirus (hCMV) promoter to drive transcription in a variety of cell types together with an intron to increase expression and the simian virus 40 (SV40) polyadenylation signal. Also included is a large polylinker for ease of cDNA insertion. The four selectable markers are for: *(i)* mycophenolic acid resistance, using the bacterial *gpt* gene (pWE 1), *(ii)* neomycin/G418 resistance, using the *neo* gene (pWE 2), *(iii)* puromycin resistance, using a puromycin acetyltransferase gene (pWE 3) and *(iv)* hygromycin resistance, using a hygromycin phosphotransferase gene (pWE 4) (2,4,6). Insertion of cDNAs into these vectors ensures that each has an identical transcription unit, while allowing different selections to be applied to different channel subunits. In addition, their identical nature in all but the selectable marker facilitates shuttling of cDNAs between one vector and another. We have introduced the cDNAs for several subunits of the N-type calcium channel into HEK293 cells using triple transfection followed by concurrent triple selection. The α1 subunit was maintained in pWE 2, the α2 subunit in pWE 3 and the β3 subunit in pWE 1. These vectors were co-transfected into HEK293 cells either by calcium phosphate co-precipitation (2) with 10 µg of each supercoiled or linear plasmid onto a 10-cm petri dish containing 10^6 cells or by electroporation with 50 µg of each linear plasmid into 10^7 cells using a single pulse of 210 V, 1180 µF in the Cell-Porator® Electroporation System (Life Technologies, Gaithersburg, MD, USA). Triple selection with mycophenolic acid, G418 (both from Life Technologies) and puromycin (Calbiochem-Novabiochem, San Diego, CA, USA) was applied to the cells 24 h posttransfection. The levels of each selection used in this combination had previously been empirically determined using a range of each drug concentration in single, double and triple selection regimes using pWE 1, 2 and 3 without inserts and with mock transfections. The levels found to be most effective at killing mock-transfected cells while allowing colonies to form in control vector transfections were 5 µg/mL mycophenolic acid (glutamate pyruvate [GPT] selection), 800 µg/mL G418 and 1 µg/mL puromycin. After three to four weeks in selection, colonies were expanded into 24-well plates. Expression of each subunit was

Cloning and Expression Vectors for Gene Function Analysis
Edited by Quinn Lu and Michael P. Weiner

Table 1. RT-PCR Analysis of Transfectant Colonies Isolated after Triple Selection

Transfection Method	Colonies Expressing: α1, α2 & β3	Any Two Subunits	Only One Subunit	Zero Subunits	Percent Triple Expression
Calcium Phosphate with Circular DNA	1	4	0	0	20%
Calcium Phosphate with Linear DNA	5	3	2	4	36%
Electroporation with Linear DNA	4	3	0	0	57%

The table shows the results obtained after RT-PCR analysis of 26 independently isolated colonies. These were isolated after co-transfection of pWE 1, 2 and 3 vectors containing β3, α1 and α2 subunits of the N-type calcium channel and triple selection (5 µg/mL mycophenolic acid, 800 µg/mL G418 and 1 µg/mL puromycin). Colonies were expanded from 3 different transfection methods, the expression of each subunit was scored, and the number of colonies expressing detectable levels of subunit RNA for: *(i)* all three, *(ii)* any combination of two, *(iii)* one or *(iv)* no subunits is presented.

assessed using RT-PCR analysis on RNA extracted from a confluent well in a 24-well plate (3), with specific pairs of primers for each channel subunit. No PCR products were detected in untransfected HEK293 cells, and the results obtained from 26 independently isolated cell lines are shown in Table 1. With each transfection method, cell lines were isolated that expressed cDNAs for all three channel subunits. This demonstrates the utility of the triple transfection/selection regime to efficiently generate cell lines expressing multiple cDNAs. Note that linearizing the vectors prior to calcium phosphate transfection increased the percentage of cell lines that expressed all three cDNAs from 20% to 36%, and electroporation with linear DNA was even more efficient, with 57% of the cell lines examined expressing all three cDNAs. Toneguzzo et al. (5) reported that electroporation of linear DNA into lymphoid cells resulted in the integration of the DNA into the host chromosomes preferentially at its ends. This may explain why linear molecules have given rise to a higher percentage of transfectants expressing all three cDNAs. Circular molecules must break somewhere to integrate into the genome and could easily break within essential sequences for expression of the inserted cDNA. Toneguzzo et al. (5) also reported that to achieve multicopy insertion of plasmid DNA into the host genome, high concentrations of

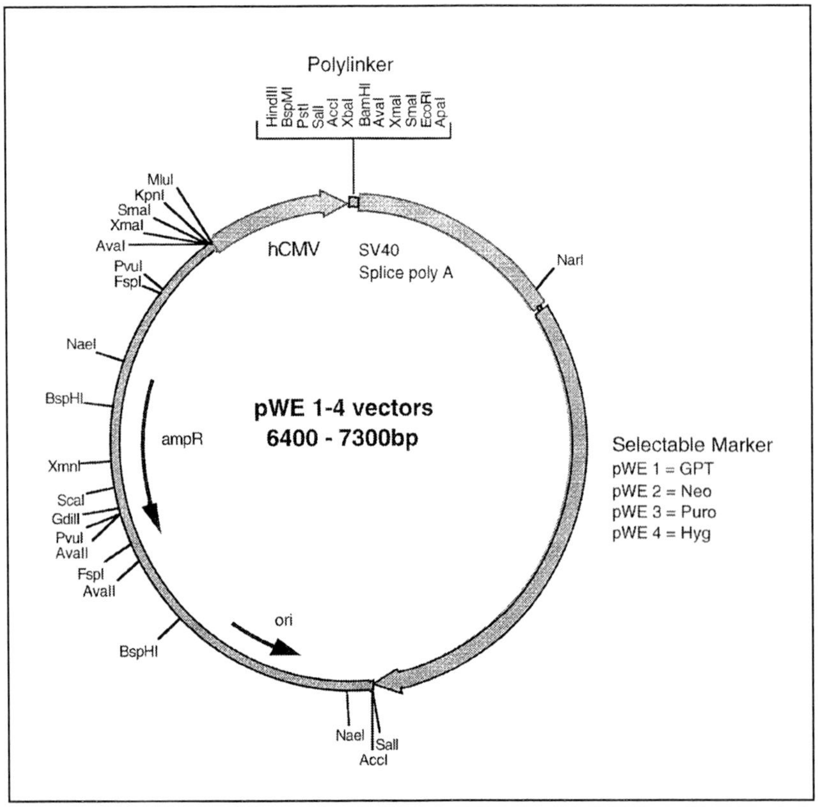

Figure 1. Structure of the pWE series of vectors used to simultaneously coexpress multiple-subunit proteins. All 4 vectors contain promoter/enhancer sequences from the immediate-early gene of the hCMV for high-level constitutive expression. This is placed immediately upstream of a convenient polylinker and the SV40 early polyadenylation signal, and small T intron ensures efficient processing of RNA. The bacterial backbone has the β-lactamase gene for ampicillin resistance and a bacterial origin of replication. Restriction enzyme sites that occur within the bacterial sequence are marked; any of these can be used for linearization of vectors before transfection. Each vector only differs in the selectable marker for selection in mammalian cells. pWE 1 has a mycophenolic acid-resistance marker, GPT; pWE 2 has a neomycin-resistance marker; pWE 3 has a puromycin-resistance marker; and pWE 4 has a hygromycin-resistance marker.

DNA were required, hence our use of 50 µg of each linearized plasmid for each electroporation.

Cell lines isolated that expressed cDNAs for all three subunits expressed different relative amounts of each subunit RNA (as determined by RT-PCR on identical amounts of cellular RNA, data not shown). This observation is useful because it allows us to examine the effect of different ratios of each subunit on the function of a multi-subunit protein, provided that the mRNA levels are proportional to the protein levels. We do not have any direct measurement of protein levels for each of the three channel subunits expressed, however we have used a ^{125}I-labeled ω-conotoxin (ω-CgTx)-binding assay (7) to assess the number of N-type calcium channels on the cell surface of transfectants expressing mRNA for all three subunits. Although ω-CgTx only binds to the $\alpha 1$ subunit of this channel, efficient cell surface expression involves cooperation among the three subunits (7). We have achieved maximal binding values of up to 34 000 sites per cell in some of our transfectants, which are similar to the levels described by Williams et al. (7), who only achieved this level of binding when all three subunits were coexpressed at levels appropriate for channel function.

In summary, we have established an efficient method for simultaneous expression of multiple cDNA sequences in mammalian cells. Use of a convenient set of mammalian expression vectors, each containing a different resistance marker, combined with electroporation of linear plasmid DNA at high concentration and a triple selection has yielded cell lines expressing three separate cDNAs with high efficiency. We only present simultaneous use of pWE 1, 2 and 3; we have however used all four vectors described in various combinations of double and triple transfections to express ion channel subunits in other cell types.

ACKNOWLEDGMENTS

We thank David Shuey for the N-type calcium channel cDNAs and Paul Spence for critical reading of this manuscript.

REFERENCES

1. **Dunlap, K., J.I. Luebke and T.J. Turner.** 1995. Exocytotic Ca^{2+} channels in mammalian central neurons. Trends Neurosci. *18*:89-98.
2. **Gorman, C.** 1985. High efficiency gene transfer into mammalian cells, p. 143-190. *In* D.M. Glover (Ed.), DNA Cloning, Volume II. Academic Press, New York.
3. **Rappolee, D.A., D. Mark, M.J. Banda and Z. Werb.** 1988. Developmental expression of PDGF, TGF-alpha, and TGF-beta genes preimplantation mouse embryos. Science *241*:708-712.
4. **Santerre, R.F., N.E. Allen, J.N. Hobbs, Jr., N. Rao and R.J. Schmidt.** 1984. Expression of prokaryotic genes for hygromycin B and G418 resistance as dominant-selection markers in mouse L cells. Gene *30*:147-156.
5. **Toneguzzo, F., A. Keating, S. Glynn and K. McDonald.** 1988. Electric field-mediated gene transfer: characterization of DNA and patterns of integration in lymphoid cells. Nucleic Acids Res. *16*:5515-5532.
6. **Vara, J., A. Portela, J. Ortin and A. Jimenez.** 1986. Expression in mammalian cells of a gene from *Streptomyces alboniger* conferring puromycin resistance. Nucleic Acids Res. *14*:4617-4624.
7. **Williams, M.E., P.F. Brust, D.H. Feldman, S. Patthi, S. Simerson, A. Maroufi, A.F. McCue, G. Velicelebi, S.B. Ellis and M.M. Harpold.** 1992. Structure and functional expression of an ω-conotoxin-sensitive human N-type calcium channel. Science *257*:389-395.

Update to: Simultaneous Expression of Multi-Subunit Proteins in Mammalian Cells Using a Convenient Set of Mammalian Cell Expression Vectors

M.I. Cockett

Bristol-Meyers Squibb, Princeton, NJ, USA

Since the original publication of this article, these four vectors have been deposited with ATCC (Manassas, VA, USA). The ATCC Accession Nos. are 87671, 87672, 87673 and 87674 for pWE1, 2, 3 and 4, respectively.

There is also one additional point that we think it is important for readers to note: with any mammalian expression vector that contains an intron downstream of an inserted cDNA, there is the possibility that cryptic splicing can occur between a 5′ cryptic splice site within the coding region of the inserted cDNA and the 3′ acceptor site of the downstream intron (2). Colleagues and we have observed this in some cases with the pWE vectors (1). Once appreciated, this can be easily screened for by accurately sizing the DNA resulting from a reverse transcription-polymerase chain reaction (RT-PCR) from RNA isolated from pWE vector transfectant cells. From our experience, in the majority of cases this is not observed, however, if aberrant splicing does occur, this can confound research and should therefore be checked for when characterizing new cell lines.

REFERENCES

1. **Bard, J.A., S. Walker, O. Bortman, R. Ochalski and S. Nawoschik.** 1999. Truncated proteins are recombinantly expressed from mammalian cell expression vectors containing introns downstream of inserted open reading frames. Soc. Neurosci. *25*:795.11.
2. **Huang, M.T. and C.M. Gorman.** 1990. The simian virus 40 small-t intron, present in many common expression vectors, leads to aberrant splicing. Mol. Cell. Biol. *10*:1805-1810.

CHAPTER 15

Isolation of Recombinant Secretory Proteins by Limited Induction and Quantitative Harvest

Helene F. Rosenberg

National Institute of Allergy and Infectious Diseases, NIH, Bethesda, MD, USA

(Reprinted from BioTechniques 24:188-192, 1998)

Vectors with features that direct recombinant proteins to the periplasm are used extensively for the expression of heterologous secretory products in *Escherichia coli*. Many of these vectors encode bacterial secretion signals that facilitate the membrane translocalization of the recombinant protein, with removal of the secretion signal occurring concomittantly. When exported to the periplasm, heterologous secretory proteins are more likely to be isolated in soluble, folded, biologically active form. Despite this advantage, yield is often limited by the capacity of the bacterial protein export system. As recombinant protein synthesis is frequently initiated by derepressing a strong promoter, the newly synthesized protein quickly overwhelms the secretory apparatus, resulting in the accumulation of cytosolic, unprocessed and often unusable forms. In this situation, isolation of the processed periplasmic fraction of recombinant protein requires time-consuming, gentle handling so as to avoid cross-contamination with the cytosolic forms. Ultimately, this procedure results in the recovery of only a small fraction of the total available yield.

Several groups have experimented with ways to amplify the rate and efficiency with which the bacterial export system handles recombinant protein. Some successful approaches include coexpression of protein disulfide isomerase (3), supplementing with bacterial chaperone proteins (6,7) or export proteins (1) and exploiting the bacterial hemolysin translocator (2). Others have had success by eliminating specific bacterial periplasmic proteins (4) and with supplementing the culture media with glycine (5). We have tried an alternative approach—rather than asking the bacteria to do more than is possible, we have modified our approach to consider how to harness the best possible outcome more effectively. We have found that by reducing the induction stimulus, we can limit recombinant protein synthesis to a rate tolerated by the native export system. By reducing the final concentration of the induction stimulus (in this case, the galactoside derivative isopropyl β-D-thiogalactopyranoside [IPTG]) from the typical 1 mM down to 1 μM, virtually 100% of the recombinant protein isolated is in mature, processed form. The absence of unprocessed cytosolic contaminants permits us to use a more rigorous, quantitative isolation procedure, improving our yield approximately 10–50-fold. Although reduction of IPTG has been suggested previously as a method to avoid formation of insoluble inclusion bodies, to the best of our knowledge there have been no previous reports describing the limitation of induction stimulus in conjunction with the processing of heterologous secretory proteins in bacteria.

Figure 1 is a schematic representation of the construct used to generate the recombinant protein described in this article.

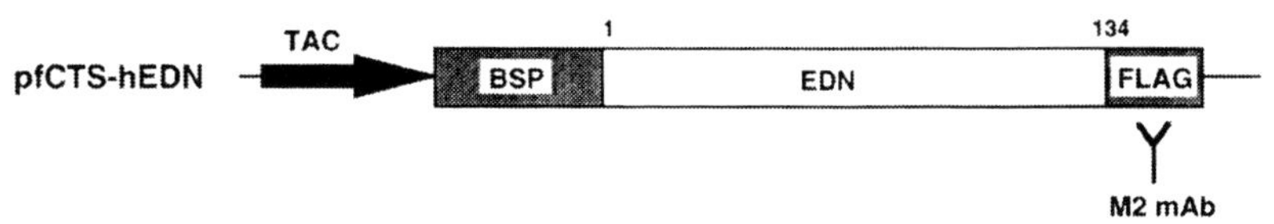

Figure 1. Schematic diagram depicting prokaryotic expression vector pfCTS-FLAG with an insert encoding the 134-bp sequence for human EDN (10). Features include *tac* promoter (responding to induction with IPTG), secretion signal (bacterial secretion piece [BSP]) directing recombinant protein to the bacterial periplasm and C-terminal FLAG octapeptide (DYKDDDDK) detected by M2 mAb.

Cloning and Expression Vectors for Gene Function Analysis
Edited by Quinn Lu and Michael P. Weiner

The plasmid pfCTS (International Biotechnologies, New Haven, CT, USA) includes a *tac* promoter, the hybrid *trp/lac* promoter that responds to induction with IPTG. The vector also encodes a carboxy-terminal FLAG® octapeptide (DYKDDDDK) (Scientific Imaging Systems [Eastman Kodak], New Haven, CT, USA) that facilitates M2 monoclonal antibody (mAb) affinity isolation of recombinant protein as described previously (8,9). The vector also includes the gene encoding β-lactamase for selection of transfectant colonies. The insert is the cDNA sequence encoding the human eosinophil-derived neurotoxin (EDN), a small, cationic protein found in eosinophils that is a member of the ribonuclease A gene superfamily (10).

As shown in Figure 2A, 1-mL cultures of bacterial transfectants were induced with varying concentrations of IPTG, harvested by brief centrifugation and resuspended in 2× sodium dodecyl sulfate (SDS) sample buffer (Novel Experimental Technology, San Diego, CA, USA) prior to electrophoresis and transfer. The Western blot demonstrates that induction with 1 mM IPTG yields multiple forms of recombinant EDN (9). These forms include both approximately 20- and 22-kDa cytosolic unprocessed forms, an approximately 18-kDa processed recombinant human (rh)EDN exported to the periplasm and a smaller (approximately 10-kDa) breakdown product also present in the periplasm. The identity of the 18-kDa periplasmic form was determined previously by amino-terminal sequencing (9). No inclusion bodies are formed when EDN is expressed in this fashion; all forms of the recombinant protein are soluble and are found in the cytosolic extract upon bacterial freeze-thaw and sonication (data not shown). Of particular interest is the fact that the 20-kDa unprocessed form accumulates in marked excess under these conditions. All four forms bind to the M2 affinity resin; the 18-kDa processed form can be isolated by gentle osmotic lysis, thereby avoiding cross-contamination with the unprocessed 20- and 22-kDa cytosolic forms (9). An analogous pattern is seen upon induction with 100 μM IPTG. However, when the concentration of IPTG is reduced to 10 μM, the relative concentrations of the 20-kDa cytosolic and 18-kDa periplasmic forms approach equality, suggesting that a greater proportion of protein synthesized is ultimately processed by the bacterial export system under conditions in which the induction stimulus is limited. The recombinant protein synthesized in response to 1 μM IPTG could not be visualized by this method.

The blot in Figure 2B depicts recombinant protein isolated from a 3-L culture of bacterial transfectants induced with 1 μM IPTG followed by quantitative cell lysis, including freeze-thaw followed by sonication; the band in lane 1 can also be detected by staining with Coomassie® blue (data not shown). The isolation of only the 18-kDa form by this method suggests that the limited induction with 1 μM IPTG permits the bacteria to process virtually all the recombinant protein produced.

The data in Figure 3 demonstrate that the recombinant protein isolated in this fashion is biologically active, with ribonuclease activity that is inhibited by placental ribonuclease inhibitor, as has been shown previously for rhEDN isolated by the osmotic lysis method (9).

In summary, we believe that this method, which involves working with rather than against the bacteria, will have wide-ranging applicability to similar recombinant protein projects. Although limited induction with 1 μM IPTG reduces the

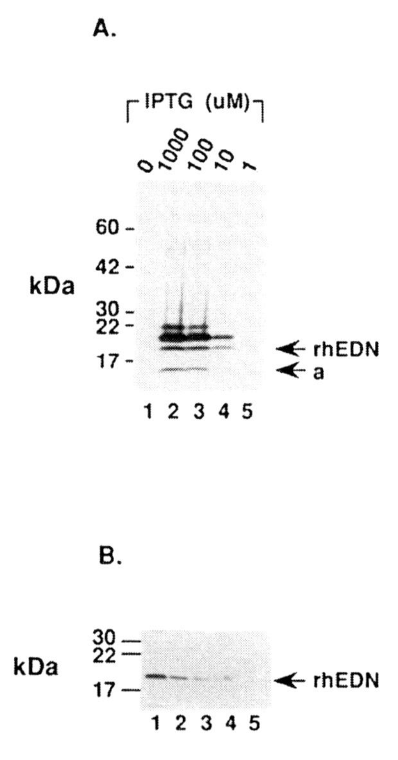

Figure 2. (A) Western blot of lysates from bacteria induced for 1 h with varying concentrations of IPTG (lanes 1–5). Recombinant protein was detected with the M2 FLAG mAb as described (9). The arrow marked "rhEDN" denotes the mobility of the processed, biologically active form of the protein isolated from the bacterial periplasm by osmotic lysis as confirmed by amino-terminal sequencing (9), and the arrow marked "a" indicates a proteolytic breakdown product; the lower-mobility bands represent cytosolic, unprocessed (BSP not removed) forms of rhEDN. (B) Western blot of processed rhEDN isolated after harvest, quantitative lysis (freeze-thaw followed by sonication) of bacterial cells induced for 3 h with 1 μM IPTG. Lanes 1–5 represent serial 3-fold dilutions of the isolated sample.

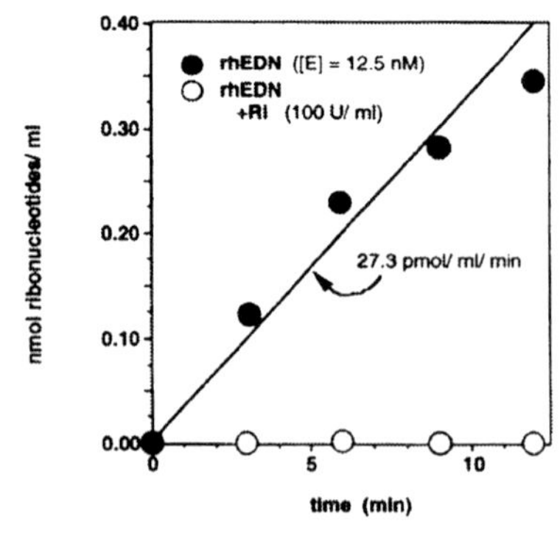

Figure 3. Ribonuclease activity [nmol/mL acid-soluble ribonucleotides generated from acid-insoluble yeast tRNA vs. time (9)] of rhEDN isolated as described. Filled circles: 12.5 nM rhEDN; open circles: 12.5 nM rhEDN + 100 U/mL placental ribonuclease inhibitor (Boehringer Mannheim, Indianapolis, IN, USA).

absolute amount of recombinant protein synthesized, the fact that all the protein was processed to the 18-kDa form has permitted us to perform quantitative lysis, resulting in the harvest of a significantly greater proportion of the recombinant protein produced. Our yield of rhEDN from 3-L cultures induced with 1 mM IPTG and harvested by osmotic lysis was routinely 5 µg or less; our yield using limited induction (1 µM IPTG) and harvest by total cell lysis has been in the range of 50–100 µg/L.

ACKNOWLEDGMENTS

FLAG is a registered trademark of Immunology Ventures. I would like to thank Drs. Harry Malech and John Gallin for their ongoing support of the work in my laboratory.

REFERENCES

1. **Blight, M.A., C. Chervaux and I.B. Holland.** 1994. Protein secretion pathway in *Escherichia coli*. Curr. Opin. Biotechnol. *5*:468-474.
2. **Blight, M.A. and I.B. Holland.** 1994. Heterologous protein secretion and the versatile *Escherichia coli* haemolysin translocator. Trends Biotechnol. *12*:450-455.
3. **Humphreys, D.P., N. Weir, A. Lawson, A. Mountain and P.A. Lund.** 1996. Co-expression of human protein disulphide isomerase (PDI) can increase the yield of an antibody Fab' fragment expressed in *Escherichia coli*. FEBS Lett. *380*:194-197.
4. **Jose, J., J. Kramer, T. Klauser, J. Pohlner and T.F. Meyer.** 1996. Absence of periplasmic DsbA oxidoreductase facilitates export of cysteine-containing passenger proteins to the *Escherichia coli* cell surface via the IgA beta autotransporter pathway. Gene *178*:107-110.
5. **Kaderbhai, N., A. Karim, W. Hankey, G. Jenkins, J. Venning and M.A. Kaderbhai.** 1997. Glycine-induced extracellular secretion of a recombinant cytochrome expressed in *Escherichia coli*. Biotechnol. Appl. Biochem. *25*:53-61.
6. **Lee, S.C. and P.O. Olins.** 1992. Effect of overproduction of heat shock chaperones GroESL and DnaK on human procollagenase production in *Escherichia coli*. J. Biol. Chem. *267*:2849-2852.
7. **Perez-Perez, J., C. Martinez-Caja, J.L. Barbero and J. Gutierrez.** 1995. DnaK/DnaJ supplementation improves the periplasmic production of human granulocyte-colony stimulating factor in *Escherichia coli*. Biochem. Biophys. Res. Commun. *210*:524-529.
8. **Rosenberg, H.F.** 1995. Recombinant human eosinophil cationic protein: ribonuclease activity is not essential for cytotoxicity. J. Biol. Chem. *270*:7876-7881.
9. **Rosenberg, H.F. and K.D. Dyer.** 1995. Eosinophil cationic protein and eosinophil-derived neurotoxin. Evolution of novel function in a primate ribonuclease gene family. J. Biol. Chem. *270*:21539-21544.
10. **Rosenberg, H.F., D.G. Tenen and S.J. Ackerman.** 1989. Molecular cloning of the human eosinophil-derived neurotoxin: a member of the ribonuclease gene family. Proc. Natl. Acad. Sci. USA *86*:4460-4464.

Update to: Isolation of Recombinant Secretory Proteins by Limited Induction and Quantitative Harvest

Helene F. Rosenberg

National Institute of Allergy and Infectious Diseases, NIH, Bethesda, MD, USA

The method presented in my original report is now our standard technique for isolating recombinant secretory RNase A ribonucleases from bacteria. We have found that, depending on which RNase A ribonuclease we are working with, we can occasionally increase the induction stimulus to as much as 10 µM isopropyl β-D-thiogalactopyyranoside (IPTG) without seeing accumulation of the unwanted, unprocessed cytoplasmic forms of these ribonucleases in our preparations. We have used this method successfully to isolate recombinant primate and rodent ribonucleases (1,2).

REFERENCES

1. **Domachowske, J.B., C.A. Bonville, K.D. Dyer and H.F. Rosenberg.** 1998. Evolution of antiviral activity in the ribonuclease A gene superfamily. Nucleic Acids Res. *26*:5327-5332.
2. **Singhania, N.A., K.D. Dyer, J. Zhang, M.S. Deming, C.A. Bonville, J.B. Domachowske and H.F. Rosenberg.** 1999. Rapid evolution of the ribonuclease A superfamily: adaptive expansion of independent gene clusters in rats and mice. J. Mol. Evol. *49*:721-728.

CHAPTER 16

Recovery of Polypeptides Cleaved from Purified Calmodulin-Binding Peptide Fusion Proteins

Peter Vaillancourt, Timothy G. Simcox and Chao-Feng Zheng

Stratagene, La Jolla, CA, USA

(Reprinted from BioTechniques 22:451-453, 1997)

Purification of recombinant proteins has been greatly simplified in recent years because of the availability of expression vectors that allow fusion of the protein coding sequence of interest to short peptide sequences or larger proteins, enabling the affinity purification of the fusion protein from crude preparations. Fusion proteins are often engineered to contain short linkers of five to six amino acids, positioned between the affinity tag and the protein of interest, which serve as recognition targets for site-specific proteases thereby allowing the proteolytic removal of the affinity tag following purification of the fusion protein. The serine protease enterokinase (EK) is particularly attractive because it cleaves after the carboxy terminus of its recognition sequence $(Asp)_4$-Lys, allowing production of cleavage products that have native amino termini following the removal of N-terminal affinity tags. In addition, the recent cloning of bovine EK and its expression and purification from *E. coli* has allowed the production of high specific activity enzyme that is virtually free of contaminating proteases (1,5). At a low enzyme-to-substrate ratio (1:1000 wt/wt), fusion protein can often be cleaved to completion in a few minutes at 37°C, whereas with other site-specific proteases such as thrombin and factor Xa, cleavage reactions are usually carried out over a period of several hours.

Following purification of fusion proteins and proteolytic removal of the affinity peptide tags, it is often desirable to recover the mature cleavage product from EK reactions that contain a mixture of the protein of interest, the free affinity tag, EK and often small amounts of uncleaved fusion protein. Here we describe a simple and efficient one-step protocol for removing both EK and EK reaction products that contain the calmodulin-binding peptide (CBP) affinity tag (6,7).

The fusion protein CBP-EK-JNK was produced by fusing the five-amino acid EK target sequence to the N-terminus of the protein c-jun N-terminal kinase (JNK) (2,4,8) by polymerase chain reaction (PCR) and subsequently inserting the PCR fragment into the expression vector pCAL-n (6,8) so that the EK target peptide-JNK fusion protein was fused in-frame downstream from the CBP purification tag. The resulting fusion protein was expressed in BL21(DE3) cells and purified from crude cell lysates to apparent homogeneity with calmodulin (CaM) affinity resin (Stratagene, La Jolla, CA, USA) (Figure 1, lane 1). Purified CBP-EK-JNK was partially digested by incubating with recombinant EK (Stratagene) at an enzyme-to-substrate ratio of 1:500 wt/wt for 3 min at 37°C. Using these conditions, slightly over half of the input fusion protein was cleaved to yield the 46-kDa JNK polypeptide and the 4 kDa CBP fragment (Figure 1, lane 2). Following the digestion, the reaction was adjusted to 4 mM $CaCl_2$ and 200 mM NaCl and mixed with 50 µL CaM-affinity resin and 20 µL soybean trypsin inhibitor-agarose (STI-agarose), the latter of which binds with high affinity to EK. The mixed resin slurry was mechanically rotated for 1 h at 4°C. Following absorption with the resin mixture, the unbound material was removed by centrifugation, and portions of the unbound supernatant and the resin pellet were analyzed by sodium dodecyl sulfate polyacrylamide gel electrophoresis (SDS-PAGE). As the results in Figure 1 indicate, the uncleaved CBP fusion protein and the cleaved CBP affinity tag were effectively removed by absorption with the resin mixture (Figure 1, compare lanes 2 and 3), while only a trace amount of cleaved JNK remained nonspecifically bound to the resin (lane 4). Protein concentration of the unbound material was assayed by the Coomassie® Plus Protein Assay (Pierce, Rockford, IL, USA), and the yield (25 µg) was determined to be approximately 83% of the JNK cleavage product present in the initial EK cleavage reaction, assuming that 60% of the input fusion protein was cleaved in the reaction.

The efficiency of EK removal with STI-agarose was determined by absorbing EK reaction mixtures, with either CaM-affinity resin alone or with a mixture of CaM-affinity resin plus STI-agarose, and assaying for EK activity remaining in the unbound material using a fluorogenic peptide substrate assay (3). In this experiment, 200 ng of EK were used to digest CBP-EK-JNK at a 1:500 enzyme-to-substrate ratio in a 5-min reaction during which over 95% of the input fusion protein was cleaved as estimated by SDS-PAGE analysis (data not shown). After EK digestion of CBP-EK-JNK, the reaction

Cloning and Expression Vectors for Gene Function Analysis
Edited by Quinn Lu and Michael P. Weiner

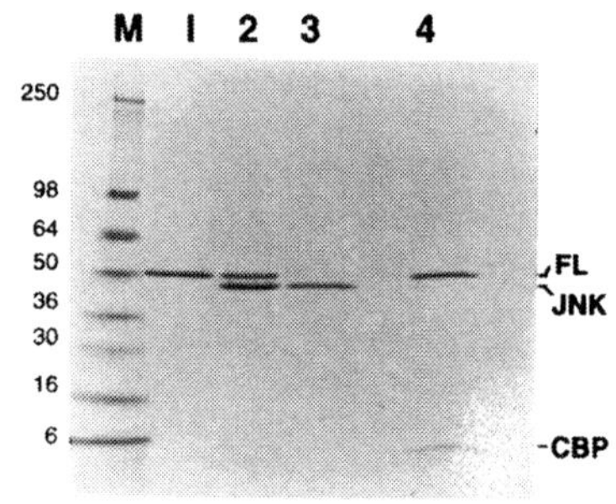

Figure 1. Removal of cleaved CBP affinity tag peptide fragment and uncleaved CBP fusion protein from EK digestion reactions. Purified CBP-EK-JNK (50 µg) (lane 1) was digested with 100 ng of EK in 100 µL of EK reaction buffer (50 mM Tris-HCl, pH 8.0, 1 mM $CaCl_2$ and 0.1% Tween® 20) for 3 min at 37°C and cooled on ice. Following removal of a portion of the reaction for SDS-PAGE analysis (lane 2), the sample was diluted to 500 µL with binding buffer (50 mM Tris-HCl, pH 8.0, 4 mM $CaCl_2$, 200 mM NaCl and 0.1% Tween 20) and added to a mixture of settled resin containing 50 µL CaM affinity resin and 20 µL STI-agarose (Sigma Chemical, St. Louis, MO, USA). The slurry was mechanically rotated at 4°C for 1 h. Following removal of unbound material (lane 3), a portion of the resin was boiled in Laemmli sample buffer (lane 4). Samples were boiled in Laemmli sample buffer and electrophoresed on a 4%–20% Tris-glycine SDS-PAGE (Novex, San Diego, CA, USA). The gel was stained with Coomassie® Brilliant Blue dye. M (molecular weight standards [kDa]), FL (full-length uncleaved CBP-EK-JNK), JNK (JNK polypeptide) and CBP (calmodulin binding peptide and EK recognition peptide fragment).

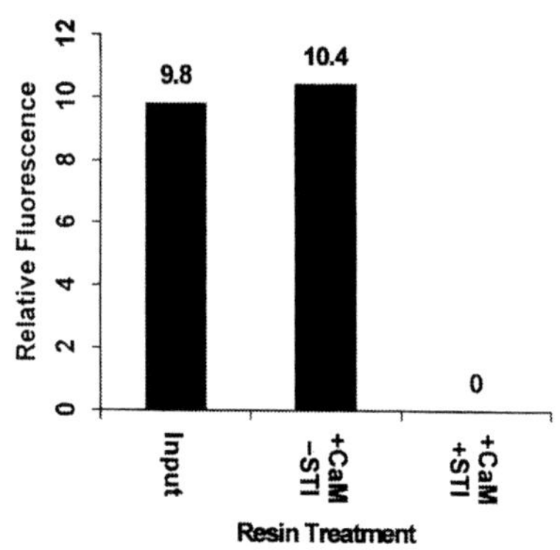

Figure 2. Determination of EK activity by fluorogenic peptide substrate assay. Purified CBP-EK-JNK (100 µg) was digested with 200 ng EK for 5 min at 37°C in a 1.0-mL reaction. The reaction was split into two 500-µL portions that were added to either 50 µL CaM-affinity resin (+CaM, -STI) or to a mixture of 50 µL CaM-affinity resin and 20 µL STI-agarose (+CaM, +STI). The resin slurries were mechanically rotated for 1 h at 4°C, and the resin was removed from the slurry by centrifugation. The unbound material was analyzed for EK activity using a fluorogenic peptide assay. The volumes of both fractions were adjusted to 420 µL with binding buffer, and the samples were mixed with 1.6 mL of a solution containing 0.5 mM of the fluorogenic EK substrate in 70 mM Tris-HCl, pH 8.0 and 10% dimethyl sulfoxide. Samples were mixed and immediately analyzed using a Model RF-1501 Spectrofluorometer (Shimadzu Scientific Instrument, Columbia, MD, USA). Enzyme activity was determined by measuring increased fluorescence (excitation 337 nm, emission 420 nm) due to the release of B-naphthylamine over a 1-min interval. For a positive control, 30 ng of EK were assayed (Input), and the increased fluorescence value was normalized to the amount of enzyme used in the experimental reactions.

was split and incubated with CaM affinity resin alone or with mixed resin as described above. To measure the amount of EK activity remaining in the supernatant, portions of the unbound material were mixed with a solution containing the fluorogenic peptide Gly-Asp-Asp-Asp-Asp-Lys-B-naphthylamide (Bachem Biosciences, Philadelphia, PA, USA). The samples were immediately analyzed in a spectrofluorometer to measure increased fluorescence due to release of B-naphthylamine over a 1-min interval. According to the results shown in Figure 2, there was no detectable EK activity remaining in the unbound fraction following incubation with STI-agarose (+STI, +CaM), while essentially all of the EK activity was recovered in the fraction that was absorbed with CaM-affinity resin alone (-STI, +CaM).

In this study, excess quantities of CaM-affinity resin and STI-agarose were used in the absorption experiments. The high affinity of both CBP fusion proteins and EK for their respective affinity resins using the mutually compatible binding conditions used in these experiments, together with the high binding capacities of the resins (both resins bind 1.5–3.0 mg of protein per mL of swollen resin) should allow this technique to be effectively used for large-scale purifications without requiring the use of large quantities of resin or excessive absorption times and should result in the rapid and efficient high yield recovery of purified EK cleavage products that have native amino acid sequence without any extraneous amino acid residues derived from the affinity tag or protease target site.

ACKNOWLEDGMENTS

The authors wish to thank Edward R. LaVallie, Mary Ellen Simcox, John C. Bauer, Alan Greener and Joseph A. Sorge for helpful discussions and Barbara McGowan, Daniel McMullan and Peter Pingerelli for preparation of calmodulin affinity resin.

REFERENCES

1. **Collins-Racie, L.A., J.M. McColgan, K.L. Grant, E.A. DiBlasio-Smith, J.M. McCoy and E.R. LaVallie.** 1995. Production of recombinant bovine enterokinase catalytic subunit in *Escherichia coli* using the novel secretory fusion partner DsbA. Bio/Technology *13*:982-987.
2. **Derijard, B., M. Hibi, I.-H. Wu, T. Barrett, B. Su, T. Deng, M. Karin and R.J. Davis.** 1994. JNK1: a protein kinase stimulated by UV light and Ha-Ras that binds and phosphorylates the c-Jun activation domain. Cell *76*:1025-1037.
3. **Grant, D.A.W. and J. Hermon-Taylor.** 1979. Hydrolysis of artificial substrates by enterokinase and trypsin and the development of a sensitive specific assay for enterokinase in serum. Biochim. Biophys. Acta *567*:207-215.
4. **Kyriakis, J.M., P. Banerjee, E. Nikolakaki, T. Dai, E.A. Rubie, M.F. Ahmad, J.U. Avruch and J.R. Woodgett.** 1994. The stress-activated protein kinase subfamily of c-jun kinases. Nature *369*:156-158.
5. **LaVallie, E.R., A. Rehemtulla, L.A. Racie, E.A. DiBlasio, C. Ferenz, K.L. Grant, A. Light and J.M. McCoy.** 1993. Cloning and functional expression of a cDNA encoding the catalytic subunit of bovine enterokinase. J. Biol. Chem. *268*:23311-23317.
6. **Simcox, T.G., C.-F. Zheng, M.E. Simcox and P. Vaillancourt.** 1995. Calmodulin binding peptide affinity tag *E. coli* protein purification system. Strateg. Mol. Cell Biol. *8*:40-43.
7. **Stofko-Hahn, R.E., D.W. Carr and J.D. Scott.** 1992. A single step purification for recombinant proteins. FEBS Lett. *302*:274-278.
8. **Zheng, C.-F. and T.G. Simcox.** 1996. Recombinant c-jun N-terminal kinase (JNK) for studying cellular stress signal transduction in mammalian cells. Strateg. Mol. Biol. *9*:20-21.

Update to: Recovery of Polypeptides Cleaved from Purified Calmodulin-Binding Peptide Fusion Proteins

Katherine A. Felts and Peter Vaillancourt

Stratagene Cloning Systems, La Jolla, CA, USA

In our original manuscript, we described the production of seamless fusion proteins by designing polymerase chain reaction (PCR) primers such that coding sequence for the desired N-terminal peptide sequence of the protein of interest resided immediately downstream of the enterokinase (EK) cleavage site. The insert was then fused downstream of and in-frame with the calmodulin-binding peptide (CBP) purification tag in the vector pCAL-n (6). We have since described the construction and testing of the CBP fusion vector pCAL-n-EK (5) in which the ligation-independent cloning (LIC) method (1) is employed such that any amino acid coding sequence can be seamlessly fused immediately downstream of the EK cleavage site, thereby allowing the recovery of proteins from EK cleavage reactions that have native N-terminal amino acid sequence. In addition to the ability to create seamless fusions between the CBP purification tag and the protein coding sequence of interest, cloning by LIC is highly efficient. Cloning efficiencies of greater than or equal to 2×10^5 cfu/µg vector are routinely achieved using as little as 25 ng PCR-generated insert, thus enough insert for several cloning reactions may be generated with as little as 10 cycles of PCR amplification (5).

There is an inherent drawback to the use of CBP fusions for certain downstream applications for which purified proteins are employed to probe protein-protein interactions. The ubiquitous presence of calmodulin in eukaryotic cells, in addition to the presence of endogenous calmodulin-binding proteins, can under some conditions contribute to background in such applications. Specifically, immunoaffinity-type "pull-down" experiments in which CBP-fusions are recovered from crude eukaryotic extracts by "immunoprecipitation" (2) with calmodulin affinity resin often result in the recovery of calmodulin binding proteins. To address this issue, we have modifed pCAL-n-EK such that the FLAG® epitope (Scientific Imaging [Eastman Kodak], New Haven, CT, USA) is positioned between the thrombin and EK cleavage sites. CBP fusion proteins expressed and purified using this vector may be cleaved with thrombin to produce FLAG fusions that may be detected and immunoprecipitated using the anti-FLAG M2 monoclonal antibody. Alternatively, the same CBP fusion may be cleaved with EK to produce proteins with natural N termini.

The pCAL-n-FLAG vector (Figure 1) is derived from the pET-11 vector series and contains the *lacI*q gene for expression of the Lac repressor protein and the hybrid T7 promoter for controlled expression of the inserted protein coding sequence (4). The vector is transformed into specialized *Escherichia coli* strains, such as BL21(DE3), which contains an expression cas-

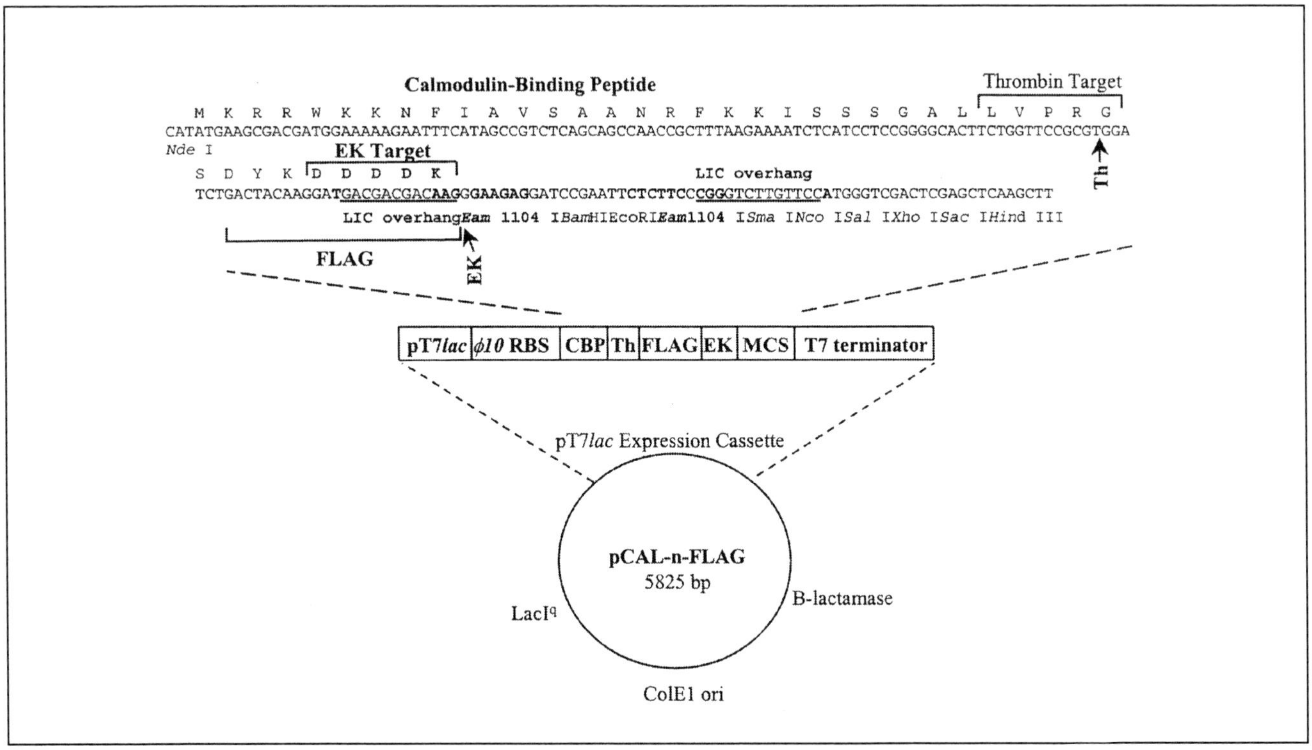

Figure 1. Map of pCAL-n-FLAG. The circular map depicts the general structure of the vector. The middle insert shows an expanded view of the pT7*lac* expression cassette, including the Lac I-repressible pT7*lac* promoter, T7 *ϕ10* ribosome binding site (RBS), the CBP fusion tag followed by a thrombin cleavage site, the FLAG epitope, an EK cleavage site, the MCS and the T7 transcriptional terminator. The top insert shows the nucleotide and amino acid sequence of the region from the CBP initiation codon through the MCS. Recognition sequences and 3′ overhang triplets for the type IIs restriction enzyme *Eam*1104I are shown in italics, the LIC 3′ terminator nucleotides are underlined.

sette for T7 RNA polymerase that is induced in the presence of isopropyl-β-D-thiogalactoside (IPTG), allowing tight control and high-level induced expression of the inserted protein coding sequence. The promoter is followed by the 26-amino acid CBP affinity tag followed by a thrombin cleavage site, the FLAG epitope, an enterokinase cleavage site, the multiple cloning site (MCS) and the T7 transcriptional terminator.

Protein coding sequence is seamlessly inserted into pCAL-n-FLAG by LIC (1), a cloning method which is universal for inserts in that it does not rely on the use of restriction sites (Figure 2). For insertion by LIC, *Eam1104 I* digested vector is treated with *Pfu* polymerase in the presence of dTTP resulting in the 3′ to 5′ exonucleolytic removal of nucleotide residues until the first dT residue is encountered, thus producing the 12- and 13-nt single-stranded overhangs shown in Figure 2A. The pCAL-n-FLAG vector is pretreated to create noncomplementary, 12- and 13-nt single-stranded tails at the two 5′ ends of the vector. These noncomplementary tails prevent the vector from recircularizing. The insert DNA is prepared by PCR amplification with gene-specific primers that include 12- and 13-nt sequences at the 5′ ends that are complementary to the pCAL-n-FLAG vector single-stranded tails, thus allowing for directional cloning. Following PCR amplification, the PCR product is purified then treated with *Pfu* DNA polymerase in the presence of dATP. The vector and insert DNA are combined, allowed to anneal at room temperature and transformed into highly competent bacterial host cells. The resultant colonies can then be screened for insert by standard methods.

The 1280-bp gene encoding c-Jun N-terminal kinase (JNK) was PCR-amplified with gene-specific primers containing LIC ends as previously described (5) and inserted into the pCAL-n-FLAG vector as described above. Mini-prep DNA was puri-

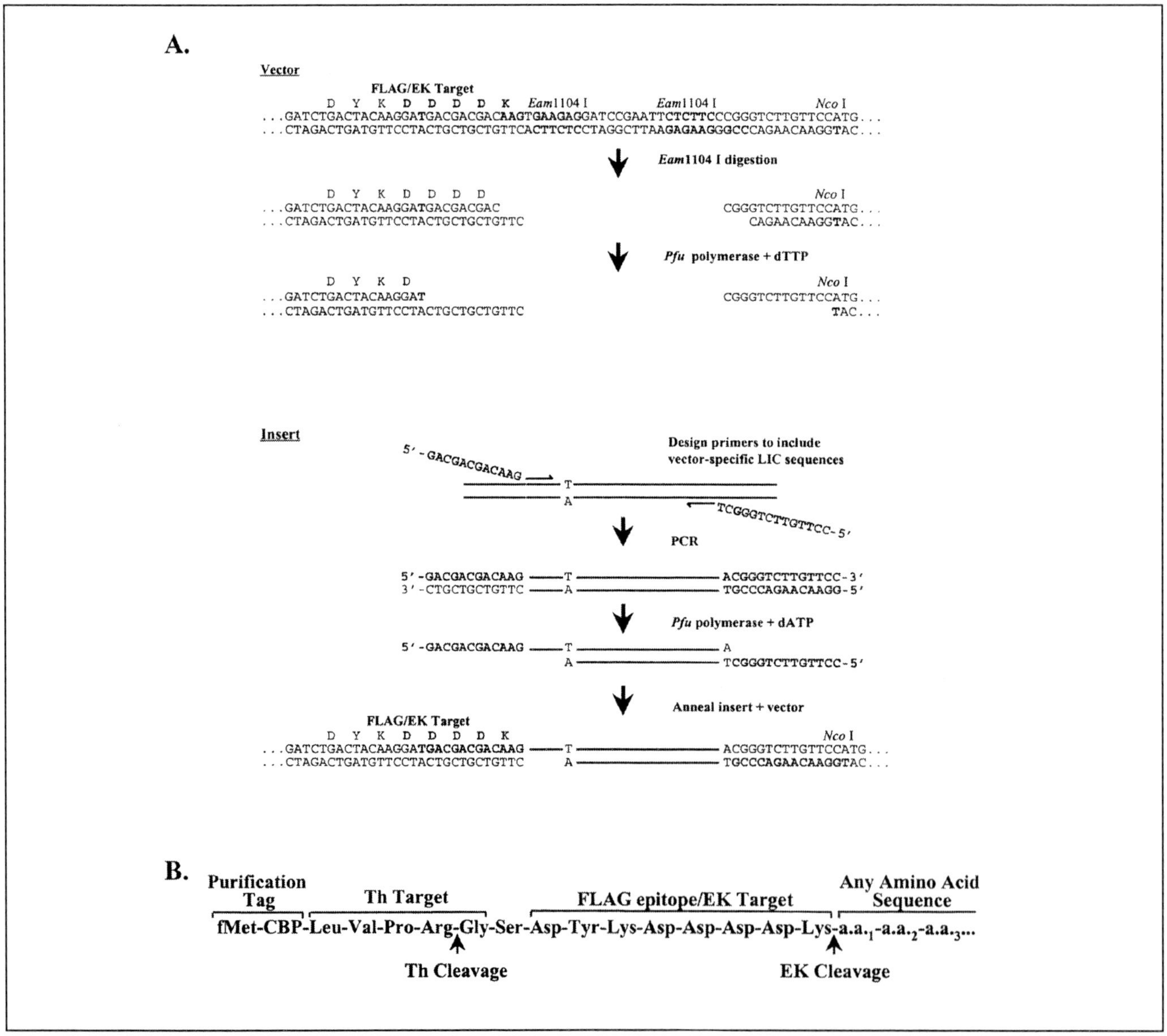

Figure 2. Seamless insertion by LIC. (A) Note that the upstream insert-specific LIC "terminator" dAMP residue need not correspond precisely with the vector-insert junction; single-stranded gaps within the annealed product (corresponding to the distance between the cloning junction and the dAMP residue) are repaired in vivo in the *E. coli* cell. (B) Predicted N-terminal amino acid sequence of pCAL-n-FLAG fusion proteins. CBP fusion proteins may be cleaved with EK to yield recombinant proteins of native amino acid sequence, or thrombin to yield FLAG fusion protein without the CBP tag.

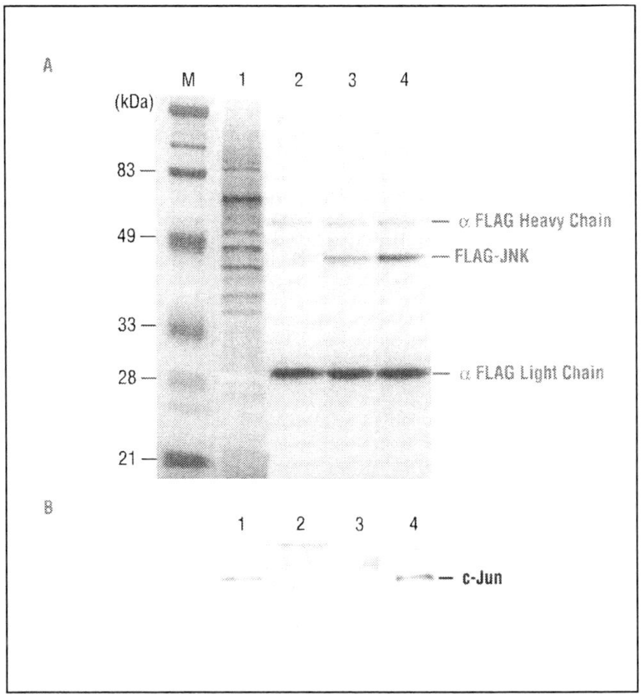

Figure 3. Immunoaffinity pull-down of c-Jun with FLAG-JNK. (A) Shown is a Coomassie Blue stain of one duplicate gel. The other gel was electroblotted onto a nitrocellulose filter and subjected to Western blot analysis with an anti-c-Jun antibody. The Western blot is shown in panel B. Lane M, markers; lane 1, c-Jun control extract; lane 2, anti-FLAG M2 affinity gel; lane 3, JNK-FLAG-anti-FLAG gel; and lane 4, JNK-FLAG-anti-FLAG gel incubated in 50 µg HeLa nuclear extract.

fied, and positive clones were subsequently confirmed by DNA sequence analysis. CBP-FLAG-JNK fusion protein was purified exactly as described in the original manuscript.

It has been shown that transcriptional activity of c-Jun is greatly enhanced in response to phosphorylation (3). JNK binds to the c-Jun transactivation domain and phosphorylates Ser-63, Ser-73 and some secondary sites. To demonstrate the utility of a FLAG-fusion protein with the CBP tag removed (Figure 2B), we performed an immunoaffinity pull-down of c-Jun from HeLa cell nuclear extract with immobilized FLAG-JNK. Prior to performing the pull-down experiment, we removed the CBP tag by digestion with thrombin and adorption of the tag with CaM-resin. CBP-FLAG-JNK in elution buffer was supplemented with $CaCl_2$ to a final concentration of 6 mM and digested at room temperature with 40 U thrombin/µg of CBP-FLAG-JNK. To evaluate the extent of digestion over time, digestion was terminated at various time points 0, 1 and 18 h. The reaction was stopped by transferring 20 µL of digest to 20 µL of 2× sodium dodecyl sulfate (SDS) sample buffer. To absorb any undigested protein and free CBP tag, calmodulin resin was added to the reaction and incubated while gently mixing at 4°C for 2 h. The resin was pelleted in a low-speed centrifuge, and 20 µL of the absorbed supernatant was transferred to 20 µL of 2× SDS sample buffer. The degree of digestion by thrombin was evaluated by performing sodium dodecyl sulfate-polyacrylamide gel electrophoresis (SDS-PAGE) (data not shown). Recognition of the FLAG epitope on the resulting FLAG-JNK fusion was verified by western blot (data not shown). JNK affinity gel was prepared by coupling FLAG-JNK to anti-FLAG M2 affinity gel (Sigma, St. Louis, MO, USA) in a binding buffer that contained 150 mM NaCl. Approximately 100 µg of FLAG-JNK was bound by 40 µL of anti-FLAG gel. The affinity gel pellet was washed and an amount equal to 5 µg FLAG-JNK was transferred to a 200-µL sample made up of 50 µg HeLa cell nuclear extract in a reaction buffer consisting of 20 mM HEPES, pH 7.7, 50 mM NaCl, 0.1 mM EDTA, 0.05% Tween® 20. The sample was gently mixed at 4°C for 5 h. Affinity gel was pelleted by low-speed centrifugation and washed extensively with binding buffer. The pellet was resuspended in 1× SDS sample buffer and boiled before being subjected to SDS-PAGE. Duplicate gels were performed. One duplicate gel was stained with Coomassie® Blue protein stain, the other gel was electroblotted onto a nitrocellulose filter and subjected to Western blot analysis with an anti-c-Jun antibody (Transduction Laboratories, Lexington, KY, USA) by standard procedures. As indicated by the results in Figure 3 (panel A, lanes 3 and 4), equivalent amounts of FLAG-JNK are observed in the Coomassie Blue-stained gel. In panel B, an anti-c-jun-specific band with a gel mobility corresponding to the expected molecular weight is specifically immunoprecipitated from HeLa nuclear extract with the M2-FLAG-JNK gel (Figure 3B, lane 4).

REFERENCES

1. **Aslanidis, C, and P.J. de Jong.** 1990. Ligation-independent cloning of PCR products (LIC-PCR). Nucleic Acids Res. *18*:6069-6074.
2. **Blanar, M.A. and W.J. Rutter.** 1992. Interaction cloning: identification of a helix-loop-helix zipper protein that interacts with c-Fos. Science. *256*:1014-1018.
3. **Derijard, B., M. Hibi, I.H. Wu, T. Barrett, B. Su, T. Deng, M. Karin and R.J. Davis.** 1994. JNK1: a protein kinase stimulated by UV light and Ha-Ras that binds and phosphorylates the c-Jun activation domain. Cell. *76*:1025-1037.
4. **Studier, F.W., A.H. Rosenberg, J.J. Dunn and J.W. Dubendorff.** 1990. Use of T7 RNA polymerase to direct expression of cloned genes. Methods Enzymol. *185*:60-68.
5. **Wyborski, D.L., J.C. Bauer, C.F. Zheng, K. Felts and P. Vaillancourt.** 1999. An *Escherichia coli* expression vector that allows recovery of proteins with native N-termini from purified calmodulin-binding peptide fusions. Protein Expr Purif. *16*:1-10.
6. **Zheng, C.F., T. Simcox, L. Xu and P. Vaillancourt.** 1997. A new expression vector for high level protein production, one step purification and direct isotopic labeling of calmodulin-binding peptide fusion proteins. Gene. *186*:55-60.

Section III

Gene Tagging and Epitope Tagging Strategies

Mini-Exon Epitope Tagging for Analysis of the Protein Coding Potential of Genomic Sequence

Desmond J. Smith

Harvard University, Cambridge, MA, USA

(Reprinted from BioTechniques 23:116-120, 1997)

ABSTRACT

A novel approach to gene discovery and analysis is described. A small exon flanked by consensus 3′ and 5′ splice sites was synthesized. The exon contains open reading frames encoding 43 amino acid peptides. There are no stop codons in any of the three reading frames, and each reading frame contains an epitope recognized by the same monoclonal antibody. The exon can be inserted into the introns of genes, and the resulting small peptide will be incorporated into the protein encoded by the host gene, regardless of the class of intron. The protein can then be recognized by the antibody, permitting functional studies.

INTRODUCTION

Knowledge of all gene products is an essential component for understanding data obtained from the Human Genome projects. Expressed sequence tags (ESTs) (1) will provide a fragment of transcribed sequence for nearly every gene in the genome, but although they provide suggestive landmarks for positional cloning projects, ESTs usually give little insight into gene function. Furthermore, it is often difficult to forecast the structure of transcripts from genomic sequence, and computational prediction of transcription initiation sites and splice sites continues to be a challenging problem. Enhancer trap experiments (14) give information about the tissue distribution of transcripts but give little data on the function or intracellular localization of protein. Fusion of a marker protein such as β-galactosidase with gene products (4,11) gives more information; however, a disadvantage is that the truncated fusion protein may possess novel properties and may, for example, travel to an inappropriate intracellular compartment.

A method is needed that easily reveals the presence of genes and their functions in a defined genomic interval. This paper presents the development of a mini-exon that can be inserted into introns, permitting the detection of gene products using the same monoclonal antibody regardless of intron class (16). The resultant protein is altered in a minor fashion and can be analyzed using functional assays such as immunofluorescence.

MATERIALS AND METHODS

Construction of the Mini-Exon Oligonucleotide and Related Plasmids

General molecular biology techniques were performed as described (16). The 170-nucleotide (nt) sequence was synthesized using a Cyclone™ Plus DNA Synthesizer (Milligen/Biosearch, Burlington, MA, USA), and because the full-length yield of such a lengthy oligonucleotide is quite low, it was amplified using the polymerase chain reaction (PCR). The reaction contained the 170-mer at 2 mM, and the primers were CTAATTCCTCTCTTCTCCTCT and CGAGATCTACTTACCTTC, which were at 20 mM each. The conditions were 30 cycles of 94°C for 1 min, 50°C for 2 min and 72°C for 3 min followed by a final extension at 72°C for 15 min.

The PCR products were 5′-phosphorylated using T4 polynucleotide kinase, the 170-bp product purified from a 3% agarose gel and cloned into *Sma*I-cut Bluescript® KS(+) (Stratagene, La Jolla, CA, USA). A recombinant clone containing the correct insert was identified using double-stranded sequencing and called pmyc2. The mini-exon was excised using *Kpn*I and *Sac*I and ligated into pUC1813 (10), which had been cut using the same enzymes, thus creating pUC1813.myc2. The mini-exon was then excised as a blunt *Sma*I fragment from pUC1813.myc2 and ligated into pBR327 (19), which had been cut with *Eco*RI and *Bsa*I, the ends having been filled in using Klenow. The resultant plasmid, pBR327.myc2 (2.6 kb), is diagrammed in Figure 1. p11-4cG contains a hybrid cDNA/genomic *p53* clone (2,12) borne on a vector containing the simian virus 40 (SV40) origin, promoter and enhancer. pBR327.myc2 was linearized just upstream of the mini-exon by partial digestion with *Hin*dIII and inserted into a *Hin*dIII site in intron 9 of the *p53* gene in p11-4cG. Since p11-4cG is ampicillin-resistant and pBR327.myc2 is tetracycline-resistant, it was easy to obtain recombinants by selecting for doubly resistant colonies. The resulting plasmid was called p11-4cG.pBR327.myc2.

Cloning and Expression Vectors for Gene Function Analysis
Edited by Quinn Lu and Michael P. Weiner

Analysis of RNA

Total RNA was prepared from COS-1 cells as described (17). Reverse transcription PCR (RT-PCR) was performed as described (18) with minor modifications using the primers recognizing exons 9 and 10 of the *p53* gene. The exon 10 primer was also used for RT. For sequencing, RT-PCR products were gel-purified and subcloned into M13mp18. The mini-exon was ligated into M13mp19 for S1 analysis. Preparation of single-stranded radiolabeled probes from the resulting M13 clone and S1 analysis using these probes were as described (17), except the primer used for probe synthesis was the SK primer (Stratagene).

Transfection and Immunofluorescence of COS Cells

The transfection and immunofluorescence of COS cells was performed as described (13) except the $NaBH_4$ treatment was omitted. Cells were fixed and permeabilized in 4% paraformaldehyde, 1% Triton® X-100, 1× phosphate-buffered saline (PBS) for 10 min, blocked in 1× PBS, 2% bovine serum albumin, 0.1% Triton X-100 for 15 min, incubated in 9E10 antibody (undiluted tissue culture supernatant; Harvard Biolabs Monoclonal Antibody Facility, Cambridge, MA, USA) for 90 min, washed in 1× PBS plus 0.1% Triton X-100, blocked again, incubated for 60 min in fluorescence isothiocyanate (FITC)-conjugated goat anti-mouse IgG (Organon-Teknika/Cappel, Durham, NC, USA) diluted 1:20 in blocking solution, washed and mounted.

RESULTS

The Mini-Exon

The mini-exon (Figure 1) has a 129-bp coding region with no stop codons in any of the three reading frames on the coding strand. In the DNA sequence, there are three direct repeats, which are separated from each other by four nucleotides. Each of the repeats encodes a myc epitope recognized by the same monoclonal antibody (9E10) in one of the three reading frames. Glycine residues flank each of the repeats. These residues are flexible and allow free rotation and mobility of the embedded repeat. Such mobility is thought to enhance antibody binding to epitopes (8). The coding region is preceeded and followed, respectively, by a consensus 3′ and 5′ splice site (3,15). The 3′ splice site includes a consensus branch point

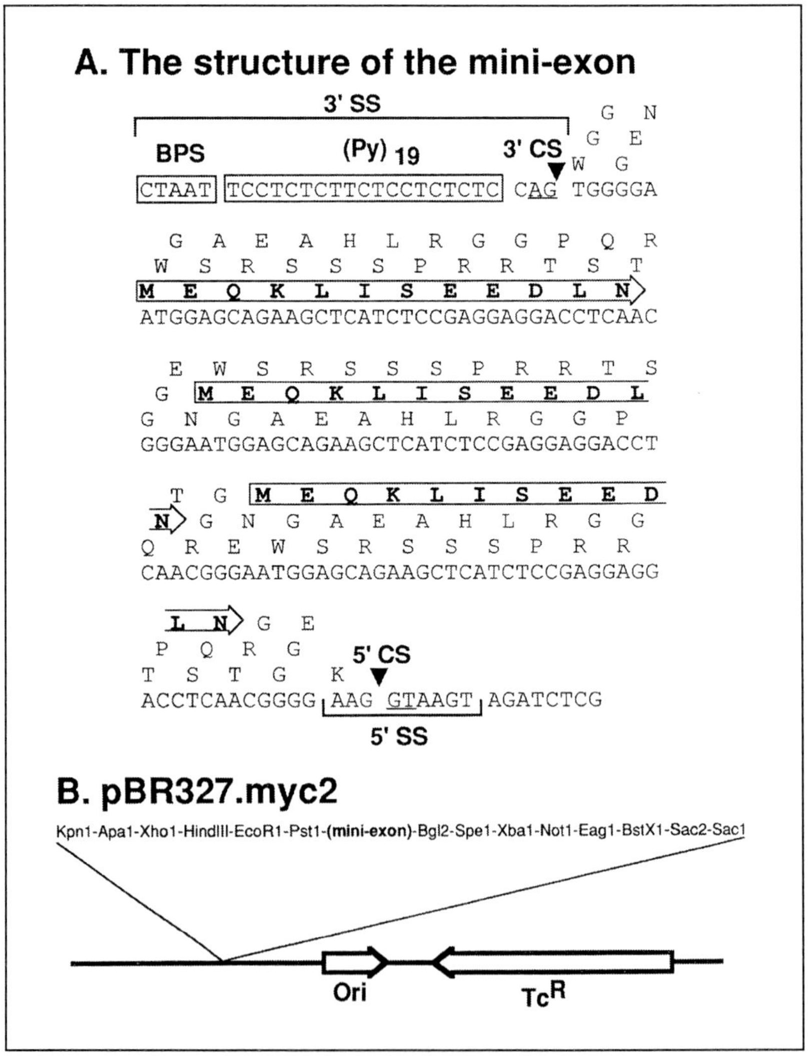

Figure 1. The mini-exon. (A) The structure of the mini-exon. The branch point sequence (BPS), the polypyrimidine tract [$(Py)_{19}$], the mandatory AG dinucleotide (underlined) and the cleavage site (3′ CS) of the 3′ splice site are shown. The three ORFs are shown, and the epitope recognized by the monoclonal antibody 9E10 is shown highlighted in each frame. The cleavage site (5′ CS) and the mandatory GT dinucleotide (underlined) of the 5′ splice site is shown. (B) pBR327.myc2. The plasmid is shown in the sense-orientation of the mini-exon, together with useful single cutter sites of the polylinker. The *Hin*dIII site was used to introduce the mini-exon into the *p53* gene, but it is not a single cutter, since there is also a *Hin*dIII site in the tetracycline-resistance (Tc^R) gene.

sequence, a polypyrimidine tract and the mandatory AG dinucleotide. The 5′ splice site includes the mandatory GT dinucleotide. Since exon/intron boundaries tend to map to the surface of final protein products (7), the mini-exon peptide will be likely to be displayed on the surface of the protein, where it will be very accessible for binding to antibody (8).

The mini-exon is cloned into a plasmid with the tetracycline-resistance marker, which facilitates further subcloning (see Materials and Methods).

Splicing of the Mini-Exon

To test the mini-exon, it was inserted into intron 9 (class 0) of a mutant allele of the *p53* tumor suppressor gene. The recombinant plasmid, p11-4cG.pBR327.myc2, was transfected into COS-1 cells, and RNA was extracted for S1 analysis using DNA complementary to the mini-exon as probe (Figure 2). The observed protected fragment was consistent with correct and 100% efficient splicing of the mini-exon into the *p53* transcript. An RT-PCR analysis was performed using primers that lie in the coding region of exons 9 and 10 of the *p53* gene and which therefore span intron 9. Figure 3 shows the results of this analysis. If the mini-exon was not included in the *p53* transcript, an RT-PCR product of 195 bp was expected; whereas, if it were spliced correctly, a product of 324 bp was predicted. The 324-bp product was observed and was again consistent with correct and 100% efficient splicing. The RT-PCR product was subcloned and sequenced. This revealed that the mini-exon had spliced exactly as predicted into the *p53* open reading frame (ORF) (data not shown).

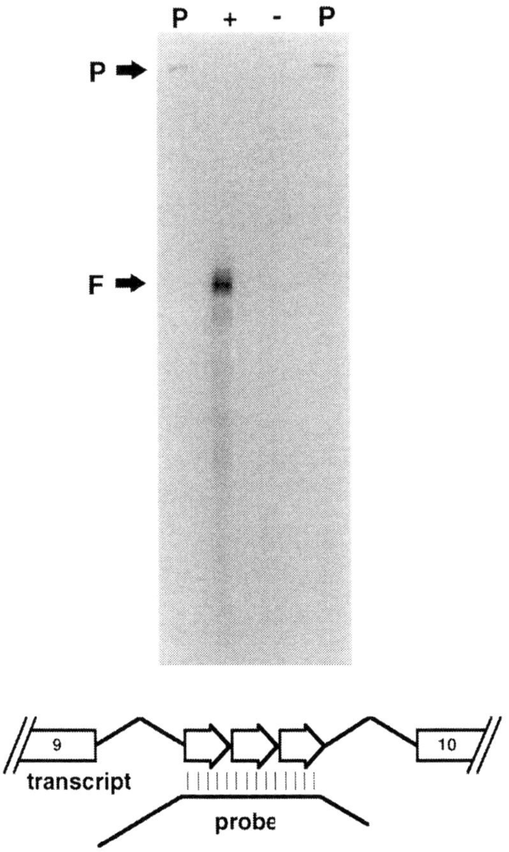

Figure 2. S1 analysis of the mini-exon splicing. The probe, indicated by the arrow marked with P, has a size of 243 nt. The expected protected fragment if there is complete splicing of the mini-exon is 131 nt and is indicated by the arrow marked with F. This equals the size of the mini-exon, 129 nt, plus the 3′-most 2 nt from exon 9 of the *p53* gene, which happen to match the invariant AG dinucleotide of the 3′ splice site of the mini-exon. Lanes: +, COS cells transfected with p11-4cG.pBR327.myc2; -, mock-transfected cells.

Expression of the Mini-Exon at the Protein Level

COS cells that had been transfected with p11-4cG. pBR327.myc2 were assayed for expression of the hybrid protein by staining the cells with a monoclonal antibody (9E10) that recognizes the repeat present in the mini-exon. This assay revealed that the fusion protein was expressed (Figure 4) and that the protein was present at roughly equal levels in the nucleus and the cytoplasm. The normal location of unmutated *p53* is nuclear, although the mutant used here is temperature-sensitive with respect to intracellular localization and is cytoplasmic at 37°C (6). The tagged version is present in the cytoplasm but has regained some of its nuclear localization.

DISCUSSION

Ultimately, computational analysis may permit analysis of the genome with complete confidence, but until then, experi-

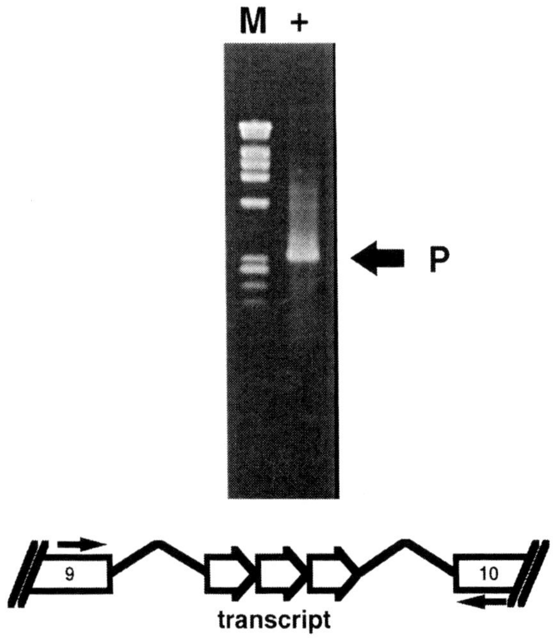

Figure 3. RT-PCR analysis of the mini-exon splicing. The arrow marked with P shows the RT-PCR product expected (324 bp) if the splicing of the mini-exon is correct. An RT-PCR was also performed on RNA that had been pretreated with DNase-free RNase and on RNA from mock-transfected COS cells. As expected, in both cases, no RT-PCR product was obtained (data not shown). Lanes: M, marker (ϕX174 digested with *Hae*III); +, COS cells transfected with p11-4cG.pBR327.myc2.

ments will be required to deduce gene function. In comparison to other techniques (4,11,14), an advantage of the present approach is that the protein is altered in a relatively minor fashion by the mini-exon and is not truncated. Placing the mini-exon between coding exons is likely to be the least-disruptive position in which to place a new epitope, since exons tend to code for quasi-independent protein domains (7). In addition, the glycines present in the mini-exon will endow the peptide with high flexibility, which may allow the adjacent exons encoded by the gene to assume their natural orientation with respect to each other. The epitope could also be used as a ligand for affinity-purification of the tagged protein using the monoclonal antibody, and this may allow biochemical characterization of the protein's properties.

Despite the relatively small perturbation of protein structure caused by the mini-exon, it is still possible that the mini-exon could influence a protein's function or site of intracellular localization, as witnessed by its effect on the mutated *p53* gene. In addition, introns do occasionally fall within functional domains. Nevertheless, a broad range of cDNAs have been tagged with the myc epitope, in general, with results faithful to the native protein. The ease of screening for genes with mini-exon epitope tagging (MEET) and the ability to tag multiple introns in the same gene should overcome any disadvantages due to sporadic effects on protein function.

As a tool for gene discovery, the mini-exon could be placed in introns in a random fashion using retroviruses or using bacterial tranposons such as γδ, permitting facile detection of proteins. As shown in this work, using the Tc^R gene, a gratuitous marker can be used, making it simple to detect insertions of the mini-exon in new locations. Since introns make up much of the mammalian genome, there is a large target size for the mini-exon, exceeding that for enhancer trapping. The simplicity of MEET should make it very amenable to automation, which has been one of the driving forces behind recent genomic advances.

Since most mutations affect the protein coding potential of genes, MEET could be particularly useful for mutation hunting by using transfection of tagged clones in conjunction with two-dimensional gel electrophoresis and immunoblotting. In addition, transfection of the mini-exon tagged DNA into cells, combined with immunofluorescence, should give information about the intracellular localization of all gene products in a mutation-critical region. This in itself may provide useful clues; for example, there was a strong belief from the biology of cystic fibrosis that the gene would turn out to be a membrane protein and this presumption proved to be correct (5).

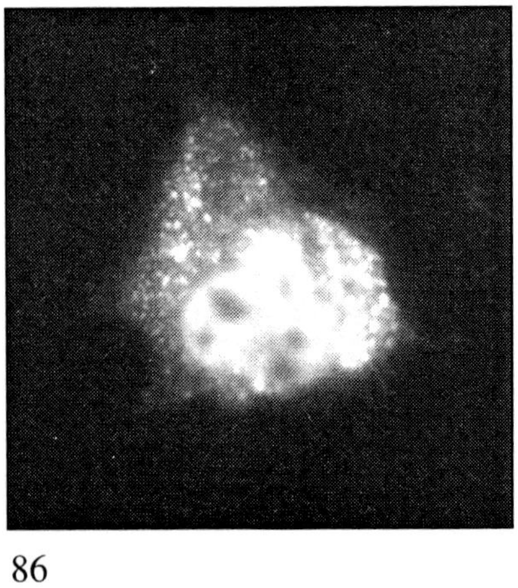

Figure 4. Immunofluorescence analysis of mini-exon protein expression. The expression of the mini-exon epitope tagged protein is evident both in the cytoplasm and nucleus. COS cells are highly transformed, and this is reflected in the unusual morphology of the cells (13). The fluoroscence is specific; mock-transfected cells show no fluorescence (data not shown).

While this manuscript was being prepared, a paper describing a similar approach to gene identification called CD-tagging was published (9). The two methods are conceptually similar, but CD-tagging uses three different vectors or three different antibodies to solve the problem of the three different reading frames by which an intron can interrupt a gene. An advantage of the approach described here is that because only one epitope and cognate antibody can be used universally, a greater speed of screening for new genes should be possible. There is already much experience with the widely used myc epitope/antibody combination, and it displays robust performance capability. Complete sequencing of both the unspliced and spliced exon (see Results) confirms that the epitope should be expressed in all three classes of introns and would be detected by the antibody. Steric hinderance should not be an obstacle to detection of the epitope in the other two frames, since the epitope in the first frame was expressed internally within the *p53* gene, but was nevertheless easily detected by the antibody. Furthermore, the use of synthetic consensus splice sequences, as demonstrated in this paper, ensures highly efficient, accurate and constitutive splicing, thus safeguarding against tissue specificity (20).

In conclusion, MEET should be a useful addition to the arsenal of techniques by which the genes uncovered by the genome product can be more accurately catalogued and understood.

ACKNOWLEDGMENTS

Many thanks to Bill Gelbart in whose laboratory this work was performed, to Nick Brown for suggesting the 5′ splice site and to Phil Hinds for p11-4cG. Thanks also to Jan-Fang Cheng, John Morrison and Chris Paszty for helpful comments on the manuscript. I especially thank Jane Iwamura and Sean Egan for encouragement. D.J.S. was supported by a SERC/NATO fellowship and the Horace Le Marquand and Dudley Bigg fellowship of the Royal Society.

REFERENCES

1. **Adams, M.D., A.R. Kerlavage, R.D. Fleischmann, R.A. Fuldner, C.J. Bult, N.H. Lee, E.F. Kirkness, K.G. Weinstock et al.** 1995. Initial assessment of human gene diversity and expression patterns based upon 83 million nucleotides of cDNA sequence. Nature *377* (Suppl):3-174.
2. **Bienz, B., R. Zakut-Houri, D. Givol and M. Oren.** 1984. Analysis of the gene coding for the murine cellular tumor antigen p53. EMBO J. *3*:2179-2183.
3. **Breathnach, R. and P. Chambon.** 1981. Organization and expression of eukaryotic split genes coding for proteins. Annu. Rev. Biochem. 50:349-383.
4. **Burns, N., B. Grimwade, P.B. Ross-Macdonald, E.Y. Choi, K. Finberg, G.S. Roeder and M. Snyder.** 1994. Large-scale analysis of gene expression, protein localization, and gene disruption in *Saccharomyces cerevisiae*. Genes Dev. *8*:1087-1105.
5. **Collins, F.S.** 1992. Cystic fibrosis: molecular biology and therapeutic implications. Science *256*:774-779.
6. **Gannon, J.V. and D.P. Lane.** 1991. Protein synthesis required to anchor a mutant p53 protein which is temperature-sensitive for nuclear transport. Nature *349*:802-806.
7. **Go, M.** 1981. Correlation of DNA exonic regions with protein structural units in haemoglobin. Nature *291*:90-92.
8. **Harlow, E. and D. Lane.** 1988. Antibodies: A Laboratory Manual. CSH

Laboratory Press, Cold Spring Harbor, NY.
9. **Jarvik, J.W., S.A. Adler, C.A. Telmer, V. Subramaniam and A.J. Lopez.** 1996. CD-tagging: a new approach to gene and protein discovery and analysis. BioTechniques *20*:896-904.
10. **Kay, R. and J. McPherson.** 1987. Hybrid pUC vectors for addition of new restriction enzyme sites to the ends of DNA fragments. Nucleic Acids Res. *15*:2778.
11. **Lynch, A.S., D. Briggs and I.A. Hope.** 1995. Developmental expression pattern screen for genes predicted in the C. elegans genome sequencing project. Nature Genet. *11*:309-313.
12. **Martinez, J., I. Georgoff, J. Martinez and A.J. Levine.** 1991. Cellular localization and cell cycle regulation by a temperature-sensitive p53 protein. Genes Dev. *5*:151-159.
13. **Munro, S. and H.R.B. Pelham.** 1987. A C-terminal signal prevents secretion of luminal ER proteins. Cell *48*:899-907.
14. **O'Kane, C.J. and W.J. Gehring.** 1987. Detection in situ of genomic regulatory elements in *Drosophila*. Proc. Natl Acad. Sci. USA *84*:9123-9127.
15. **Reed, R.** 1989. The organization of 3′ splice-site sequences in mammalian introns. Genes Dev. *3*:2113-2123.
16. **Sharp, P.A.** 1981. Speculations on RNA splicing. Cell *23*:643-646.
17. **Smith, D.J.** 1988. The complete sequence of a frog α-tubulin gene and its regulated expression in mouse L-cells. Biochem. J. *249*:465-472.
18. **Smith, D.J., Y. Zhu, J.-L. Zhang, J.-F. Cheng and E.M. Rubin.** 1995. Construction of a panel of transgenic mice containing a contiguous 2-Mb set of YAC/P1 clones from human chromosome 21q22.2. Genomics *27*:425-434.
19. **Soberon, X., L. Covarrubias and F. Bolivar.** 1980. Construction and characterization of new cloning vehicles. IV. Deletion derivatives of pBR322 and pBR325. Gene *9*:287-305.
20. **Youderian, P., S. Bouvier and M.M. Susskind.** 1982. Sequence determinants of promoter activity. Cell *30*:843-853.

Update to: Mini-Exon Epitope Tagging for Analysis of the Protein Coding Potential of Genomic Sequence

Desmond J. Smith

UCLA School of Medicine, Los Angeles, CA, USA

The field of genomics has advanced with unsettling speed since the previous publication. In the area of structural genomics, the databases currently boast complete genome sequences from 35 organisms, including one mammal (*Homo sapiens*), two invertebrates (the worm, *Caenorhabditis elegans* and the fruit fly, *Drosophila melanogaster*), one unicellular eukaryote (the yeast *Saccharomyces cerevisiae*) and 25 bacterial species and strains (http://www.ncbi.nlm.nih.gov/). Of course, this tally is only temporary and will continue to increase for the forseeable future. The developments responsible for these dramatic increases in sequencing speed have relied upon a number of incremental advances in technology, but surely prominent among these has been the advent of commercially available capillary sequencing machines. On another front, there has been the increasing realization that single nucleotide polymorphisms (SNPs) will be very powerful for genetic mapping, because of their abundance and the prospects of rapid scoring of these markers using technologies such as gene chips (9). In addition, it has been realized that SNPs may play an important role in individual variation and susceptibility to disease (10).

The power of functional genomics has been driven home by the use of microarrays and gene chips, which have allowed gene expression profiling for entire or substantial portions of genomes. These technologies have given profound insights into the coordinated control of genes involved in various physiological responses such as metabolic shifts due to carbon source changes (4,13,14), sporulation (3,13), wound healing (7) and cell cycle progression (2).

It is likely that this ever-increasing pace of technology development will continue to feature prominently in genomics. In 1996, I pointed out in my original report that computational investigation might one day "...permit analysis of the genome with complete confidence, but until then, experiments will be required to deduce gene function". This dilemma has only deepened with the completion of the human genome, where the inadequacy of computational analysis is becoming painfully clear. There even exists controversy over such an apparently elementary topic as the number of genes in the human genome (12). This dilemma is exemplified by the difficulties in distinguishing between coding and noncoding regions of the genome. Even very deep expressed sequence tag (EST) databases from a variety of tissues fail to cover all genes and all splicing variants. In addition, EST databases will not detect posttranslational modifications of proteins, which are frequently dynamic and are of undeniable importance in phenomena such as signal transduction and cancer (6,8).

As a result of these conundrums, there is a movement within the field of functional genomics towards proteomics, the study of how the genome coordinates and regulates the sets of proteins that carry out the myriad complex reactions of living matter (11). One technology which shows increasing promise in this endeavor is mass spectrometry (MS) (15). With a complete knowledge of all transcript sequences, MS can allow comprehensive cataloging of all the proteins and their posttranslational modifications that constitute a proteome. However, this technology does not directly address the distinction between exons and introns, and, for higher eukaryotes, this information is necessary for proper interpretation of MS data. In addition, MS does not allow the intracellular location of proteins to be deduced. Fusion of transcript sequences to reporter genes, such as β-galactosidase (1) or green fluorescent protein (GFP) (5), does allow intracellular localization of proteome constituents, but there is the risk that protein fusions will be misfolded because of their large size, resulting in the routing of protein cargo to inappropriate intracellular compartments.

Thus, the need addressed by mini-exon epitope tagging remains paramount; namely how to identify transcribed regions of the genome from "junk" and how best to assess the properties of the encoded proteins. There is little doubt that mini-exon epitope tagging, and similar technologies, will assume increasing importance in deciphering the molecular logic of the genome. The plasmid containing the mini-exon is available from the ATCC (Accession No. ATCC 87574; Manassas, VA, USA).

REFERENCES

1. **Burns, N., B. Grimwade, P.B. Ross-Macdonald, E.Y. Choi, K. Finberg, G.S. Roeder and M. Snyder.** 1994. Large-scale analysis of gene expression, protein localization, and gene disruption in *Saccharomyces cerevisiae*. Genes Dev. *8*:1087-1105.
2. **Cho, R.J., M.J. Campbell, E.A. Winzeler, L. Steinmetz, A. Conway, L. Wodicka, T.G. Wolfsberg, A.E. Gabrielian et al.** 1998. A genome-wide transcriptional analysis of the mitotic cell cycle. Mol. Cell *2*:65-73.
3. **Chu, S., J. DeRisi, M. Eisen, J. Mulholland, D. Botstein, P.O. Brown and I. Herskowitz.** 1998. The transcriptional program of sporulation in budding yeast. Science *282*:699-705.
4. **DeRisi, J.L., V.R. Iyer and P.O. Brown.** 1997. Exploring the metabolic and genetic control of gene expression on a genomic scale. Science *278*:680-686.
5. **González, C. and L.A. Bejarano.** 2000. Protein traps: using intracellular localization for cloning. Trends Cell Biol. *10*:162-165.
6. **Hunter, T.** 2000. Signaling—2000 and beyond. Cell *100*:113-127.
7. **Iyer, V.R., M.B. Eisen, D.T. Ross, G. Schuler, T. Moore, J.C.F. Lee, J.M. Trent, L.M. Staudt et al.** 1999. The transcriptional program in the response of human fibroblasts to serum. Science *283*:83-87.
8. **Kouzarides, T.** 2000. Acetylation: a regulatory modification to rival phosphorylation? EMBO J. *19*:1176-1179.
9. **Kruglyak, L.** 1997. The use of a genetic map of biallelic markers in linkage studies. Nat. Genet. *17*:21-24.
10. **Kwok, P.Y. and Z. Gu.** 1999. Single nucleotide polymorphism libraries: why and how are we building them? Mol. Med. Today *5*:538-543.
11. **Pandey, A. and M. Mann.** 2000. Proteomics to study genes and genomes. Nature *405*:837-846.
12. **Pennisi, E.** 2000. Human Genome Project. And the gene number is...? Science *288*:1146-1147.
13. **Roth, F.P., J.D. Hughes, P.W. Estep and G.M. Church.** 1998. Finding DNA regulatory motifs within unaligned noncoding sequences clustered by whole-genome mRNA quantitation. Nat. Biotechnol. *16*:939-945.
14. **Wodicka, L., H. Dong, M. Mittmann, M.H. Ho and D.J. Lockhart.** 1997. Genome-wide expression monitoring in *Saccharomyces cerevisiae*. Nat. Biotechnol. *15*:1359-1367.
15. **Yates, J.R., III.** 2000. Mass spectrometry. From genomics to proteomics. Trends Genet. *16*:5-8.

CHAPTER 18

Multiple Epitope Tagging of Expressed Proteins for Enhanced Detection

Ron Hernan, Ken Heuermann and Bill Brizzard

Sigma, St. Louis, MO, USA

(Reprinted from BioTechniques 28:789-793, 2000)

ABSTRACT

Three FLAG® epitopes have been incorporated into the mammalian expression vector pCMV-5 to create a transient expression vector, p3XFLAG-CMV-7. The vector was designed to express FLAG fusion proteins that can be detected at tenfold lower expression levels than the current FLAG fusion protein expression system. The usefulness of this expression and detection system was demonstrated by expression of bacterial alkaline phosphatase in COS-7 cells. In addition, 3X FLAG bacterial alkaline phosphatase was expressed in Escherichia coli, *purified on anti-FLAG M2 affinity gel, and detection of 500 pg of purified protein by Western blot analysis is demonstrated.*

INTRODUCTION

Epitope tagging has become a powerful tool for the detection and purification of expressed proteins. This methodology has been used for protein localization, immunoprecipitation and protein-protein interaction. Many types of tags have been used, with c-*myc* and FLAG® tags (Scientific Imaging [Eastman Kodak], New Haven, CT, USA) being two of the most popular epitopes used (5). Generally, these epitopes are fused to the N or C terminus of the expressed protein, which makes them more accessible to the antibody for detection and less likely to cause severe structural or functional perturbations.

The original FLAG sequence, Asp-Tyr-Lys-Asp-Asp-Asp-Asp-Lys, is recognized by two monoclonal antibodies, M1 and M2 (6,14). In addition, the FLAG sequence with an initiator methionine attached is recognized by the M2 antibody and a third antibody, M5 (2). The last five amino acids of the FLAG sequence are the recognition site for the protease enterokinase, which allows the removal of the FLAG epitope.

The FLAG epitope has been used in expression systems for detection and purification of heterologous proteins in *Escherichia coli* (3), *Saccharomyces cerevisiae* (10,14) *Drosophila* (20), baculovirus (4,15) and mammalian systems (13, 16). For mammalian expression systems, expression levels are low, and the effective detection of expressed foreign proteins using established methods can be difficult. We describe a mammalian expression plasmid containing multiple FLAG epitopes in tandem, p3XFLAG CMV-7, designed for intracellular expression with increased sensitivity of detection. This vector contains the cytomegalovirus (CMV) promoter (19) and simian virus 40 (SV40) origin of replication for efficient expression in COS-7 cells (12). Moreover, the detection of triple FLAG-tagged BAP that was expressed and purified from *E. coli* was compared to single FLAG-tagged BAP.

MATERIALS AND METHODS

All materials were supplied by Sigma (St. Louis, MO, USA) unless otherwise stated.

p3XFLAG-CMV-7 Construction

p3XFLAG-CMV-7 was constructed from the mammalian expression vector, pCMV-5 (1). The triple FLAG sequence was constructed from two pairs of complementary oligonucleotides. The first pair of oligonucleotides was synthesized as follows: 5′-GAAGAATTCACCATGGACTACAAAGACCATGACGG-TGATTATAAAGATCATGAT-3′ and 5′-ATCATGATCTTTAT-AATCACCGTCATGGTCTTTGTAGTCCATGGTGAATTC-TTC-3′. The second pair was synthesized with the following sequence: 5′-GAAGATATCGATTACAAGGATGACGATGA-CAAGCTTGGG-3′ and 5′-CCCAAGCTTGTCATCGTCATC-CTTGTAATCGATATCTTC-3′. The first pair of oligonucleotides were annealed together and digested with *Eco*RI and *Eco*RV. The second pair of oligonucleotides were annealed together and digested with *Eco*RV and *Hin*dIII. The two pairs of digested nucleotide cassettes were ligated into CMV-5, which had been double-digested with *Eco*RI and *Hin*dIII. The nucleotide sequence was verified by automated DNA sequencing.

Cloning and Expression Vectors for Gene Function Analysis
Edited by Quinn Lu and Michael P. Weiner

Construction of pFLAG-CMV7-BAP

A modified version of the *E. coli phoA* gene (7) (for which the leader sequence and the N-terminal four amino acids of the mature enzyme were deleted) was subcloned into the vector p3XFLAG-CMV-7 to express BAP. The modified sequence was cut from pFLAG-ATS-BAP by double digestion with *Hin*dIII and *Bgl*II. The fragment was then cloned into p3XFLAG CMV-7, which had been double-digested with *Hin*dIII and *Bam*HI to generate p3XFLAG-CMV-7-BAP. The nucleotide sequence at the N terminus of the *phoA* coding region was verified by automated DNA sequencing.

Triple FLAG-ATS-BAP Construction

Two oligonucleotides encoding the sense and antisense strand for the triple flag sequence were synthesized (sequences not shown), 5′-phosphorylated with T4 polynucleotide kinase and annealed together to generate a cassette with an *Nde*I and a *Hin*dIII restriction enzyme overhang. The pFLAG-ATS-BAP was digested with *Nde*I and *Hin*dIII, and the vector was purified by gel electrophoresis. The annealed cassette was ligated to the double-digested pFLAG-ATS-BAP vector with T4 DNA ligase, and the reaction was carried out overnight at 16°C for 16 h. The ligation was enriched by digestion with *Nru*I to linearize the parent vector that was not ligated with the triple FLAG sequence, and then transformed into *E. coli* DH5α. Clones were isolated, and the nucleotide sequence was verified by automated DNA sequencing.

Bacterial Expression and Protein Purification

E. coli BL21 (DE3) cells were transformed with the expression plasmid containing the triple FLAG BAP construct. Cells were grown in Terrific broth containing 100 µg/mL ampicillin at 37°C with agitation. The culture was grown to an $A_{600} = 4.0$ and then induced with IPTG at a final concentration of 1 mM. The cell culture was grown for an additional 3 h at 37°C and then harvested by centrifugation. The cell pellet was resuspended in 50 mM Tris-HCl, pH 8.0, and the cells were disrupted by sonication. Then, the cellular debris was removed by centrifugation. BAP was affinity-purified on M2 antibody-conjugated sepharose 4B resin equilibrated with 50 mM Tris-HCl, pH 8.0, 150 mM NaCl (TBS). The resin was washed with

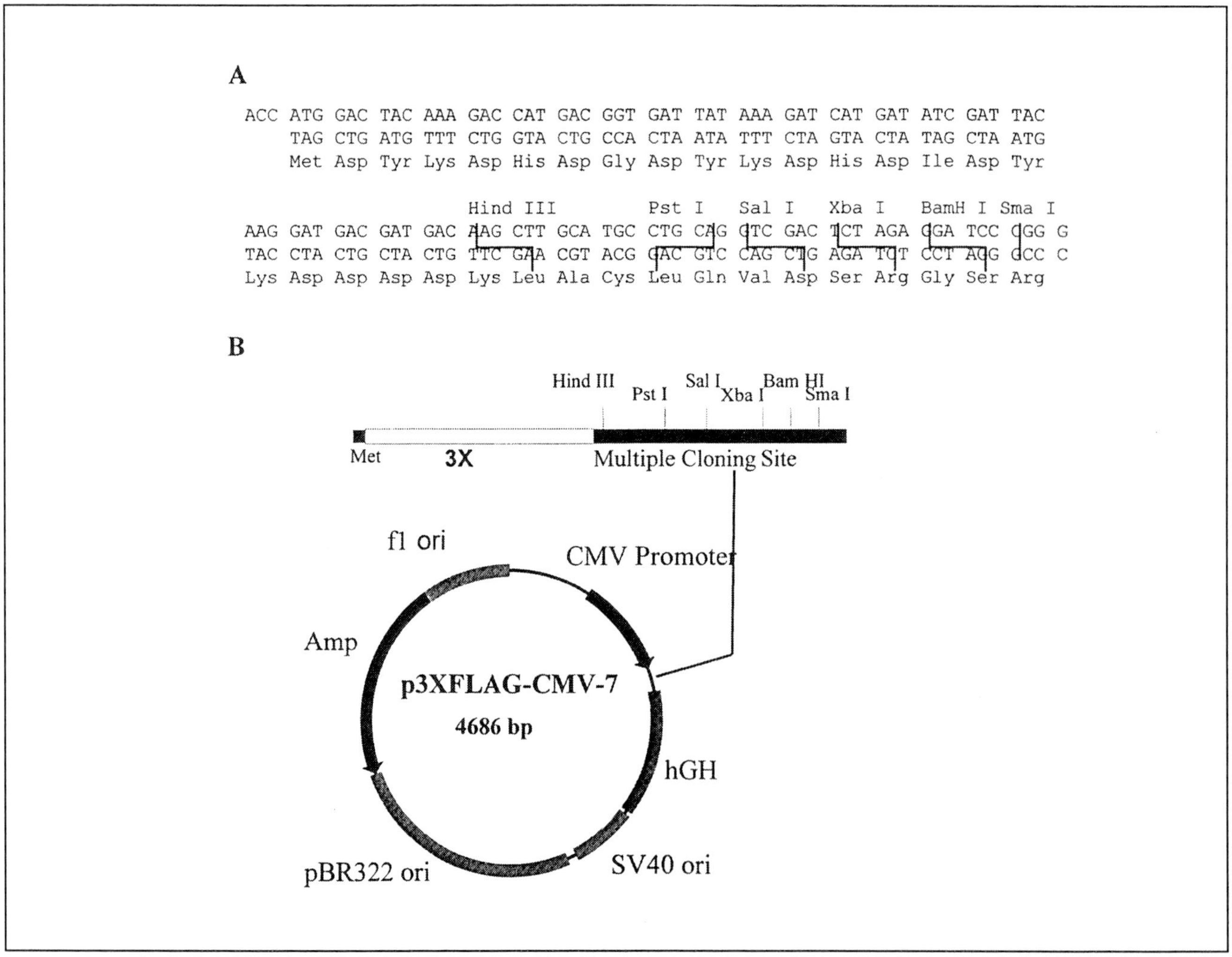

Figure 1. p3XFLAG-CMV-7 expression vector. (A) DNA and protein sequence of 3X FLAG-CMV-7 multiple cloning site and FLAG sequences. (B) Plasmid map of the p3XFLAG-CMV-7 showing the CMV promoter, human growth hormone transcription termination and polyadenylation site, SV40 origin of replication, Col E1 origin of replication and β-lactamase gene.

20 bed volumes of TBS, and then the triple FLAG BAP was eluted with five column volumes of 0.1 M Glycine, pH 3.5. The eluted protein was pooled and adjusted to pH 7.5 with 1 M Tris-HCl, pH 8.0. Protein content was determined by the Bradford method and by absorbance using ε_{280} = 0.7 mL/mg; both methods yielded the same value.

Western Blot

Purified 3X FLAG-BAP and N-BAP were diluted with to 2× Laemlli buffer (9), boiled for 5 min and then placed on ice. Samples were resolved on a 15% SDS-PAGE using the method of Laemlli (9) and then transferred to nitrocellulose membranes. The membrane was blocked with PBS containing 3% non-fat dry milk for 1 h and then rinsed 3× in TBS, 0.05% Tween® 20 (TBS-T). The membrane was incubated with M2 antibody at a final concentration of 10 µg/mL for 30 min in TBS-T and then rinsed 3× in TBS-T. The membrane was then incubated for 30 min with a goat anti-rabbit IgG (whole molecule) horseradish peroxidase (HRP) conjugate diluted 1:10000 in TBS-T, then rinsed 3× in TBS-T. The FLAG-tagged proteins were detected with the HRP conjugate and visualized by chemilumenescent detection using an ECL™ kit (Amersham Pharmacia Biotech, Piscataway, NJ, USA) with exposures from 1–30 min on Kodak X-Omat® MR film (Eastman Kodak, Rochester, NY, USA).

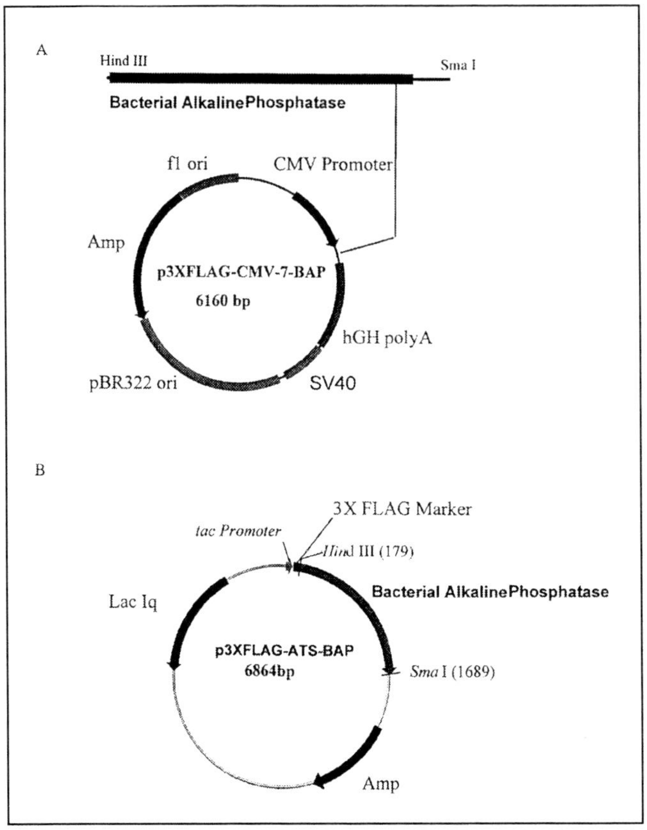

Figure 2. Expression vectors p3XFLAG-CMV-7-BAP and p3XFLAG-ATS-BAP. (A) Vector map of p3XFLAG-CMV-7-BAP showing insertion of the phoA coding region into p3XFLAG-CMV-7. (B) Vector map of p3XFLAG-ATS-BAP showing insertion of the *phoA* coding region into pFLAG-ATS-BAP.

Transfection of COS-7 Cells with p3XFLAG-CMV-7-BAP

COS-7 cells were cultured on 35 mm^2 plates in DMEM containing 10% fetal bovine serum, 4 mM L-glutamine, 5 µg/mL gentamycin. Cells were grown at 37°C in a humidified CO_2 incubator with 5% CO_2. Transfection of the p3XFLAG-CMV-7-BAP plasmid was accomplished using LIPOFECTAMINE™ (Life Technologies, Gaithersburg, MD, USA) according to manufacturer's directions. Two micrograms of vector DNA per plate were used for the transfection. Immunostaining was done 72 h after transfection.

Immunostaining

At 72 h after transfection, the cells were washed with 50 mM Tris-HCl, pH 7.4, 150 mM NaCl (TBS). The cells were fixed with 1:1 (vol/vol) methanol-acetone mixture for 1 min. The fixed cells were washed 4× with TBS and then incubated with 10 µg/mL M2 antibody-HRP conjugate in TBS for 1 h. Cells were washed with TBS 5×, and the M2 antibody-HRP conjugate was visualized with freshly prepared 0.01 mg/mL *o*-dianisidine, 0.015% hydrogen peroxide in TBS. Cells were stained for approximately 15 min and then visually inspected.

RESULTS

Vector Construction

We have constructed a vector for expression of proteins in mammalian host cells using a modified version of the FLAG expression system, which contains 3X FLAG sequences in tandem (Figure 1). This construct was designed to improve the detection limit of expressed proteins in mammalian host cells. The first two flag peptides are modified FLAG sequences. The original FLAG epitope is Asp-Tyr-Lys-Asp-Asp-Asp-Asp-Lys, while the first two flag recognition sequences are Asp-Tyr-Lys-Asp-His-Asp with a Gly and a Ile spacer between the two sequences. These alternative sequences arise from phage display studies in which a different binding motif was determined (11). This allows the introduction of additional FLAG antibody binding sites without the addition of extra enterokinase recognition/cleavage sites.

The p3XFLAG-CMV-7 expression vector contains the promoter region of the human CMV major immediate early gene, which allows for constitutive expression of cloned genes in mammalian cell lines. The Kozak consensus sequence (8) is provided in the vector along with a multiple cloning site, which allows for a variety of cloning strategies. The multiple cloning site is compatible with the other existing CMV mammalian expression vectors. In addition, the expression vector contains the SV40 origin of replication for efficient high-level transient expression (12) and a DNA segment from the human growth hormone containing transcriptional termination sequence and polyadenylation signals (17). p3XFLAG-CMV-7 contains the β-lactamase gene for selection of the plasmid in *E. coli*.

Expression of 3XFLAG-BAP in COS-7 Cells

p3XFLAG-CMV-7-BAP (Figure 2) was transfected into COS-7 cells as described in Materials and Methods. At 72 h after transfection, the cells were analyzed by immunostaining using an anti-FLAG M2-HRP conjugate. Figure 3 shows light microscopy of cells detected with M2 antibody and visualized with *o*-dianisidine.

BAP Expression

To address whether a triple FLAG fusion protein produces a more sensitive response then the traditional FLAG epitope, a triple FLAG version of BAP was constructed for expression in *E. coli* (Figure 2). The vector p3XFLAG-ATS-BAP was transformed into *E. coli* and the 3X FLAG-BAP expressed and purified as described in Materials and Methods. An N-FLAG-BAP containing the traditional epitope was also expressed and purified (3). A comparison of the sensitivity of the single and the triple FLAG-BAP was demonstrated by Western blot analysis as described above. Figure 4 shows the Western blot of purified single and triple FLAG that was probed with M2 anti-FLAG antibody and detected by chemiluminescence. The results clearly indicate a tenfold increase in the detection limit of the triple FLAG-BAP as compared to the single FLAG-BAP fusion protein. We were able to detect 500 pg of purified 3X FLAG-BAP with exposures as short as 1 min. With increased exposure time, detection as low as 100 pg has been achieved but with increased background (data not shown). We have also demonstrated a tenfold increase detection in both dot blot and ELISA (R. Hernan, personal communication).

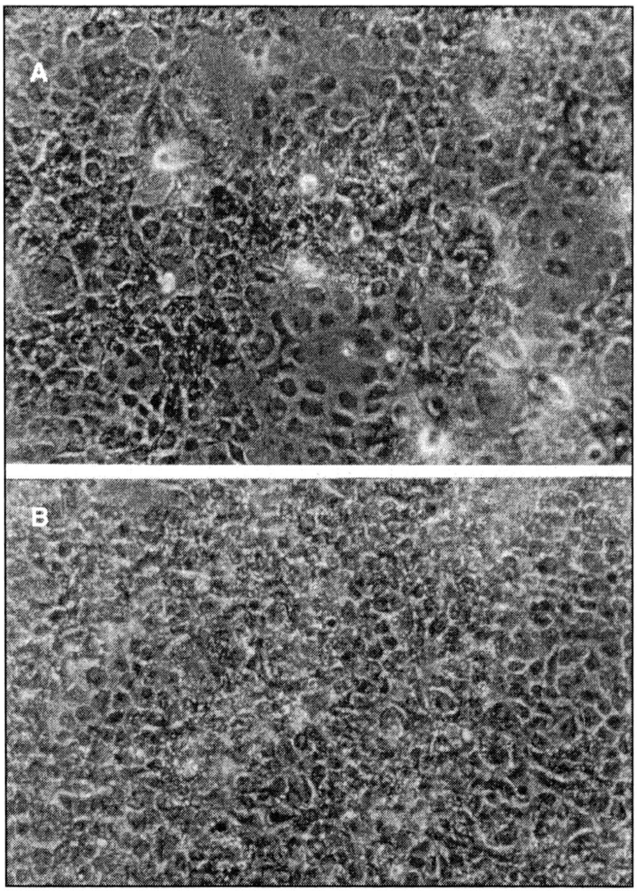

Figure 3. Light microscopy of immunostained cells. (A) COS-7 cells transfected with p3XFLAG-CMV-7-BAP. (B) Control COS-7 cells.

DISCUSSION

The FLAG epitope tag has been effectively used to detect and purify proteins (3) in mammalian and bacterial systems. We have demonstrated that the presence of three FLAG epitopes significantly increases the detection limit of purified BAP. Moreover, we have found that 3X FLAG-BAP cannot be eluted from anti-FLAG M2 affinity gel by competition with the original FLAG peptide. However, 3X FLAG-BAP and the 1X FLAG-BAP can be competitively eluted from the anti-FLAG M2 affinity gel using 3X FLAG peptide (data not shown). The p3XFLAG-CMV-7 vector was designed for expression and detection of heterologous proteins in mammalian cells and is compatible with existing pFLAG-CMV vectors. This allows easy subcloning between vectors containing the single FLAG and the triple FLAG (Additional expression vectors encoding the triple FLAG sequence are in development). The immunostaining results show that expression of the *phoA* gene in COS-7 cells is not significantly perturbed by addition of the 3X FLAG sequence.

The M2 antibody reacts with the alternate epitope in the 3X FLAG sequence. In contrast, the M5 antibody fails to show the increased sensitivity of the M2 antibody (results not shown). Recent results using phage display (18) have demonstrated that the critical residues for M2 binding and M5 binding are slightly different. The M2 antibody prefers the sequence Asp-Tyr-Lys-Xxx-Xxx-Asp-Xxx-Xxx, while the M5 prefers Asp-Tyr-Xxx-Xxx-Asp-Asp-Xxx-Xxx. The triple FLAG sequence, Asp-Tyr-Lys-Asp-His-Asp, clearly favors the binding of M2 over that of the M5 or even the M1 antibody. This expression system allows for increased sensitivity and detection of the FLAG epitope tagging system with many similarities to the existing FLAG mammalian expression system.

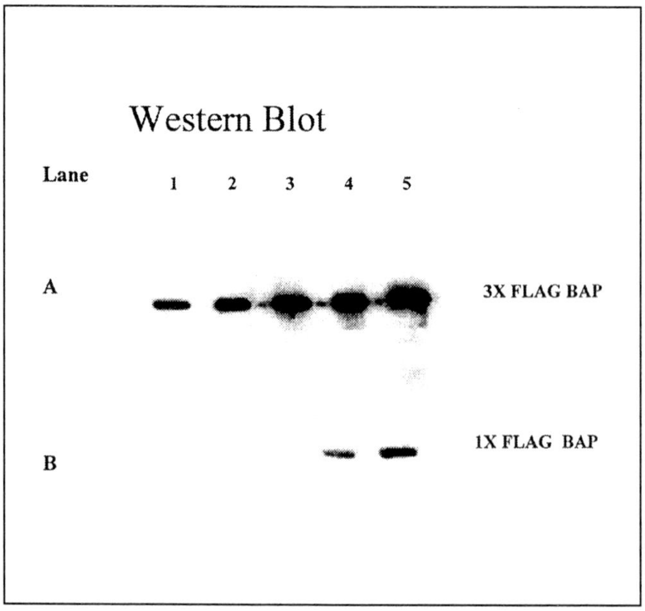

Figure 4. Detection of purified 3XFLAG-BAP by Western blot. (A) Western Blot of purified 3X FLAG-BAP using anti-FLAG M2 antibody. Lane 1, 0.5 ng; lane 2, 1 ng; lane 3, 2 ng; lane 4, 5 ng; and lane 5, 10 ng. (B) Western blot of purified N-FLAG-BAP using anti-FLAG M2 antibody. Lane 1, 0.5 ng; lane 2, 1 ng; lane 3, 2 ng; lane 4, 5 ng; and lane 5, 10 ng.

ACKNOWLEDGMENTS

The authors thank Richard Chubet and Fred Redeaux for their help in the beginning of this work, Hank Hwang for the M2-HRP conjugate and Kelly Prusheik for technical assistance and critical review of the manuscript.

REFERENCES

1. **Andersson, S., H. Davis, H. Dahlbach, H. Jornvall and D. Russell.** 1989. Cloning, structure and expression of the mitochondrial cytochrome P-450 sterol 26-hydroxylase, bile biosynthetic enzyme. J. Biol. Chem. *264*:8222-8229.
2. **Brizzard, B. and R. Chubet.** 1999. Epitope tagging of recombinant proteins, p. 5.8.1-5.8.11. *In* J.N. Crawley, C.R. Gerfen, R. McKay, M.A. Rogawski, D.R. Sibley and P. Skolnick (Eds.), Current Protocols in Neuroscience. John Wiley & Sons, New York.
3. **Brizzard, B., R. Chubet and D. Vizard.** 1994. Immunoaffinity purification of FLAG epitope-tagged bacterial alkaline phosphatase using a novel monoclonal antibody and peptide elution. BioTechniques *16*:730-735.
4. **Dent, P., D. Reardon, D. Morrison and T. Sturgill.** 1995. Regulation of raf-1 and raf-1 mutants by ras dependent and ras independent mechanisms in vitro. Mol. Cell Biol. *15*:4125-4135.
5. **Evan, G., G. Lewis, G. Ramsay and J. Bishop.** 1985. Isolation of monoclonal antibodies specific for human C-myc proto-oncogene product. Mol. Cell Biol. *5*:3610-3616.
6. **Hopp, T., V. Prickett, R. Price, R. Libby, D. March, D. Cerretti, D. Urdal and P. Conlon.** 1988. A short polypeptide marker sequence useful for recombinant protein identification and purification. BioTechnology. *6*:1204-1210.
7. **Inouye, H., S. Michaelis, A. Wright and J. Beckwith.** 1981. Cloning and restriction mapping of the alkaline phosphatase structural gene (phoA) of *Escherichia coli* and generation of deletion mutants in vitro. J. Bacteriol. *146*:668-746.
8. **Kozak, M.** 1984. Point mutations close to the AUG initiator codon affect the efficiency of translation of rat preproinsulin in vivo. Nature *308*:241.
9. **Laemmli, V.** 1970. Cleavage of the structural proteins during the assembly of the head of Bacteriophage T4. Nature *227*:680-685.
10. **Lee, J., J. Laydon, P. McDonnell, T. Gallagher, S. Kumar, D. Green, D. McNulty, M. Blumenthal et al.** 1994. A protein kinase involved in the regulation of imflammatory cytokine biosynthesis. Nature *372*:739-746.
11. **Miceli, R., M. DeGraaf and H. Fischer.** 1994. Two-stage selection of sequences from a random phage display library delineates both core residues and permitted structural range within an epitope. J. Immunol. Methods *167*:279-287.
12. **Okayama, H. and P. Berg.** 1983. A cDNA cloning vector that permits expression of cDNA inserts in mammalian cells. Mol. Cell Biol. *3*:280-289.
13. **Overholt, S., T. Liu, D. Taylor, M. Wang, A.K. El-Naggar, E. Shillitoe, K. Adler-Storthz et al.** 1997. Head and neck squamous cell growth suppression using adenovirus-p53-FLAG: a potential marker for gene therapy trials. Clin. Cancer Res. *3*:185-191.
14. **Prickett, K., D. Amberg and T. Hopp.** 1989. A calcium dependent antibody for identification and purification of recombinant proteins. BioTechniques *7*:580-589.
15. **Ritchie, P., A. Decout, J. Amey, C. Mann, J. Read, M. Rosseneu, J. Scott and C. Shoulders.** 1999. Baculovirus expression and biochemical characterization of the human microsomal triglyceride transfer protein. Biochem. J. *338*:305-310.
16. **Schulte am Esch, J., II, J. Sevigny, E. Kaczmarek, J. Siegel, M. Imai, K. Koziak, A. Beaudoin and S. Robson.** 1999. Structural elements and limited proteolysis of CD39 influence ATP diphosphohydrolase activity. Biochemistry *38*:2248-2258.
17. **Seeburg, P.** 1982. The human growth hormone gene family: nucleotide sequences show recent divergence and predict a new polypeptide hormone. DNA *1*:239-249.
18. **Slootstra, J., D. Kuperus, A. Pluckthun and R. Meloen.** 1996. Identification of new tag sequences with differential and selective recognition properties for the anti-FLAG monoclonal M1, M2 and M5. Mol. Divers. *2*:156-164.
19. **Thomsen, D., R. Stenberg, W. Goins and M. Stinski.** 1984. Promoter-regulatory region of the major immediate early gene of human cytomegalovirus. Proc. Natl. Acad. Sci. USA *81*:659-663.
20. **Xu, T. and G. Rubin.** 1993. Analysis of genetic mosaics in developing and adult Drosophila tissues. Development *117*:1223-1237.

Update to: Multiple Epitope Tagging of Expressed Proteins for Enhanced Detection

Ron Hernan, Ken Heuermann, Lin Zhang, and Bill Brizzard

Sigma, St. Louis, MO, USA

Recently, a series of new mammalian expression vectors containing the 3X FLAG® epitope (Scientific Imaging [Eastman Kodak], New Haven, CT, USA) have been developed. The new vector series offers a number of additional features (see Table 1). Transient expression vectors p3XFLAG-cytomegalovirus (CMV)-7.1 and 8 have the 3X FLAG at the N terminus. Vectors for stable transfection, p3XFLAG-CMV-9 and 10, contain a neomycin marker, and transfectants can be maintained by selection with G418 (Geneticin®; Life Technologies, Rockville, MD, USA). We have also generated vectors for the stable expression of 3X FLAG sequence at the C terminus of the expressed protein, p3XFLAG-CMV-13 and 14. Vectors with and without a secretion leader sequence (designed for secreted versus intracellular expression) are provided. In addition, the multiple cloning region of the new vectors is compatible with the existing pFLAG-CMV-1, 2, 3, and 4 vectors.

We have also generated a series of dual epitope vectors containing the 3X FLAG sequence and a c-myc tag. This series includes transient expression vectors, p3XFLAG-CMV-23 and 24, and stable expression vectors, p3XFLAG-CMV-25 and 26. These vectors are useful for protein-protein interaction by co-immunoprecipitation, sequential purification, western blots, immunofluorescent and also for the study of membrane topology.

Table 1. The p3XFLAG Expression Vectors

	N-Terminal 3X FLAG	C-Terminal 3X FLAG	C-Terminal Myc tag	Transient	Stable	Intracellular	Secreted
p3XFLAG-CMV-7	X			X		X	
p3XFLAG-CMV-7.1	X			X		X	
p3XFLAG-CMV-8	X			X			X
p3XFLAG-CMV-9	X				X		X
p3XFLAG-CMV-10	X				X	X	
p3XFLAG-CMV-13		X			X		X
p3XFLAG-CMV-14		X			X	X	
p3XFLAG-CMV-23	X		X	X			X
p3XFLAG-CMV-24	X		X	X		X	
p3XFLAG-CMV-25	X		X		X		X
p3XFLAG-CMV-26	X		X		X	X	

CHAPTER 19

Versatile Epitope Tagging Vector for Gene Expression in Mammalian Cells

Tanya Hosfield and Quinn Lu

Stratagene Cloning Systems Inc., La Jolla, CA, USA

(Reprinted from BioTechniques 25:306-309, 1998)

ABSTRACT

We have constructed an epitope-tagging vector, pCMV-Tag 1, for gene expression in mammalian cells. This vector, which allows for N-terminal, C-terminal and internal tagging of the gene product of interest with the FLAG® and/or c-myc epitopes, enables researchers to rapidly and efficiently characterize gene products in vivo.

INTRODUCTION

Epitope tagging is a convenient technique in which a known peptide epitope is fused to a target protein of interest, allowing expression of the fusion protein to be monitored using a tag-specific antibody. Thus, a new protein can be easily studied without having to generate new, specific antibodies to that protein. Epitope tagging can be used to localize gene products in living cells, identify associated proteins, track the movement of fusion proteins within the cell or characterize new proteins by immunoprecipitation (3,5,7–9). Epitope tags are generally added at the N (amino) or C (carboxy) terminus, as the ends of proteins are more likely to be accessible to the antibody, and the addition is less likely to affect the normal function of proteins (1). However, if the functional portions of the protein are located at either terminus, an internal tagging option is also available using the FLAG® epitope (Scientific Imaging Systems [Eastman Kodak], New Haven, CT, USA). In past experiments, the addition of an epitope coding sequence has normally been performed by individually adding the tag sequence to each construct. More recently, the preferred way to obtain an epitope-tagged protein is to fuse the coding sequence for the target protein to a vector that contains an epitope tag. To date, most epitope-tagging vectors constructed have only one peptide tag and only allowed for tagging at one terminus, which limited the general utility of the vector (3). In this report, we describe a new epitope-tagging vector, pCMV-Tag 1, which contains sequences for both the FLAG and the c-myc epitopes. In addition, pCMV-Tag 1 was specifically designed for versatile cloning purposes, allowing for both internal and terminal cloning with a choice of either epitope tag. Advantageous properties of the FLAG and c-myc epitopes include possessing a strong, specific detection signal with limited background and being innocuous to transfected cells so that they do not interfere with the function of the tagged protein (5,8,10,11). The versatility of pCMV-Tag 1 enables researchers to find a suitable tagging strategy that will greatly assist in biochemical studies of their protein of interest.

MATERIALS AND METHODS

Construction of pCMV-Tag 1

pCMV-Tag 1 is derived from pCMV-Script™ (Stratagene, La Jolla, CA, USA). Briefly, 5 μg of pCMV-Script were digested with *Eco*RV/*Bam*HI and then ligated with an oligonucleotide fragment coding for the sequence of the FLAG epitope, DYKDDDDK (4). This construct was given the name pCMV-Tag 1F and was subsequently digested with *Xho*I/*Kpn*I and then ligated with another oligonucleotide fragment coding for the c-myc epitope, EQKLISEEDL (2), resulting in pCMV-Tag 1. All cloning junctions were verified by DNA sequencing. pCMV-Tag 1 is 4319 bp in size, and the compiled nucleotide sequence is available from GenBank® (Accession No. AF 025668).

Cloning of the Luciferase Gene into the pCMV-Tag 1 Vector

The expression level of the pCMV-Tag 1 vector was tested by cloning the reporter gene luciferase into the construct and then assaying for function of the gene. To do so, pCMV-Tag 1 was digested with *Bgl*II/*Xho*I, treated with shrimp alkaline phosphatase (Amersham Pharmacia Biotech, Piscataway, NJ,

Cloning and Expression Vectors for Gene Function Analysis
Edited by Quinn Lu and Michael P. Weiner

USA) and then ligated with a *Bam*HI/*Xho*I fragment containing the firefly luciferase gene produced by polymerase chain reaction (PCR). The ligated DNA was transformed into XL10-Gold cells (Stratagene), and the correct clones were identified by restriction analysis.

Cell Culture Transfection Assays of pCMV-Tag 1

Chinese hamster ovary (CHO) cells were grown from a stock culture in Dulbecco's modified Eagle medium (DMEM) complete media (Life Technologies, Gaithersburg, MD, USA) containing 10% fetal bovine serum (FBS) and 1% glutamine. Transfection of the "pCMV-Tag 1 plus luciferase" construct into CHO cells was performed using the LIPOFECTAMINE™ method (Life Technologies). Two micrograms of DNA were used to transfect approximately 5.0×10^5 cells. The cells were harvested approximately 48 h after the initial transfection.

Stable Line Generation

Stable CHO cell lines containing pCMV-Tag 1 and derived constructs were generated using the $CaPO_4$ Mammalian Transfection Kit (Stratagene). A kill curve experiment was performed before the transfection to optimize the selection concentration of the G418 sulfate (Stratagene). This demonstrated that the optimal amount of G418 was 600 µg/mL for 14 days. Twelve micrograms of DNA were transfected into approximately 5.0×10^5 cells on 100-mm tissue culture plates. The stable line was generated approximately two weeks after the initial transfection.

Western Analysis of Epitope-Tagged Luciferase

Western blots were performed using Xcell II products, reagents and protocols (Novex, San Diego, CA, USA). Approximately 20 µL of cell lysate containing 20 µg of protein were electrophoresed on a 4%–20% Tris-glycine acrylamide gel at 120 V for 1.5 h. Protein was transferred from the gel onto nitrocellulose at 30 V for 1 h, probed with the appropriate monoclonal or polyclonal antibodies and detected with a horseradish peroxidase (HRP)-conjugated anti-mouse or anti-rabbit, secondary antibody. The signal was detected by chemiluminescence (SuperSignal™; Pierce Chemical, Rockford, IL, USA), using an exposure time of 1 min. Primary antibodies used were as follows: rabbit polyclonal anti-luciferase (Promega, Madison, WI, USA), mouse monoclonal anti-FLAG® M2 (Stratagene) and mouse monoclonal anti-c-myc (Boehringer Mannheim, Indianapolis, IN, USA). The secondary antibody used was anti-mouse/rabbit conjugated HRP (Amersham Pharmacia Biotech).

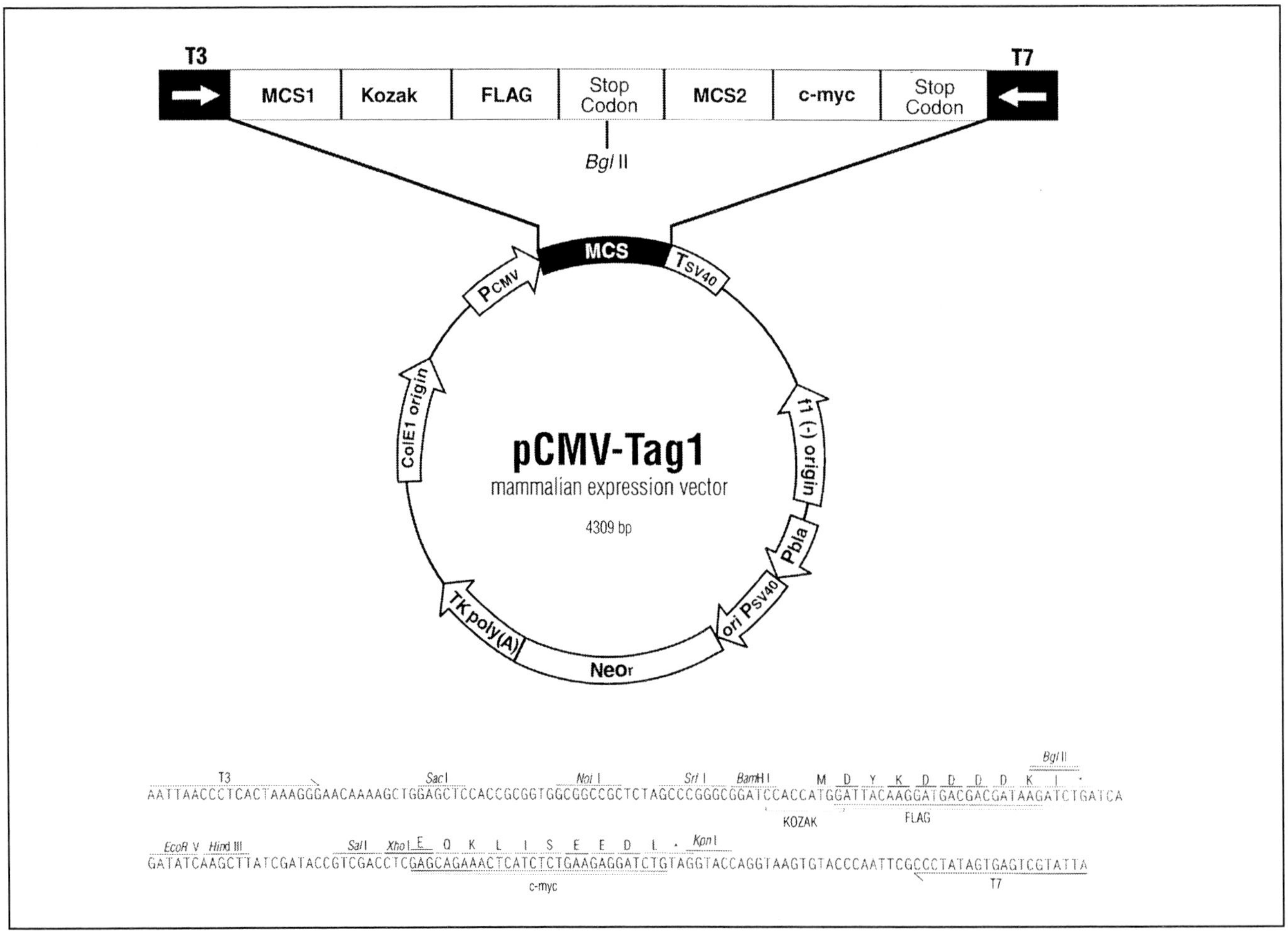

Figure 1. Plasmid map of pCMV-Tag 1. The nucleotide sequence around the MCS with the corresponding amino acid sequences for the epitope tags are shown as well as the unique restriction sites in the MCS region and the T3 and T7 promoters.

Table 1. Cloning Strategies for Epitope Tagging in the pCMV-Tag 1 Vector

Epitope Tag	Tag Location	Cloning Site	Kozak Sequence
FLAG	C terminus	MCS 1	Insert supplied
FLAG	N terminus	*Bgl*II	Vector supplied
FLAG and c-myc	N and C termini	*Bgl*II and MCS 2	Vector supplied
c-myc	C terminus	MCS1 and 2	Insert supplied
FLAG	Internal	MCS1 and *Bgl*II	Insert supplied
No tag	N/A	MCS 1 or MCS 1 and 2	Insert supplied

N/A = not applicable

Enzyme Assays

Luciferase expression levels in CHO cells were determined using a Luciferase Assay Kit (Stratagene). Approximately 1 μg of total cell protein in a volume of 1 μL was added to 100 μL of luciferase assay reagent, and the resulting light was monitored by a luminometer. Protein concentration was determined using a standard Bradford assay (Pierce Chemical).

RESULTS AND DISCUSSION

The pCMV-Tag 1 Epitope Tagging Vector

The pCMV-Tag 1 vector (Figure 1), derived from pCMV-Script, contains sequences for the FLAG and the c-myc epitopes. These specific epitope tags were chosen for their high immuno-reactivity, their small size and their low likelihood to interfere with the function of a protein of interest. The FLAG epitope is a synthetic epitope that consists of 8 amino acid (aa) residues, DYKDDDDK (4), while the c-myc epitope, found in the human c-*myc* gene, is 10-aa-residues long, EQKLISEEDL (4). In addition to the epitope tag sequences, pCMV-Tag 1 contains additional features that enable it to propagate and express fusion proteins in eukaryotic cells. The human cytomegalovirus (CMV) promoter allows for constitutive expression of the cloned gene in a wide variety of mammalian cell lines. Stable selection of clones in eukaryotic cells is made possible by the presence of the neomycin-resistance gene, which is driven by the simian virus 40 (SV40) early promoter and terminated by the thymidine kinase (TK) transcription stop signal. The neomycin-resistance gene is also driven by the β-lactamase promoter to provide selection of the plasmid in *E. coli* cells using kanamycin. The multiple cloning site (MCS) of pCMV-Tag 1 is specifically designed to allow a variety of cloning strategies to be used, resulting in C-terminal, N-terminal and internal tagging of a protein of interest. A KOZAK consensus sequence of GCC(A or G)CC ATGG (6) is provided for N-terminal FLAG epitope tagging.

Table 1 summarizes the cloning strategies for various tagging choices. A gene of interest can be efficiently cloned to achieve a terminal tag by using the *Bgl*II, MCS 1 or MCS 2 sites. For internal tagging, initially the 5′ portion of the gene

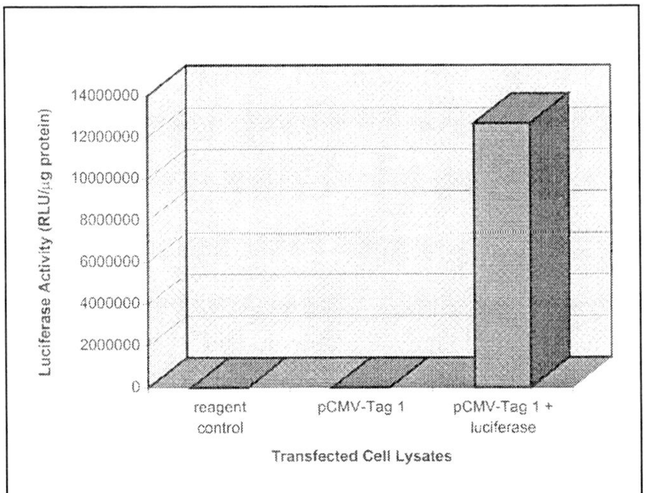

Figure 2. Luciferase assay of epitope-tagged luciferase. CHO cells were transfected with the pCMV-Tag 1 plus luciferase construct and the pCMV-Tag 1 vector alone. One microliter (1 μg protein) of cell lysate was mixed with 100 μL of luciferase assay reagent (Stratagene), and detection was made using a luminometer. The results were normalized using protein concentration. Luciferase activity was measured in relative light units (RLU) per μg of protein.

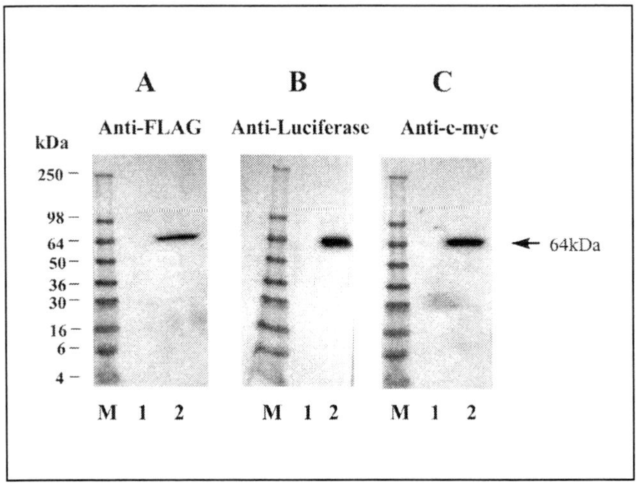

Figure 3. Western analysis of the FLAG-luciferase-c-myc fusion protein. Lane 1 is the pCMV-Tag 1 vector alone, and lane 2 is pCMV-Tag 1 plus luciferase. Twenty microliters of CHO cell lysate (containing 20 μg protein) were electrophoresed on each 4%–20% Tris-glycine gel for 1.5 h at 120 V, transferred to a nitrocellulose membrane for 1 h at 30 V and probed with anti-FLAG (A), anti-luciferase (B) and anti-c-myc (C) antibodies. The signal was detected by chemiluminescence.

should be cloned within the MCS 1, and then the 3′ portion within the *Bgl*II site. The variety of tagging locations enables researchers to optimize the tagging choices for their proteins.

Expression Analysis of pCMV-Tag 1

We have tested the pCMV-Tag 1 vector by using the firefly luciferase gene as a reporter. The luciferase gene was chosen because it can be assayed both enzymatically and immunologically. The firefly luciferase gene was cloned into the *Bgl*II/*Xho*I site of pCMV-Tag 1 such that the luciferase protein was tagged with the FLAG epitope at the N terminus and with the c-myc epitope at the C terminus. This construct was transiently transfected into CHO cells, and the cell lysates were assayed for luciferase activity. Figure 2 shows the results of transiently expressed luciferase within the pCMV-Tag 1 vector. The data demonstrate that the doubly tagged luciferase is biologically active. The activity of the tagged luciferase is over 6000-fold higher than that of the reagent control and the pCMV-Tag 1 vector alone. This activity is comparable to the activity of untagged firefly luciferase (data not shown).

To demonstrate the ease of detection of the epitope tags, Western analyses were performed. Figure 3 shows the Western analysis of cell lysates derived from CHO cells transfected with the pCMV-Tag 1 vector alone and the vector containing the doubly tagged luciferase. The samples were simultaneously loaded, electrophoresed in triplicate and probed with their respective antibodies. The results clearly indicate that the fusion protein, composed of FLAG-luciferase-c-myc, can be easily detected by antibodies against FLAG, luciferase and c-myc, respectively. Stable CHO cell lines were generated with pCMV-Tag 1 and pCMV-Tag 1 plus luciferase, using the $CaPO_4$ Mammalian Transfection Kit (Stratagene).

CONCLUSIONS

The pCMV-Tag 1 expression vector contains the small and highly immunoreactive FLAG and c-myc epitopes for both N-terminal, C-terminal and internal tagging. These tags eliminate the need for raising specific antisera to study a target gene, and both can be easily detected in transfected cells using well-characterized commercially available antibodies. The pCMV-Tag 1 vector offers a fast, versatile and reliable method for analyzing the function of gene products in vivo.

ACKNOWLEDGMENTS

FLAG and Anti-FLAG are trademarks of Immunex Corporation. We would like to thank the members of the Research and Development Department at Stratagene Cloning Systems, Inc., Phyllis Frosst and the Gerace laboratory at TSRI for suggestions, discussion and materials.

REFERENCES

1. **Cravchik, A. and A. Matus.** 1993. A novel strategy for the immunological tagging of cDNA constructs. Gene *137*:139-143.
2. **Evan, G.I., G.K. Lewis, G. Ramsay and J.M. Bishop.** 1985. Isolation of monoclonal antibodies specific for human C-myc proto-oncogene product. Mol. Cell. Biol. *5*:3610-3616.
3. **Georgiev, O., J. Bourquin, M. Gstaiger, L. Knoepfel, W. Schaffner and C. Hovens.** 1996. Two versatile eukaryotic expression vectors permitting epitope-tagging, radiolabelling, and nuclear localization. Gene *168*:165-167.
4. **Hopp, H.P., K.S. Prickett, V.L. Price, R.T. Libby, C.J. March, D.P. Cerretti, D.L. Urda and P.J. Conlon.** 1988. A short polypeptide marker sequence useful for recombinant protein identification and purification. Bio/Technology *6*:1204-1210.
5. **Kolodziej, P.A. and R.A. Young.** 1991. Epitope tagging and protein surveillance. Methods Enzymol. *194*:508-519.
6. **Kozak, M.** 1991. Structural features in eukaryotic mRNAs that modulate the initiation of translation. J. Biol. Chem. *266*:19867-19870.
7. **Sells, M. and J. Chernoff.** 1995. Epitope-tag vectors for eukaryotic protein production. Gene *152*:187-189.
8. **Shiio, Y., M. Itoh and J. Inoue.** 1995. Epitope tagging. Methods Enzymol. *254*:487-502.
9. **Simonsen, H. and H. Lodish.** 1994. Cloning by function: expression cloning in mammalian cells. Curr. Techniques *15*:437-441.
10. **Surdej, P. and M. Jacobs-Lorena.** 1994. Strategy for epitope tagging the protein-coding of any gene. BioTechniques *17*:560-565.
11. **Uttam, P.K.** 1992. Novel vectors for expression of cDNA encoding epitope-tagged proteins in mammalian cells. Gene *114*:285-288.

Update to: Versatile Epitope Tagging Vector for Gene Expression in Mammalian Cells

Tanya Hosfield, Carsten Carstens, and Quinn Lu[1]

Stratagene, La Jolla, CA; and [1]*SmithKline Beecham Pharmaceuticals, King of Prussia, PA, USA*

ADDITIONAL MAMMALIAN EXPRESSION VECTORS FOR CONVENIENT EPITOPE TAGGING AND ANTIBIOTIC SELECTION

We have previously developed the mammalian epitope tagging vector pCMV-Tag 1 for single or double tagging with the c-myc epitope and/or the FLAG® epitope (Scientific Imaging Systems [Eastman Kodak], New Haven, CT, USA). However, the unique restriction sites in the multiple cloning sites are limited, and it generally requires PCR amplification of the target gene for generating in-frame fusions with the epitope tags. To circumvent these limitations and to extend this expression vector line, we have developed pCMV-Tag 2, 3, 4 and 5 series of vectors. Each vector is available in three versions, A, B, and C, which differ only by one or two bases. Thus, each pCMV-Tag vector has the capacity to clone a test gene into one version of the vector so that it is fused in frame with the epitope tag.

All of the pCMV-Tag vectors are derived from the pCMV-Script™ vector (Stratagene, LA Jolla, CA, USA) and contain sequences for either the FLAG or c-myc epitope at either the N or C terminus, or both (Figure 1). In addition to the epitope tag sequences, the pCMV-Tag vectors contain features for expression of fusion proteins in eukaryotic cells. The cytomegalovirus (CMV) promoter allows constitutive expression of the cloned DNA in a wide variety of mammalian cell lines. The neomycin resistance gene is under control of both the prokaryotic β-lactamase promoter to provide kanamycin resistance in bacteria and the SV40 early promoter to provide G418 resistance in mammalian cells. The multiple cloning site (MCS) of the pCMV-Tag vectors allows for a variety of cloning strategies, resulting in either C-terminal or N-terminal fusions with either FLAG or c-myc. A Kozak consensus

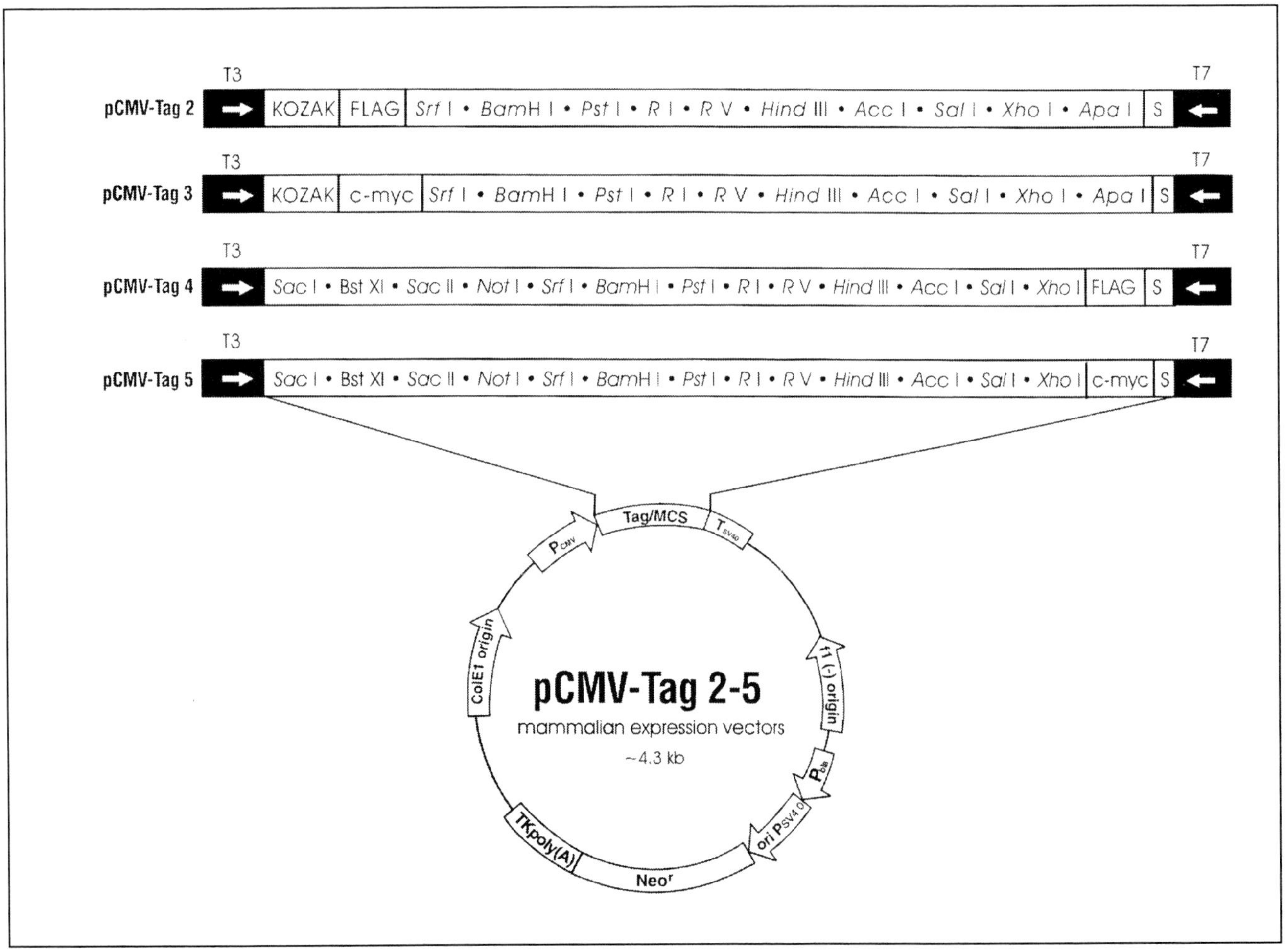

Figure 1. Circular map of the pCMV-Tag 2-5 mammalian expression vectors. The pCMV-Tag mammalian expression vectors feature eukaryotic expression, epitope tagging options, neomycin and kanamycin resistance, T3 and T7 RNA promoters, and single-stranded rescue.

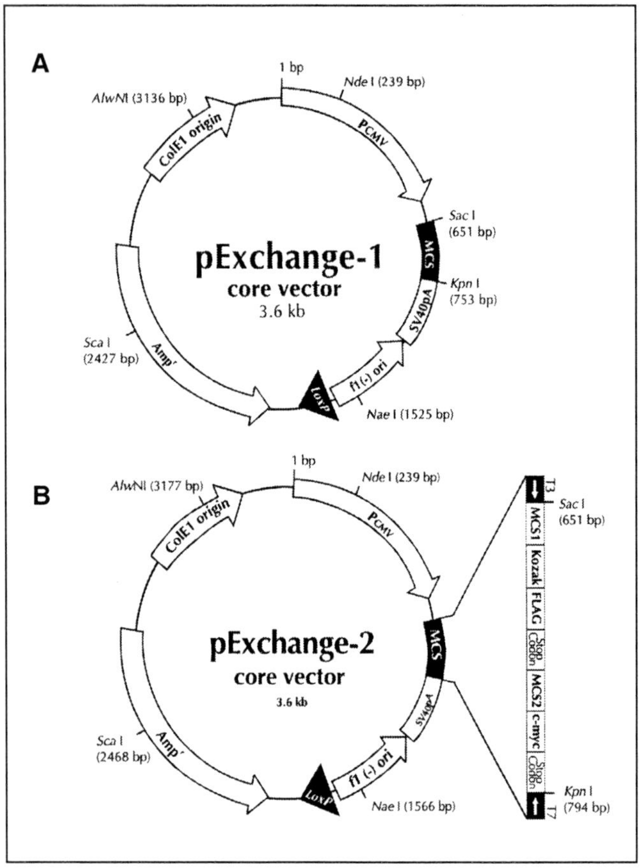

Figure 2. (A) pExchange-1 and (B) pExchange-2 core vectors.

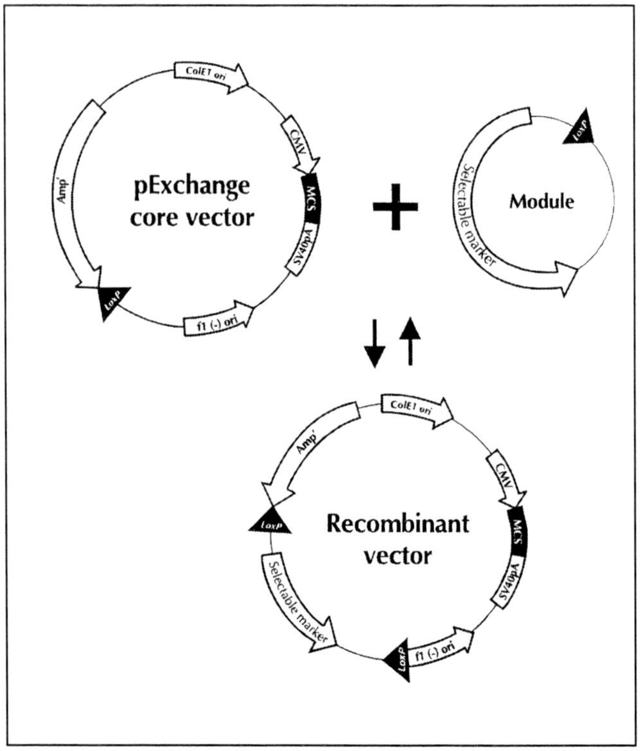

Figure 3. Site-specific recombination catalyzed by *Cre* recombinase results in the insertion of a module into a core vector.

sequence (1) is included to provide optimal expression of the fusion protein when N-terminal FLAG epitope is used. Other cloning options, which require fusion proteins to include their own translational start sequence, are also possible. These epitope tagging vectors offer fast, versatile, and reliable methods to generate tagged constructs for use in analyzing the function of gene products in vivo.

To further increase the flexibility and functionality of the pCMV-Tag vectors, we created a vector system (called the Exchanger™ System) (2) that allows convenient exchange of the antibiotic selectable marker in pCMV-Script and pCMV-Tag1. The Exchanger System includes two core vectors that contain a single *LoxP* site in each basic mammalian expression vector; (*i*) the pExchange-1 core vector (Figure 2), which is derived from the pCMV-Script mammalian expression vector, and (*ii*) the pExchange-2 core vector, which is derived from the pCMV-Tag 1 vector (Figure 2). The Exchanger System utilizes *Cre*-mediated site-specific recombination to allow quick and efficient directional insertion of prefabricated marker modules (hygromycin, puromycin, and neomycin) into the core vectors (Figure 3).

We have also made additional core pExchange vectors based on pCMV-Tag 2, 3, 4, and 5 vectors. pExchange-3 and 4 vectors provide an N-terminal FLAG or c-myc epitope tags, pExchange 5 and 6 provide C-terminal FLAG and c-myc epitope tags, respectively (vector maps not shown). Analogous to the pCMV-Tag series of vectors, three versions of the vectors indicated by A, B, and C were constructed to allow adjustment of the reading frames of the inserts to the epitope tags.

pCMV-Tag vectors and pExchange vectors are commercially available from Stratagene Cloning Systems (La Jolla, CA).

REFERENCES

1. **Kozak, M.** 1991. Structural features in eukaryotic mRNAs that modulate the initiation of translation. J. Biol. Chem. *266*:19867-19870.
2. **Marsh, L. and C. Carstens.** 1999. Mammalian expression vectors with interchangeable selectable markers. Strategies *13-2*:54-55.

CHAPTER 20

Epitope Tag-Antibody Combination Useful for the Detection of Protein Expression in Prokaryotic and Eukaryotic Cells

Manfred Rüdiger, Brigitte M. Jockusch and Martin Rothkegel

Technical University of Braunschweig, Braunschweig, Germany

(Reprinted from BioTechniques 23:96-97, 1997)

Epitope tagging has proven a useful method for monitoring recombinant proteins expressed in either prokaryotic or eukaryotic cells. For this purpose, the epitope should not be present on any of the host cell's proteins, it should not interfere with either structure or function of the tagged protein, and the corresponding antibody must be of high specificity and affinity. To this day, only a few epitope-antibody combinations have been found satisfactory (5).

We describe the successful application of a new epitope-antibody combination. The epitope consists of a 10-amino acid sequence motif present in profilin from birch pollen. It is recognized by the monoclonal anti-birch profilin antibody 4A6, which has been characterized as an immunoglobulin IgG2a (1). Epitope recognition by 4A6 is restricted to profilins of the birch tree family (4) and has been demonstrated in immunoblotting, immunoprecipitation and immunofluorescence (1,3).

A DNA cassette coding for the 4A6 epitope (S F P Q F K P Q E I) (4) with an *Eco*RI-compatible 5′ end and an *Xba*I-compatible 3′ overhang was generated by annealing the synthetic oligonucleotides BiPro1 (5′-AAT TCC TTC CCA CAG TTT AAG CCT CAG GAA AT-3′) and BiPro2 (5′-CTA GAT TTC CTG AGG CTT AAA CTG TGG GAA GG-3′). The resulting double-stranded DNA fragment was cloned into the *Eco*RI/*Xba*I-digested vectors pMal/c (New England Biolabs, Schwalbach, Germany) and pcDNA3 (Invitrogen, Leek, Netherlands) to produce a prokaryotic and a eukaryotic expression vector, respectively. The resulting constructs pMal/BiPro and pcDNA3/BiPro were verified by DNA sequencing. Using the C-terminal birch profilin tag, one can monitor: *(i)* the expression in transformed *E. coli* of any protein as a fusion protein with the maltose-binding protein (MBP); *(ii)* its purity after amylose affinity chromatography; and *(iii)* its integrity after cleaving the N-terminal MBP by factor Xa (according to the instructions by the manufacturer).

We tested the usefulness of pMal/BiPro in this context by cloning into it the cDNA of mouse adenylyl cyclase-associat-

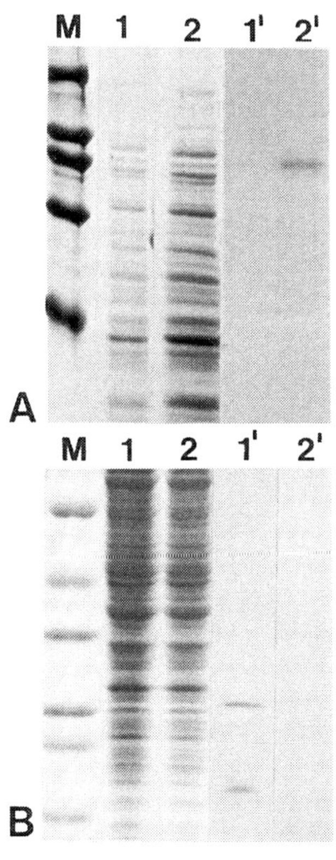

Figure 1. Detection of tagged proteins in prokaryotic and eukaryotic cells. (A) Bacterial lysates of *E. coli* carrying the vector pMal-CAPBiPro were analyzed before (lanes 1 and 1′) and after IPTG induction of the bacteria (lanes 2 and 2′) by SDS-PAGE/Coomassie® blue staining (lanes 1 and 2) and by immunoblotting with 4A6 (lanes 1′ and 2′). Lane M: molecular weight markers at 45, 66, 97.5, 116 and 205 kDa. (B) Detection of the tagged vinculin tail in lysates of transfected PtK_2 cells. Coomassie blue-stained gels are shown of lysates from PtK_2 cells transfected either with pcDNA3-VTBiPro (lane 1) or with pcDNA3/BiPro (lane 2, control). Lanes 1′ and 2′: corresponding immunoblots with 4A6. The arrowhead indicates the expected position of the vinculin tail fragment at 29 kDa. The lower band corresponds to a degradation product, as identified by a chicken vinculin tail-specific antibody (2). Lane M: molecular weight markers at 20.1, 24, 29, 36, 45 and 66 kDa.

Cloning and Expression Vectors for Gene Function Analysis
Edited by Quinn Lu and Michael P. Weiner

ed protein (CAP; GenBank® Accession No. L12367). The 1.4-kb DNA fragment was fused in-frame to the MBP and the epitope tag as an *Eco*RI fragment. The resulting construct (pMal-CAPBiPro) was verified by restriction analysis and DNA sequencing. Expression of the tagged fusion protein was induced with isopropyl-β-D-thiogalactopyranoside (IPTG) and monitored by immunoblotting after sodium dodecyl sulfate polyacrylamide gel electrophoresis (SDS-PAGE; Figure 1A). The tagged MBP-CAP fusion protein (99 kDa) was faithfully and specifically recognized by 4A6 (Figure 1A, lane 2′).

To test the eukaryotic expression vector pcDNA3/BiPro, we inserted a sequence encoding amino acids 858–1066 of chicken vinculin (GenBank Accession No. J04126) as a *Hin*dIII/*Eco*RI fragment into the *Hin*dIII/*Eco*RI-digested pcDNA3/BiPro to generate pcDNA3-VTBiPro. The expression of the tagged polypeptide was monitored in a mammalian epithelial cell line (PtK_2) and transfected by the $CaPO_4$ precipitation method. Immunoblot analysis with 4A6 proved the presence of the tagged avian vinculin tail in PtK_2, seen at the expected position of 29 kDa, together with a proteolytic degradation product (Figure 1B). Both proteins were also recognized by a monoclonal antibody specific for the avian vinculin tail (Reference 2; data not shown). In immunofluorescence analysis of the transfected cells, 4A6 identified the vinculin tail fragment localized to actin filament bundles and in large actin filament aggregates (Figure 2). This association of the vinculin tail domain with actin filaments in transfected cells has been described previously by our group (unpublished).

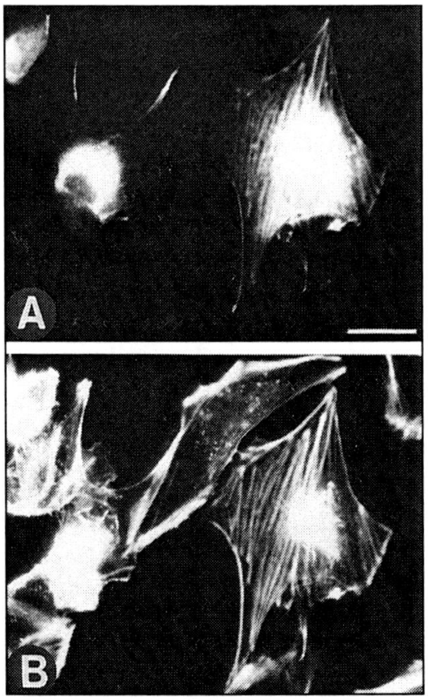

Figure 2. Immunofluorescence of PtK_2 cells transfected with pcDNA3-VTBiPro. Cells were stained for the tagged vinculin tail with 4A6 and a rhodamine-coupled secondary antibody (A) and for F-actin with fluorescein isothiocyanate (FITC)-phalloidin (B). Note the co-localization of the expressed vinculin tail in stress fibers and in large, perinuclear aggregates with filamentous actin. Bar, 20 μm.

In addition, we found that 4A6 can be faithfully used to immunoprecipitate BiPro-tagged proteins and as first antibody in enzyme-linked immunosorbent assay (ELISA) and microinjection studies (data not shown). Thus, it should be applicable in a wide range of studies where proteins have to be identified.

ACKNOWLEDGMENTS

We thank Dr. C. Schwienbacher and M. Lederer for their support and stimulating discussions, C. Wucherpfennig and M. Bock for expert technical assistance and the Deutsche Forschungsgemeinschaft for financial support.

REFERENCES

1. **Giehl, K., R. Valenta, M. Rothkegel, M. Ronsiek, H.G. Mannherz and B.M. Jockusch.** 1994. Interaction of plant profilin with mammalian actin. Eur. J. Biochem. *226*:681-689.
2. **Menkel, A.R., M. Kroemker, P. Bubeck, M. Ronsiek, G. Nikolai and B.M. Jockusch.** 1994. Characterization of an F-actin-binding domain in the cytoskeletal protein vinculin. J. Cell Biol. *126*:1231-1240.
3. **Rothkegel, M., O. Mayboroda, M. Rohde, C. Wucherpfennig, R. Valenta and B.M. Jockusch.** 1996. Plant and animal profilins are functionally equivalent and stabilize microfilaments in living animal cells. J. Cell Sci. *109*:83-90.
4. **Wiedemann, P., K. Giehl, S.C. Almo, A.A. Fedorov, P. Steinberger, M. Rüdiger, M. Ortner, M. Sippl et al.** 1996. Molecular and structural analysis of a continous birch profilin epitope defined by a monoclonal antibody. J. Biol. Chem. *271*:29915-29921.
5. **Zentgraf, H., M. Frey, S. Schwinn, C. Tessmer, B. Willemann, Y. Samstag and I. Velhagen.** 1995. Detection of histidine-tagged fusion proteins by using a high-specific mouse monoclonal anti-histidine tag antibody. Nucleic Acids Res. *23*:3347-3348.

Update to: Epitope Tag-Antibody Combination Useful for the Detection of Protein Expression in Prokaryotic and Eukaryotic Cells

Manfred Rüdiger, Brigitte M. Jockusch and Martin Rothkegel

Cell Biology, Zoological Institute, Technical University of Braunschweig, Braunschweig, Germany

In the last few years further vectors for the generation of BiPro-tagged proteins have been created by inserting the BiPro-tag cDNA cassette into common expression plasmids. The general use of the BiPro-tag/4A6 antibody combination has been shown by expression and detection of recombinant proteins in *Escherichia coli* (3,4), vertebrate cells (2,5) and insect cell lines (Rothkegel, unpublished). Furthermore, the BiPro-tag has been successfully combined with the 6xHis-tag to enable a simple purification protocol for BiPro-tagged proteins by Ni-affinity chromatography (2,3). In addition, the number of applications of this epitope-tag antibody combination is continuously growing. Recently, the antibody 4A6 has not only been demonstrated to immunoprecipitate BiPro-tagged proteins from cell lysates, but also to provide a powerful tool for protein-ligand interaction analysis (1,2,5). Erratum: The monoclonal antibody 4A6 has been characterized as an immunoglobulin IgG1 (6).

REFERENCES

1. **Giesemann, T., S. Rathke-Hartlieb, M. Rothkegel, J.W. Bartsch, S. Buchmeier, B.M. Jockusch and H. Jockusch.** 1999. A role for polyproline motifs in the spinal muscular atrophy protein SMN. Profilins bind to and colocalize with smn in nuclear gems. J. Biol. Chem. *274*:37908-37914.
2. **Harbeck, B., S. Hüttelmaier, K. Schlüter, B.M. Jockusch and S. Illenberger.** 2000. Phosphorylation of the vasodilator-stimulated phosphoprotein (VASP) regulates its interaction with actin. J. Biol. Chem. *275*:30817-30825.
3. **Hüttelmaier, S., B. Harbeck, O. Steffens, T. Meßerschmidt, S. Illenberger and B.M. Jockusch.** 1999. Characterization of the actin binding properties of the vasodilator-stimulated phosphoprotein VASP. FEBS Lett. *451*:68-74.
4. **Pandjaitan, B., P.B. Eibensteiner, S. Vrtala, B. Hayek, M. Grote, R. Reichelt, H. Rumpold, R. Valenta and S. Spitzauer.** 1999. pET-prof, a plasmid for high-level expression of recombinant peptides fused to a birch profilin-derived hexadecapeptide tag: a system for the detection and presentation of recombinant antigens. Gene *237*:333-342.
5. **Weiss, E.E., M. Kroemker, A.H. Rüdiger, B.M. Jockusch and M. Rüdiger.** 1998. Vinculin is part of the cadherin-catenin junctional complex: complex formation between alpha-catenin and vinculin. J. Cell Biol. *141*:755-764.
6. **Wiedemann, P., K. Giehl, S.C. Almo, A.A. Fedorov, M. Girvin, P. Steinberger, M. Rüdiger, M. Ortner et al.** 1996. Molecular and structural analysis of a continuous birch profilin epitope defined by a monoclonal antibody. J. Biol. Chem. *271*:29915-29921.

Use of *Staphylococcus aureus* Protein-A Subdomains as a Tag for the Sensitive Detection of Recombinant Fusion Proteins

Mark M. Gosink and Richard D. Vierstra

University of Wisconsin-Madison, Madison, WI, USA

(Reprinted from BioTechniques 25:374-378, 1998)

The use of translationally fused peptide tags has greatly facilitated the detection and purification of recombinant proteins. The first tags included narrowly defined epitopes (e.g., c-myc, hemagglutinin and FLAG®) that could be detected by a corresponding monoclonal antibody (1,6). More recently, tags have been developed that also facilitate purification of the corresponding fusion with an appropriate affinity matrix. Examples of this type include poly-histidine (11), glutathione S-transferase (10), maltose-binding protein (3) and cellulose-binding peptide (9). However, each of these tags requires a specific antibody to detect the tagged molecule.

One tag that eliminates the need to prepare specific antibodies/antisera has been derived from from *Staphylococcus aureus* Protein A (SPA). SPA has a natural affinity for a wide range of immunoglobulins (Ig) due to the presence of an Ig-binding motif repeated five times within the polypeptide [domains D, A, B, C and E (7,8)]. This affinity has been widely exploited for the purification of recombinant proteins on nonspecific Ig matrices. More recently, we and others have noted that the Ig-binding domains of SPA are remarkably stable and retain their Ig-binding activity even after the fused protein is subjected to sodium dodecyl sulfate polyacrylamide gel electrophoresis (SDS-PAGE) and Western blot transfer (2,5). This stability suggested that SPA also could be used for immunoblot detection of tagged proteins and for immunopurification.

We show that SPA tags can provide a highly sensitive method to detect recombinant proteins subjected to immunoblot analysis. In fact, the limits of detection using nonspecific IgGs to recognize the SPA tag were comparable to those obtained with antisera specific for the rest of the polypeptide. Moreover, if a nonspecific secondary antibody conjugated to a reporter molecule was used alone, a less sensitive but easy one-step immunoblot procedure was possible. Thus, SPA tags can be convenient and affordable for use in recombinant protein detection.

We originally isolated the coding regions for the Ig-binding domains D and DABC from SPA by polymerase chain reaction (PCR) amplification of *S. aureus* genomic DNA as described in Gosink and Vierstra (2). The SPA_D domain added 64 amino acids (aa) (7.1 kDa) to the target, whereas the SPA_{DABC} added 238 aa (27 kDa). These coding regions were cloned into a pET Vector (Novagen, Madison, WI, USA) harboring the wheat ubiquitin conjugating enzyme (*UBC4*) gene to create the pET-UBC4-Protein A_D (UBC4-SPA_D) and pET-UBC4-Protein A_{DABC} (UBC4-SPA_{DABC}) vectors (2). To construct the SPA_D- and SPA_{DABC}-β-galactosidase (SPA_D-LACZ and SPA_{DABC}-LACZ, respectively) expression vectors, the Ig-binding domains, D and DABC, were PCR-amplified at low stringency from the pET-UBC4-SPA_{DABC} using the primers, 5′-CTTAATGACCATATGGCTCCAAAAGCTGATGCGCAAC-3′ and 5′-CTTTGTTGAATTTAGATCTTTTGGTGCTTGAGC-3′. The primers added an *Nde*I site and an ATG codon before codon 90 and added a *Bgl*II site after either codon 153 or 327 of SPA. The products encompassing the D and DABC domains were gel-purified, digested with *Nde*I/*Bgl*II and ligated into *Nde*I/*Bgl*II-digested pET29b to create the vectors pET-SPA_D and pET-SPA_{DABC}, respectively. These vectors were digested with *Bgl*II and ligated in-frame with the 5′ end of the *Lac*Z, which had been released from pET control F plasmid (Novagen) by *Bam*HI digestion. Recombinant proteins were synthesized in and extracted from the *Escherichia coli* strain BL21(DE3) as described (2). All four SPA-tagged proteins were soluble following extraction, suggesting that the tags did not interfere with protein folding or stability (data not shown).

Total protein extracts from the bacteria were separated by SDS-PAGE and electrophoretically transferred onto an Immobilon®-P polyvinylidene fluoride (PVDF) membrane (Millipore, Bedford, MA, USA). When a primary antibody was used, membranes were first rinsed 2× for 5 min in TBST (TBST buffer contained 0.1% [vol/vol] Tween® 20 in TBS

Cloning and Expression Vectors for Gene Function Analysis
Edited by Quinn Lu and Michael P. Weiner

[TBS contained 137 mM NaCl, 2.7 mM KCl, 25 mM Tris-HCl, pH 8.0]), once for 15 min in HST (TBST plus 0.5 M NaCl) and hybridized for 1 h with the primary reagent diluted with HST. Nonspecific rabbit IgGs (Sigma Chemical, St. Louis, MO, USA) were used at 10 μg/mL. Rabbit antiserum against wheat UBC4 was diluted 1:1000. Following incubation with the primary antibody or in cases when only a secondary antibody was used, the membrane was rinsed 2× for 5 min in TBST followed by one 15-min rinse in HST. Membranes were then hybridized for 1 h with the indicated secondary reagent in HST. Concentrations of alkaline-phosphatase-labeled goat anti-rabbit IgGs (AP-GAR) and horseradish-peroxidase-labeled rabbit anti-goat IgGs (HRP-RAG) (Kirkegaard & Perry Laboratories, Gaithersburg, MD, USA) were 1 μg/mL. Membranes were rinsed once for 15 min in HST, followed by two 5-min rinses in TBST and one 1-min rinse in distilled water. Conjugated antibodies bound to the membrane were visualized using either HRP color development reagent (Bio-Rad, Hercules, CA, USA) or nitroblue tetrazolium and 5-bromo-4-chloro-3-indolyl phosphate as described (4).

As shown in Figure 1, the SPA-tagged versions of UBC4 and LACZ were easily visualized by a one-step immunoblot protocol using either AP-GAR or HRP-RAG as the sole antibody probes. In this case, AP-GAR and HRP-RAG bound directly to the SPA sequences. Interestingly, the SPA-tagged proteins were detected with approximately the same sensitivity with the goat antibodies (AP-GAR) as with the rabbit antibodies (HRP-RAG) despite previous reports demonstrating that intact SPA has little affinity for goat IgGs. Either the SPA_D or the SPA_{DABC} tag was effective, although the latter appeared to provide higher sensitivity (Figure 2). From immunoblot analysis of serial dilutions of SPA-tagged UBC4, <74 ng of protein could be detected easily if a single Ig-binding domain was added (UBC4-SPA_D) (Figure 2A); this limit decreased by a factor of 10 (ca. 2.7 ng) if all four Ig-binding domains were used (UBC4-SPA_{DABC}) (Figure 2A).

As compared to a two-step conventional immunoblot protocol using UBC4 antibodies in conjunction with AP-GAR, the one-step procedure using just AP-GAR was approximately 1/30 as sensitive in detecting UBC4-SPA_D (Figure 2, A and C). However, near equal sensitivity could be achieved simply by using a two-step protocol with nonspecific rabbit IgGs as the primary antibody followed by AP-GAR as the secondary antibody (Figure 2B). When UBC4 antisera or nonspecific rabbit IgGs were used as the primary antibodies, we could detect >0.9 ng of UBC4-SPA_D. Even less protein was detected when tagged with four Ig-binding domains (UBC4-SPA_{DABC}).

In contrast to other tags currently available [e.g., c-myc, poly-histidine or glutathione S-transferase (6,10,11)], SPA tags offer several advantages. First, SPA tags can be detected by nonspecific Igs, thus obviating the need to prepare specific antisera either to the tag or to the modified protein. In our hands, the sensitivity of detection and the level of nonspecific staining produced with a two-step immunoblot procedure using nonspecific-rabbit IgGs as the primary antibody rivaled that obtained with a specific antiserum. We were successful with various IgGs (rabbit, mouse and goat), enabling researchers to use a variety of nonspecific Igs without the need to obtain a specific antibody for each protein or application. As shown here, fusion of the SPA tag to either the N or C terminus was possible.

A second advantage is that SPA-tagged proteins also can be detected easily using a simple one-step immunoblot protocol having a single antibody probe (e.g., AP-GAR). Although not

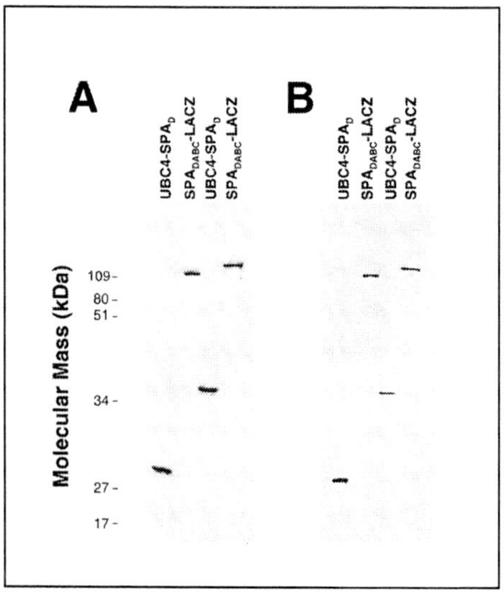

Figure 1. Detection of SPA-tagged proteins with a one-step immunoblot procedure using AP-GAR or HRP-RAG. UBC4 and LacZ proteins were tagged with either one or four Ig-binding domains from SPA (SPA_D or SPA_{DABC}, respectively) and expressed in *E. coli*. Bacterial extracts were subjected to SDS-PAGE, Western blot transfer onto PVDF membranes and immunoblot analysis using only a secondary antibody for detection: AP-GAR (A) or HRP-RAG (B). Identical protein loads and immunoblot development times were used in both panels.

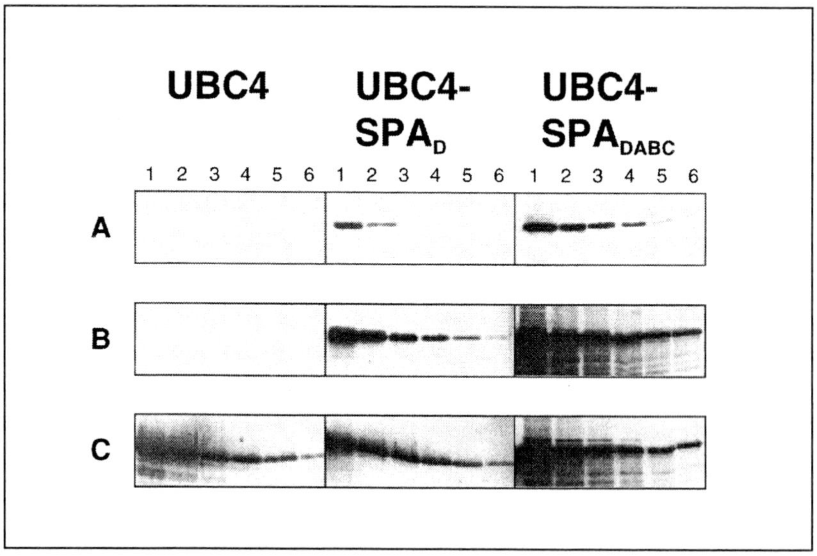

Figure 2. Sensitivity of detection for SPA-tagged proteins. Purified UBC4 and UBC4-SPA_D and partially purified UBC4-SPA_{DABC} were serially diluted in threefold increments, subjected to SDS-PAGE and transferred onto a PVDF membrane. Lanes 1–6 for UBC4 and UBC4-SPA_D contained 220, 74, 25, 8.2, 2.7 and 0.9 ng of pure protein, respectively. Lanes 1–6 for UBC4-SPA_{DABC} contained 660, 220, 74, 25, 8.2 and 2.7 ng of total protein, respectively. Wild-type or tagged UBC4 were detected with either a one-step hybridization using just AP-GAR (A), a two-step hybridization using nonspecific rabbit IgGs followed by AP-GAR (B) or a two-step hybridization using rabbit antisera against UBC4 followed by AP-GAR (C). The proteins were subjected simultaneously to SDS-PAGE and immunoblotting, and the membranes were both incubated with the color reagents for 5 min.

as sensitive as the two-step protocol, the one-step method enabled us to detect, in less than 1 h, recombinant proteins subjected to Western blot transfer. A third advantage is that, unlike most other peptide tag systems, the same reagent can be used for detection and purification of the SPA-tagged protein. Whereas the poly-histidine tag requires a separate reagent for detection and purification [poly-histidine antisera and chelated Ni^{2+}, respectively (11)], SPA-based tags can use IgGs in both applications.

The sensitivity and ease of use of the SPA-based tags in protein detection clearly demonstrate their usefulness as versatile markers for protein analysis. Although their size is greater than poly-histidine (64 aa for domain D of SPA vs. 6 aa for poly-histidine), it is comparable or smaller than other polypeptide tags (e.g., glutathione S-transferase and maltose-binding protein). SPA-based tags do not seem to adversely affect the stability or activity of the tagged proteins and provide several convenient ways to detect and purify recombinant proteins using inexpensive readily available reagents.

ACKNOWLEDGMENTS

This work was supported by grants from The Procter & Gamble Company to M.M.G. and R.D.V. and from the USDA-NRICGP (Grant No. 97-35301-4218) to R.D.V.

REFERENCES

1. **Brizzard, B.L., R.G. Chubet and D.L. Vizard.** 1994. Immunoaffinity purification of FLAG® epitope-tagged bacterial alkaline phosphatase using a novel monoclonal antibody and peptide elution. BioTechniques *16*:730-732.
2. **Gosink, M.M. and R.D. Vierstra.** 1995. Redirecting the specificity of ubiquitination by modifying ubiquitin-conjugating enzymes. Proc. Natl. Acad. Sci. USA *92*:9117-9121.
3. **Guan, C., P. Li, P.D. Riggs and H. Inouye.** 1988. Vectors that facilitate the expression and purification of foreign peptides in *Escherichia coli* by fusion to maltose-binding protein. Gene *67*:21-30.
4. **Harlow, E. and D. Lane (Eds.).** 1988. Antibodies: A Laboratory Manual. CSH Laboratory Press, Cold Spring Harbor, NY.
5. **Jang, Y.-J., J.-M. Lecerf and B.D. Stollar.** 1996. Heavy chain dominance in the binding of DNA by a lupus mouse monoclonal autoantibody. Mol. Immunol. *33*:197-210.
6. **Kolodziej, P.A. and R.A. Young.** 1991. Epitope tagging and protein surveillance. Methods Enzymol. *194*:508-519.
7. **Nilsson, B. and L. Abrahmsén.** 1990. Fusions to staphylococcal protein A. Methods Enzymol. *185*:144-161.
8. **Nilsson, B., L. Abrahamsén and M. Uhlén.** 1985. Immobilization and purification of enzymes with staphylococcal protein A gene fusion vector. EMBO J. *4*:1075-1080.
9. **Ramírez, C., J. Fung, R.C. Miller, Jr., R. Antony, J. Warren and D.G. Kilburn.** 1993. A biofunctional affinity linker to couple antibodies to cellulose. Bio/Technology *11*:1570-1573.
10. **Smith, D.B. and K.S. Johnson.** 1988. Single-step purification of polypeptides expressed in *Escherichia coli* as fusions with glutathione S-transferase. Gene *67*:31-40.
11. **Van Dyke, M.W., M. Sirito and M. Sawadogo.** 1992. Single-step purification of bacterially expressed polypeptides containing an oligo-histidine domain. Gene *111*:99-104.

Update to: Use of *Staphylococcus aureus* Protein-A Subdomains as a Tag for the Sensitive Detection of Recombinant Fusion Proteins

Mark M. Gosink and Richard D. Vierstra

University of Wisconsin-Madison, Madison, WI, USA

The utility of Protein-A (SPA) subdomains for tagging both prokaryotic and eukaryotic polypeptides has been demonstrated by a number of researchers (2,4). However, the versatility of SPA is not limited to detection. We and others have found that SPA fusions can provide a near one-step purification of recombinant proteins from crude extracts of bacterial, insect and mammalian origin, using inexpensive IgG-Sepharose® columns (Amersham Pharmacia Biotech, Piscataway, NJ, USA) (1,5,6). In these cases, the tagged proteins are bound and eluted from the IgG columns using the conditions optimized for intact Protein A. Because the acidic elution conditions required to elute fusions generated with native SPA domains can be detrimental to some proteins, researchers have created modified version of the SPA domains, which can be effectively released using milder elution conditions (3).

REFERENCES

1. **Aoki, T., M. Satoh and H. Watabe.** 1998. Detection of blotted antigen using a fusion protein between protein A and pepsinogen C. J. Biochem. Tokyo *124*:572-577.
2. **Fahrenkrog, B., J.P. Aris, E.C. Hurt, N. Pante and U. Aebi.** 2000. Comparative spatial localization of protein-A-tagged and authentic yeast nuclear pore complex proteins by immunogold electron microscopy. J. Struct. Biol. *129*:295-305.
3. **Gulich, S., M. Uhlen and S. Hober.** 2000. Protein engineering of an IgG-binding domain allows milder elution conditions during affinity chromatography. J. Biotechnol. *76*:233-244.
4. **Pool, M.R., E. Lopez-Huertas and A. Baker.** 1998. Characterization of intermediates in the process of plant peroxisomal protein import. EMBO J. *17*:6854-6862.
5. **Shinoda, K., Y. Morishita, K. Sasaki, Y. Matsuda, I. Takahashi and T. Nishi.** 1997. Enzymatic characterization of human alpha1,3-fucosyltransferase Fuc-TVII synthesized in a B cell lymphoma cell line. J. Biol. Chem. *272*:31992-31997.
6. **Zhou, D., M. Malissard, E.G. Berger and T. Hennet.** 2000. Secretion and purification of recombinant beta1-4 galactosyltransferase from insect cells using pFmel-protA, a novel transposition-based baculovirus transfer vector. Arch. Biochem. Biophys. *374*:3-7.

CHAPTER 22

Monitoring and Purification of Proteins Using Bovine Papillomavirus E2 Epitope Tags

Niilo Kaldalu, Dina Lepik, Arnold Kristjuhan and Mart Ustav

Tartu University, Tartu, Estonia

(Reprinted from BioTechniques 28:456-462, 2000)

ABSTRACT

We describe here the use of two newly mapped bovine papillomavirus type 1 (BPV-1) E2 protein epitopes as tags. We constructed several vector plasmids for overexpression as well as for moderate expression of single- or double-tagged proteins in either Escherichia coli *or eukaryotic cells. The new tags were fused to several proteins, and the activity of the tagged proteins was tested in different assays. The tags were shown not to interfere with the function of these proteins in vivo and in vitro. Interaction of the monoclonal antibodies 3F12 and 1E2 with their respective epitopes was specific and had high affinity in a variety of conditions. We have demonstrated that the 3F12 antibody-epitope interaction tolerates high salt concentrations up to 2 M. This permits immunoprecipitation and immunopurification of the tagged proteins in high-salt buffers and reduction of the nonspecific binding of the contaminating proteins. We also provide a protocol for DNA binding and DNase I footprinting assays using the tagged, resin-bound DNA-binding proteins. The BPV-1 E2-derived tags can be recommended as useful tools for detection and purification of proteins.*

INTRODUCTION

Epitope tagging is a recombinant DNA technique by which a protein is made immunoreactive to a preexisting antibody. This technique simplifies detection, characterization, purification and in vivo localization of proteins and has become a standard method of molecular genetics (3). However, some tags are not useful in certain applications due to high background binding. In some cases, affinity purification and immunoprecipitation of a tagged protein is problematic due to co-precipitation of contaminating proteins.

Here, we describe the use of two recently mapped bovine papillomavirus type 1 (BPV-1) E2 protein epitopes as tags. We constructed several vector plasmids for over-expression as well as for moderate-level expression of either single- or double-tagged proteins in *Escherichia coli* and eukaryotic cells. The new tags were fused to functionally different proteins: a bacterial transcriptional activator, XylS, that agregates and becomes nonfunctional at high levels of expression, several mutants of the tumor suppressor protein p53 for overexpression in *E. coli* and rat glyceraldehyde-3-phosphate dehydrogenase (GAPDH) for expression in eukaryotic cells. Hereby, we report the use of BPV-1 E2-derived tags and respective monoclonal antibodies for the identification of proteins on western blots, in the immunofluorescence staining of cells and DNA band-shift assay. In addition, we describe an application of the resin-bound tagged protein in DNA binding and DNase I footprinting assays.

MATERIALS AND METHODS

Plasmid Constructions

For the construction of pBR-3F12 and pBR-1E2, we inserted the coding sequence of XylS with the N-terminally fused influenza virus hemagglutinin (HA) epitope (2) from the pET11c-based parent plasmid pETSN117 (4) between the *Nhe*I and *Bam*HI sites of pBR322. The subcloned fragment contained in addition to a ribosome binding site, an extra *Nde*I site and a start codon preceding the tag sequence. The resultant plasmid pBRSN 117 contains *Xba*I and *Bam*HI sites for cloning of a recombinant coding sequence (4). Then, the coding sequence of the HA epitope between *Nde*I and *Xba*I sites was replaced with double-stranded synthetic oligonucleotides encoding peptides GVSSTSSDFRDR and TTGHYSVRD, recognized by anti-BPV-1 E2 monoclonal antibodies 3F12 and 1E2, respectively (7).

For the construction of pBR-NC, we amplified *xylS* sequence by PCR using the 3′-end primer containing a coding sequence for the peptide TSSDFRDR. The peptide recognized by 3F12 Mab was fused in frame to the C terminus of

Cloning and Expression Vectors for Gene Function Analysis
Edited by Quinn Lu and Michael P. Weiner

XylS and flanked by *Kpn*I and *Bam*HI sites. The resultant PCR fragment was cloned into pBR-1E2 generating the expression plasmid with cloning sites *Xba*I and *Kpn*I.

pET-3F12 was generated by cloning the *Bam*HI/*Nde*I fragment from pBR-3F12-*xylS* into the corresponding sites in pET-11c. Then, the *XylS* gene was removed by cleavage with *Xba*I and *Bam*HI and replaced with coding sequences for different mutant p53 proteins bearing *Xba*I and *Bgl*II sites at the ends.

For the construction of pCG-3F12, double-stranded synthetic oligonucleotide encoding 3F12 epitope and incorporating multicloning sites was inserted between the *Xba*I and *Bgl*II sites of pCG (9). Plasmid pCG-3F12-GAPDH contains rat glyceraldehyde-3-phosphate dehydrogenase (GAPDH) inserted between *Bam*HI and *Bgl*II sites of pCG-3F12. Plasmid pCG-3F12-GAPDH-NLS contains nuclear localization signal of human p53 (amino acids 305–327) fused to the C-terminus of the GAPDH protein.

Immunoaffinity Binding of 3F12-XylS

pBR-3F12-*xylS* bearing *E. coli* DH5α cells were grown in LB medium supplemented with 100 µg/mL ampicillin at 20°C to an A_{600} of approximately 1.0. Cells were harvested, washed with TBS and resuspended in 1/10 volume of high-salt lysis buffer containing 100 mM Tris-HCl, pH7.5, 1.5 M NaCl and 5 mM EDTA, 20% (wt/vol) glycerol. Cell suspension was frozen in liquid nitrogen and stored at -70°C.

For the preparation of crude extracts, cells were thawed and dithiothreitol (DTT; 10 mM), PMSF (100 µg/mL), aprotinin (1 µg/mL), and CHAPS (0.2%) were added. Cells were incubated with lysozyme (0.5 mg/mL) on ice for 20 min and disrupted by sonication. The lysate was clarified by centrifugation at 40 000× *g* at 4°C for 30 min.

Affinity beads were prepared by coupling 3F12 anti-BPV E2 monoclonal antibody to divinylsulfon-activated Toyopearl® HW65 TSK-gel (TOSOH, Tokyo, Japan). The 3F12-TSK affinity beads were incubated with crude lysate at 4°C for 1 h with end-over-end agitation. Beads were washed extensively with high-salt lysis buffer on glass filter. These 3F12-XylS beads were stored at -20°C in approximately 10 gel volumes of storage buffer containing 10 mM Tris-HCl, pH 7.5, 100 mM KCl, 10 mM $MgCl_2$, 80 µM EDTA, 80 µM DTT, 50% (vol/vol) glycerol, 100 µg/mL PMSF and 1 µg/mL aprotinin. These beads were used for DNA binding and "footprinting" assays.

DNA Precipitation Assay

End-labeled restriction fragments of pUPM190 *Hpa*II/*Hin*fI digest and 50 µL of the suspension of 3F12-XylS beads were mixed with 0.3 mL of DNA binding buffer containing 50 mM Tris-HCl, pH 7.5, 100 mM KCl, 2 mM EDTA, 10% (wt/vol) sucrose, 0.1% CHAPS, 700 µg/mL BSA, 1 µg/mL aprotinin and 1 mM *meta*-toluate. The mixture was incubated for 30 min at room temperature with end-over-end agitation. Beads were washed 3× with 1 mL of DNA binding buffer and incubated with Proteinase K (50 µg/mL) at 37°C for 30 min in stop solution containing 200 mM NaCl, 5 mM EDTA and 1% SDS. The retained DNA was released by phenol-extraction, ethanol-precipitated, and identified in 5% PAGE under non-denaturing conditions.

DNase I Footprinting

The *Om*-containing *Eco*RI/*Xho*I fragment of pUPM190 was end-labeled at *Xho*I site in the lower strand with Klenow fragment. The labeled probe was incubated with 50 µL of the suspension of 3F12-XylS beads in DNA binding buffer as for the DNA precipitation. The beads were washed 3× and suspended in 100 µL of DNase I buffer containing 10 mM Tris-HCl, pH 7.5, 100 mM KCl, 5 mM $MgCl_2$, 1 mM $CaCl_2$, 100 µM DTT, 100 µM EDTA, 100 µg/mL BSA and 2 µg/mL sonicated salmon-sperm DNA. DNase I was added to concentrations of 0.6–3.0 µg/mL, and the reaction mixture was incubated for 1 min at 30°C. The reaction was stopped by 200 µL of stop solution containing 600 mM NaCl, 120 mM EDTA, 3% SDS and 150 µg/mL dextran, and the beads were treated with Proteinase K (50 µg/mL) at 37°C for 30 min. DNA was extracted from the beads with phenol/chloroform, precipitated with ethanol and analyzed on sequencing gel. DNase I cleavage of the unbound template was carried out in the same buffer at DNase I concentrations of 30–300 ng/mL.

Gel-Shift Assay

Gel-shift assay was performed as previously described (1). We used the crude lysate preparations of p53 mutants. The double-stranded p53 con oligo (5′-GAT CCG GAC ATG CCC GGG CAT GTC CGG ATC-3′) was used as a probe. Protein-DNA complexes were separated from the unbound DNA on 5% PAGE (55:1).

Immunodetection of Proteins

For western blot analysis, cells expressing the tagged proteins were lyzed in SDS sample buffer, total protein extracts were separated on 12% SDS-PAGE and electroblotted onto nitrocellulose membrane filters. Blots were blocked with 1% nonfat dry milk for 1 h, incubated with anti-BPV E2 1E2, 3F12 or anti-p53 pAb 240 monoclonal antibodies, followed by incubation with anti-mouse IgG alkaline phosphatase-conjugated secondary antibody.

For immunofluorescence staining, Saos-2 cells were transfected with plasmids expressing 3F12-GAPDH-NLS or 3F12-GAPDH proteins. Cells were grown on microscopy cover glasses and fixed 24 h after transfection with methanol at -20°C. Proteins were detected with 3F12 and anti-mouse Ig antibody conjugated with FITC as a primary and secondary antibody, respectively.

RESULTS AND DISCUSSION

Moderate Level Bacterial Expression of Epitope-Tagged Proteins

The XylS protein, a transcriptional activator from the TOL plasmid pWWO of the soil bacterium *Pseudomonas putida*, like some other AraC/XylS family transcription factors, does not tolerate high level of over-expression and is prone to aggregation both inside the cell as well as in the solution, in the course of purification (4,5). We expressed XylS and several truncated variants of the protein at a near to native level in

Table 1. Main Characteristics of Expression Vectors

Plasmid	Promoter	Tag location	Tag sequence
pBR-1E2	*Ptet*	N-terminus	M* T T G H Y S V R D S R... 5′-CAT ATG ACA ACA GGG CAT TAC TCT GTA AGA GAT TCT AGA GGA TCC -3′ *NdeI* **XbaI** **BamHI**
pBR-3F12	*Ptet*	N-terminus	M* G V S S T S S D F R D R S R... 5′-CAT ATG GGT GTC TCA TCC ACC TCT TCT GAT TTT AGA GAT CGC TCT ACA GGA TCC -3′ *NdeI* **XbaI** **BamHI**
pBR-NC	*Ptet*	N-terminus 1E2 C-terminus 3F12	M* T T G H Y S V R D S R... 5′-CAT ATG ACA ACA GGG CAT TAC TCT GTA AGA GAT TCT AGA-3′... *NdeI* **XbaI** ...G T S S D F R D R STOP ...5′-GGT ACC TCT TCT GAT TTT AGA GAT CGC TGA GGA TCC-3′ **KpnI** **BamHI**
pET-3F12	T7	N-terminus	M* G V S S T S S D F R D R S R... 5′-CAT ATG GGT GTC TCA TCC ACC TCT TCT GAT TTT AGA GAT CGC TCT AGA-3′ **NdeI** **XbaI**
pCG-3F12	CMV IE	N-terminus	M* G V S S T S S D F R D R... 5′-TCT AGA ATG GGT GTC TCA AGT ACT TCT TCT GAT TTT AGA GAT CGC... **XbaI** *ScaI* ...G S K L P G L Q G T R S ...GGA TCC AAG CTT CCC GGG CTG CAG GGT ACC AGA TCT-3′ **BamHI** **HindIII** **SmaI** **PstI** **KpnI** **BglII**

In amino acid sequences, the epitopes are underlined and the initiator methionine codons are marked with asterisks
In nucleotide sequences, cleaving sites for restriction endonucleases are underlined.
Restriction enzymes which cut once per plasmid are marked in bold. pET-3F12 contains an additional *Xba*I site which is sensitive to *dam* methylation and is not cleaved when the plasmid is isolated from *dam*+ strains

E. coli and tested their physiological activities in vivo. For that, we constructed vectors for the moderate level of expression of epitope-tagged fusion proteins. The vectors were based on pBR322, and the *tet* promoter of this plasmid was used for expression of recombinant proteins. The vectors pBR-3F12 and pBR-1E2 were constructed for the expression of proteins with N-terminally fused peptides GVSSTSSDFRDR and TTGHYSVRD, recognized by anti-BPV E2 monoclonal antibodies 3F12 and 1E2, respectively (7). Both vectors contain *Xba*I and *Bam*HI sites for cloning of the recombinant sequence. We also constructed the vector pBR-NC for the expression of proteins with different epitope tags in both N and C termini. The N-terminal tag was TTGHYSVRD as in pBR-1E2 and the C-terminal tag was TSSDFRDR, a shorter version of the epitope recognized by 3F12 Mab. The cloning sites for a coding sequence in pBR-NC are *Xba*I and *Kpn*I (Table 1).

We transformed *E. coli* DH5α with plasmids expressing the tagged versions of XylS and analyzed the expression of the tagged XylS proteins by western blotting. The 3F12 antibody recognized both 3F12-XylS and NC-XylS proteins, while the 1E2 antibody recognized 1E2-XylS and NC-XylS proteins as single bands on the western blot (Figure 1A, lanes 1–6). No cross-reaction with cellular proteins was observed. However, when 3F12 Mab was used for the detection of the double-tagged NC-XylS protein, a much weaker signal was detected when compared with the 1E2 Mab signal. (Figure 1A, lanes 4 and 6). This could be explained by the use of the shorter version of the 3F12-specific epitope in the double-tagged protein. We studied the effect of these tags on the activity of XylS protein in *E. coli* strain CC118*Pm-lacZ*, which carries a chromosomal copy of the XylS responsive *Pm* promoter fused to the *lacZ* gene (6). The tags had no effect on the transcriptional activation by XylS (data not shown).

Study of Specific DNA Binding by the Matrix-Attached Epitope-Tagged Protein

XylS is a DNA-binding protein, which specifically binds to the *Om* operator sequence and activates the responsive *Pm* promoter. However, suitable conditions to study the soluble XylS in vitro have not been found, as the protein tends to aggregate. To demonstrate the site-specific DNA binding of epitope-tagged XylS in vitro, we used immunobound 3F12-

XylS protein that was attached to the TSK beads through the N-terminal tag and 3F12 antibody. Western blot analysis showed that high salt concentrations, up to 2 M NaCl, do not hinder the interaction of 3F12 MAb with the specific epitope (data not shown). Therefore, to avoid aggregation of XylS and coimmunoprecipitation of contaminating proteins, the crude lysate was prepared in a high-salt lysis buffer, containing 1.5 M NaCl, and 3F12-XylS was bound to the affinity resin carrying 3F12 Mab by a single-step, batchwise procedure.

The DNA binding assay was carried out by mixing the 3F12-XylS beads with a mixture of end-labeled restriction fragments of the *Om*-containing plasmid pUPM190 (4). After several washes, the bound DNA was released and identified by gel electrophoresis, using the input mixture of fragments as a marker. Figure 2A shows that only a single *Om*-containing fragment of the pUPM190 digest was retained on the beads, and binding of any other fragment could not be detected.

Further, we analyzed the interaction of the immobilized 3F12-XylS protein with *Om* by DNase I footprinting. The specific complexes of the *Om*-containing DNA fragment and TSK-bound 3F12-XylS were prepared identically as for DNA precipitation and were treated with DNaseI. After the cleavage, DNA was extracted from the beads and analyzed on a sequencing gel. As a control, the unbound template was cleaved at lower concentrations of DNaseI to obtain the equal rate of cleavage. Figure 2B shows that 3F12-XylS protects a 44 bp area on the lower strand and four DNase I hypersensitive sites occur within the protected region. We obtained an identical DNaseI footprint earlier with HA-epitope tagged XylS, using similar technical approach (4). DNA precipitation and DNase I footpriting with a resin-bound epitope-tagged protein (10) are the methods of choice for proteins that are prone to aggregation and are difficult to purify.

High Level Bacterial Expression of Epitope-Tagged Proteins

For high-level bacterial expression of N-terminally tagged proteins, we constructed the vector pET-3F12, a derivative of pET-11c that contains the tag-encoding sequence and the cloning sites identical to pBR-3F12 (Table 1). The coding sequences for mutant p53 proteins were cloned into the vector. Mutant proteins ΔN39ΔC362 and ΔN61ΔC362 had their transactivation and regulatory parts deleted, but maintained the ability to bind DNA, while the mutant protein ΔN39ΔC362trp248 had lost its DNA-binding ability due to the point mutation in its DNA-binding domain (8). Plasmids, generated for the T7 promoter-directed expression of 3F12-tagged p53 fusion proteins were transformed into *E. coli* strain BL21 (DE3).

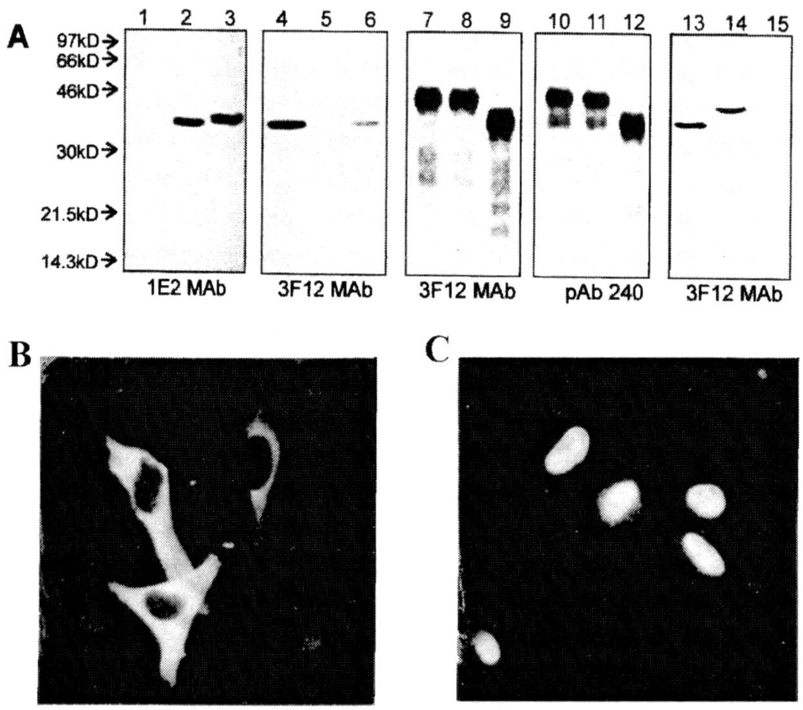

Figure 1. Detection of the tagged proteins. (A) Western blot analysis of the tagged proteins. Total protein extracts, separated by 12% SDS-PAGE and electroblotted, were analyzed using 1E2 (lanes 1–3), 3F12 (lanes 4–9 and 13–15) or anti-p53 pAb 240 (lanes 10–12) primary antibodies and alkaline phosphatase-conjugated secondary antibody. Lanes 1–6: *E. coli* DH5α cells producing tagged XylS protein, bearing pBR-3F12 (lanes 1 and 4), pBR-1E2 (lanes 2 and 5) and pBR-NC (lanes 3 and 6) derived expression plasmids. Lanes 7–12: *E. coli* BL21 (DE3) cells bearing pET-3F12 derived expression plasmids, producing 3F12-tagged p53 variants ΔN39ΔC362 (lanes 7 and 10), ΔN39ΔC362trp248 (lanes 8 and 11) and ΔN61ΔC362 (lanes 9 and 12). Lanes 13 and 14: Saos-2 cells transfected with pCG-3F12 derived expression plasmids producing 3F12-tagged GAPDH (lane 13) and GAPDH-NLS (lane 14). Saos-2 cells expressing untagged p53 were used for a negative control (lane 15). (B and C) Subcellular localization of 3F12-GAPDH and 3F12-GAPDH-NLS proteins. Saos-2 cells were transfected with pCG-3F12-GAPDH (B) or pCG-3F12-GAPDH-NLS (C) and tagged proteins were detected with immmunofluorescence analysis.

Expression of the tagged p53 proteins was monitored by western blot analysis. The tag-specific 3F12 antibodies and p53-specific pAb240 antibodies were used for the detection of the proteins. Using this vector system resulted in enormous overexpression of the protein detected with 3F12 Mab as well as with pAb240 antibodies (Figure 1A, lanes 7–12).

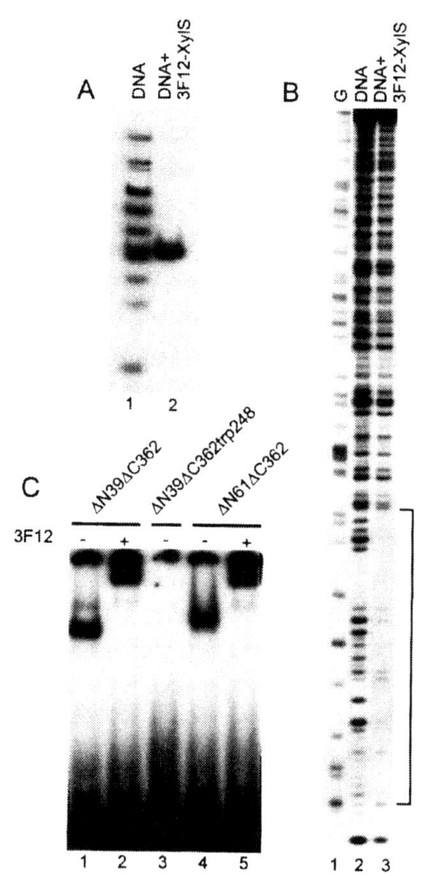

Figure 2. Use of the tagged proteins in DNA binding assays. (A) DNA precipitation by 3F12-tagged XylS. Radiolabeled *Hpa*II/*Hin*fI digest of the *Om*-containing plasmid pUPM190 was incubated with 3F12-beads containing 3F12-XylS. Unbound DNA was removed by washing. Bound DNA was extracted from the beads and analyzed on nondenaturing TBE/PAGE (5%), using the fragment mixture as a marker (lane 1). Only a single fragment, which contained the XylS binding site *Om*, was retained on the beads (lane 2). (B) DNase I footprinting by resin-bound 3F12-tagged XylS. *Xho*I/*Eco*RI *Om* containing fragment from pUPM190 was end-labeled in the lower strand at 3′ terminus and incubated with 3F12-beads containing 3F12-XylS. Unbound DNA was removed by washing. Both free and protein-bound templates were subjected to DNase I cleavage. Lane 1, G-specific DNA sequence marker; lane 2, DNase I digest of the unbound DNA fragment; lane 3, DNase I digest of the fragment bound to 3F12-XylS. Brackets indicate the region protected from the DNase I cleavage by the DNA-bound XylS. (C) Band-shift assay of the 3F12-tagged p53 proteins. The DNA-binding activity of the p53 proteins was studied by separating the protein-DNA complexes from the unbound DNA on 5% PAGE (55:1). Crude *E. coli* lysates containing p53 mutants ΔNΔC362 and ΔNΔC362 produce a shifted band (lanes 1 and 4). Supershift with tag-specific monoclonal antibody 3F12 was used to show that the protein-DNA complexes contain p53 (lanes 2 and 5). The ΔN39ΔC362trp248 p53 protein carrying a point mutation in its DNA-binding domain was used as a negative control (lane 3).

The DNA-binding activity of the tagged p53 proteins was studied in a band-shift assay (Figure 2C). The p53 mutant ΔN39ΔC362trp248, carrying a point mutation in its DNA-binding domain, was used as a negative control. The mutants ΔN39ΔC362 and ΔN61ΔC362 are functional in DNA binding and produced a shifted band. To show that the produced complex really contains p53, the protein-DNA complexes were supershifted with the tag-specific monoclonal antibody 3F12. This way, epitope tags can be used to verify specificity of the shifted complex in a DNA band-shift assay without purification of the protein.

Expression of Epitope-Tagged Proteins in Eukaryotic Cells

For eukaryotic expression, two coding sequences of rat glyceraldehyde-3-phosphate dehydrogenase (GAPDH) gene were cloned into pCG-3F12 plasmid (Table 1). The first had 3F12 epitope fused in frame to GAPDH amino acids 2–333 (pCG-3F12-GAPDH) and the second also contained a nuclear localization signal of p53 (amino acids 305–327) fused to the C terminus of the GAPDH protein (pCG-3F12-GAPDH-NLS).

Saos-2 cells were transfected with 1 μg of expression plasmids and analyzed 24 h after transfection. Expression and localization of the proteins were determined by western blotting and immunofluorescence analysis, respectively. The 3F12 antibody recognized both proteins as single bands on the western blot (Figure 1A, lanes 13 and 14) and no cross-reaction with cellular proteins was observed (Figure 1A, lane 15). Immunofluorescence staining of transfected cells with 3F12 antibody indicated that both proteins were localized in the appropriate compartment of the cell; the 3F12-GAPDH in the cytoplasm and 3F12-GAPDH-NLS in the nucleus (Figure 1, B and C). These results indicate that 3F12 epitope-tag can be used for the detection and determination of the localization of proteins expressed in eukaryotic cells.

Advantages of the BPV E2-Derived Epitope Tags

We analyzed the expression of proteins tagged with the BPV E2-derived epitopes in *E. coli* and eukaryotic cells. Detection of the tagged proteins both on immunoblots and by immunofluorescence staining of cells indicates low background activity, sensitivity and good signal-to noise ratio of the used epitope-antibody combinations. We did not observe any cross-reaction with cellular proteins. Because of the high specificity of the epitope-antibody interaction, our tagging system is especially useful for the studies of protein localization in the cells. In addition, we have shown that interaction of 3F12 MAb with the specific epitope is not hindered by high salt concentrations. That allows us to immunoprecipitate and immunopurify the tagged proteins in high-salt conditions and to avoid coimmunoprecipitation of contaminating proteins as well as to avoid aggregation of the protein of interest in the course of purification.

ACKNOWLEDGMENTS

The study was supported by the Estonian Science Foundation Grant Nos. 2496, 2497, 2315 and 2316.

REFERENCES

1. **Abroi, A., R. Kurg and M. Ustav.** 1996. Transcriptional and replicational activation functions in the bovine papillomavirus type 1 E2 protein are encoded by different structural determinants. J. Virol. *70*:6169-6179.
2. **Field, J., J. Nikawa, D. Broek, B. MacDonald, L. Rodgers, I.A. Wilson, R.A. Lerner and M. Wigler.** 1988. Purification of a RAS-responsive adenylyl cyclase complex from *Saccharomyces cerevisiae* by use of an epitope addition method. Mol. Cell. Biol. *8*:2159-2165.
3. **Jarvik, J.W. and C.A. Telmer.** 1998. Epitope tagging. Annu. Rev. Genet. *32*:601-618.
4. **Kaldalu, N., T. Mandel and M. Ustav.** 1996. TOL plasmid transcription factor XylS binds specifically to the *Pm* operator sequence. Mol. Microbiol. *20*:569-579.
5. **Kessler, B., M. Herrero, K.N. Timmis and V. de Lorenzo.** 1994a. Genetic evidence that the XylS regulator of the Pseudomonas TOL meta operon controls the Pm promoter through weak DNA-protein interactions. J. Bacteriol. *176*:3171-3176.
6. **Kessler, B., K.N. Timmis and V. de Lorenzo.** 1994b. The organization of the *Pm* promoter of the TOL plasmid reflects the structure of its cognate activator protein XylS. Mol. Gen. Genet. *244*:596-605.
7. **Kurg, R., J. Parik, E. Juronen, T. Sedman, A. Abroi, I. Liiv, U. Langel and M. Ustav.** 1999. Effect of bovine papillomavirus E2 protein-specific monoclonal antibodies on papillomavirus DNA replication. J. Virol. *73*:4670-4677.
8. **Lepik, D., I. Ilves, A. Kristjuhan, T. Maimets and M. Ustav.** 1998. p53 protein is a suppressor of papillomavirus DNA amplificational replication. J. Virol. *72*:6822-6831.
9. **Tanaka, M. and W. Herr.** 1990. Differential transcriptional activation by Oct-1 and Oct-2: interdependent activation domains induce Oct-2 phosphorylation. Cell *60*:375-386.
10. **Ustav M., E. Ustav, P. Szymanski and A. Stenlund.** 1991. Identification of the origin of replication of bovine papillomavirus and characterization of the viral origin recognition factor E1. EMBO J. *10*:4321-4329.

Update to: Monitoring and Purification of Proteins Using Bovine Papillomavirus E2 Epitope Tags

Reet Kurg and Mart Ustav

Tartu University and Estonian Biocentre, Tartu, Estonia

In the article, we characterize further the use of Bovine papillomavirus type 1 (BPV-1) E2 protein epitopes as tags. We have mapped more precisely the essential sequence of the epitope for antibody 3F12 and show the possibility of using this epitope as internal tag in the case of proteins where both terminus are functionally important.

We described the use of two BPV-1 E2 protein epitopes as tags in our earlier publication (2). The new tags were fused to the N and C terminus of the several proteins of bacterial and eukaryotic origin, and the activities of the tagged proteins were tested in different assays, including western blot assay immunofluorescence staining of cells, DNA band-shift assay, as well as the DNA binding and Dnase I footprinting assays (2–4). We showed that 3F12 antibody interaction with its respective epitope was specific and had high affinity, and in addition, it tolerated high salt concentrations up to 2 M. This interesting characteristic allows to detect the tagged proteins under high salt conditions to lower the background signals. In this study, we have mapped the epitope for monoclonal antibody 3F12 more precisely. We generated several truncated 3F12 tags (Figure 1A) in pCG-3F12 vector between the *Xba*I and *Bam*HI sites instead of the previously used 3F12 epitope. As a marker gene, the DNA binding domain (DBD) of E2 was incorporated into *Bgl*II site of multicloning site of pCG vector (2). Figure 1B shows the results of western blot analysis obtained with COS-7 cell lysates transfected with 500 ng of expression plasmids and probed with 3F12 antibody. From these data, we concluded that the high affinity 3F12 antibody epitope consists of 10 amino acid residues, SSTSSDFRDR. All these ten amino acids are required and are sufficient for strong interaction with monoclonal antibody 3F12. The first half of the sequence consists polar amino acids, and the second half contained charged amino residues resulting in very hydrophilic peptide. Such highly hydrophilic sequences have been proven to have strong antigenicity and are correspondingly likely to adopt a highly exposed conformation in the three-dimensional folding of a protein (1).

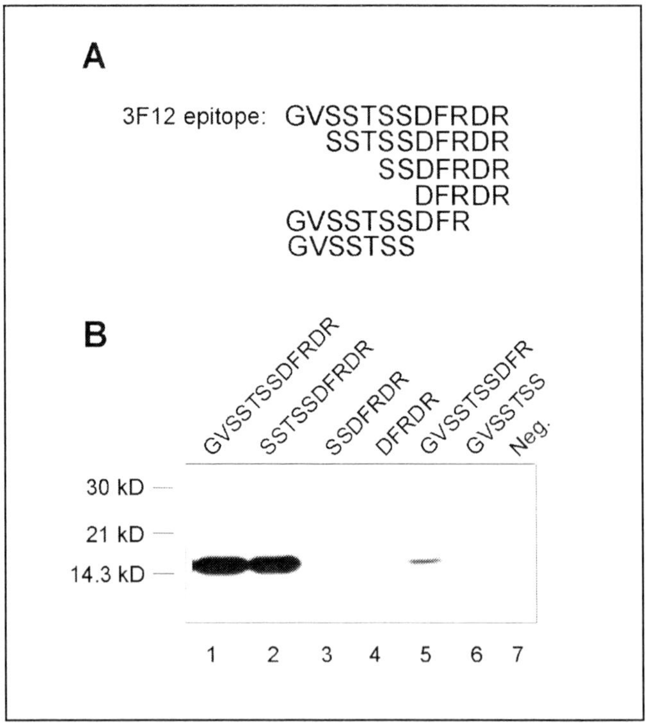

Figure 1. Mapping of the 3F12-epitope. (A) The sequences of peptides used in this study. (B) Western blot analysis of the tagged proteins. COS-7 cells were transfected with pCG-3F12-DBD derived expression plasmids containing 3F12 epitopes of different size. Lane 7 is the vector alone. 48 h after electroporation, the cells were lysed in Laemmli buffer, separated by 15% sodium dodecyl sulfate-polyacrylamide gel electrophoresis (SDS-PAGE) and electroblotted to polyvinylidene difluoride (PVDF) filter and probed with Mab 3F12. The signal was detected by chemoluminescence.

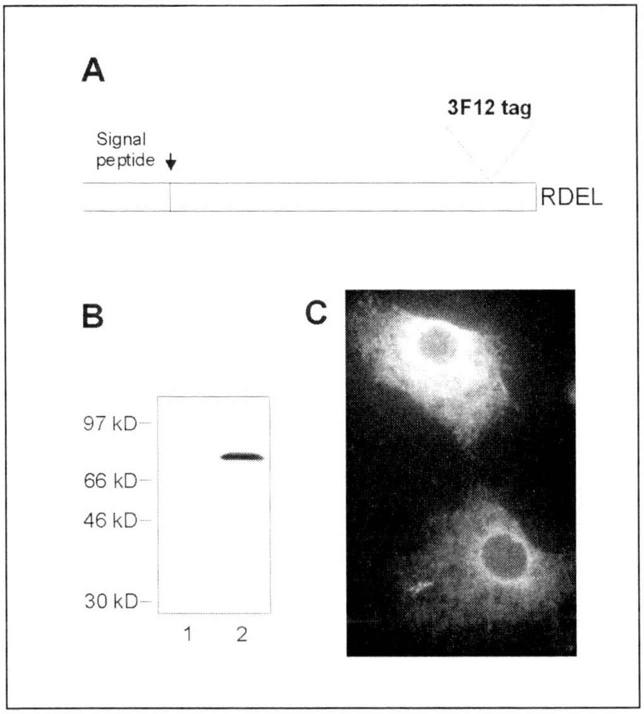

Figure 2. The 3F12-epitope as an internal tag. (A) Schematic representation of the cyclooxygenase and the position of the 3F12-epitope tag within the protein. (B) Western blot analysis of the tagged protein. Cos-7 cells were transfected with expression vector containing the 3F12 tagged cyclooxygenase. 48 h after electroporation, the cells were lysed in Laemmli buffer, separated by 8% SDS-PAGE and electroblotted to PVDF filter and probed with Mab 3F12. The signal was detected by chemoluminescence. COS-7 cells expressing untagged Cox was used as a negative control (lane 1). (C) Immunofluorescence staining of COS-7 cells transfected with the 3F12 tagged cox expressing vector. Cells were grown on microscopy cover glasses and fixed 24 h after transfection with acetone:methanol (1:1) at -20°C. Protein was detected with Mab 3F12 and anti-mouse Ig antibody conjugated with fluorescein isothiocyanate (FITC) as a primary and secondary antibody, respectively.

Epitope tags are usually added at the extreme N or C terminus of the protein, since the ends of proteins are more likely to be accessible to the antibody, and the addition of tag is less likely to affect the function of the protein, and because it is more convenient. However, when the extreme ends of the protein are important for its function or when processing is taking place at these ends, then an internal tagging must be used. Next, we decided to determine whether the 3F12 epitope is also recognizable and usable in the middle of the protein. Cyclooxygenases are glycoproteins associated with membranes of the endoplasmic reticulum (ER) and nuclei and contain a signal peptide in the amino acid sequence, which is cleaved to yield the mature protein (5). The C terminus of the protein contains the sequence RDEL, which probably helps to retain the protein in the ER. We inserted the 3F12 epitope into the open reading frame of cyclooxygenase of Arctic coral (6) into the C-terminal 22-amino acid region, which is known to reside in the lumen of ER and has been shown to be dispensable for Cox enzymatic activity. As shown on Figure 2, the 3F12 epitope was readily detectable by western blot analysis of the lysates of the transfected COS-7 cells and by immunofluorescence analysis of the intact cells. Immunofluorescence staining of transfected COS-7 cells with 3F12 antibody exhibited signal at the ER and the nuclear membrane, similar to results obtained with green fluorescent protein (GFP)-tagged Cox and cox-specific antibodies (5). In conclusion, these results show that 3F12 epitope-tag is usable in the extreme N and C terminus as well as in the middle of the protein. This tag can be used as a tool for the detection and determination of the localization of proteins expressed in eukaryotic cells and also for expression and immunopurifying proteins from bacteria.

The 3F12 epitope tags and respective antibodies are available from: Quattromed Ltd., Nooruse 9, 50411 Tartu, ESTONIA. E-mail:erki.molder@quattromed.ee. http//:www.quattromed.ee

REFERENCES

1. **Hopp, T.** 1986. Protein surface analysis: methods for identifying antigenic determinants and other interaction sites. J. Immunol. Methods. *88*:1-18.
2. **Kaldalu, N., D. Lepik, A. Kristjuhan and M. Ustav.** 2000. Monitoring and purification of proteins using bovine papillomavirus E2 epitope tags. BioTechniques. *28*:456-462.
3. **Kaldalu, N., T. Mandel and M. Ustav.** 1996. TOL plasmid transcription factor XylS binds specifically to the Pm operator sequence. Mol. Microbiol. *20*:569-579.
4. **Kaldalu, N., U. Toots, V. de Lorenzo and M. Ustav.** 2000. Functional domains of the TOL plasmid transcription factor XylS. J. Bacteriol. *182*:1118-1126.
5. **Trifan, O., R. Smith, B. Thompson and T. Hla.** 1999. Overexpression of cyclooxygenase-2 induces cell cycle arrest. J. Biol. Chem. *274*:34141-34147.
6. **Varvas, K., I. Järving, R. Koljak, K. Valmsen, A. Brash and N. Samel.** 1999. Evidence of a cyclooxygenase-related prostaglandin synthesis in coral. J. Biol. Chem. *274*:9923-9929.

Section IV

Reporter Gene Technologies

Dominant Positive and Negative Selection Using Luciferase, Green Fluorescent Protein and β-Galactosidase Reporter Gene Fusions

Craig A. Strathdee, Marilyn R. McLeod and T. Michael Underhill

The John P. Robarts Research Institute and University of Western Ontario, London, ON, Canada

(Reprinted from BioTechniques 28:210-214, 2000)

The introduction of exogenous DNA sequences into cultured cells is a cornerstone in the study of mammalian gene expression. For most cell types, the efficiency of stable gene transfer is usually quite low, which mandates the use of selectable markers to isolate a population of transfected cells with the desired phenotype. In simple gene expression experiments, only one type of selectable marker is required. Positive selectable markers such as the bacterial *neo* and *hph* genes facilitate the direct selection for cells that are resistant to the cytocidal effects of the antibiotics, G418 and hygromycin B, respectively (1,12). Conversely, negative selectable markers facilitate the elimination of the cells in which they are expressed.

The most popular negative selectable marker is the herpes simplex virus thymidine kinase gene, *tk*, which sensitizes cells to the toxic effects of the nucleoside analogue gancyclovir (11,15). In more complex gene transfer experiments such as in the use of gene, promoter or enhancer traps, it becomes necessary to include multiple marker genes. For example, fusions between the bacterial *lac*Z reporter gene and the selectable neo marker (2), the *hph* and *tk* selectable markers (7) or the *neo* and *tk* selectable markers have been developed for use in enhancer or gene trapping experiments. Here, we describe the further development of gene fusions between the *tk* and *neo* selectable markers and the bacterial *lac*Z, jellyfish *gfp* and firefly *luc* reporter genes.

The gene fusions were constructed using cDNAs derived from the previously described pPNT (14), pβgeo (2), pGL3-Control (Promega, Madison, WI, USA), and pEGFP-C1 (Clontech Laboratories, Palo Alto, CA, USA) vector constructs, which encode the HSV *tk* and bacterial *neo* genes, a *lac*Z-*neo* fusion gene, the enhanced *luc* and *gfp* reporter genes, respectively. A *tkneo* gene fusion was generated using PCR amplification in conjunction with vector-specific primers to eliminate the endogenous stop codon in the *tk* gene of pPNT through the incorporation of a *Bam*HI site. The amplified product was cloned as an *Eco*RI-*Bam*HI fragment into the corresponding *Eco*RI-*Bam*HI sites of pβgeo, which effectively replaces the *lac*Z-portion of the *lac*Z-*neo* gene fusion. The *tkneo* fusion gene was then subcloned as a *Hind*III-*Not*I fragment into the pEFG3 expression vector, and this construct was designated pEF-tkneo (Figure 1). We incorporated the *gfp* cDNA by subcloning the corresponding reading frame as a blunt-ended *Nhe*I-*Bgl*II fragment from pEGFP-C1 into the blunt-ended *Bam*HI site to generate pEF-tkgfneo.

The *tklucneo* gene fusion was generated using PCR amplification in conjunction with vector-specific primers to introduce *Pac*I restriction endonuclease sites into the 5′-UTR and to eliminate the endogenous stop codon in the *luc* gene of pGL3-Control, which was then subcloned into the *Bam*HI site of pEF-tkneo to generate pEF-tkβgeo (Figure 1). The *lac*Z gene was incorporated into the *tkneo* gene fusion by subcloning the corresponding reading frame as a *Bam*HI fragment from pβgeo into the *Bam*HI site of pEF-tkneo to generate pEF-tkβgeo (Figure 1). The *gfp* cDNA was incorporated into the *Bam*HI site of this construct to generate pEF-TβGN (Figure 1). As positive control plasmids for our experiments, we subcloned the individual *luc*, *gfp* and *lac*Z genes into the pEF1 vector (Invitrogen, Carlsbad, CA, USA) to generate pEF-Luc, pEF-GFP and pEF-βGal, respectively (Figure 1).

The activity of the reporter gene components in each of the five fusion genes was assayed following transient transfection into HEK293 (human embryonic kidney) cells. Cells were grown in DMEM supplemented with 10% FCS under standard culture conditions and were transfected with 1 μg of plasmid DNA in a single well of a 6-well culture dish using cationic liposomes according to the manufacturer's protocols (LipofectAMINE Plus™; Life Technologies, Gaithersburg, MD, USA). The results indicate that all four gene fusions that contain a reporter gene component retain their respective

Cloning and Expression Vectors for Gene Function Analysis
Edited by Quinn Lu and Michael P. Weiner

activities, although to a different degree in each case (Table 1). Specifically, the fusions that contain a β-galactosidase (β-gal) reporter gene retain their full activity in comparison to the pEF-βGal non-fusion control, and the fusions that contain a *luc* or *gfp* reporter gene retain only approximately 5% of their activity in comparison to the respective pEF-Luc and pEF-GFP non-fusion controls.

The activity of the selectable marker components in each of the five fusion genes was assayed using stable transfection experiments. Each linearized expression construct was transfected into HEK 293 cells as described above, and two days later the cells were divided evenly into duplicate 100-mm culture dishes. One dish was selected in 400 μg/mL G418 to assay for function of the *neo* marker in the fusion gene, while the other dish was selected in 400 μg/mL G418 and 2 μM gancyclovir to assay for the function of the *tk* marker. Those transfectants selected in G418 yielded equivalent numbers of resistant colonies for each construct, including the pEF-Luc, pEF-GFP and pEF-βGal controls, indicating that the neo component of each of the fusion proteins retained its function (Table 1). With the exception of a single "breakthrough", the transfectants selected in G418 and gancyclovir yielded no colonies for any of the fusion genes. Because there was no effect on the number of colonies obtained in the pEF-Luc, pEF-GFP and pEF-βGal control transfections that do not express *tk*, this indicates that the *tk* component of each of the fusion proteins also retained its function (Table 1).

The stable clones were assayed for β-gal activity by histochemical staining, which showed that the reporter gene activity of each fusion construct was similarly retained in all of the clones examined (Figure 2). However, only about 10% of the clones examined expressed detectable green fluorescent protein (GFP). This finding is consistent with our previous data from the transient assays for GFP, which suggests that although every clone may express the fusion gene, the reduced GFP signal will only be detectable in those that express it at high levels. Interestingly, the subcellular localization of the fusion proteins differs somewhat, with the *tkgfneo* fusion yielding a diffuse cytoplasmic signal similar to the non-fusion GFP control, and the TβGN fusion yielding an intense punctate signal suggestive of incorporation into inclusion bodies (Figure 2 and References 3 and 13). Although expression levels are similar between the clones, GFP from the TβGN fusion is easier to detect because of a significant enhancement of signal intensity.

The selectable reporter fusion genes we have developed are well suited to serve as qualitative and quantitative probes of promoter or enhancer strength. They also provide a simple mechanism to eliminate false positive clones in gene, promoter, and enhancer trapping strategies (2) by virtue of their ability to be both positively and negatively selected. In a related application, we are currently using a "floxed" version of the

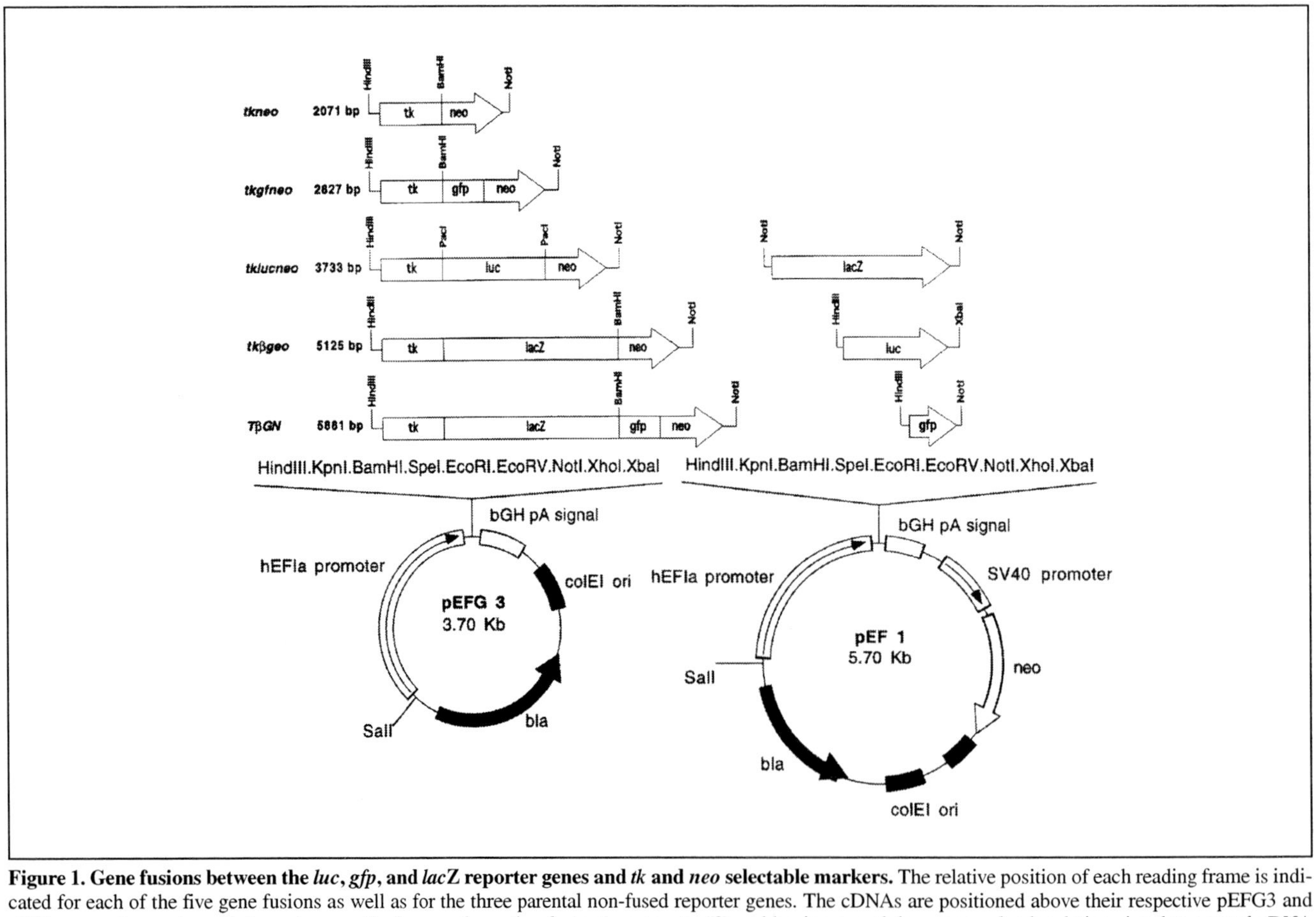

Figure 1. Gene fusions between the *luc*, *gfp*, and *lacZ* reporter genes and *tk* and *neo* selectable markers. The relative position of each reading frame is indicated for each of the five gene fusions as well as for the three parental non-fused reporter genes. The cDNAs are positioned above their respective pEFG3 and pEF1 expression vectors. Both vectors use the human elongation factor 1a promoter (9) and bovine growth hormone polyadenylation signal to control cDNA expression (10), with the pEF1 vector containing a separately expressed *neo* marker to provide selection for stable clones following transfection.

Table 1. Analysis of Selectable Reporter Gene Fusions in HEK293 Cells

Construct	G418^R Colonies[a]	G418^{R+} Gancyclovir^R Colonies[a]	Relative β-gal Activity[b,c]	Relative Luciferase Activity[b,c]	Relative GFP[b,c]
pEF-GFP	220	150	1.2	-[d]	720
pEF-βGal	164	147	180	-	0
pEF-tkneo	286	0	0.7	-	0
pEF-tkβgeo	174	0	200	-	0
pEF-tkgfneo	187	0	0.6	-	39
pEF-TβGN	185	0	103	-	16
pEF-Luc	118	64	-	920	-
pEF-tkneo	98	0	-	0	-
pEF-tklucneo	63	1	-	52	-

[a]Independent colonies per P100 culture dish.

[b]Cells were washed once in PBS, harvested, and lysed using a proprietary detergent formulation according to manufacturers protocols (Reporter Lysis Buffer; Promega, Madison, WI, USA). The luciferase and β-gal activities in the lysates were determined by assaying duplicate dilutions using commercially available luminescent (Luciferase Assay System; Promega) and chemiluminescent (Galacto-Light Plus; Tropix, Bedford, MA, USA) substrates. Light yield from the reactions was detected using a tube format luminometer with autoinjector (Lumat LB9507; EG&G Wallac, Gaithersburg, MD, USA). The GFP in the lysates was determined by assaying duplicate dilutions directly in a fluorescent plate reader using an FITC filter set (Fluorescence Concentration Analyzer; IDEXX Laboratories, Westbrook, ME, USA).

[c]Corrected for differences in transfection efficiency by including 5 ng of either pGL3-Control (top series) or pCMVβ (bottom series) (Clontech Laboratories) in each reaction.

[d]-, not done.

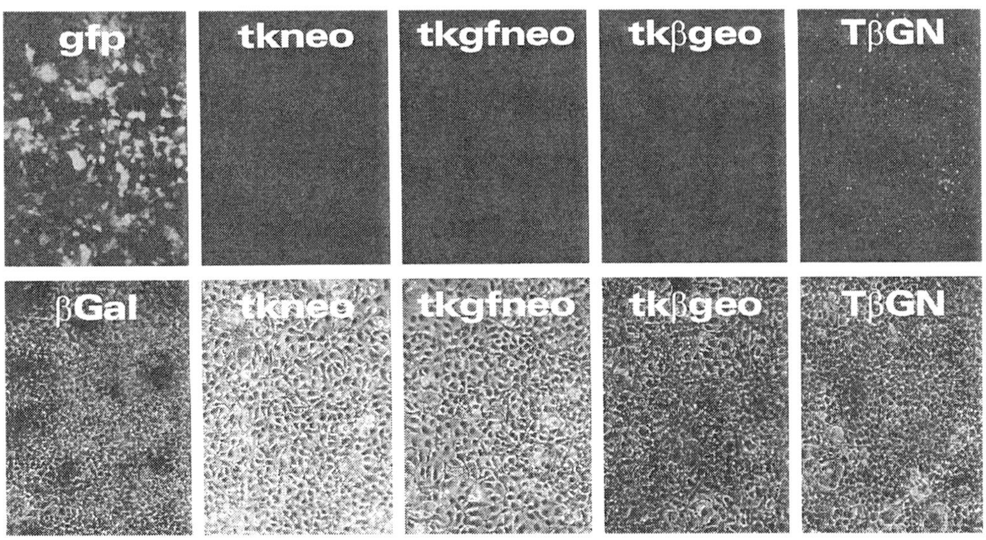

Figure 2. Detection of GFP and β-galactosidase activity in stable HEK293 clones expressing the reporter gene fusions. Cells were fixed in paraformaldehyde/glutaraldehyde using standard methods (8). GFP was detected directly using fluorescence microscopy (top row), whereas β-galactosidase activity was detected using X-gal histochemical staining (bottom row) (8). The respective fusion proteins are indicated for each panel.

tklucneo fusion gene as a selectable marker in conjunction with the tetracycline-regulated transactivator expression system (4–6), where it has proven useful in the identification of stable transfectants that have a high-level of "on" and low-level "off" expression following modulation of promoter activity by doxycycline (manuscript in preparation).

ACKNOWLEDGMENTS

We thank Chris Pászty and Phil Soriano for generously providing the pPNT and pβgeo constructs, respectively. This research was supported by grants from the ALS Society of Canada (C.A.S.) and Medical Research Council of Canada (T.M.U.). C.A.S. is a Cancer Research Society/Medical Research Council of Canada Scholar.

REFERENCES

1. **Blochlinger, K. and H. Diggelmann.** 1984. Hygromycin B phosphotransferase as a selectable marker for DNA transfer experiments with higher eucaryotic cells. Mol. Cell Biol. *4*:2929-2931.
2. **Friedrich, G. and P. Soriano.** 1991. Promoter traps in embryonic stem cells: a genetic screen to identify and mutate developmental genes in mice. Genes Dev. *5*:1513-1523.
3. **Friedrich, V.L., Jr., G.R. Holstein, X. Li, A. Gow, K.A. Kelley and R.A. Lazzarini.** 1993. Intracellular distribution of transgenic bacterial β-galactosidase in central nervous system neurons and neuroglia. J. Neurosci. Res. *36*:88-98.
4. **Gossen, M. and H. Bujard.** 1992. Tight control of gene expression in mammalian cells by tetracycline-responsive promoters. Proc. Natl. Acad. Sci. USA *89*:5547-5551.
5. **Gossen, M., S. Freundlieb, G. Bender, G. Muller, W. Hillen and H. Bujard.** 1995. Transcriptional activation by tetracyclines in mammalian cells. Science *268*:1766-1769.
6. **Gu, H., J.D. Marth, P.C. Orban, H. Mossmann and K. Rajewsky.** 1994. Deletion of a DNA polymerase β gene segment in T cells using cell type-specific gene targeting. Science *265*:103-106.
7. **Lupton, S.D., L.L. Brunton, V.A. Kalberg and R.W. Overell.** 1991. Dominant positive and negative selection using a hygromycin phosphotransferase-thymidine kinase fusion gene. Mol. Cell Biol. *11*:3374-3378.
8. **MacGregor, G.R., A.E. Mogg, J.F. Burke and C.T. Caskey.** 1987. Histochemical staining of clonal mammalian cell lines expressing *E. coli* β-galactosidase indicates heterogeneous expression of the bacterial gene. Somat. Cell Mol. Genet. *13*:253-265.
9. **Mizushima, S. and S. Nagata.** 1990. pEF-BOS, a powerful mammalian expression vector. Nucleic Acids Res. *18*:5322.
10. **Pfarr, D.S., L.A. Rieser, R.P. Woychik, F.M. Rottman, M. Rosenberg and M.E. Reff.** 1986. Differential effects of polyadenylation regions on gene expression in mammalian cells. DNA *5*:115-122.
11. **Smith, K.O., K.S. Galloway, W.L. Kennell, K.K. Ogilvie and B.K. Radatus.** 1982. A new nucleoside analog, 9-[[2-hydroxy-1-(hydroxymethyl)ethoxyl]methyl]guanine, highly active in vitro against herpes simplex virus types 1 and 2. Antimicrob. Agents Chemother. *22*:55-61.
12. **Southern, P.J. and P. Berg.** 1982. Transformation of mammalian cells to antibiotic resistance with a bacterial gene under control of the SV40 early region promoter. J. Mol. Appl. Genet. *1*:327-341.
13. **Tu, P.H., K.A. Robinson, F. de Snoo, J. Eyer, A. Peterson, V.M. Lee and J.Q. Trojanowski.** 1997. Selective degeneration of Purkinje cells with Lewy body-like inclusions in aged NFHLACZ transgenic mice. J. Neurosci. *17*:1064-1074.
14. **Tybulewicz, V.L., C.E. Crawford, P.K. Jackson, R.T. Bronson and R.C. Mulligan.** 1991. Neonatal lethality and lymphopenia in mice with a homozygous disruption of the c-abl proto-oncogene. Cell *65*:1153-1163.
15. **Wei, C.M., M. Gibson, P.G. Spear and E.M.** Scolnick. 1981. Construction and isolation of a transmissible retrovirus containing the src gene of Harvey murine sarcoma virus and the thymidine kinase gene of herpes simplex virus type 1. J. Virol. *39*:935-944.

Update to: Dominant Positive and Negative Selection Using Luciferase, Green Fluorescent Protein and β-Galactosidase Reporter Gene Fusions

Craig A. Strathdee, Marilyn R. McLeod and T. Michael Underhill

The John P. Robarts Research Institute and University of Western Ontario, London, ON, Canada

Since this article was published, we have developed a *gfluc* reporter gene based on a fusion between the jellyfish *gfp* and firefly *luc* reporter genes. Briefly, the *gfp* cDNA was fused in-frame with the *luc* cDNA by cloning it as a polymerase chain reaction (PCR)-derived *Nco*I-*Bsp*HI fragment into the *Nco*I site of pGL3 (Promega, Madison, WI, USA). The resulting fusion protein retains the properties of both parental reporter genes, namely that cells expressing *gfluc* are easily visualized through fluorescence microscopy by virtue of the *gfp* component, and that expression levels between different cell populations can easily be quantitated by virtue of the *luc* component.

CHAPTER 24

Quantification of Gene Expression with a Secreted Alkaline Phosphatase Reporter System

Te-Tuan Yang, Parisa Sinai, Paul A. Kitts and Steven R. Kain

CLONTECH Laboratories, Palo Alto, CA, USA

(Reprinted from BioTechniques 23:1110-1114, 1997)

ABSTRACT

The cDNA encoding secreted alkaline phosphatase (SEAP) is a useful tool for investigating the function of known or putative enhancer/promoter elements. SEAP has the unusual properties of extreme heat stability and resistance to the phosphatase inhibitor L-homoarginine. Therefore, endogenous alkaline phosphatase activity in transfected cells can be minimized by pretreatment of samples at 65°C and incubation with the inhibitor. With the use of the chemiluminescent substrate CSPD®, 10^{-13} g of enzyme can be detected in culture medium, and the enzyme activity can be detected as early as 24 h after transfection. The chemiluminescence-based SEAP assay is about 10-fold more sensitive than similar assays using firefly luciferase as the reporter enzyme. The SEAP activity can also be assayed with a fluorescent substrate MUP, which provides sensitivity comparable to luciferase. Since the enzyme is secreted to culture medium, the enzyme assay can be performed on small samples of the culture supernatant. Because preparation of cell lysates is not required, assaying for SEAP activity is faster and more convenient than assaying for intracellular reporters. Furthermore, because the transfected cells are not disturbed by the sampling procedure, the same cultures can be repeatedly sampled for time-course studies or used for further investigations.

INTRODUCTION

Studies of transcription control elements have been greatly facilitated by linking them to reporter genes, which are readily assayed; expression of the reporter gene is used as an indication of transcriptional activity (10). The key features of a useful reporter gene are that its activity be easily assayed and that its endogenous activity in the target cells be low. β-Galactosidase (β-gal) is perhaps the best example of such a gene and has proven to be invaluable in studies with many cell types (12,14). The bacterial chloramphenicol acetyltransferase (*cat*) gene (1,8) has been extensively used in the analysis of transcription control elements in mammalian tissue culture cells. The genes encoding the human growth hormone (13) and firefly luciferase (LUC) (6,9) have also been presented as useful alternatives. However, each of these genes has disadvantages, such as the presence of a similar endogenous enzyme activity or the requirement of a cumbersome assay.

Berger et al. described a novel reporter gene, secreted alkaline phosphatase (SEAP) (2,5), which avoids several of these disadvantages. As a reporter of gene expression, SEAP differs from CAT, β-gal and LUC in that SEAP is secreted from transfected cells and can thus be assayed using a small aliquot of the culture medium. Levels of SEAP activity detected in culture medium are directly proportional to changes in intracellular concentrations of SEAP mRNA and protein (15). The enzyme can withstand temperatures as high as 65°C and is resistant to the phosphatase inhibitor L-homoarginine (3). Therefore, endogenous alkaline phosphatase activity can be eliminated by pretreatment of samples at 65°C and incubation with the inhibitor.

The secreted nature of SEAP provides several advantages for its use as a genetic reporter: *(i)* preparation of cell lysates is not required; *(ii)* the kinetics of gene expression can be easily studied by repeated collection of medium from the same cultures; *(iii)* transfected cells are not disturbed during measurement of SEAP activity and remain intact for further investigations; *(iv)* background from endogenous alkaline phosphatase (AP) activity in the culture medium is low; and *(v)* sample collection and assay can be automated using multiwell plates. This latter characteristic makes SEAP reporter systems particularly well-suited to high-throughput applications.

MATERIALS AND METHODS

Materials

L-homoarginine, human placental alkaline phosphatase (PLAP) and calf intestinal alkaline phosphatase (CIAP) were obtained from Sigma Chemical (St. Louis, MO, USA). All

Cloning and Expression Vectors for Gene Function Analysis
Edited by Quinn Lu and Michael P. Weiner

media, serum and other tissue culture supplements were purchased from Life Technologies (Gaithersburg, MD, USA). The CalPhos Maximizer™ Transfection Kit was obtained from CLONTECH Laboratories (Palo Alto, CA, USA).

Cell Culture and Supplements

CHO-K1 cells were cultured in Ham's F-12 medium supplemented with 10% fetal bovine serum (FBS). BHK cells were cultured in Dulbecco's modified Eagle medium (DMEM) supplemented with 10% FBS. All cell lines used in the studies were obtained from ATCC (Rockville, MD, USA). Media for all cultures routinely included 100 U/mL of penicillin and 100 µg/mL of streptomycin. Cultures were maintained at 37°C with 5% CO_2/95% air.

pSEAP2 Plasmid Vectors

The pSEAP2 series is an integrated set of plasmids that differ only by the presence or absence of simian virus 40 (SV40) promoter and/or enhancer sequences. pSEAP2-basic has a multiple cloning site (MCS) upstream of the SEAP gene, into which putative promoters can be inserted. The SEAP coding sequence is followed by the SV40 late polyadenylation signal to ensure proper, efficient processing of the transcript. The vector backbone also provides an F1 origin for single-stranded DNA production, a pUC19 origin of replication (ori) and an ampicillin-resistance (*amp*^{r}) gene for propagation and selection in *E. coli*. The sequences around the SEAP ATG have been modified to generate a strong Kozak consensus translation initiation site (11). A synthetic transcription blocker (TB), composed of adjacent polyadenylation and transcription pause sites, reduces background transcription (7). pSEAP2-control (Figure 1) is pSEAP2-basic with the SV40 early promoter inserted into the MCS upstream of the SEAP gene and the SV40 enhancer inserted downstream. pSEAP2-enhancer and pSEAP2-promoter are identical to pSEAP2-control except that they lack the SV40 promoter or enhancer sequences, respectively.

Transfections

Transfections were performed using duplicate cultures by seeding 5×10^5 cells per well in 6-well tissue culture plates one day before transfections in the appropriate growth medium. Cells were transfected with 3 µg pSEAP2 vectors using the CalPhos Maximizer Transfection Kit according to the manufacturer's instructions.

SEAP Activity Assay

All materials required for the detection of SEAP activity are available in the Great EscAPe™ SEAP Genetic Reporter System (CLONTECH). Both the chemiluminescence and fluorescence assays for SEAP activity were performed according to the manufacturer's instructions using 25 µL of conditioned medium from the transfected cells. Chemiluminescence signals were detected using a Model TD-20e Luminometer (Turner Designs, Sunnyvale, CA, USA) with a 15-s integration time. Fluorescence signals were detected with a CytoFluor™ II Fluorometer (PerSeptive Biosystems, Framingham, MA, USA) with excitation and emission at 360 and 449 nm, respectively.

RESULTS AND DISCUSSION

Assay of SEAP Activities in Cell Culture Medium Allows Kinetic Analysis of Gene Expression

There are several criteria to consider when selecting a transcription reporter gene: *(i)* expression of the gene must not alter the physiology of the cells used for transfection; *(ii)* the reporter protein must be either absent or easily distinguishable from endogenous proteins; *(iii)* the reporter protein should be detectable over a wide linear range to facilitate analysis of both weak and strong changes in gene expression; and *(iv)* a simple, rapid and sensitive assay should be available to detect the reporter protein. Many of these criteria are fulfilled by the commonly used *cat*, β-gal and *luc* genes. However, these reporters are intracellular enzymes and require harvesting and lysing the transfected cells before quantifying activities. In contrast, the secreted nature of SEAP allows the transfected cells to remain intact, and hence the kinetics of gene expression can be easily studied using the same cultures by repeatedly sampling the medium (Figure 2). Furthermore, after samples have been taken for the SEAP assay, the same cells can be used for further analysis.

Detection of SEAP Activity by Chemiluminescence and Fluorescence

The original assay for SEAP activity uses the chromogenic AP substrate ρ-nitrophenyl phosphate (5). The sensitivity of this assay is poor and has a narrow linear dynamic range. A bioluminescence assay using a reaction between SEAP and LUC using the substrate luciferin-*O*-phosphate provides improved sensitivity, but is cumbersome because of the use of two reporter proteins (2). We assayed SEAP activities directly using a chemiluminescent AP substrate, CSPD® (Tropix, Bedford, MA, USA) (4). The sensitivity of SEAP detection with this substrate is illustrated in Figure 3A, where the luminometer assay can detect as little as 10^{-13} g of purified PLAP. The

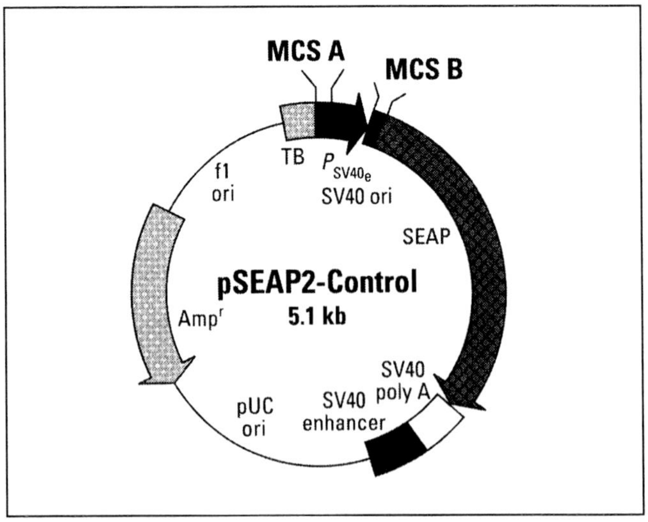

Figure 1. Map of pSEAP2-control. The MCS region is identical in all four vectors except for a 209-bp promoter fragment that has been inserted between the *Bgl*II and *Hin*dIII sites in pSEAP2-control and pSEAP2-promoter. pSEAP2-control and pSEAP2-enhancer also have the SV40 enhancer inserted downstream of the SEAP gene.

sensitivity of chemiluminescence detection with CSPD is 80–100-fold greater than the colorimetric procedure. Moreover, these results show that the SEAP assay has a linear dynamic range of over four orders of magnitude. In addition, the CSPD substrate produces a stable chemiluminescence signal that rises rapidly, reaches a maximum in approximately 10 min and remains constant for more than 1 h (Figure 3B). Little background was detected using conditioned medium lacking purified enzyme. The stability of the signal allows its measurement in luminometers not equipped with an automatic injection device and by exposure to X-ray film. The sensitivity is consistently 10-fold greater than assays for LUC, as shown in Figure 4, and 10^4–10^5-fold more sensitive than CAT assays (2).

Similar results were obtained for fluorescence detection of SEAP activity using MUP (Molecular Probes, Eugene, OR, USA) as the substrate. Although 10-fold less sensitive than its chemiluminescence-based counterpart, the fluorescence-based SEAP assay is comparable to LUC and is suitable for most reporter applications. Furthermore, like the chemiluminescence assay, the fluorescence SEAP assay is linear over a 10^4-fold range of enzyme concentrations (data not shown).

SEAP Expression in Transfected Mammalian Cells

CHO-K1 cells were transfected separately with each of three SEAP vectors, and samples of the conditioned media

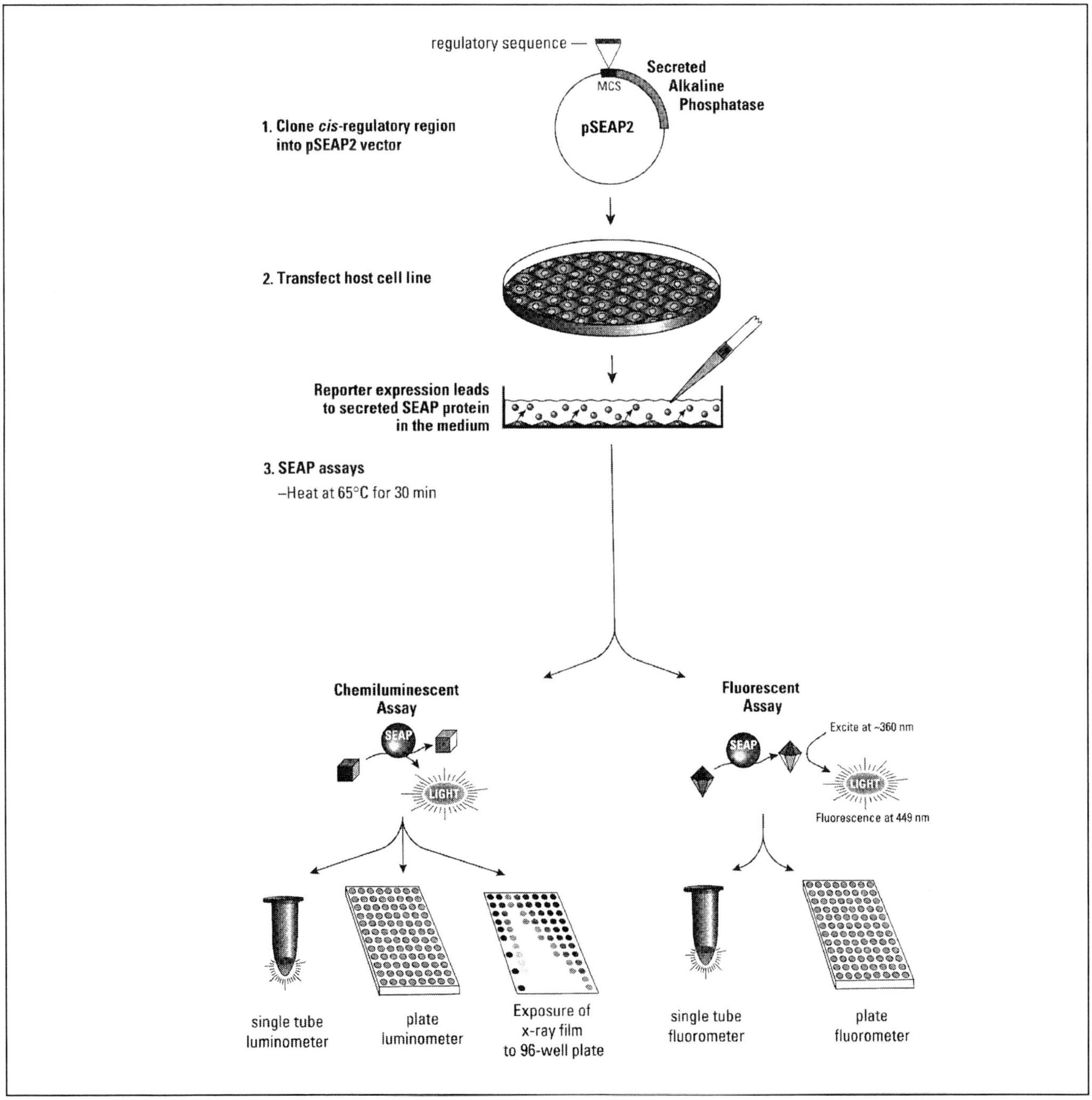

Figure 2. Flow diagram for the SEAP assay in transfected mammalian cells.

were assayed for SEAP activity using chemiluminescence- or fluorescence-based detection at various time points. The results of the chemiluminescence assay (Figure 5A) indicated an approximately linear increase in SEAP activity between 24 and 72 h after transfecting the cells with the pSEAP2-control or pSEAP2-promoter vectors. Negligible activity was observed when cells were transfected with pSEAP2-basic, which lacks a functional promoter. The basal level of activity from the pSEAP2-promoter vector permits the clear observa-

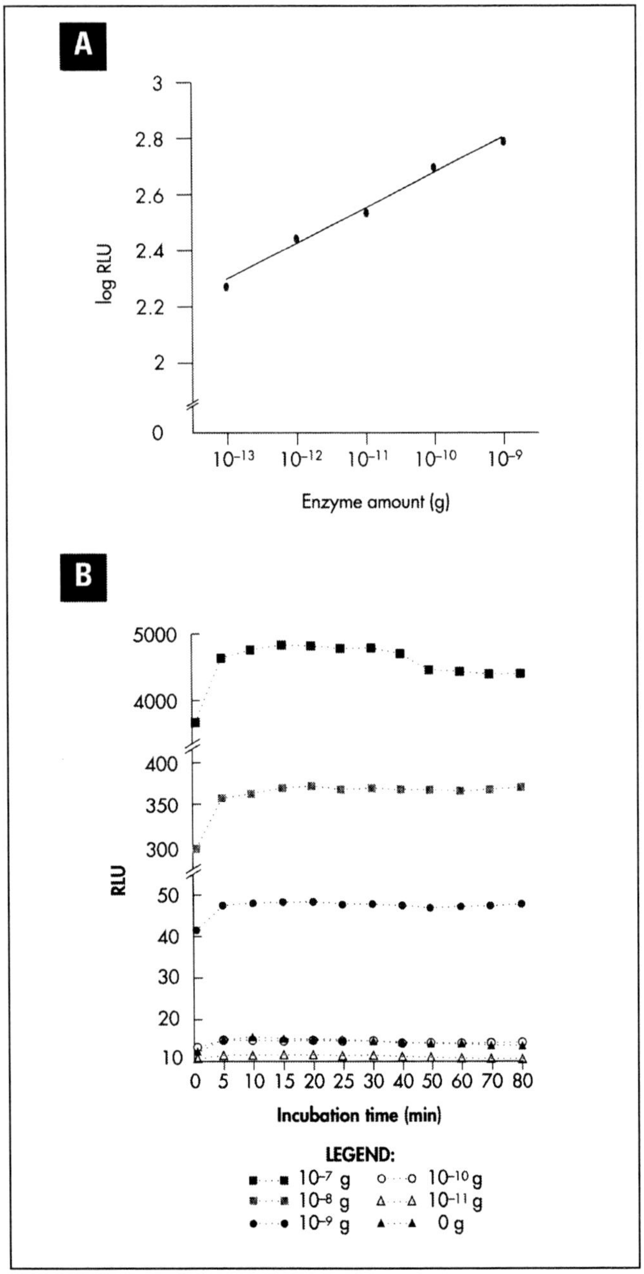

Figure 3. Chemiluminescence detection of human PLAP activity. (A) Sensitivity of chemiluminescence detection of PLAP. Known amounts of purified PLAP were added to conditioned medium from NIH3T3 cell cultures containing 10% calf serum. Samples of the AP/medium mixtures were diluted in buffer and assayed as described in Materials and Methods. (B) Stability of the chemiluminescence signal produced in an AP assay using CSPD. The indicated amounts of purified PLAP were added to conditioned medium from NIH3T3 cell cultures containing 10% calf serum and assayed as described in Materials and Methods.

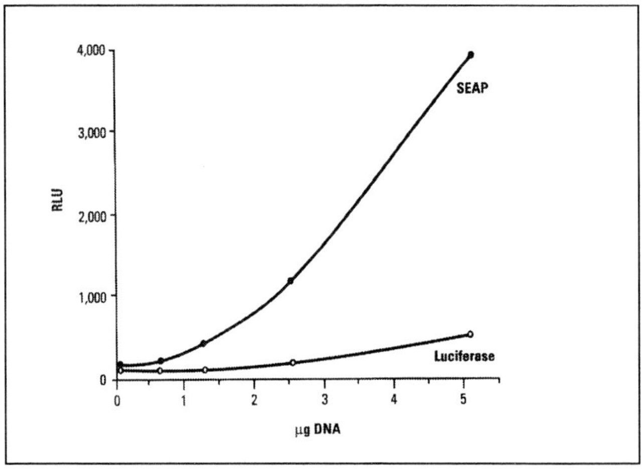

Figure 4. Comparison of SEAP and LUC sensitivities. Parallel culture of 1.5×10^6 BHK cells were plated in 35-mm dishes, grown overnight and transiently transfected with the indicated amounts of either pSEAP2-control vector or a similar vector containing the *luc* gene. After 24 h, SEAP activity was assayed in the media of the SEAP-transfected cultures using the chemiluminescence detection. Similarly, after 24 h, cell lysates were prepared from the LUC cultures, and LUC activity was assayed using a commercial kit. RLU = relative light units.

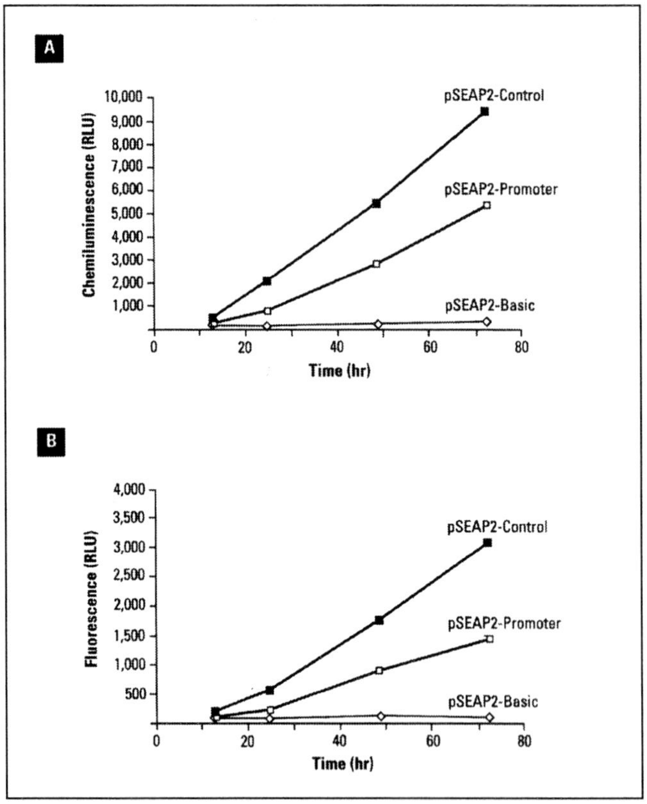

Figure 5. Time course of SEAP expression in mammalian cells transfected with pSEAP2-basic, pSEAP2-promoter and pSEAP2-control vectors. Monolayer cultures of CHO-K1 cells were grown in 35-mm plates in medium containing 10% calf serum. Duplicate cultures were transfected with 3.0 µg of the indicated SEAP vector using the CalPhos Maximizer transfection procedure. Samples of the conditioned media were collected from duplicate plates at 16, 24, 48 and 72 h posttransfection. (A) SEAP activities were assayed by chemiluminescence detection as described in Materials and Methods. Signals were measured on a luminometer. Background values obtained with conditioned medium from untransfected cells were subtracted from all values. (B) SEAP activities were assayed by fluorescence detection as described in Materials and Methods. Signals were measured using the CytoFluor II Fluorometer.

tion of further increases in SEAP expression brought about by cloning enhancer sequences into this vector upstream of the SV40 promoter. Similar results were obtained when SEAP activities were detected by the fluorescence assay using MUP (Figure 5B).

REFERENCES

1. **Alam, J. and J.L. Cook.** 1990. Reporter genes: application to the study of mammalian gene transcription. Anal. Biochem. *188*:245-254.
2. **Berger, J., J. Hauber, R. Hauber, R. Geiger and B.R. Cullen.** 1988. Secreted placental alkaline phosphatase: a powerful new quantitative indicator of gene expression in eukaryotic cells. Gene *66*:1-10.
3. **Berger, J., A.D. Howard, L. Gerber, B.R. Cullen and S. Udenfriend.** 1987. Expression of active, membrane-bound human placental alkaline phosphatase by transfected simian cells. Proc. Natl. Acad. Sci. USA *84*:4885-4889.
4. **Bronstein, I., J. Fortin, P.E. Stanley, G.S.A.B. Stewart and L.J. Kricka.** 1994. Chemiluminescent and bioluminescent reporter gene assays. Anal. Biochem. *219*:169-181.
5. **Cullen, B.R. and M.H. Halim.** 1992. Secreted placental alkaline phosphatase as a eukaryotic receptor gene. Methods Enzymol. *216*:362-368.
6. **de Wet, J.R., K.V. Wood, M. DeLuca, D.R. Helinske and S. Subramani.** 1987. Firefly luciferase gene: structure and expression in mammalian cells. Mol. Cell. Biol. *7*:725-737.
7. **Eggermont, J. and N. Proudfoot.** 1993. Poly(A) signals and transcriptional pause sites combine to prevent interference between RNA polymerase II promoters. EMBO J. *12*:2539-2548.
8. **Gorman, C.M., L.F. Moffat and B.H. Howard.** 1982. Recombinant genomes which express chloramphenicol acetyl transferase in mammalian cells. Mol. Cell. Biol. *2*:1044-1051.
9. **Gould, S.J. and S. Subramani.** 1988. Firefly luciferase as a tool in molecular and cell biology. Anal. Biochem. *175*:5-13.
10. **Kain, S.R. and S. Ganguly.** 1995. Overview of genetic reporter systems. *In* F.M. Ausubel, R. Brent, R.E. Kingston, D.D. Moore, J.A. Smith, J.G. Seidman and K. Struhl (Eds.), Current Protocols in Molecular Biology. John Wiley & Sons, New York.
11. **Kozak, M.** 1987. An analysis of 5′-noncoding sequences from 699 vertebrate messenger RNAs. Nucleic Acids Res. *15*:8125-8148.
12. **Rose, M., M.J. Casadaban and D. Botstein.** 1981. Yeast genes fused to β-galactosidase in *E. coli* can be expressed normally in yeast. Proc. Natl. Acad. Sci. USA *78*:2460-2464.
13. **Selden, R.F., K. Burke-Howie, M.E. Rowe, H.M. Goodman and D.D. Moore.** 1986. Human growth hormone as a reporter gene in regulation studies employing transient gene expression. Mol. Cell. Biol. *6*:3173-3179.
14. **Simon, J.A., C.A. Sutton, R.B. Lobell, R.L. Glaser and J.T. Lis.** 1985. Determinants of heat shock-induced chromosome puffing. Cell *40*:805-817.
15. **Tate, S.S., R. Urade, R. Micanovic, L. Gerber and S. Udenfriend.** 1990. Secreted alkaline phosphatase: an internal standard for expression of injected mRNA in the Xenopus oocyte. FASEB J. *4*:227-231.

CHAPTER 25

Application of the Green Fluorescent Protein as a Reporter for Ace1-Based, Two-Hybrid Studies

Gerd Mayer, Heike Launhardt and Thomas Munder

Hans-Knöll-Institut für Naturstoff-Forschung e.V., Jena, Germany

(Reprinted from BioTechniques 27:86-94, 1999)

ABSTRACT

The two-hybrid system in Saccharomyces cerevisiae *is a genetic approach for the detection of protein-protein interactions in vivo. This technology relies on the activity of separated DNA-binding and transactivation domains of specific transcription factors to reconstitute an active transcription factor complex if interacting proteins are fused to these domains. Interactions are consequently detected through the activity of reporter genes. The two-hybrid technology has been successfully applied for the determination of interactions between numerous proteins of several organisms. Conventional reporter systems, such as the β-galactosidase from* Escherichia coli, *suffer from a variety of drawbacks, including the requirement for external substrates. In this report, we describe an alternative version of the two-hybrid system using the combined advantages of the copper-inducible transcription factor Ace1 together with the yeast metallothionein gene* CUP1 *and the green fluorescence protein from aquatic invertebrates as reporters. This technique allows the copper-dependent monitoring of protein-protein interactions in living yeast cells.*

INTRODUCTION

Since its introduction in 1989, the two-hybrid system in *Saccharomyces cerevisiae* has become a widely distributed technology for the detection of protein-protein interactions in vivo (7). Recently, we developed a two-hybrid approach that is based on the activity of the yeast transcription factor Ace1 (14). Ace1 stimulates transcription of the *CUP1* gene through binding to regulatory sequences within the *CUP1* promoter (8,16). *Cup1* encodes the yeast metallothionein, which tightly chelates intracellular copper and other metal ions, protecting the cell against copper toxicity (9). The binding of Ace1 to DNA is dependent on the presence of copper ions, because copper can induce a conformational change in the amino terminus of Ace1, allowing the protein to interact with DNA (8).

To monitor protein-protein interactions with the Ace1-based, two-hybrid system, the coding sequences of the respective proteins were fused to those of the Ace1 DNA-binding domain and the transactivation domain. These hybrid proteins were expressed in yeast reporter strains. Positive interactions were detected by the survival of cells on copper-containing media due to the expression of genomic CUP1 or through a copper-dependent expression of the β-galactosidase (β-gal) linked to the CUP1 promoter (14).

For the elimination of false positives that occur frequently in the two-hybrid system, several reporter genes under control of different promoters have been developed (1). Recently, the green fluorescent protein (GFP) from *Aequorea victoria* has been described as a novel reporter for two-hybrid studies (4,13). The GFP makes yeast cells strongly fluorescent without causing toxicity (15). Light-stimulated GFP fluorescence does not require any cofactors, external substrates or additional gene products. Detection of GFP can be performed in living cells instead of fixed samples and without cell lysis.

In this report, we describe an improved Ace1-based, two-hybrid system, which is more efficient than the existing one (14). In contrast to the conventional two-hybrid systems, the reporter gene activity is inducible by copper ions, which is of value for certain applications, e.g., drug screening. Moreover, we used the advantages of the GFP as a readout for monitoring protein-protein interaction. This combination allows a copper-dependent detection of interacting proteins in living yeast cells.

MATERIALS AND METHODS

Yeast Strains, Transformation and Gene Replacement of ACE1

Transformation of yeast cells followed the protocol of Klebe and co-workers (11). Yeast transformants were selected and cultivated onto SD synthetic medium (2% glucose and 0.67% yeast nitrogen base without amino acids) supplemented with essential amino acids and nucleotides. The copper-resistant yeast strain DBY747 (*MAT***a** *ura3-52 his3-171 trp1-289 leu2-3,-112 CUP1*^{r}) was used as a host for the two-hybrid analyses. A gene-replacement cassette was constructed, which contained the ACE1 coding sequence substituted by the TRP1 selection marker. The genomic ACE1 allele was

Cloning and Expression Vectors for Gene Function Analysis
Edited by Quinn Lu and Michael P. Weiner

disrupted in strain DBY747 by homologous recombination at the *Nco*I site within the ACE1 promoter and the *Nhe*I site within the ACE1 terminator regions present on the disruption cassette. Colonies were selected on SD medium lacking tryptophan. This gave rise to strain TFY118-1 (*MAT***a** *ura3-52 his3-171 trp1-289 ace1::TRP1 leu2-3,-112 CUP1*[r]). Correct gene-replacement events were verified by loss of copper resistance and by polymerase chain reaction (PCR) with primers flanking the insertion sites. Amplified sequences were analyzed by agarose gel electrophoresis. The plasmid pTY99, carrying a CUP1 promoter-GFP reporter (cycle 3 mutant gene) (5) construct, was linearized with *Bst*EII and integrated into the *leu2* locus of TFY118-1. The transformants were selected on medium that lacked leucine. This strain was named ITH5 (*MAT***a** *ura3-52 his3-171 trp1-289 ace1::TRP1 leu2-3,-112 LEU2::pTY99 CUP1*[r]).

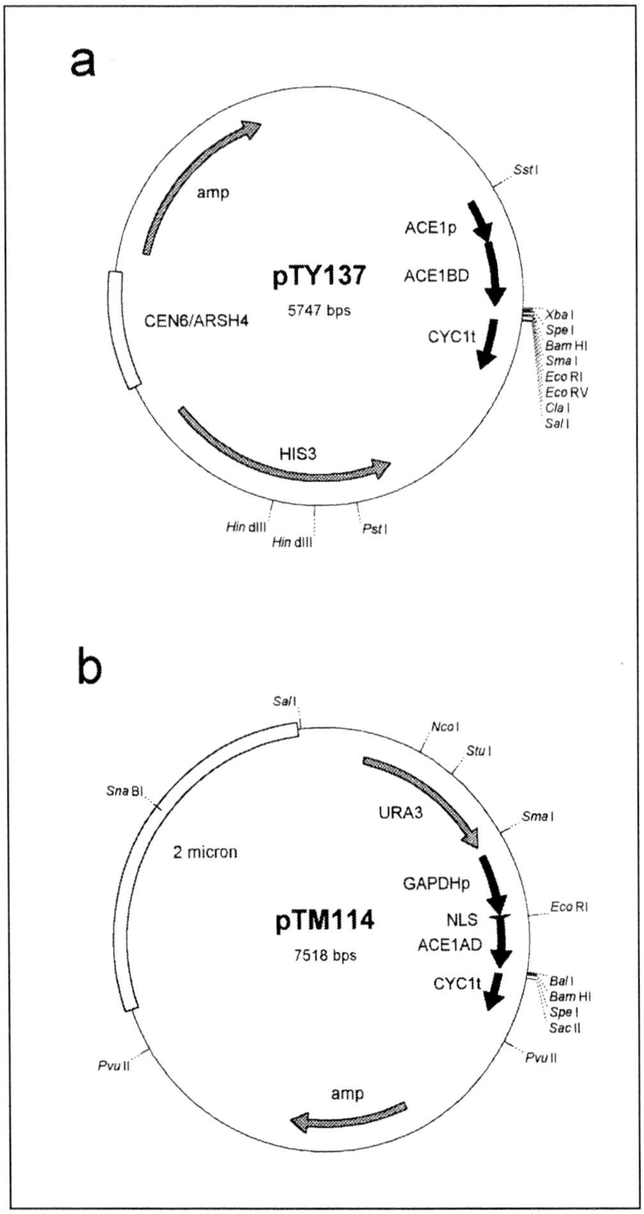

Figure 1. Key vectors for the two-hybrid system carrying the coding sequences of the (a) Ace1BD and (b) Ace1AD. GAPDHp, promoter of the glyceraldehyde-3-phosphate dehydrogenase gene; NLS, nuclear localization sequence of the SV40 large T antigen; CYC1t, terminator sequences of the cytochrome c_1 oxidase gene; and ACE1p, promoter of the *ACE1* gene. Detailed description of the expression plasmid constructions are available from the corresponding author.

GFP Assay

The derivatives of the GFP reporter strain ITH5 carrying the two-hybrid plasmids were grown overnight at 30°C in SD medium. Cells were diluted with SD medium to a cell density of $OD_{600nm} = 0.5$, and 100 µL of the suspension were transferred to opaque 96-well plates together with various $CuSO_4$ concentrations. The microplates were covered by sticking strips and a lid and incubated for several hours on a shaking platform at 30°C. The on-line registration of GFP activity was monitored upon excitation at 405 nm, and the whole cell fluorescence was measured at 510 nm using the Victor 1420 Multilabel Counter Luminescence Spectrophotometer (EG&G Wallac, Turku, Finland). The copper-induced GFP fluorescence was expressed as arbitrary units regarding the cellular background fluorescence, which was detected before the addition of copper.

RESULTS AND DISCUSSION

Development of a Novel Ace1-Based, Two-Hybrid System and Verification by Metallothionein Expression

As an alternative to conventional two-hybrid systems, we have developed a two-hybrid approach based on the activity of the yeast transcriptional activator Ace1 (14). This system suffers from inconveniences due to the necessity of a genomic integration of the Ace1 DNA-binding domain (Ace1BD) plasmid. We now show that this can be avoided using the Ace1-based, two-hybrid vectors pTY137 (carrying Ace1BD) and pTM114 [carrying the Ace1 activation domain (Ace1AD)], which contain their own origins of replication (see Figure 1). Both plasmids can be transformed simultaneously into yeast cells. To enhance the probability of nuclear

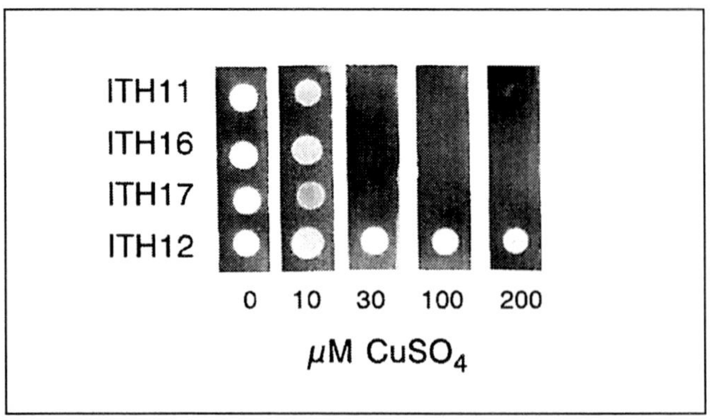

Figure 2. Induction of metallothionein expression through homodimerization of Bcy1 in the Ace1-based, two-hybrid system. Equal numbers of cells from yeast strains ITH11 (expressing Ace1BD and Ace1AD), ITH16 (expressing Ace1BD-Bcy1 and Ace1AD), ITH17 (expressing Ace1BD and Ace1AD-Bcy1) and ITH12 (expressing Ace1BD-Bcy1 and Ace1AD-Bcy1) were spotted onto SD plates containing the indicated $CuSO_4$ concentrations and incubated for 3 days at 30°C.

import of the prey hybrid protein, the Ace1AD was fused to the nuclear localization sequence of the simian virus 40 (SV40) large T antigen (10).

The system was tested with the homodimerization of the yeast regulatory subunit of the cyclic AMP-dependent protein kinase encoded by the *BCY1* gene (3,17). The coding sequences for the amino terminal localized dimerization part of Bcy1 (17) were fused to the sequences of Ace1BD and Ace1AD using plasmids pTY137 and pTM114 (Figure 1). Plasmids were co-transformed in the yeast strain ITH5, which carries several tandemly amplified copies of the *CUP1* gene. The strain ITH12, expressing both Ace1BD-Bcy1 and Ace1AD-Bcy1 fusion proteins, was resistant to 200 µM copper, whereas the strains ITH11 (expressing Ace1BD and Ace1AD alone), ITH16 (containing Ace1BD-Bcy1 and Ace1 AD) and ITH17 (containing Ace1BD and Ace1AD-Bcy1) did not grow at concentrations of more than 10 µM copper (Figure 2). Thus, the dimerization of Bcy1 reconstituted a functional Ace1 protein, which then induced the transcription of CUP1, detected by an enhanced copper resistance of yeast cells. This was in accordance with our previously published data (14). The Ace1 two-hybrid system described in this report did not require a genomic integration of the bait vector and therefore enabled a co-transformation of both plasmids.

We propose that our system will also be suitable for a screening with activation domain tagged libraries to identify proteins interacting with a given bait. The transformants carrying Ace1BD-bait and Ace1AD-prey plasmids, or optionally, the diploid cells obtained by the mating-selection method of a two-hybrid screening (2), can be spread directly onto agar plates containing $CuSO_4$. Only those cells expressing interacting proteins can endure the toxicity of copper ions, detected by the growth monitoring on copper-containing plates. To test this, we transformed ITH5 with the Ace1-Bcy1 containing plasmids and spread the cells onto plates supplemented with 30–100 µM copper. Only those cells expressing both fusion proteins grew on copper medium (not shown). Because interacting clones are identified by a growth selection, this system is highly sensitive as stated by Le Douarin et al. (12). Modulating the stringency of the CUP1 selection by different concentrations of copper might allow the discrimination between strong or weak interactions in a screening procedure at an early step.

The GFP Is a Reporter for Ace1-Based, Two-Hybrid Analyses

The detection of the enzymatic activity of classical reporters for quantitative two-hybrid studies, such as the β-gal from *Escherichia coli*, usually requires cell lysis and the addi-

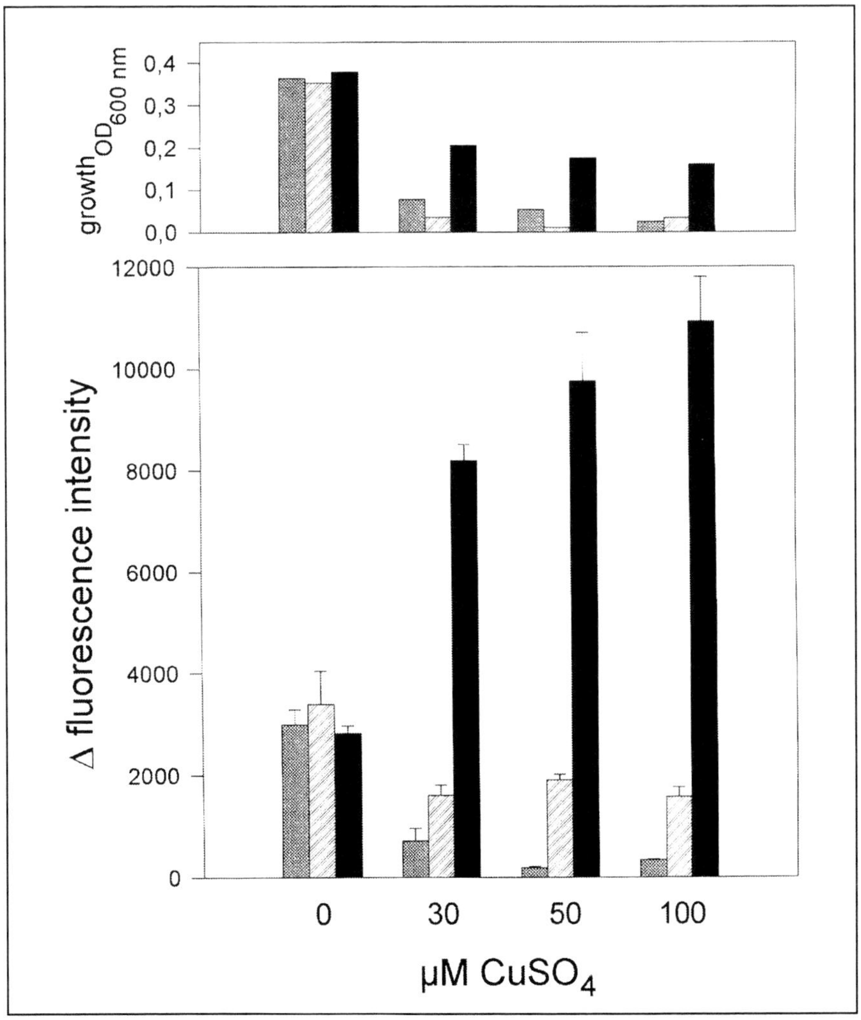

Figure 3. (Lower panel) GFP activity in the yeast strains ITH17 (expressing Ace1BD and Ace1AD-Bcy1, dark grey bars), ITH16 (expressing Ace1BD-Bcy1 and Ace1AD, bright gray bars) and ITH12 (expressing Ace1BD-Bcy1 and Ace1AD-Bcy1, black bars). Cells were incubated with the indicated $CuSO_4$ concentrations in microplates. The GFP activity was determined as the difference of yeast background fluorescence and the fluorescence obtained after a 4-h incubation with copper. The means and standard deviations (error bars) from at least 6 experiments are shown. (Upper panel) Growth increase of yeast strains ITH17, ITH16 and ITH12 during a 4-h incubation in microplates. Following the fluorescence detection, cells of each strain were collected from the wells and transferred into micro-square cuvettes. The cell density was measured photometrically at 600 nm. Concentration at t = 0 was OD_{600nm} = 0.5.

tion of substrates. Furthermore, reporter activities can only be monitored at a given endpoint. In this report, we describe the application of a copper-inducible GFP expression as an alternative reporter system allowing the measurement of protein-protein interaction directly in living yeast cells.

For our studies, we used the strain ITH5 containing a stable genomic integrated copy of a CUP1 promoter-GFP reporter fusion. The copper-regulated functional GFP expression was validated by overexpression of the entire ACE1 coding sequence, as present on a 2-µm-derived vector (not shown).

The application of the copper-dependent GFP expression in two-hybrid analyses, using the homodimerization of Bcy1 as an example, was tested with the plasmids described above. Plasmids containing the Ace1BD-Bcy1 and Ace1AD-Bcy1 hybrids were co-transformed together or with plasmids carrying the Ace1BD and the Ace1AD alone into ITH5. Figure 3 shows that copper induced an enhanced GFP signal up to four-fold in strain ITH12 expressing both Ace1-Bcy1 proteins compared to the noninduced culture. In contrast, the strains ITH16 and ITH17 expressing either the Ace1BD-Bcy1 hybrid or the Ace1AD-Bcy1 hybrid alone showed a reduced fluorescence after addition of copper as compared to the untreated cells. These results reflect the relationship between the GFP expression and the survival of yeast cells under copper stress. Due to the lack of CUP1 expression, the cells containing non-interacting Ace1 fusion proteins were sensitive to copper treatment, implying that these cells entered a cell cycle arrest, which also contributed to a reduced GFP signal. Cells expressing interacting Ace1 fusion proteins produced metallothionein and are therefore growth-resistant to copper. Figure 3 shows that the cell growth was more reduced in copper-treated cells of strain ITH12 compared to untreated cells. Therefore, the strongly enhanced GFP signal in these cells was not dependent on a growth increase but on the copper-dependent stimulation of GFP expression.

The combination of CUP1 and GFP expression should also allow for distinguishing strong from weak interactors. As reported earlier, it is inappropriate to use the quantitative readout of only one reporter to determine the strength of interaction (6). In our system, the growth of yeast strains expressing low-affinity interacting proteins on copper will be much more depressed than that of cells containing strongly interacting pairs. Furthermore, the GFP signal in these cells is greatly reduced compared to strains expressing proteins with a high-binding affinity.

Our data demonstrate that the copper-dependent GFP reporter expression allowed the measurement of Bcy1 dimerization using the Ace1-based, two-hybrid system. The detection of GFP fluorescence does not require lysis of yeast cells. This allows the monitoring of reporter gene activity as a consequence of time. To prove this, ITH12 cells were treated with 100 µM $CuSO_4$, and fluorescence was measured after different time points. The GFP signal in these cells increased with the incubation time, whereas the untreated cells produced only a slightly enhanced GFP fluorescence (not shown).

Taken together, our data have shown that the GFP and the endogenous CUP1 reporter in combination with the copper-inducible Ace1 transcription factor, allowed an efficient detection of protein-protein interaction in living yeast cells. Our system is a simple and fast approach avoiding cell lysis and the addition of substrates for detection of enzymatic activities. The commercially available displayGREEN™ Two-Hybrid System (Display Systems Biotechnology, Vista, CA, USA) uses the GFP reporter in combination with lexA and the acidic activator peptide B42. In this system, the bait, prey and reporter genes are located on three different 2- µm-derived plasmids. To avoid recombination events, our system uses vectors from different origins of replication (see Figure 1).

Finally, we would like to stress that two-hybrid systems can be setup to screen for compounds interfering in the interaction between therapeutic relevant proteins, which has been demonstrated recently (18). We propose that the described copper-inducible, Ace1-based, two-hybrid system is a very useful approach for the development of high-throughput screening assays. The expression of the reporter gene in our system is solely initiated by the addition of copper. Therefore, the incubation of cells with an inhibitory drug followed by the addition of copper should result in a clearly reduced GFP expression. Conversely, the reporter gene in non-inducible systems is already expressed during the whole fermentation process of yeast cells before the addition of inhibiting compounds. Due to the already existing level of the reporter proteins, the inhibitory effects of such compounds could hardly be detected. Work is in progress to examine the qualification of the Ace1 system for drug screening purposes.

ACKNOWLEDGMENTS

We are grateful to Martin Funk for the gift of plasmid p413GPD. We thank Alan Riboldi-Tunnicliffe and Stefan Wölfl for critically reading the manuscript. This work was supported by a grant from the Thüringer Ministerium für Wissenschaft, Forschung und Kultur (Contract Grant No. B301-95039). The components of the Ace1-based, two-hybrid system are available upon request to T.M.

REFERENCES

1. **Bartel, P., C.-T. Chien, R. Sternglanz and S. Fields.** 1993. Elimination of false positives that arise in using the two-hybrid system. Bio/Technology *14*:920-924.
2. **Bendixen, C., S. Gangloff and R. Rothstein.** 1994. A yeast mating-selection scheme for detection of protein-protein interactions. Nucleic Acids Res. *22*:1778-1779.
3. **Cannon, J.F. and K. Tatchell.** 1987. Characterization of *Saccharomyces cerevisiae* genes encoding subunits of cyclic AMP-dependent protein kinase. Mol. Cell Biol. *7*:2653-2663.
4. **Cormack, R.S., K. Hahlbrock and I.E. Somssich.** 1998. Isolation of putative plant transcriptional coactivators using a modified two-hybrid system incorparting a GFP reporter gene. Plant J. *14*:685-692.
5. **Crameri, A., E.A. Whitehorn, E. Tate and W.P.C. Stemmer.** 1996. Improved green fluorescent protein by molecular evolution using DNA shuffling. Nature Biotech. *14*:315-319.
6. **Estojak, J., R. Brent and E. Golemis.** 1995. Correlation of two-hybrid affinity data with in vitro measurements. Mol. Cell Biol. *15*:5820-5829.
7. **Fields, S. and O.-K. Song.** 1989. A novel genetic system to detect protein/protein interactions. Nature *340*:245-246.
8. **Fürst, P., S. Hu, R. Hackett and D. Hamer.** 1988. Copper activates metallothionein gene transcription by altering the conformation of a specific DNA binding protein. Cell *55*:705-717.
9. **Hamer, D.H.** 1986. Metallothionein. Annu. Rev. Biochem. *55*:913-951.
10. **Kalderon, D., B.L. Roberts, W.D. Richardson and A.E. Smith.** 1984. A short amino acid sequence able to specify nuclear location. Cell *39*:499-509.
11. **Klebe, R.J., J.V. Harriss, Z.D. Sharp and M.G. Douglas.** 1983. A gener-

al method for polyethylene-glycol-induced genetic transformation of bacteria and yeast. Gene *25*:333-341.

12. **Le Douarin, B., B. Pierrat, E. vom Baur, P. Chambon and R. Losson.** 1995. A new version of the two-hybrid assay for detection of protein-protein interactions. Nucleic Acids Res. *23*:876-878.
13. **Mancini, A., R. Niedenthal, H. Joos, A. Koch, S. Trouliaris, H. Niemann and T. Tamura.** 1997. Identification of a second Grb2 binding site in the v-Fms tyrosine kinase. Oncogene *15*:1565-1572.
14. **Munder, T. and P. Fürst.** 1992. The *Saccharomyces cerevisiae* Cdc25 gene product binds specifically to catalytically inactive Ras proteins *in vivo*. Mol. Cell Biol. *12*:2091-2099.
15. **Niedenthal, R.K., L. Riles, M. Johnston and J.H. Hegemann.** 1996. Green fluorescent protein as a marker for gene expression and subcellular localization in budding yeast. Yeast *12*:773-786.
16. **Thiele, D.J.** 1988. ACE1 regulates expression of the *Saccharomyces cerevisiae* metallothionein gene. Mol. Cell Biol. *8*:2745-2752.
17. **Toda, T., S. Cameron, P. Sass, M. Zoller, J.D. Scott, B. McMullen, M. Hurwitz, E. Krebs and M. Wigler.** 1987. Cloning and characterization of BCY1, a locus encoding a regulatory subunit of the cyclic AMP-dependent protein kinase in *Saccharomyces cerevisiae*. Mol. Cell Biol. *7*:1371-1377.
18. **Young, K., S. Lin, L. Sun, E. Lee, M. Modi, S. Hellings, M. Husbands, B. Ozenberger and R. Franco.** 1998. Identification of a calcium channel modulator using a high throughput yeast two-hybrid system. Nature Biotechnol. *16*:946-950.

Update to: Application of the Green Fluorescent Protein as a Reporter for Ace1-Based, Two-Hybrid Studies

Gerd Mayer, Heike Launhardt and Thomas Munder

Hans-Knöll-Institut für Naturstoff-Forschung e.V., Jena, Germany

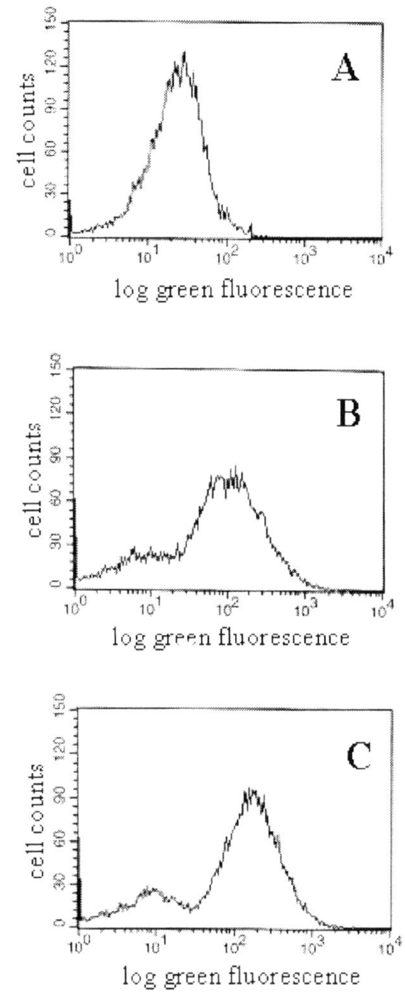

Figure 1. Flow cytometric analysis of yeast strain ITH12 grown under different induction conditions. Cells were incubated overnight on an orbital shaker at 30°C in the absence (A) or presence of 10 µM CuSO4 (B) or 30 µM $CuSO_4$(C). Cells were washed once with phosphate-buffered saline (PBS) and resuspended in the same buffer. The GFP fluorescence was detected by a FACStar^{PLUS} cytometer. Excitation wavelength was 488 nm, and fluorescence was monitored at 530 ± 15 nm. Typical FACS profiles of the cells are shown.

Detection of Protein-Protein Interaction with the Ace1-Based, Two-Hybrid System by Flow Cytometry

The fluorescence of yeast cells expressing the green fluorescent protein (GFP) was also monitored by flow cytometry using a FACStar$^{PLUS TM}$ cytometer (Becton-Dickinson, Heidelberg, Germany). Cells of yeast strain ITH12, expressing the Ace1BD-Bcy1 and Ace1AD-Bcy1 fusion proteins, were grown overnight under inducing (with copper) and noninducing (without copper) conditions. Figure 1A shows the autofluorescence of the cells in the absence of copper. Addition of $CuSO_4$ strongly enhanced their fluorescence intensity, even at low copper concentrations (Figure 1, B and C). This indicated that the homodimerization of Bcy1 reconstituted an active Ace1 protein, which can be detected by a copper-dependent GFP expression measured by flow cytometry.

The fluorescence-activated cell sorting (FACS) profiles show that the fluorescence of the cellular population, which did not express GFP in presence of copper, shifted to a lower intensity. This probably points to the fact that copper has a deleterious effect to those cells not expressing the endogenous *CUP1* gene. This may reflect the slow growth phenotype of ITH12 cells in liquid cultures supplemented with $CuSO_4$ (not shown).

These data demonstrate that flow cytometry is a useful tool for the detection of protein-protein interaction with the Ace1-based, two-hybrid system. This is of particular importance for screening experiments with Ace1-AD-tagged cDNA molecules. We postulate that yeast cells expressing interacting proteins of certain baits may be sorted and differentiated from those cells containing noninteracting protein pairs.

The Ace1-based, two-hybrid system is now available from MobiTec (Göttingen, Germany).

CHAPTER 26

Use of GFP as a Reporter for the Facile Analysis of Sequence-Specific Proteases

J.C. Pehrson, A. Weatherman, J. Markwell, G. Sarath and S.D. Schwartzbach

University of Nebraska-Lincoln, Lincoln, NE, USA

(Reprinted from BioTechniques 27:28-32, 1999)

A large number of sequence- and structure-specific peptidases are recognized to play important roles in cell biology, although they are relatively difficult to characterize (1,3). A major factor limiting the characterization of sequence- and structure-specific peptidases is the time-consuming nature of their assay. The requirement for specific amino acid sequences or peptide structures generally precludes the use of the small chromogenic peptide analog substrates, which have proven so useful with catabolic peptidases. The most common assay for sequence- and structure-specific proteases requires in vitro synthesis of radiolabeled protein substrates, incubation of the substrate with an enzyme, separating the reaction products by sodium dodecyl sulfate polyacrylamide gel electrophoresis (SDS-PAGE) and quantitating the amount of proteolytic product with a phosphor imager or by densitometry of fluorographs. The time and labor-intensive nature of this assay represents a major impediment to the purification and characterization of these interesting enzymes.

Development of a rapid and convenient assay with general applicability to structure- and sequence-specific proteases will greatly assist in the purification and analyses of these important enzymes. Figure 1 portrays a simple and rapid fluorescent assay that can be adapted for the assay of most site- and sequence-specific proteases. The basis of this assay is a protease substrate composed of an N-terminal His tag and a polypeptide containing the peptidase cleavage site fused to green fluorescent protein (GFP). Proteolysis releases GFP from the fusion protein, and batch metal-chelate affinity chromatography is used to separate the fluorescent uncleaved fusion protein (pellet) from GFP liberated by proteolysis (Figure 1). The amount of liberated GFP in the supernatant is then measured in a spectrofluorometer.

A plasmid, pETEGFP, encoding a protein containing an N-terminal His tag and a 23-amino acid peptide containing a thrombin cleavage site and EGFP (CLONTECH Laboratories, Palo Alto, CA, USA), was constructed by excising the EGFP coding region from the pEGFP vector as a *Bam*HI/*Not*I fragment and cloning it into the *Bam*HI/*Not*I site of the bacterial expression vector pET28b (Novagen, Madison, WI, USA). The fluorescent thrombin substrate was produced by growing *Escherichia coli* BL-21 (DE-3) cells transformed with the pETEGFP plasmid in 3 L of LB media containing 50 µg/mL kanamycin at 37°C. Protein expression was induced at a culture density of 1.8 A_{600} by addition of 0.4 mM isopropyl-β-D-thiogalactopyranoside (IPTG) for 6 h. Cells were harvested by centrifugation at 6000× *g* for 10 min, resuspended in TALON™-Binding Buffer (CLONTECH) and lysed by sonication. The soluble cell fraction obtained after centrifugation at 32 500× *g* for 15 min was incubated at 70°C for 5 min. Under these conditions, most cellular proteins were denatured with no loss of EGFP fluorescence. Denatured proteins were removed by centrifugation at 32 500× *g* for 15 min, and the EGFP fusion product in the supernatant was purified using metal-chelate chromatography on commercially available TALON Metal Affinity Resin (CLONTECH) as described by the manufacturer. In the case of heat-labile fusion proteins, the soluble cell fraction can be applied directly to the TALON Column (CLONTECH). Omission of the heat treatment has no effect on the subsequent purification of the fusion protein substrates, but this does result in longer column chromatographic times due to increased protein load. After elution from the metal-chelate column, the purified fusion protein substrate was buffer-exchanged into thrombin buffer (20 mM Tris-HCl, 150 mM NaCl, 2.5 mM $CaCl_2$, pH 8.4; Novagen) and adjusted to a fluorescence of approximately 46 000 fluorescence U (Flu)/mL (Ex = 484 nm; Em = 509 nm). Thrombin was assayed by incubation at 37°C in thrombin buffer containing varying or fixed amounts of the purified EGFP fusion protein. Aliquots of 200 µL were removed from the reaction mixtures at specific time points (during the linear phase of the assay) and heat-treated at 70°C for 5 min to stop the reaction by denaturation of thrombin. The heat-treated sample was mixed with 200 µL of 5× binding buffer, 200 µL of a 1:1 aqueous slurry of TALON Beads (CLONTECH) and 400 µL of water and shaken at 200 rpm for 30 min at room temperature. After a brief centrifugation at 13 000× *g* for 30 s to sediment the beads containing unreacted substrate and His-tag fusion peptides, the supernatant was analyzed for released GFP fluores-

Cloning and Expression Vectors for Gene Function Analysis
Edited by Quinn Lu and Michael P. Weiner

cence in a Model LS 50 Luminescence Spectrofluorimeter (Perkin-Elmer, Norwalk, CT, USA). TALON was the most effective resin tested, removing greater than 97% of the fluorescent protein from solution in the absence of thrombin.

Preliminary experiments indicated that the crude bacterial cell-free extract containing the EGFP fusion protein substrate contained inhibitors of thrombin. Purification of the substrate by metal-chelate chromatography removed the inhibitory materials producing a substrate suitable for protease assays. At saturating substrate concentrations, the rate of thrombin-dependent release of EGFP from purified EGFP fusion protein substrate was linear for at least 60-min over a thrombin concentration range of 0.57–4.6 U/μL reaction (Figure 2A). At the highest thrombin concentration tested, approximately 23% of the EGFP added to the reaction was released from the fusion protein. By sampling at multiple time points over a 60-min period and using a least-squares linear regression analysis to fit a line to the experimental points, differences in background levels do not interfere with the accurate measurement of reaction rates. Lines fitted to the experimental points (Figure 2A) by least-squares linear regression analysis had correlation coefficients >0.95, indicative of the accuracy of the rate determinations. Over approximately a tenfold concentration difference of thrombin, the reaction rates, as determined by linear regression analysis, were directly proportional to the amount of thrombin added to the assay (Figure 2B).

A useful enzyme assay requires a simple method for measuring the amount of substrate converted to product. Conversion must also be linear with time and proportional to the amount of enzyme added, and sufficient purified substrate must be readily available to perform reactions at saturating substrate concentrations. Using thrombin as a model protease, release of EGFP from an EGFP fusion protein was shown to be proportional to time and the amount of thrombin present in the assay. The assay was sensitive and had a large dynamic range capable of measuring differences of at least an order of magnitude in thrombin concentration. The assay is also rapid, requiring a minimum number of steps, thus making it suitable for assaying the large number of samples generated during enzyme purification.

The major limitation to the general use of this assay to measure sequence-specific proteases will be the ability to produce large quantities of a soluble His-tagged EGFP fusion protein having the necessary protease recognition sequence. EGFP has been fused to a large number of proteins producing soluble biologically active fluorescent proteins (5). Conditions for soluble expression of foreign proteins in bacterial cells is becoming better understood. The ability to easily engineer EGFP fusions with exotic protease recognition sequences and inexpensively express them as soluble proteins in bacterial cells makes this assay ideally suited for the characterization of a large variety of sequence- and domain-specific peptidases.

An assay for site-specific proteases has been published based on measuring fluorescent resonance energy transfer (FRET) between two GFP variants linked by the protease target sequence (2). Cleavage of the substrate separates the two GFPs, thus preventing resonance energy transfer. Although very elegant, FRET is a very complex function of the distance between the chromophores and the angle formed by their dipole moments (4), requires specialized instrumentation and is limited to assaying one sample at a time. Measuring the direct

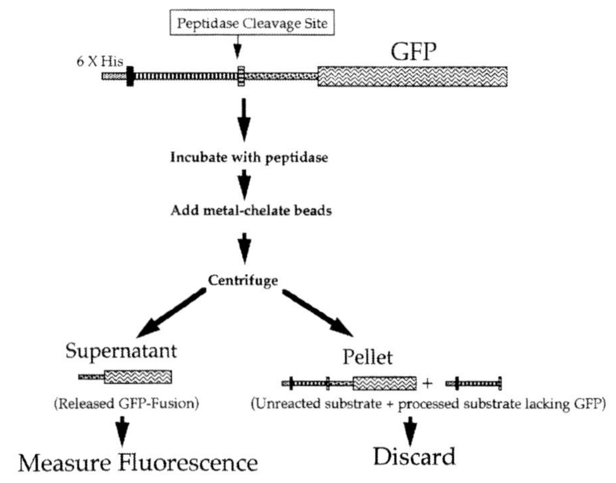

Figure 1. Schematic representation of the fluorescent assay for site-specific proteases.

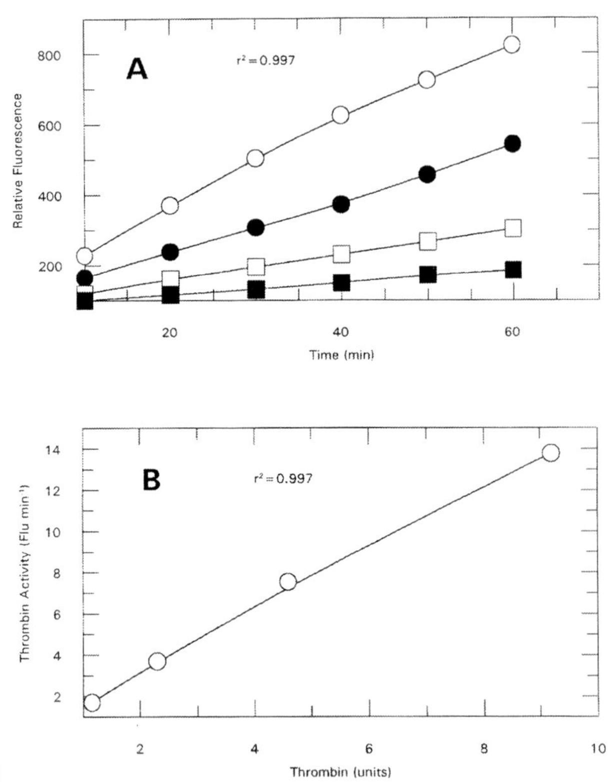

Figure 2. Validation of the fluorescent assay using thrombin. All assays were performed as indicated in the text. (A) Supernatant relative fluorescence as a function of time at 4 different concentrations of added thrombin (■ = 0.57 U; □ = 1.15 U; ● = 2.30 U and ○ = 4.60 U). (B) Supernatant relative fluorescence as a function of added thrombin.

release of EGFP from a fusion protein as described in this paper is simpler, a more suitable method for the rapid analysis of many samples for enzyme kinetics and provides a good complementation to other fluorescence-based proteolytic assays.

ACKNOWLEDGMENTS

This work was supported in part by a grant from the Center for Biotechnology, University of Nebraska-Lincoln funded through the Nebraska Research Initiative (G.S., J.M. and S.D.S.), The National Science Foundation Grant No. MCB-9630817 (S.D.S.) and by the Howard Hughes Medical Institute for Undergraduate Research Support. This paper is No. 12481 in a journal series from the Nebraska Agricultural Research Division.

REFERENCES

1. **Dalbey, R.E. and G. von Heijne.** 1992. Signal peptidase in prokaryotes and eukaryotes—a new protease family. Trends Biochem. Sci. *17*:474-478.
2. **Mitra, R.D., C.M. Silva and D.C. Youvan.** 1996. Fluorescence resonance energy transfer between blue-emitting and red-shifted excitation derivatives of the green fluorescent protein. Gene *173*:13-17.
3. **Smeekens, S.P.** 1993. Processing of protein precursors by a novel family of subtilisin-related mammalian endoproteases. Bio/Technology *11*:182-186.
4. **Stryer, L.** 1978. Fluorescence energy transfer as a spectroscopic ruler. Annu. Rev. Biochem. *47*:819-846.
5. **Tsien, R.Y.** 1998. The green fluorescent protein. Annu. Rev. Biochem. *67*:509-544.

Update to: Use of GFP as a Reporter for the Facile Analysis of Sequence-Specific Proteases

G. Sarath, J. Markwell and S.D. Schwartzbach

University of Nebraska-Lincoln, Lincoln, NE, USA

The fluorometric assay described in this paper has been successfully used to measure the activity of the polyprotein processing protease found in *Euglena* chloroplasts (2,4). Some *Euglena* chloroplast proteins, including the small subunit of ribulose-bis-phosphate carboxylase (SSU), are synthesized as polyprotein precursors containing 8 mature proteins covalently joined by a decapeptide, A**MFA**XX**G**X**K**X (1,3) containing 4 invariant (bold) and 2 conserved positions (5).

Cleavage at both the N and C termini of the decapeptide by the polyprotein processing protease is required for release of mature proteins during precursor maturation within the chloroplast. Using a His-tagged-SSU-decapeptide-green fluorescent protein (GFP) as substrate, the activity of the polyprotein processing protease could be measured in crude membrane-free chloroplast extracts (4). The activity of the polyprotein processing protease with a decapeptide-containing substrate was linear for 18 min and 3-fold higher than the activity with an identical substrate lacking the decapeptide (4). Immunoblots of the protease reaction products demonstrated that GFP was released from the decapeptide-containing substrate confirming the reaction specificity (4). The product of the reaction with the substrate lacking the decapeptide was similar in size to the uncleaved substrate, suggesting that it is cleaved within the N-terminal region at a site other than the polyprotein processing peptidase cleavage site. Cleavage is most likely due to a protease present in the crude chloroplast lysate. Polyprotein processing protease activity, with a substrate containing a decapeptide in which the conserved K was mutated to T, was the same as the activity with the substrate lacking the decapeptide (S. Schwartzbach, unpublished). A detailed protocol for adapting this assay to the measurement of any site-specific protease, including a discussion of potential pitfalls and strategies for production of GFP fusion substrates, has appeared (4).

REFERENCES

1. **Chan, R.L., M. Keller, J. Canaday, J.H. Weil and P. Imbault.** 1990. Eight small subunits of *Euglena* ribulose 1-5 bisphosphate carboxylase/oxygenase are translated from a large mRNA as a polyprotein. EMBO J. 9:333-338.
2. **Enomoto, T., C. Sulli and S.D. Schwartzbach.** 1997. A soluble chloroplast protease processes the *Euglena* polyprotein precursor to the light harvesting chlorophyll a/b binding protein of Photosystem II. Plant Cell Physiol. *38*:743-746.
3. **Muchhal, U. and S.D. Schwartzbach.** 1992. Characterization of a *Euglena* gene encoding a polyprotein precursor to the light harvesting chlorophyll a/b binding protein of photosystem II. Plant Mol. Biol. *18*:287-299.
4. **Sarath, G. and S.D. Schwartzbach.** 2000. Use of GFP as a reporter for the analysis of sequence-specific proteases. *In* J.E. Coligan, B.M. Dunn, H.L. Ploegh, D.W. Speicher and P.T. Wingfield (Eds.), Current Protocols in Protein Science. John Wiley & Sons, New York (In press).
5. **Schiff, J.A., S.D. Schwartzbach, T. Osafune and E. Hase.** 1991. Photocontrol and processing of LHCPII apoprotein in *Euglena*: possible role of golgi and other cytoplasmic sites. J. Photochem. Photobiol. B *11*:219-236.

CHAPTER 27

Applications of Green Fluorescent Protein in Plants

Staci M. Leffel, Stephen A. Mabon and C. Neal Stewart, Jr.

University of North Carolina, Greensboro, NC, USA

(Reprinted from BioTechniques 23:912-918, 1997)

ABSTRACT

Green fluorescent protein (GFP) is increasingly being used in plant biology from the cellular level to whole plant level. At the cellular level, GFP is being used as an in vivo reporter to assess frequency of transient and stable transformation. GFP has also proven to be an invaluable tool in monitoring trafficking and subcellular localization of protein. At the organ level and up, many exciting applications are rapidly emerging. The development of brighter GFP mutants with more robust folding properties has enabled better macroscopic visualization of GFP in whole leaves and plants. One interesting example has been the use of GFP to monitor virus movement in and among whole plants. GFP is also emerging as a powerful tool to monitor transgene movement and transgenic plants in the field. In a proof-of-concept study, tobacco was transformed with a modified version of the GFP gene controlled by a constitutive (35S) promoter. GFP expression in progeny plants ranged from 0% to 0.5%, and approximately 0.1% GFP was the minimal amount needed for unambiguous macroscopic detection. GFP is the first truly in vivo reporter system useful in whole plants, and we project its usefulness will increase even further as better forms of GFP genes become available.

INTRODUCTION

Green fluorescent protein (GFP) has emerged as a powerful new tool in plant biotechnology research and applications. The purpose of this review is to briefly chronicle GFP research in plants and to summarize some of the bold directions this relatively new tool is taking plant biotechnology.

GFP was first cloned and sequenced from the cnidarian, *Aequorea victoria*, in the early nineties (32). Since then, it has been expressed in a variety of organisms, ranging from bacteria (19), fungi (41), invertebrates (30), vertebrates (49) and plants (12,34,37,44). Wild-type GFP is a 27-kDa monomer consisting of 238 amino acids that has the unique characteristic of emitting green light when excited with UV (360–400 nm) or blue (440–480 nm) light. GFP is the only well-characterized example of a protein that displays strong, visible fluorescence without any additional substrates or co-factors (14). It maintains this fluorescence even after prolonged incubation in strong denaturing agents such as 6 M guanidine HCl, 8 M urea or 1% sodium dodecyl sulfate (SDS) (4). GFP has a broad range of pH stability, retaining conformation from pH 5.5–12.0, and is extremely thermostable, surviving temperatures up to 65°C (4).

In the jellyfish *A. victoria*, its role is to transduce the blue chemiluminescence of another protein (aequorin) into green light (22,23). The light absorbing agent of the protein, the chromophore, is covalently attached to the protein and is formed by a post-translational modification in which a tripeptide within the protein is cyclized and oxidized (7,13,27,38). This chromophore is protected inside the protein, whose molecular structure has been recently elucidated (27,46). The overall shape of the protein consists of eleven strands of β-sheet forming the shape of a cylinder, capped with α-helices on the top and bottom (46). These α-helices also form a scaffold for the chromophore, which is located near the geometric center of the protein (46). This structure represents a new protein fold nicknamed the β-can (46). Oxygen has a paradoxical relationship with the protein, for it is believed to be necessary for correct protein folding, but its contact with the chromophore will quench fluorescence (46).

GFP IN CELLS

GFP Optimization

Green fluorescent protein retains fluorescence when fused to another protein both on the N- and C-terminal ends, which makes it an attractive fluorescent tag to monitor subcellular activities such as gene expression, protein-protein interactions and protein trafficking and localization (5,10,15, 37). These applications have been improved by the development of chro-

Cloning and Expression Vectors for Gene Function Analysis
Edited by Quinn Lu and Michael P. Weiner

mophore-mutated versions of GFP, in which single amino-acid substitutions have produced forms of the protein with shifted excitation peaks. Wild-type GFP has two excitation points at about 395 and 470 nm, and it emits the most amount of green light at 508–509 nm (5,45). A blue-shifted variant has been produced by substituting histidine (Y66H) or tryptophan (Y66W) for tyrosine at position 66 in the chromophore (14). These mutants, called blue fluorescent proteins (BFPs) (14) were created to enhance GFP as a molecular tool by providing a second visibly distinct color for in vivo visualization of subcellular activities (13). Other mutations made by exchanging amino acid serine 65 to cysteine (S65C) and threonine (S65T) have increased the magnitude of green fluorescence, which has significantly improved the detection levels of the GFP (34). In addition, other modifications, such as F64L and S65T, have improved the folding in EGFP, an enhanced version of the protein (8,47).

Though the wild-type gene has been expressed in both monocot and dicot cells (34), expression has generally been low or unstable in plants (16). Therefore, modified versions of the gene, such as *mgfp4*, were produced when an 84-bp sequence necessary for proper protein folding and expression was found to be mis-spliced in *Arabidopsis* (11). The sequence similarity to known plant introns caused it to be spliced-out during transcription, so Haseloff and colleagues reduced its AU content and reduced the chance of the sequence being excised, thereby restoring proper expression in the plant (12). To help improve visible fluorescence from GFP, Haseloff et al. has also recently transformed *Arabidopsis* with a modified GFP construct (*mgfp5-er*) that targets the protein to the endoplasmic reticulum, resulting in high, in vivo, expression in plants (12). In addition, this construct also has V163A and S175T mutations that enhance folding, stability and hence in vivo fluorescence (39).

Figure 1. Photographs of green GFP transgenic and red non-transgenic tobacco plants under 365-nm UV illumination. Plants were characterized by visual inspection for green fluorescence using a hand-held, long-wave UV spotlight (Model B-100AP 100 W: 365-nm wavelength; UVP, San Gabriel, CA, USA). Each plant was classified as high, medium, low or not fluorescent according to "greenness" of the leaves under UV light compared to control plants. (A) Highly fluorescent (right) and nonfluorescent (left) influorescences and (B) leaves are shown. Plants were illuminated by two UVP spotlights at 1 m distance. Photos were taken using a 35-mm single reflex camera with Fuji 100 ASA color film, at *f* stop 11 and 60-s exposures with a yellow filter on a macro lens.

Transformation Tool

The fact that GFP requires no additional substrates or co-factors to fluoresce and its ability to be expressed in a variety of organisms has made it a valuable universal reporter gene for scientific research (2,5,31). One obvious use for GFP is as a β-glucuronidase (GUS) substitute in transient and stable expression assays. For example, in the first studies using GFP in plant cells, it was shown that wild-type GFP was fluorescent in orange (25) and *Arabidopsis* (16) protoplasts. Modified versions have been utilized to a greater extent in maize (6,29,34) and alfalfa protoplasts (34), and the cells of *Arabidopsis* (11, 12,28,37), corn (28), rice (24,48), soybean (30), spruce and pine (44). It is conceivable that GFP can be used in place of selectable markers, such as those conferring antibiotic or herbicide resistance to recover transformed cells, tissues, organs and ultimately plants. For example, in transformation systems in which in vivo technology is used, such as bombardment of meristems (20,21), it would be much more cost-efficient to use an in vivo marker such as GFP rather than GUS. Chimeric transformants can be identified easily and nondestructively and en masse. Thus, GFP may reduce the cost of transformation and make transformation systems of economically less important crops, model and wild plants more feasible.

Cell Biology

Beyond transformation studies as noted above, GFP has been used in plants mainly as a subcellular targeting tool and as a marker gene (11). GFP has been targeted to the endoplasmic reticulum (3,12) and fused to proteins containing nuclear localization sequences (9) and to mitochondrial tar-

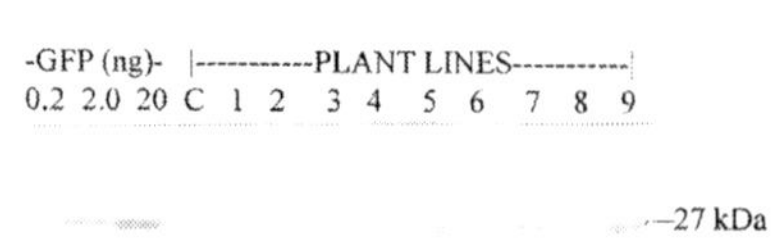

Figure 2. Immunostained protein blot analysis of plants synthesizing GFP. Protein standards are wild-type GFP from CLONTECH Laboratories (Palo Alto, CA, USA). Plant samples identities follow: lane C, non-transgenic control; lanes 1 and 2, nonfluorescing transgenics; lanes 3 and 4, low-fluorescing transgenics; lanes 5 and 6 medium-fluorescing transgenics; lanes 7–9, high-fluorescing transgenics. Sample 8 represents expression of 0.1% *mgfp4*, and 10 represents expression of approximately 0.5%.

geting signals (17) to study intracellular protein targeting. In a cell cycle study, GFP has been fused to phragmoplastin, a protein targeted to the cell plate (10). In this case, the chimeric gene proved valuable to monitor early events in cell plate formation. These studies, especially the last, demonstrate the power of GFP in cell biology inasmuch that cells do not have to be destroyed during assay, and that GFP may potentially be fused to any peptide.

WHOLE PLANT FLUORESCENCE

Disease Movement

While most of the research to date with GFP in plants has been performed at the cellular level, perhaps the most exciting applications will be at the organ level and higher. For all these studies, only modified versions of GFP have been used because of the requisite high expression. Especially exciting is the prospect of using GFP in monitoring and studying pathogen infection of plants. For example, viral infections in plants have been studied using GFP by replacing coat-protein genes from potato virus X (PVX) with the fluorescent protein and monitoring virus-infected tissue with microscopy to study virus movements (2) and by fusing the protein-to-movement proteins to determine how a virus spreads through and among host cells (15,17). This approach can also be used macroscopically. For example, researchers at the Scottish Crop Research Institute have fused coat protein with GFP and a protein of interest to use in a production system of foreign proteins in plants, a GFP overcoat (26,42). Under UV illumination, viral movement can be easily seen in whole plants; hence, bioproduction of the protein of interest can be monitored in real time.

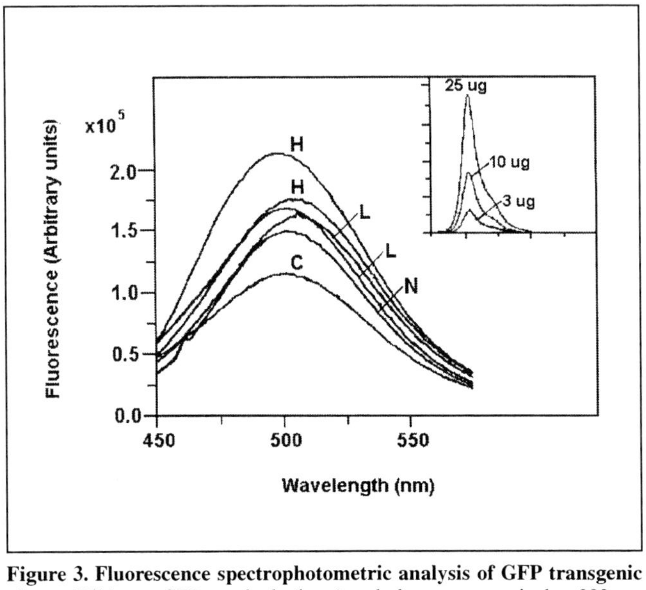

Figure 3. Fluorescence spectrophotometric analysis of GFP transgenic plants. Wild-type GFP standards (inset) and plants were excited at 380 nm with the emission scan measured from 400–575 nm, with excitation peak at 502 nm. C = control non-transgenic plants; N = GFP-transgenic plants that were indistinguishable visually from C; L = low fluorescence-group plant by visual observation; H = high fluorescence-group plant by visual observation.

Ecological Monitoring

One use for GFP in whole plants is to use it as an in vivo marker to monitor transgene spread in the environment. Since some engineered crop species have weedy relatives growing nearby, researchers and environmentalists are concerned that fitness-enhancing transgenes will be introgressed into wild populations (35,40), which could foreseeably alter natural and agricultural ecosystems, a likely situation (1,18,36). One ramification of large-scale releases is the possibility of increased invasiveness and competition of transgenic weeds containing a transgene conferring an increment of fitness. No tracking system is in place to monitor transgene introgression into unintended hosts; however, GFP is the best candidate for this application (43).

To test the practicality of using GFP as a transgene monitoring tool, we have engineered tobacco (*Nicotiana tabacum* cv. Xanthi) for whole-plant fluorescence using *mgfp4*, a mutagenized version of the gene that retains UV excitation under the control of the 35S promoter. The goal is to provide an in vivo, real-time, scorable marker gene to monitor transgenic plants in the field. When a plant transgenic for the GFP gene is excited with a hand-held, long-wave UV light (365 nm), the plant appears green compared to non-transgenic plants, which appear red due to the red autofluorescence of chlorophyll (Figure 1). To visibly detect GFP expression in the plant, protein expression must be high enough to mask this red autofluorescence (43) (Figure 1). Several practical problems present themselves in visualizing stably transformed plants for GFP and engineering a transgene tracking system. Two of these are addressed below.

Is fluorescence stable across generations? We have produced over 200 independent lines of transgenic tobacco. In analyzing approximately 20 of these lines representing a range of fluorescence and growing out progeny, we have determined that the transgene is stably expressed in whole-plants-based visual assessment, fluorescence spectrophotometry and Western blot analysis (Table 1). We currently have fourth-generation plants that are fluorescent. There do seem to be environmental interactions associated with fluorescence phenotypes, but these are independent of transgenic line effects.

How much GFP must be produced to be unambiguously visualized in whole plants? Protein analysis was conducted on transgenic plants to quantify GFP protein expression and to compare these data to visual fluorescence data (Table 1).

Approximately 0.1% mGFP4 expression is sufficient to yield highly fluorescent plants, assuming all species are in the active (fluorescent) form on the basis of visual and immunostain data comparisons (Table 1; Figure 2). Fluorescence was positively correlated with protein expression (P >0.05; Figure 2). Fluorescence spectrophotometry confirmed the visible fluorescence data (Figure 3). The fluorescence spectrophotometry was valuable for comparative purposes, but was not useful to quantify the number of green fluorescent molecules per unit leaf area because the extinction coefficient of the wild-type standard was presumed different than the recombinant GFP in plants. Fluorescence spectrophotometric scans also reveal endogenous fluorescent compounds in leaves. Also, since leaves were frozen before extraction, there may have been protein degradation before measurement.

In conclusion, green fluorescent protein is stably inherited

Table 1. Summary of GFP Transgenic Plant Progeny Data: Means and Standard Deviations

Primary Transgenic Fluorescence Class[a]	Progeny Transgenic Fluorescence Class[a]	Progeny Fluorescence Spectrometry Data[b]	Progeny GFP Expression[c]	Seed Yield[d]
High (4)	3.14 ± 1.16	1.57 ± 0.50	0.19 ± 0.19	1.03 ± 1.38
Low (1)	1.92 ± 0.86	1.43 ± 0.30	0.12 ± 0.18	2.65 ± 2.07
Control (0)	0 ± 0	1.35 ± 0.28	0 ± 0	1.80 ± 0.56

[a]Plants were grouped according to relative visual observation and a scale from 0–4 (in parentheses), with 0 being red and 4 being green under UV illumination; e.g., highly fluorescing plants appeared green under UV light (365 nm) as in the green plant in Figure 1, low fluorescence was intermediate between red and green, etc. Control plants were non-transgenic and appeared red.

[b]Fluorescence spectrophotometry was performed as another assay for visual fluorescence of GFP in plant extract samples as prepared below. Extraction for fluoresence spectrometry were performed from leaves that were harvested on dry ice and stored at -80°C. Samples and GFP standards were excited at 380 nm with the emission scan measured from 400–575 nm. Fluorescence in arbitrary units ($\times 10^5$) with excitation at 380 nm and detection at 502 nm.

[c]Gene expression is represented as % GFP of total extractable protein as determined by immunostained protein blot analysis. A 0.1-g (1.5 cm^2) portion of the fourth leaf from the apex leaf was homogenized with 600 µL extraction buffer (50 mM Tris-HCl, pH 8.0, 10 mM EDTA, pH 8.0, 10 mM 2-mercaptoethanol, 18% glycerol) on ice. Samples were clarified by centrifugation at 16400× *g* for 5 min. Protein blot analysis was performed using 2 µg of total protein in each sample lane and was analyzed according to published methods (33). Recombinant GFP protein (CLONTECH) was used for standards. Immunostaining was performed by applying each of the following antibodies sequentially using methods published elsewhere (33): rabbit polyclonal anti-GFP serum (1:3000 dilution; CLONTECH), goat-anti rabbit (Sigma Chemical, St. Louis, MO, USA) and rabbit anti-goat/alkaline phosphatase (Sigma Chemical). GFP protein was detected by exposing the nitro-cellulose membrane to nitro blue tetrazolium/bromochloroindolyl phosphate substrate for approximately 4 min. Total protein was quantified using the Bradford assay. Densitometric analysis was used to quantify immunostained GFP.

[d]Seed weight per plant at the end of the growing season of plants grown in a greenhouse and common garden in 1-gallon plastic pots.

in progeny and is useful as a tag to mark transgenic plants in vivo. We (43) and others (12,28) have shown that whole plant fluorescence with GFP is possible. This, to our knowledge, is the first report to show how much GFP must be synthesized in plants to be reliably visualized.

There are several other questions that must be answered before GFP can be used reliably in whole-plant applications. *What is the relationship between light quality and light quantity that plants used for growth and fluorescence?* We have noticed that plant growth conditions, namely irradiance, can cause apparent switch-on/switch-offs of fluorescence. Under prolonged periods of cloudiness, we observe that previously highly fluorescent plants are scored low-to-no fluorescence. While it seems that light quantity is the major factor, we cannot rule out light quality factors at this point. *Is there any yield drag associated with GFP fluorescence and expression?* Field experiments need to be performed to test whether GFP decreases plant growth, seed yield and photosynthesis. Likewise, GFP dynamics in the field need to be documented. While much more needs to be done in biotechnology, much of the needed technology for commercial applications has to be developed in remote sensing, optics, computer technology and spectral analysis. For example, at least one company is in the process of developing a portable fluorometer designed to measure GFP in intact leaves in the field.

It is foreseeable that in 21st century agriculture, GFP will be a valuable tool in several arenas. It is envisaged that GFP could be used to monitor pathogen spread across whole fields. Pathogens, insects and other stresses might trip inducible promoter-fused GFPs in transgenic plants that would serve as an early-warning system for crop failure, enabling the grower to take timely action. Different colored fluorescence proteins might be used to code for various pests and stresses. GFP biotechnology should be useful to monitor other agricultural biotechnologies and guard against transgene flow to weedy, wild relatives that might disrupt ecosystems. These systems would certainly increase crop yields, decrease chemical inputs and protect natural ecosystems, thereby increasing sustainability of agriculture. We tip our hats to the jellyfish.

ACKNOWLEDGMENTS

This research is supported by NSF award ISB-9604528 to C.N.S. Also funds from USDA Grant No. 9166384 were used in early aspects of the research. We thank Jacque Bailey, Sheila Branch, Ed Hellen, Mark Hens, Lisa Bruneau and Bryan Yacono for technical assistance. We especially thank Jim Haseloff for the gift of the *mgfp4* gene.

REFERENCES

1. **Arias, D.M. and L.H. Reiseberg.** 1994. Gene flow between cultivated and wild sunflowers. Theor. Appl. Genet. *89*:655-660.
2. **Baulcombe, D.C., S. Chapman and S. Santa Cruz.** 1995. Jellyfish green fluorescent protein as a reporter for virus infections. Plant J. *7*:1045-1053.
3. **Boevink, P., S. Santa Cruz, C. Hawes, N. Harris and K.J. Oparka.** 1996. Virus-mediated delivery of the green fluorescent protein to the endoplasmic reticulum of plant cells. Plant J. *10*:935-941.
4. **Bokman, S.H. and W.W. Ward.** 1981. Renaturation of *Aequorea* green fluorescent protein. Biochem. Biophys. Res. Commun. *101*:1372-1380.
5. **Chalfie, M., Y. Tu, G. Euskirchen, W.W. Ward and D.C. Prasher.** 1994. Green fluorescent protein as a marker for gene expression. Science *263*:725-888.
6. **Chiu, W.L., Y. Niwa, W. Zeng, T. Hirano, H. Kobayashi and J. Sheen.** 1996. Engineered GFP as a vital reporter in plants. Curr. Biol. *6*:325-330.
7. **Cody, C.W., D.C. Prasher, W.M. Westler, F.G. Prendergast and W.W. Ward.** 1993. Chemical structure of the hexapeptide chromophore of the *Aequorea* green fluorescent protein. Biochemistry *32*:1212-1218.
8. **Cormack, B.P., R.H. Valdivia and S. Falkow.** 1996. FACS optimized mutants of the green fluorescent protein (GFP). Gene *173*:33-38.
9. **Grebenok, R.J., E. Pierson, G.M. Lambert, F. Gong, C.L. Afonso, R. Haldeman-Cahill, J.C. Carrington and D.W. Galbraith.** 1997. Green-fluorescent protein fusions for efficient characterization of nuclear targeting. Plant J. *11*:573-586.
10. **Gu, X. and D.P.S. Verma.** 1997. Dynamics of phragmoplastin in living cells during cell plate formation and uncoupling of cell elongation from the plane of cell division. Plant Cell 9:157-169.
11. **Haseloff, J. and B. Amos.** 1995. GFP in plants. Trends Genet. *11*:328-329.
12. **Haseloff, J., K.R. Siemering, D.C. Prasher and S. Hodge.** 1997. Removal of a cryptic intron and subcellular localization of green fluorescent protein are required to mark transgenic *Arabidopsis* plants brightly. Proc. Natl. Acad. Sci. USA *94*:2122-2127.
13. **Heim, R., D.C. Prasher and R.Y. Tsien.** 1994. Wavelength mutations and posttranslational autoxidation of green fluorescent protein. Proc. Natl. Acad. Sci. USA *91*:12501-12504.
14. **Heim, R. and R.Y. Tsien.** 1996. Engineering green fluorescent protein for improved brightness, longer wavelengths and fluorescence resonance energy transfer. Curr. Biol. *6*:178-182.
15. **Heinlein, M., B.L. Epel, H.S. Padgett and R.N. Beachy.** 1995. Interaction of tobamovirus movement proteins with the plant cytoskeleton. Science *270*:1983-1985.
16. **Hu, W. and C.-L. Cheng.** 1995. Expression of *Aequorea* green fluorescent protein in plant cells. FEBS Lett. *369*:331-334.
17. **Kohler, R.H., W.R. Zipfel, W.W. Webb and M.R. Hanson.** 1997. The green fluorescent protein as a marker to visualize plant mitochondria *in vivo*. Plant J. *11*:613-621.
18. **Langevin, S.A., S.A. Clay and J.B. Grace.** 1990. The incidence and effects of hybridization between cultivated rice and its related weed, red rice, *Oryza sativa* L. Evolution *44*:1000-1008.
19. **Leff, L.G. and A.A. Leff.** 1996. Use of green fluorescent protein to monitor survival of genetically engineered bacteria in aquatic environments. Appl. Environ. Microbiol. *62*:3486-3488.
20. **McCabe, D.E. and B.J. Martinell.** 1993. Transformation of elite cotton cultivars via particle bombardment of meristems. Bio/Technology *11*:596-598.
21. **McCabe, D.E., W.F. Swain, B.J. Martinell and P. Christou.** 1988. Stable transformation of soybean (*Glycine max*) by particle acceleration. Bio/Technology *6*:923-926.
22. **Morin, J.G. and J.W. Hastings.** 1971. Energy transfer in a bioluminescent system. J. Cellular Physiol. *77*:313-318.
23. **Morise, H., O. Shimomura, F.H. Johnson and J. Winant.** 1974. Intermolecular energy transfer in the bioluminescent system of *Aequorea*. Biochemistry *13*:2656-2662.
24. **Nagatani, N., S. Takumi, M. Tomiyama, T. Shimada and E. Tamiya.** 1997. Semi-real time imaging of the expression of a maize polyubiquitin promoter-GFP gene in transgenic rice. Plant Sci. *124*:49-56.
25. **Niedz, R.P., M.R. Sussman and J.S. Satterlee.** 1995. Green fluorescent protein: an *in vivo* reporter of plant gene expression. Plant Cell Rep. *14*:403-406.
26. **Oparka, K.J., P. Boevink and S. Santa Cruz.** 1996. Studying the movement of plant viruses using green fluorescent protein. Trends Plant Sci. *1*:412-418.
27. **Ormo, M., A.B. Cubitt, K. Kallio, L.A. Gross, R.Y. Tsien and J. Remington.** 1996. Crystal structure of the *Aequorea victoria* green fluorescent protein. Science *273*:1392-1395.
28. **Pang, S.-Z., D.L. DeBoer, Y. Wan, G. Ye, J.G. Layton, M.K. Neher, C.L. Armstrong, J.E. Fry, M.A. Hinchee and M.E. Fromm.** 1996. An improved green fluorescent protein gene as a vital marker in plants. Plant Physiol. *112*:893-900.
29. **Paterson, A.H., K.F. Schertz, Y.F. Lin, S.-C. Liu and Y.-L. Chang.** 1995. The weediness of wild plants: molecular analysis of genes influencing dispersal and persistence of johnsongrass, *Sorghum halepense* (L.) Pers. Proc. Natl. Acad. Sci. USA *92*:6127-6131.
30. **Plautz, J.D., R.N. Day, G.M. Dailey, S.B. Welsh, J.C. Hall, S. Halpain and S.A. Kay.** 1996. Green fluorescent protein and its derivatives as versatile markers for gene expression in living *Drosophila melanogaster*, plant and mammalian cells. Gene *173*:83-87.
31. **Prasher, D.C.** 1995. Using GFP to see the light. Trends Genet. *11*:320-323.
32. **Prasher, D.C., V.K. Eckenrode, W.W. Ward, F.G. Prendergast and M.J. Cormier.** 1992. Primary structure of the *Aequorea victoria* green fluorescent protein. Gene *111*:229-233.
33. **Pratt, L.H., D.W. McCurdy, Y. Shimazaki and M.-M. Cordonnier.** 1986. Immunodetection of phytochrome: immunocytochemistry, immunoblotting, and immunoquantitation, p. 50-74. *In* H.F. Linskens and J.F. Jackson (Eds.), Modern Methods of Plant Analysis. Springer, New York.
34. **Reichel, C., J. Mathur, P. Eckes, K. Langenkemper, C. Koncz, J. Schell, B. Reiss and C. Maas.** 1996. Enhanced green fluorescence by the expression of an *Aequorea victoria* green fluorescent protein mutant in mono- and dicotyledonous plant cells. Proc. Natl. Acad. Sci. USA *93*:5888-5893.
35. **Rissler, J. and M. Mellon.** 1993. Perils Amongst the Promise. Union of Concerned Scientists, MIT Press, Cambridge, MA.
36. **Sawahel, W.A.** 1994. Transgenic plants: performance, release and containment. W.J. Microbiol. Biotechnol. *10*:139-144.
37. **Sheen, J., S.B. Hwang, Y. Niwa, H. Kobayashi and D.W. Galbraith.** 1995. Green fluorescent protein as a new vital marker in plant cells. Plant J. *8*:777-784.
38. **Shimomura, O.** 1979. Structure of the chromophore of *Aequorea* green fluoroscent protein. FEBS Lett. *104*:220-222.
39. **Siemering, K.R., R. Golbik, R. Sever and J. Haseloff.** 1996. Mutations that suppress the thermosensitivity of green fluorescent protein. Curr. Biol. *6*:1653-1663.
40. **Snow, A.A. and P.M. Palma.** 1997. Commercialization of transgenic plants: potential ecological risks. Bioscience *47*:86-96.
41. **Spellig, T., A. Bottin and R. Kahmann.** 1996. Green fluorescent protein (GFP) as a new vital marker in the phytopathogenic fungus *Ustilago maydis*. Mol. Gen. Genet. *252*:503-509.
42. **Sterling, J.** 1997. Novel production systems. Genet. Eng. News *17*:8.
43. **Stewart, C.N., Jr.** 1996. Monitoring transgenic plants using in vivo markers. Nature Biotechnology *14*:682.
44. **Tian, L., A. Sequin and P.J. Charest.** 1997. Expression of the green fluorescent protein gene in conifer tissues. Plant Cell Rep. *16*:267-271.
45. **Ward, W.W., H.J. Prentice, A.F. Roth, C.W. Cody and S.C. Reeves.** 1982. Spectral perturbations of the *Aequorea* green fluorescent protein. Photochem. Photobiol. *35*:803-808.
46. **Yang, F., L.G. Moss and G.N. Phillips, Jr.** 1996. The molecular structure of green fluorescent protein. Nature Biotechnology *14*:1246-1251.
47. **Yang, T.T., L.Z. Cheng and S.R. Kain.** 1996. Optimized codon usage and chromophore mutations provide enhanced sensitivity with the green fluorescent protein. Nucleic Acid Res. *24*:4592-4593.
48. **Yue, H., X. Xin-Ping and L. Bao-Jian.** 1997. Improved green fluorescent protein as a fast reporter of gene expression in plant cells. Biotechnol. Tech. *11*:133-136.
49. **Zolotukhin, S., M. Potter, W.W. Hauswirth, J. Guy and N. Muzyczka.** 1996. A "humanized" green fluorescent protein cDNA adapted for high level expression in mammalian cells. J. Virol. *70*:4646-4654.

Update to: Applications of Green Fluorescent Protein in Plants

C. Neal Stewart, Jr.

University of North Carolina, Greensboro, NC, USA

NEW GFPs EXPRESSED IN PLANTS

Since 1997, only a few new green fluorescent protein (GFP) variants have been developed and used in plants. A DNA shuffled mutant that has greater solubility and fluorescence called mut3GFP (2) has the V164A (V163A) mutation that improves fluorescence and folding at higher temperatures (15). It also includes the F100S (F99S) and M154T (m153T) mutations (2). Davis and Vierstra (3) further modified mGFP4 to include the mut3 mutations and called it smGFP. In addition, S65T and Y66H mutation variants were produced (3). They called these smRS-GFP (sm, soluble modified; RS, red-shifted) and sm blue fluorescent protein (BFP), respectively. When these transgenes were expressed in *Arabidopsis*, no difference in expression was found between smGFP and smRS-GFP (3). However, both provided improvements over mGFP4 fluorescence. As with all other BFPs, smBFP had little fluorescence compared with BFPs (3). Like mut3, smGFP had primarily UV excitation (with a minor blue peak), and smRS-GFP had an excitation maximum of 495 nm and emission maximum at 507 nm.

While these new variants are available, my laboratory still primarily uses mGFP5-er for most GFP applications. This variant is valuable because of the heat-stability mutation (V163A) and dual blue/UV excitation that provides flexibility in excitation modes.

DETECTION OF GFP IN TRANSGENIC PLANTS

In 1997, the state-of-the-art photonic devices that use blue light excitation were better suited for microscopic detection of GFP (epifluorescence microscopes using fluorescein isothiocyanate [FITC] filter sets) while UV devices were primarily used for macroscopic detection (a bright 100 W UV lamp). This situation is largely changed. There are, however, blue light/emission filter macroscopic detection systems now available for blue light-excited GFPs, such as the Little Luma LT 9700 (Lightools, Encinitas, CA, USA). The greatest technological improvement that we currently use is a scanning fluorescence spectrophotometer for the quantification of GFP directly on the plant: Spex Fluorescence Fluoromax (Jobin Yvon & Glen Spectra, Edison, NJ, USA). This instrument is equipped with a fiber optic probe that can excite and collect emissions on intact leaves or other tissue. This instrument allows for the standardization of measurements by anchoring scans of different samples to an average nontransgenic sample outside of the GFP emission peak, between 440 and 470 nm (Figure 1). In the 1997 paper, we used homogenized leaf tissue for crude protein isolates and subsequent fluorescence spectrophotometry. In doing so, we also extracted phenolic compounds that interfered with accurate measurements. The on-the-plant fluorescence spectrometry approach controls for autofluorescence and negates the need for tissue homogenization.

ANSWERED QUESTIONS

There were several predictions about the utilization of GFP in plants in the 1997 paper. The one that is most quickly coming to fruition is the use of GFP as an aid in plant transforma-

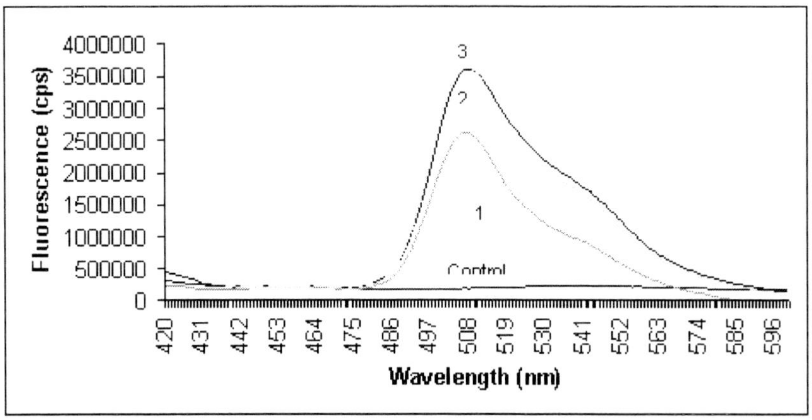

Figure 1. Fluorescence spectrophotometric analysis of GFP transgenic plants. Single leaves from three different transgenic *Brassica napus* (canola) events producing mGFP5-er and a nontransgenic control plant were analyzed. The leaves were excited with 390 nm and fluorescent scans were obtained using a Spex Fluorescence Fluoromax fluorescence spectrophotometer. Excitation and emission light was collected using a probe at the end of a 2-m fiber optic bundle, which was placed on each intact leaf. The plots were standardized by transposing GFP transgenic scans onto a control leaf scan between 440 and 470 nm, so that all baseline emission scans had equivalent counts per second.

tion procedures (1,4–6,8,10–13,14,17). A full review of these applications is forthcoming (16).

Two questions that we posed in the 1997 paper that have now been answered are:

1. What is the relationship between light quality and light quantity that plants use for growth and fluorescence? In the "early days" we noticed that if plants transgenic for mGFP4 were in low light, then the GFP fluorescence would be attenuated. We performed some experiments (unpublished), in which we masked off portions of leaves so that parts of leaves would be in shade and parts would be in light. One could clearly see lines of demarcation. The conclusion that was tentatively drawn was that light interacted with GFP fluorescence. When we later performed the same experiments with plants transgenic for mGFP5-er and sGFP (S65T), this phenomenon did not occur. And, when we compared the fluorescence of these three GFP variants in plants in the field, mGFP4-trasngenic plants had, by far, the lowest fluorescence, even though all three genes were expressed in the respective plants (7). Therefore, we have come to the conclusion that the wild-type (or near wild-type) GFPs simply do not stay folded as well under field conditions and are generally less stable in plants. However, we have noticed that, when plants are stressed and protein synthesis is compromised, GFP is a good molecular indicator at the organ level (lights-on when plants are healthy; lights-out when plants are stressed, unpublished data).

2. Is there any yield drag associated with GFP fluorescence and expression? There has long been anectdotal evidence that GFP might be cytotoxic. In a recent study in certain mammalian cell lines, the overexpression of GFP caused apoptosis (9). We addressed the cytotoxicity issue by 2 years of field experiments using plants expressing various levels of mGFP4, mGFP5-er or sGFP (S65T). There was no association between any fitness parameters such as yield or biomass with GFP expression or fluorescence (7). While in isolated cell lines or plant types and conditions, there may be some affect of overexpressing GFP, but clearly, GFP is not cytotoxic in mature plants. Its overexpression seems to have a neutral effect on plant fitness.

ACKNOWLEDGMENTS

Thanks to Reggie Millwood for providing Figure 1. This work was supported by USDA Grant No. 9804086 and NSF Grant No. IBN-96-04528.

REFERENCES

1. **Ahlandsberg, S., P. Sathish, C. Sun and C. Jansson.** 1999. Green fluorescent protein as a reporter system in the transformation of barley cultivars. Physiol. Plant. *107*:194-200.
2. **Crameri, A., E.A. Whitehorn, E.Tate and W.P.C. Stemmer.** 1996. Improved green fluorescent protein by molecular evolution using DNA shuffling. Nat. Biotechnol. *14*:315-319.
3. **Davis, S.J. and R.D. Vierstra.** 1998. Soluble, highly fluorescent variants of green fluorescent protein (GFP) for use in higher plants. Plant Mol. Biol. *36*:521-528.
4. **Elliot, A.R., J.A. Campbell, I.S. Brettell and P.L. Grof.** 1998. *Agrobacterium*-mediated transformation of sugarcane using GFP as a screenable marker. Aust. J. Pl. Physiol. *25*:739-743.
5. **Elliot, A.R., J.A. Campbell, B. Dugdale, I.S. Brettell and C.P.L. Grof.** 1999. Green-fluorescent protein facilitates rapid in vivo detection of genetically transformed plant cells. Plant Cell Rep. *18*:707-714.
6. **Ghorbel, R., J. Juarez, L. Navarro and L. Pena.** 1999. Green fluorescent protein as a screenable marker to increase the efficiency of generating transgenic woody fruit plants. Theor. Appl. Genet. *99*:350-358.
7. **Harper, B.K., S.A. Mabon, S.M. Leffel, M.D. Halfhill, H.A. Richards, K.A. Moyer and C.N. Stewart, Jr.** 1999. Green fluorescent protein as a marker for expression of a second gene in transgenic plants. Nat. Biotechnol. *17*:1125-1129.
8. **Kaeppler, H.F., G.K. Menon, R.W. Skadsen, A.M. Nuutila and A.R. Carlson.** 2000. Transgenic oat plants via visual selection of cells expressing green fluorescent protein. Plant Cell Rep. *19*:661-666.
9. **Liu, H.-S., M.-S. Jan, C.-K. Chou, P.-H. Chen and H.-J. Ke.** 1999. Is green fluorescent protein toxic to the living cells? Biochem. Biophys. Res. Comm. *260*:712-717.
10. **Maximova, S.N., A.M. Dandekar and M.J. Guiltinan.** 1998. Investigation of *Agrobacterium*-mediated transformation of apple using green fluorescent protein: high transient expression and low stable transformation suggest that factors other than T-DNA transfer are rate-limiting. Plant Mol. Biol. *37*:549-559.
11. **McCormac, A.C., H. Wu, M. Bao, Y. Wang, R. Xu, M.C. Elliot and D.-F. Chen.** 1998. The use of visual marker genes as cell-specific reporters of *Agrobacterium*-mediated T-DNA delivery to wheat (*Triticum aestivum* L.) and barley (*Hordeum vulgare* L.). Euphytica *99*:17-25.
12. **Molinier, J., C. Himber and G. Hahne.** 2000. Use of green fluorescent protein for detection of transformed shoots and homozygous offspring. Plant Cell Rep. *19*:219-223.
13. **Nehlin, L., C. Möllers, P. Bergman and K. Glimerlius.** 2000. Transient *B-gus* and *gfp* gene expression and viability analysis of microprojectile bombarded microspores of *Brassica napus* L. J. Plant Physiol. *156*:175-183.
14. **Ponappa, T., A.E. Brzozowski and J.J. Finer.** 2000. Transient expression and stable transformation of soybean using jellyfish green fluorescent protein (GFP). Plant Cell Rep. *19*:6-12.
15. **Siemering, K.R., R. Golbik, R. Sever and J. Haseloff.** 1996. Mutations that suppress the thermosensitivity of green fluorescent protein. Curr. Biol. *6*:1653-1663.
16. **Stewart, C.N., Jr.** The utility of green fluorescent protein in transgenic plants. Plant Cell Rep. (In press).
17. **Vain, P., B. Worland, A. Kohli, J. Snape and P. Christou.** 2000. The green fluorescent protein (GFP) as a vital screenable marker in rice transformation. Theor. Appl. Genet. *96*:164-169.

CHAPTER 28

Green Fluorescent Protein as a Reporter of Gene Expression and Protein Localization

S.R. Kain, M. Adams, A. Kondepudi, T.-T. Yang, W.W. Ward[1] and P. Kitts

CLONTECH Laboratories, Palo Alto, CA, and [1]Cook College, Rutgers University, New Brunswick, NJ, USA

(Reprinted from BioTechniques 19:650-655, 1995)

ABSTRACT

The green fluorescent protein (GFP) from the jellyfish Aequorea victoria *is rapidly becoming an important reporter molecule for monitoring gene expression and protein localization in vivo, in situ and in real time. GFP emits bright green light (λ_{max} = 509 nm) when excited with UV or blue light (λ_{max} = 395 nm, minor peak at 470 nm). The fluorescence excitation and emission spectra of GFP are similar to those of fluorescein, and the conditions used to visualize this fluorophore are also suitable for GFP. Unlike other bioluminescent reporters, the chromophore in GFP is intrinsic to the primary structure of the protein, and GFP fluorescence does not require a substrate or cofactor. GFP fluorescence is stable, species-independent and can be monitored non-invasively in living cells and, in the case of transparent organisms, whole animals. Here we demonstrate GFP fluorescence in bacterial and mammalian cells and introduce our Living Colors™ line of GFP reporter vectors, GFP protein and anti-GFP antiserum. The reporter vectors for GFP include a promoterless GFP vector for monitoring the expression of cloned promoters/enhancers in mammalian cells and a series of six vectors for creating fusion proteins to either the N or C terminus of GFP.*

INTRODUCTION

Light is produced in several bioluminescent species as a result of energy transfer to green fluorescent proteins (GFPs). One such GFP from the jellyfish *Aequorea victoria* fluoresces in vivo after receiving energy from the Ca^{2+}-activated photoprotein aequorin (9,18). Purified GFP has similar spectral properties to the protein expressed in vivo, absorbing blue light and emitting green light, which is detectable using a fluorescence microscope, UV light box or by fluorescence-activated cell sorting (FACS; Reference 1). Full-length GFP appears to be required for fluorescence; however, the minimal chromophore responsible for light emission can be assigned to a hexapeptide within the protein (2). This region of the protein contains a Ser-dehydroTyr-Gly trimer, which cyclizes and yields a chromophore that emits light by an as yet unknown mechanism. Mutagenesis in the chromophore region has revealed a dehydroTyr-Gly consensus, with mutations at additional sites in the hexapeptide producing functional GFP mutants with altered fluorescence spectra (3,7,8).

The cloning and sequencing of the GFP gene from *A. victoria* (14) has allowed expression of this protein in heterologous systems. GFP expressed in both prokaryotic and eukaryotic cells yields a bright green fluorescence when the cells are excited by blue or UV light. GFP fluorescence does not require additional gene products from *A. victoria* and occurs in a species-independent fashion. Recent studies using the nematode *Caenorhabditis elegans*, an organism often used for developmental studies, have demonstrated the utility of GFP as a reporter of gene expression in vivo (1). GFP has also been expressed in yeast, mammalian cells and *Drosophila melanogaster* (1,4,13,15,17). Moreover, both N- and C-terminal protein fusions with GFP have been constructed and have shown to maintain the fluorescence properties of native GFP (4,13,15,17).

The utility of GFP as a marker of gene expression and protein localization is facilitated by the introduction of our Living Colors™ product line, which includes a series of eight GFP vectors, purified GFP protein and anti-GFP antiserum. The comprehensive set of GFP vectors provides a ready source of GFP cDNA for the construction of specialized vectors by the researcher, a promoterless transcription reporter vector for analysis of promoter/enhancer elements and a set of protein fusion vectors for fusion of heterologous protein and peptide sequences to the N or C terminus of GFP. The GFP protein and antiserum provide important control reagents, allowing the researcher to correlate levels of GFP expression with fluorescence intensity.

Cloning and Expression Vectors for Gene Function Analysis
Edited by Quinn Lu and Michael P. Weiner

MATERIALS AND METHODS

Materials

All mammalian cell lines used in these studies were obtained from ATCC (Rockville, MD, USA). All media, serum and other reagents used in the culture of mammalian cells are from Life Technologies (Gaithersburg, MD, USA). *Escherichia coli* strain BL21(DE3) (16) was obtained from Novagen (Madison, WI, USA). The *E. coli* GFP expression vector, TU#58 (1), was obtained from Dr. Martin Chalfie (Columbia University). pET5ahead, a pET5a derivative containing the head portion of the fiber gene from adenovirus 2, was generously provided by Jadwiga Chroboczek (CNRS Institut de Biologie Structurale). The Western Exposure ™ Chemiluminescent Detection System, Living Colors™ GFP Reporter Vectors, GFP protein and anti-GFP antiserum are each from CLONTECH Laboratories (Palo Alto, CA, USA).

Mammalian Cells and the Method of Culture

CHO-K1 cells were cultured in Ham's F-12 medium containing 10% fetal bovine serum (FBS), COS-1 cells were cultured in Dulbecco's modified Eagle medium (DMEM) containing 10% FBS, NIH 3T3 cells were cultured in DMEM containing 4.5 g/L glucose and 10% newborn calf serum and Hela cells were cultured in minimum essential medium (MEM) with Earle's balanced salt solution, nonessential amino acids and 10% FBS. HEp-2 cells, an epidermoid carcinoma cell line derived from human larynx, were grown in RPMI 1640 medium supplemented with 100 mM L-glutamine and 5% FBS. Each cell line was propagated as monolayer cultures at 37°C in a humidified atmosphere of 95% air, 5% CO_2.

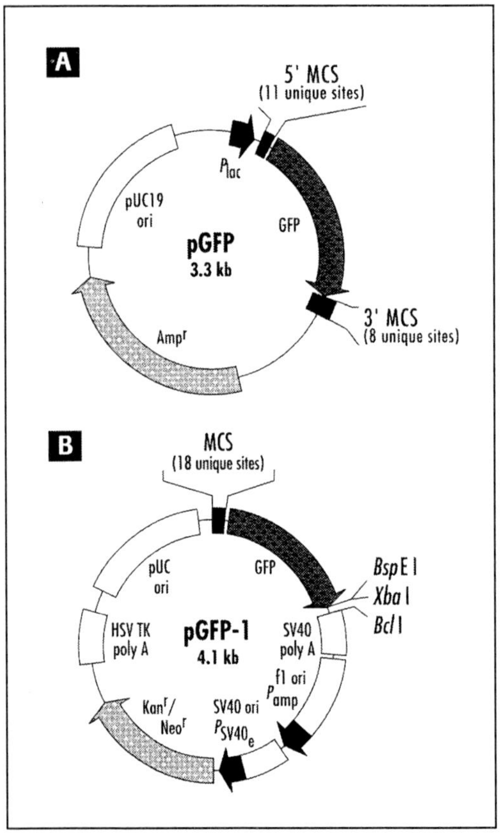

Figure 1. Plasmid map of pGFP cDNA vector and pGFP-1 promoter reporter vector. Panel A, pGFP cDNA vector. The GFP coding sequence in pGFP is flanked by separate MCS at the 5′ and 3′ ends. The 5′ MCS consists of *Hin*dIII, *Sph*I, *Pst*I, *Sal*I, *Xba*I, *Bam*HI, *Xma*I, *Sma*I, *Kpn*I, *Asp*718I and *Age*I. The 3′ MCS consists of *Bsm*I, *Eco*RI, *Stu*I, *Spe*I, *Eag*I, *Bsi*WI, *Bsp*120I and *Apa*I. The pUC19 backbone of pGFP provides a high copy number origin of replication and ampicillin-resistance gene for propagation in *E. coli*. pGFP was previously described as TU#60 (1). Panel B, pGFP-1 promoter reporter vector. pGFP-1 contains an MCS upstream of the promoterless GFP gene. The MCS contains the same sites as the MCS in pGFP-N1 (Figure 2), except pGFP-1 does not contain the *Nhe*I site. A simian virus 40 (SV40) polyadenylation signal downstream of the GFP gene directs proper processing of the mRNA in eukaryotic cells. A neomycin-resistance cassette allows for G418 selection of stably transfected eukaryotic cells.

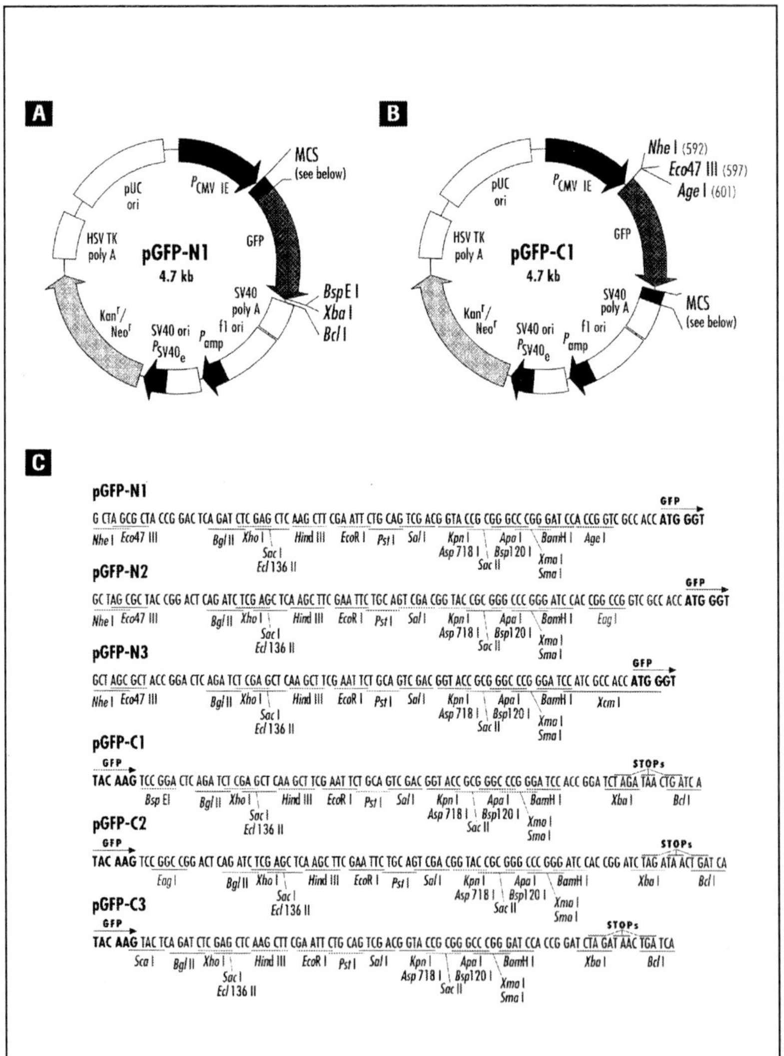

Figure 2. Plasmid map of pGFP N- and C-terminal protein fusion vectors. pGFP-N1 (Panel A), -N2 and -N3 allow genes cloned into an MCS upstream of the GFP coding sequences to be expressed as fusions to the N terminus of GFP. pGFP-C1 (Panel B), -C2 and -C3 allow genes cloned into an MCS at the end of the GFP coding sequences to be expressed as fusions to the C terminus of GFP. For both the pGFP-N and pGFP-C vectors, the fusion proteins are expressed under control of the immediate early promoter of CMV ($P_{CMV\ IE}$). The remainder of the vector backbone is identical to pGFP-1 (see Figure 1). Panel C shows the MCS of GFP protein fusion vectors.

Transfection of Mammalian Cells with GFP Reporter Vectors

For each of the mammalian cell lines used in these studies, 1–2 $\times 10^5$ cells were plated on 60-mm tissue-culture plates containing a glass coverslip. The cells were cultured for 18–24 h in the appropriate growth medium supplemented with serum. Cells were transfected with 4.0 µg of GFP plasmid DNA in Opti-MEM® I reduced-serum medium using the Lipofectin® reagent, both from Life Technologies, according to the manufacturer's instructions. The cells were incubated in the transfection solution containing DNA/Lipofectin for 8 h at 37°C in a humidified atmosphere of 95% air, 5% CO_2. The cells were allowed to recover for a period of 36 h in complete medium, and GFP fluorescence was visualized by fluorescent microscopy.

Fluorescence Microscopy

Microscopy was performed using a Zeiss AxioScope™ Model 50 fluorescent microscope (Carl Zeiss, Thornwood, NY, USA) to generate the data in Figure 4. The filter set used was a standard fluorescein isothiocyanate (FITC) set (Chroma Technology, Brattleboro, VT, USA). The maximum excitation and emission wavelengths for this filter set were 485 nm (range 460–510 nm) and 540 nm (range 515–565 nm), respectively. Photographs were taken using Kodak Ektachrome 400 ASA film (Eastman Kodak, Rochester, NY, USA). In Figure 6, microscopy was performed using an MRC®-600 confocal fluorescence microscope (Bio-Rad, Hercules, CA, USA) with standard FITC and rhodamine filter sets.

Electrophoresis, Western Blotting and Chemiluminescent Detection

Prior to electrophoresis, samples were solubilized by boiling for 5 min in sample buffer containing 62.5 mM Tris-HCl (pH 6.8), 2% (wt/vol) sodium dodecyl sulfate (SDS), 10% (vol/vol) glycerol, 0.1% bromophenol blue and 80 mM 1,4-dithiothreitol (Sigma Chemical, St. Louis, MO, USA). Electrophoresis was performed using the method of Laemmli (12).

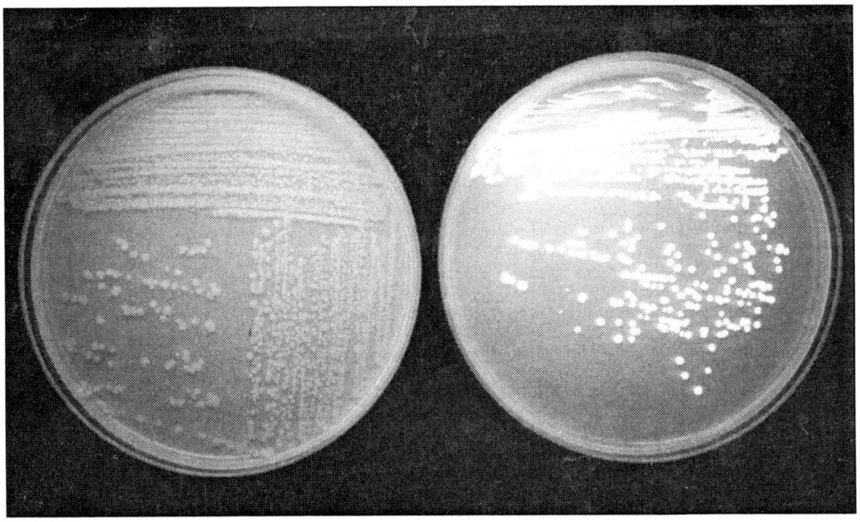

Figure 3. Expression of GFP in *E. coli*. The *E. coli* expression host, BL21(DE3), was transformed with TU#58, a pET3a derivative containing the GFP gene (Reference 1; right plate) or pET5ahead, a pET5a derivative containing the head portion of the fiber gene from adenovirus 2 as a negative control (left plate). Streaks of the transformed bacteria were grown at 37°C, without IPTG induction, and photographed under illumination from a long-wave UV lamp (365 nm) using Kodak Ektachrome film.

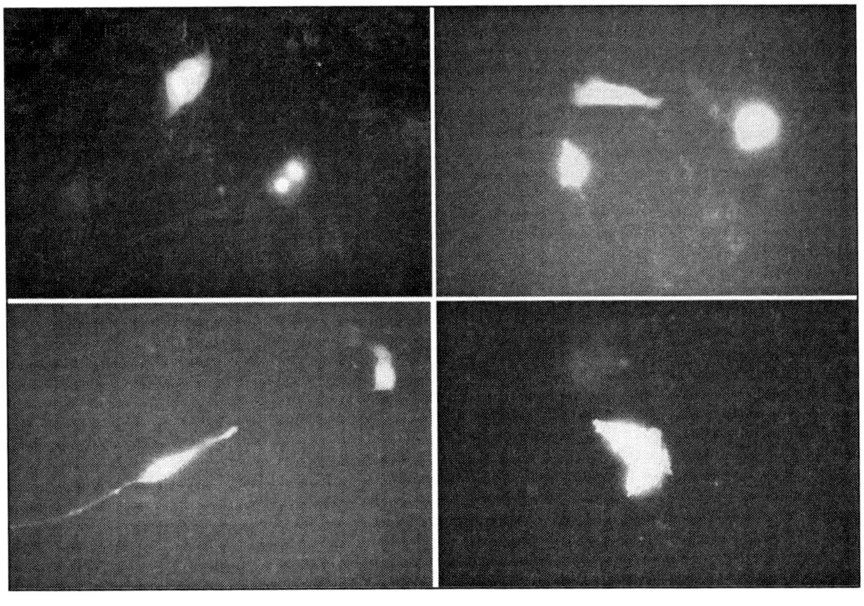

Figure 4. Transient expression of pGFP-C1 in mammalian cells. Monolayer cultures of COS-1, CHO-K1, NIH 3T3 and HeLa cells were transfected separately with pGFP-C1 as described in Materials and Methods. Transfection with pGFP-1 was used as a negative control to discern signals arising from autofluorescence in each cell type. Cells were fixed in 3.7% formaldehyde/PBS at 24 h post-transfection and prepared for fluorescence microscopy as described in Materials and Methods. Photographs were taken with a 40× oil immersion objective.

Electrophoretic transfer of proteins to polyvinylidene fluoride (PVDF; CLONTECH) membranes was performed using a Mini-trans-Blot® apparatus (Bio-Rad) in cold buffer containing 25 mM Tris-HCl (pH 8.3), 192 mM glycine, 20% (vol/vol) methanol according to the manufacturer's instructions. Following transfer of proteins to PVDF, the antigen-specific signals were visualized using the Western Exposure Chemiluminescent Detection System (CLONTECH; Reference 10) with a 1:2000 dilution of rabbit anti-GFP antibody and a 1:5000 dilution of goat anti-rabbit IgG conjugated to alkaline phosphatase according to the manufacturer's instructions.

Bacterial Invasion and Fluorescence Actin Staining

The GFP expression vector pFPV.1 used to transform *Salmonella typhimurium* was constructed by insertion of a Plac-gfp fusion into the *Bam*HI site of the plasmid pBR322. In this construct, the GFP gene is placed under the control of the *lac* promoter. Approximately 10^5 HEp-2 cells were seeded on glass coverslips and infected with *S. typhimurium* (strain SL1344) containing pFPV.1 at a multiplicity of infection of 10:1 for 5 min as described previously (5,6). The cells were fixed with 3.7% formaldehyde in phosphate-buffered saline (PBS), washed again and permeabilized with 0.1% Triton® X-100 (Sigma Chemical) for 10 min at room temperature. The detergent was washed away with PBS, and the cells were stained with rhodamine-phalloidin (Molecular Probes, Eugene, OR, USA) using a 1:500 dilution of 300 U stock for 20 min at room temperature. The coverslips were washed a final time in PBS, air-dried and mounted on glass slides for microscopy.

RESULTS AND DISCUSSION

The Living Colors line of reporter vectors provides researchers with a comprehensive set of GFP vectors suitable for a variety of applications. The vector pGFP (Figure 1A) is primarily intended as a convenient source of the GFP cDNA (1,14), which can be readily excised using restriction enzyme sites in the two multiple cloning sites (MCS) flanking the GFP coding sequence. Alternatively, the GFP coding sequences can be amplified by PCR. pGFP-1 (Figure 1B) is a versatile transcription reporter vector for monitoring the function of promoter/enhancer elements cloned into the MCS upstream of the promoterless GFP gene. The sequences around the GFP start codon have been converted to a Kozak consensus translation initiation site to increase translation efficiency in eukaryotic cells (11). The pGFP-1 vector also contains a neomycin/kanamycin-resistance cassette for selection of transformed bacteria and eukaryotic cells. pGFP-1 is suitable for the in vivo analysis of promoter/enhancer activities in live cells or organisms, thereby allowing the researcher to observe real-time changes in gene expression.

It has been shown that chimeric genes encoding both N- and C-terminal fusions to GFP retain the normal biological activity of the heterologous partner, as well as maintaining fluorescent properties similar to native GFP (4,13,15,17). The use of GFP in this capacity provides a "fluorescent tag" on the protein, which allows for in vivo localization of the fusion protein. The use of GFP fusions can provide enhanced sensitivity and resolution in comparison with standard antibody staining techniques (17), and the GFP tag eliminates the need for fixation, cell permeabilization and antibody incubation steps normally required when using antibodies tagged with chemical fluorophores. Lastly, use of the GFP tag permits kinetic studies of protein localization and trafficking due to the resistance of the GFP chromophore to photobleaching (1,15,17).

Studies using GFP fusion proteins are facilitated by the Living Colors protein fusion vectors shown in Figure 2. pGFP-N1 (Panel A) is one of three vectors for fusing heterologous proteins to the N-terminus of GFP. The same MCS is present in a different reading frame in pGFP-N1, -N2 and -N3. Similarly, pGFP-C1, -C2 and -C3 (Panel B) can be used to create in-frame fusions to the C terminus of GFP. All six GFP protein fusion vectors contain the Kozak consensus sequence, a neomycin/kanamycin-resistance cassette and the immediate early promoter of cytomegalovirus (CMV) for constitutive expression of GFP fusions in mammalian cells.

An important advantage of GFP as an in vivo marker of gene expression is the ability of this protein to generate a fluorescent signal without the requirements for additional factors from *A. victoria*. Crude preparations of recombinant GFP from *E. coli* yield fluorescence excitation and emission spectra that are indistinguishable from that produced by purified GFP from

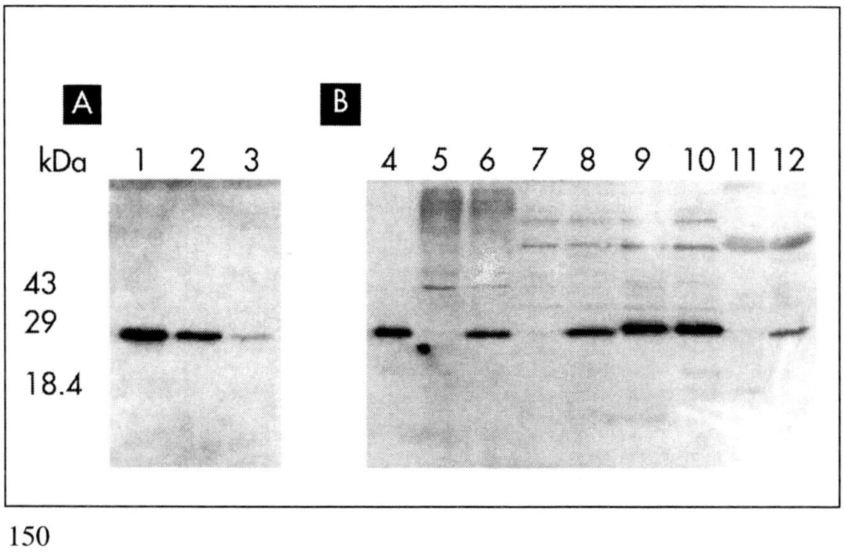

Figure 5. Detection of purified and expressed GFP on Western blots. Samples were electrophoresed on an SDS-polyacrylamide gel and electroblotted to a PVDF membrane. Antigen-specific signals are detected with a rabbit anti-GFP antiserum and a goat anti-rabbit IgG conjugated to alkaline phosphatase. The chemiluminescent signal is imaged on Kodak X-OMAT film (Eastman Kodak) with a 30-min exposure. Lanes 1–3: 4 ng, 400 pg and 40 pg of GFP protein, respectively. Lane 4: 4 ng GFP protein; lane 5: *S. cerevisiae* Y187 lysate; lane 6: *S. cerevisiae* Y187 lysate plus 4 ng GFP; lane 7: *E. coli* lysate; lane 8: *E. coli* lysate plus 4 ng GFP; lane 9: expression of GFP from the plasmid pGFP without IPTG stimulation in *E. coli*; lane 10: expression of GFP from the plasmid pGFP with IPTG stimulation in *E. coli*; lane 11: mammalian cell CHO-K1 lysate; lane 12: mammalian cell CHO-K1 lysate plus 4 ng GFP.

A. victoria (1). The identity of these fluorescence spectra suggests that soluble GFP expressed in bacteria undergoes equivalent protein-folding and chromophore-forming reactions as those occurring in the jellyfish (see Figure 3 for a simple illustration of GFP fluorescence in transformed *E. coli*). These observations explain, in part, why GFP is functional as a fluorescent marker in a variety of cell types and in different species. Figure 4 illustrates this fact, where GFP is expressed in four different mammalian cell lines from the plasmid pGFP-C1. The fluorescence intensity produced by GFP is roughly equivalent in each cell type, with little background fluorescence observed from untransfected cells. Similar results were obtained in live cells and in fixed cells counterstained with propidium iodide to visualize untransfected cells (data not shown). The observed lack of background staining represents another advantage of the GFP tag over antibody-based detection methods, as the latter technique is subject to background contributed by nonspecific binding of antiserum.

The use of GFP as a genetic marker has not met without problems in certain experimental systems. For example, high-level expression of GFP in *E. coli* may result in incorporation of GFP into inclusion bodies (8). This population of GFP protein is indistinguishable from soluble GFP on denaturing polyacrylamide gels, but is completely nonfluorescent. The time required for posttranslational maturation of the GFP chromophore can be lengthy and appears to vary in different cell types. Therefore, a fluorescent signal below the threshold of detection may simply be due to an inadequate time interval for chromophore formation. Lastly, several groups have noted that mutagenesis targeted at the chromophore region (3,7,8) can alter the fluorescence spectra of GFP such that detection of the fluorescence signal requires alternate excitation and/or emission conditions.

Each of the above difficulties can be addressed by the use of reagents to directly quantify expression of GFP protein distinct from GFP fluorescence. The results in Figure 5 show how anti-GFP antiserum can be used in Western blotting applications to quantify levels of GFP protein in lysates prepared from bacteria, yeast (*Saccharomyces cerevisiae*) and mammalian cells. As little as 40 pg of purified GFP can be detected on Western blots (Figure 5, lane 3) using a chemiluminescence detection method (10). Purified recombinant GFP added to cell lysates is evident as a prominent 27-kDa band against minor background signals in the corresponding cell lysate (compare Figure 5, lane 5 vs. lane 6; lane 7 vs. lane 8; and lane 11 vs. lane 12). GFP protein was also detected in bacteria transformed with pGFP (lanes 9 and 10). Western blotting experiments of this type can be used to correlate levels of GFP protein expression with fluorescence intensity or as a means to confirm expression of GFP in cases where detection of the fluorescence signal is problematic. Additional applications of purified GFP include its use as a standard on protein gels, to calibrate fluorometers and flow cytometers and for microinjection of GFP into cells and tissues.

The application of GFP as a reporter goes beyond the analysis of gene expression and protein localization and can cover diverse fields such as cell-population dynamics, evolution, ecology and pathogenesis. Figure 6 illustrates the results of one example in which GFP is used as a tool for the analysis of bacterial pathogenesis. Infection of human HEp-2 cells by invasive *Salmonella typhimurium* leads to formation of plasma membrane ultrastructures referred to as ruffles (5,6). The resulting change in host cell morphology following bacterial uptake is revealed by staining with rhodamine-phalloidin to visualize rearranged actin filaments, which appear as a red "splash" at the cell surface. *S. typhimurium* expressing GFP fluoresce green when outside the host cell and appear yellow when internalized because of the overlapping signals of GFP and rhodamine. The use of GFP as the fluorescent tag in *S. typhimurium* allows for real-time analysis of bacterium entry into host cells and provides the means for a simple color screen to quantify the extent of bacterial invasiveness.

Future directions for improvement of GFP as a genetic reporter include efforts to increase the intensity of the fluorescent signal and thereby enhance detection sensitivity, and to

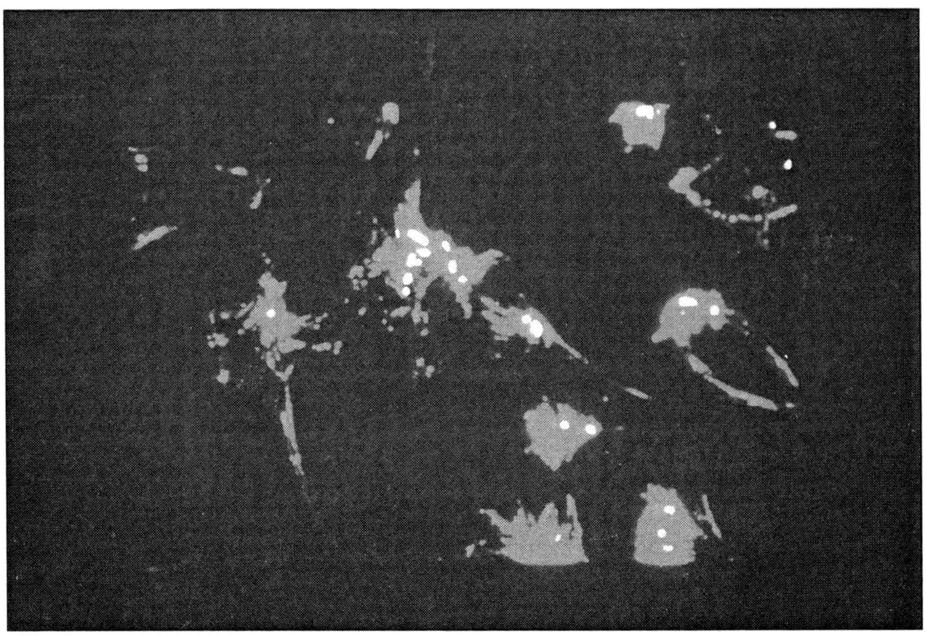

Figure 6. *S. typhimurium*-dependent induction of ruffles in HEp-2 cells. HEp-2 cells are infected with *S. typhimurium*, which constitutively express GFP. The cells are further stained with rhodamine-phalloidin to visualize actin filaments and prepared for confocal fluorescence microscopy as described in Materials and Methods. The image was photographed and merged as described previously (5,6) to allow for the simultaneous detection of the fluorescent signals from GFP (green) and rhodamine (red).

reduce the variability of the signal between replicate samples observed by some researchers. The availability of variant GFPs with shifted emission wavelengths may allow the use of this reporter to study two or more transcriptional events in the same cell population. Multiple fluorescent proteins derived from wild-type GFP, such as those described previously (3,7,8), will facilitate not only the simultaneous detection of several transcription reporters, but will also permit a similar analysis of protein trafficking events using GFP-fusion proteins.

ACKNOWLEDGMENTS

The data shown in Figure 6 was generously provided by Raphael H. Valdivia and Stanley Falkow, Department of Microbiology and Immunology, Stanford University School of Medicine, Stanford, CA, and Rocky Mountain Laboratory-NIH, Hamilton, MT. We express our appreciation to David Gunn and Anne Scholz for their critical comments and helpful suggestions during the course of this work. We also thank Yoshimi Munch and Marcie Jordan for preparation of the figures.

REFERENCES

1. **Chalfie, M., Y.T. Tu, G. Euskirchen, W.W. Ward and D.C. Prasher.** 1994. Green fluorescent protein as a marker for gene expression. Science *263*:802-805.
2. **Cody, C.W., D.C. Prasher, W.M. Westler, F.G. Prendergast and W.W. Ward.** 1993. Chemical structure of the hexapeptide chromophore of the *Aequorea* green fluorescent protein. Biochemistry *32*:1212-1218.
3. **Delagrave, S., R.E. Hawtin, C.M. Silva, M.M. Yang and D.C. Youvan.** 1995. Red-shifted excitation mutants of the green fluorescent protein. Bio/Technology *13*:151-154.
4. **Flach, J., M. Bossie, J. Vogel, A. Corbett, T. Jinks, D.A. Willins and P.A. Silver.** 1994. A yeast RNA-binding protein shuttles between the nucleus and the cytoplasm. Mol. Cell. Biol. *14*:8399-8407.
5. **Francis, C.L., T.A. Ryan, B.D. Jones, S.S. Smith and S. Falkow.** 1993. Ruffles induced by *Salmonella* and other stimuli direct macropinocytosis of bacteria. Nature *364*:639-642.
6. **Francis, C.L., M.N. Starnbach and S. Falkow.** 1992. Morphological and cytoskeletal changes in epithelial cells occur immediately upon interaction with *Salmonella typhimurium* grown under low-oxygen conditions. Mol. Microbiol. *6*:3077-3087.
7. **Heim, R., A.B. Cubitt and R.Y. Tsien.** 1995. Improved green fluorescence. Nature *373*:663-664.
8. **Heim, R., D.C. Prasher and R.Y. Tsien.** 1994. Wavelength mutations and posttranslational autoxidation of green fluorescent protein. Proc. Natl. Acad. Sci. USA *91*:12501-12504.
9. **Inouye, S. and F.I. Tsuji.** 1994. *Aequrea* green fluorescent protein: expression of the gene and fluorescence characteristics of the recombinant protein. FEBS Lett. *341*:277-280.
10. **Kain, S.R., K. Mai and P. Sinai.** 1994. Human multiple tissue western blots: a new immunological tool for the analysis of tissue-specific protein expression. BioTechniques *17*:982-987.
11. **Kozak, M.** 1987. An analysis of 5′-coding sequences from 699 vertebrate messenger RNAs. Nucleic Acids Res. *15*:8125-8148.
12. **Laemmli, U.K.** 1970. Cleavage of structural proteins during the assembly of the head of bacteriophage T4. Nature *227*:680-685.
13. **Marshall, J., R. Molloy, G.W.J. Moss, J.R. Howe and T.E. Hughes.** 1995. The jellyfish green fluorescent protein: a new tool for studying ion channel expression and function. Neuron *14*:211-215.
14. **Prasher, D.C., V.K. Eckenrode, W.W. Ward, F.G. Prendergast and M.J. Cormier.** 1992. Primary structure of the *Aequorea victoria* green fluorescent protein. Gene *111*:229-233.
15. **Stearns, T.** 1995. The green revolution. Curr. Biol. *5*:262-264.
16. **Studier, F.W., A.H. Rosenburg, J.J. Dunn and J.W. Dubendorff.** 1990. Use of T7 RNA polymerase to direct expression of cloned genes. Methods Enzymol. *185*:60-89.
17. **Wang, S. and T. Hazelrigg.** 1994. Implications for *bcd* mRNA localization from spatial distribution of *exu* protein in *Drosophila* oogenesis. Nature *369*:400-403.
18. **Ward, W.W. and M.J. Cormier.** 1978. Energy transfer via protein-protein interaction in *Renilla* bioluminescence. Photobiochem. Photobiol. *27*:389-396.

Section V

Special Purpose Vectors

GCN4-Based Expression System (pGES): Translationally Regulated Yeast Expression Vectors

Avishai Mimran, Irit Marbach and David Engelberg

The Hebrew University of Jerusalem, Jerusalem, Israel

(Reprinted from BioTechniques 28:552-560, 2000)

ABSTRACT

The expression of foreign proteins in Saccharomyces cerevisiae *is a powerful tool for basic research and the biotechnological industry. In spite of the potential of* S. cerevisiae, *only a few useful expression vectors have been developed for this yeast. These vectors are based on an increasing transcription rate in combination with an increase in gene dosage. Most vectors are maintained as plasmids, which forces growth of cultures on poor selective media. Expression of the yeast Gcn4 protein is regulated at the translational level and increases strongly under amino acid starvation. Because under these conditions protein synthesis in general ceases, it is conceivable that regulatory elements that control Gcn4 expression could support selective expression of foreign genes. We cloned DNA fragments residing upstream from the* GCN4 *coding sequence (including the 5′ untranslated region [UTR]) and ligated them to a cDNA that encodes the human serum albumin* (HSA) *gene. These* GCN4 *regulatory elements induced efficient HSA expression at the translational level under amino acid starvation. The* GCN4/HSA *cassette promoted efficient, inducible expression on either a multicopy or integrative plasmid. The integrated cassette induced a high level of HSA in dense cultures grown on rich media. Thus, the* GCN4-*based expression system (pGES) provides high protein quantities. pGES is the first expression vector to be induced at the translational level.*

INTRODUCTION

The expression of proteins in microorganisms is a powerful tool for basic research and for various medical and industrial needs. A variety of advanced expression systems have been developed in prokaryotes, in particular in the bacterium *Escherichia coli* (9,20,28). In this organism, a foreign protein can be expressed efficiently, reaching a level of 10%–30% of its total protein content (7,9,20,28). However, many eukaryotic proteins expressed in bacteria are not active. For various reasons, proteins expressed in bacteria must be extensively purified and assayed in rigorous and expensive quality-control systems. Therefore, although powerful, the bacterial systems are disadvantageous in many cases (3,27).

An attractive alternative is the baker's yeast, *Saccharomyces cerevisiae*. Most foreign proteins expressed in this eukaryote are accurately processed posttranslationally and are biologically active (3,10,21,27). *S. cerevisiae* has been widely used in the food industry and is regarded as a safe organism; the risk of side effects and the odds of the presence of pathogenic agents are minimal (3,10,27). Although *S. cerevisiae* was used successfully for the production of several proteins for therapeutic and diagnostic uses (10), and in spite of its great potential as a protein factory, this organism is not widely used for the production of proteins. The major reason for this seems to be that the level of expression obtained in yeast is significantly lower than that obtained in bacteria (3,10,21,23,24).

Only a few yeast expression vectors are available. All vectors (including bacterial and mammalian vectors) contain strong promoters (usually a promoter of a glycolitic gene; e.g., *GPD1*, *ADH1* or *PGK1*) (21,23,24). In addition, they contain the 2 μm origin of replication or a multicopy integration system that promotes maintenance of the vector in 60–100 copies per cell (10,22). Some of the vectors contain an inducible promoter, the most widely used being the *GAL1-10* promoter (21,23). The strategy of all yeast expression vectors is therefore based on a high transcription rate and the presence of many copies per cell.

Most yeast expression vectors are present in the cell as nonintegrated plasmids. This forces the growth of expressing cultures on synthetic media under selective conditions. When grown in laboratory flasks, these cultures enter the stationary phase at a low cell density, and the quantities of proteins obtained are low. When grown in industrial fermentors, high cell concentrations could be obtained but would require special treatments and long, expensive incubation time.

Here, we describe the development and use of a novel type of expression vector that is fully inducible and can be used in rich media. This vector is based on the regulatory elements of the *GCN4* gene. *GCN4* encodes a transcriptional activator with target genes that encode amino acids and nucleotides biosynthetic enzymes (15,16). Under optimal growth conditions on media supplemented with amino acids and pyrimi-

Cloning and Expression Vectors for Gene Function Analysis
Edited by Quinn Lu and Michael P. Weiner

dine and purine bases, Gcn4 is not expressed. When cells are starved for amino acids, Gcn4 expression is dramatically induced, and consequently, transcription of biosynthetic genes commences. The expression of Gcn4 is mainly regulated at the translational level. The mRNA levels of *GCN4* are high under most growth conditions, but the translation rate of this mRNA is extremely low (17). The low translation rate is maintained by four short open reading frames (ORFs) located at the 5′ noncoding leader of the message (upstream ORFs 1–4; 4uORF) (4,15–17) (Figure 1). Ribosomes recognize uORF1 and begin translation, yet they dissociate from the mRNA at the stop codon and reinitiate translation at another short ORF. In most cases, ribosomes do not reach the fifth ORF, which is the *GCN4* coding sequence (17). The inhibition of *GCN4* translation depends on the high rate of reinitiation. Under amino acid starvation, the rate of reinitiation is reduced to a minimum, and consequently, translation activity in the cell pauses, which is an obvious protective response. Reduction in reinitiation activity is mediated by the Gcn2 kinase that phosphorylates, and thereby inhibits, the activity of initiation factor 2α (eIF2α) (4,17). Although translation in the cell is generally ceased under these conditions, translation of *GCN4* mRNA is specifically induced (15–17). This activity seems a logical response because the Gcn4 protein is required for the induction of amino acid biosynthesis. Following the translation of Gcn4 and the synthesis of amino acids, general translation activity restarts. In the laboratory, amino acid starvation could be easily mimicked by the treatment of yeast cells with inhibitors of biosynthetic enzymes. The most widely used compound is 3-amino-1,2,4-triazole (3-AT), a competitive inhibitor of His3. Cells combat the competitive inhibition effect simply by producing more His3. The *GCN4* translation rate, which leads to *HIS3* transcription, dramatically increases in response to 3-AT treatment (12,15).

We proposed that the unusual pattern of Gcn4 expression could be advantageous for the expression of a foreign protein. We reasoned that if the mRNA of a given foreign protein is selectively translated when translation of most cellular mRNAs is halted, its concentration should be high. To test this idea, we have cloned the regulatory elements of *GCN4* (from -1067 to -1; +1 is the initiation of translation) and ligated this fragment to the coding sequence of the human serum albumin (*HSA*) gene. We show that under the control of *GCN4* regulatory elements, expression of HSA is rapidly and efficiently induced by 3-AT. We further show that induction occurs at the translational level and could be finely controlled by using various 3-AT concentrations. Importantly, this vector induces efficient expression as an integrated copy, which allows growth on rich media to high cell density.

MATERIALS AND METHODS

Yeast Strains and Growth Conditions

The SP1 strain (*MATa*, *his3*, *leu2*, *ura3*, *trp1*, *ade8*, *can*^{r}; obtained from M. Wigler) (18) was used as a source of genomic DNA from which the *GCN4* upstream sequences were cloned. Strain BJ2168 (*MATa*, *prc1-407*, *prb1-1122*, *pep4-3*, *leu2*, *trp1*, *ura3-52*; obtained from the yeast genetic stock center at the University of California-Berkeley) (29) was used as a host for the expression plasmids. Yeast cultures were grown on either 1% yeast extract, 2% bactopeptone, 2% glucose (YPD medium) or 0.17% yeast nitrogen base without amino acids and ammonium sulfate (Difco, Detroit, MI, USA), 0.5% $NH_4(SO_4)_2$, 2% glucose and the required amino acids and uracil (40 mg/L) (SD medium). Strains that harbor plasmids were grown on SD(-uracil). For expression experiments, cultures were harvested at the logarithmic phase (unless mentioned otherwise) and resuspended in SD(-uracil, -histidine) [or in SD(-histidine) for integrated vectors], supplemented with 3-AT (see details of each experiment).

PCR Cloning of *GCN4* Fragments and Plasmid Constructions

PCRs were performed using genomic DNA of the yeast strain SP1 as a template and the following primers. Primer 3′ (5′-GCGCCGGATCCTTTATTTGTATTTAATTTATTTTCTTGAGC-3′) contains a *Bam*HI site and was used in all PCRs (Figure 1). The long promoter (5′-GCGCCGACTAGTTTGCCACGTACATGACATTA-3′) was used to amplify the sequence from -1067. The middle-sized promoter (5′-GCGCCGACTAGTGCCGAACACCACCGGCATCTTG-3′) was used to amplify the sequence from -904. The short promoter (5′-GCGCCGACTAGTCGGAAGATAAATACTCCAAC-3′) was used to amplify the sequence from -706. These three primers contain *Spe*I sites. They were used at the 5′ side of the *GCN4* promoter (Figure 1). PCR products were digested with *Spe*I and *Bam*HI and ligated to pBluescript® (Stratagene, La Jolla, CA, USA) to obtain pBSG-Short (pBSG-S), pBSG-Middle (pBSG-M) and pBSG-Long (pBSG-L). The cloned DNAs were sequenced and found to be identical to the sequence of this region as it appears in the *Saccharomyces* genome database.

To ligate the coding sequence of *HSA* downstream from the *GCN4* sequences, the cDNA of *HSA* was cloned as a *Bam*HI-*Eco*RI fragment to pBSG-S, pBSG-M and pBSG-L. These chimeric genes contained about 70 bp of 5′ UTR sequences of *HSA*, which were found to reduce the expression

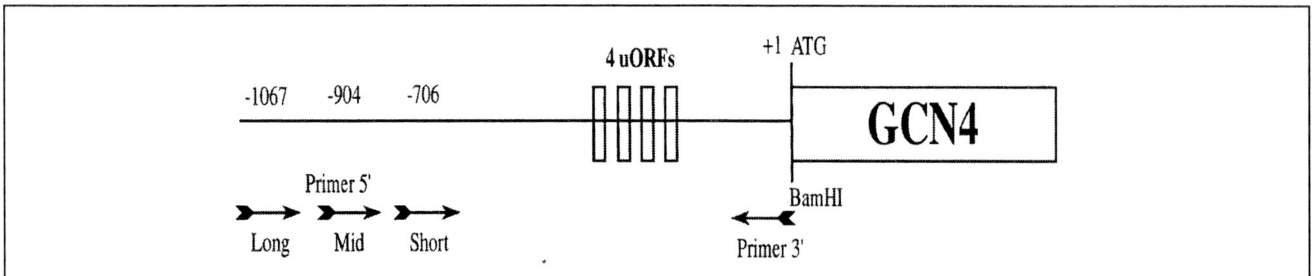

Figure 1. Schematic description of the *GCN4* gene and the primers used for promoter cloning. ORFs are shown as boxes and primers as arrows.

level (data not shown). To remove this sequence from each of the three plasmids, a 885 bp *Bam*HI-*Nco*I fragment that contains the entire 5′ UTR and 815 bp of the *HSA* coding sequence was replaced with a 815 bp PCR product that contains only the first 815 bp of the *HSA* coding sequence. Primers used for this PCR were HSA-ATG (5′-GCGCGCGGATCCATGAAGTGGGTAACCTTTATTTCCC-3′) and HSA-815 (5′-GCAGATCTCCATGGCAGCATTCC-3′).

The plasmids obtained were designated pBSG-S-HSA, pBSG-M-HSA and pBSG-L-HSA. The *GCN4-HSA* moiety of each of these plasmids was removed as a *Spe*I-*Eco*RI fragment and inserted into pRS316 (a centromeric yeast plasmid) (25), pRS306 (an integrated yeast vector) (25) and pRS426 (a 2 µm-based vector) (2). The terminator of *ADHI* (0.6 kb *Eco*RI fragment) was ligated to the *Eco*RI site of each of the nine plasmids obtained. The final constructs were designated as follows: *(i)* pGES306-S-HSA (for *GCN4*-based expression system, derived from pRS306, which contains the short *GCN4* promoter and the *HSA* cDNA; *(ii)* pGES306-M-HSA, which is similar to pGES306-S-HSA but contains the medium-sized *GCN4* promoter); *(iii)* pGES306-L-HSA, which contains the long *GCN4* promoter; *(iv)* pGES316-S-HSA; *(v)* pGES316-M-HSA; *(vi)* pGES316-L-HSA; *(vii)* pGES426-S-HSA; *(viii)* pGES426-M-HSA; and *(ix)* pGES426-L-HSA. Figure 2 depicts pGES306-L-HSA, pGES316-L-HSA and pGES426-L-HSA.

Integrated vectors were digested with *Stu*I before yeast transformation. This digestion directs the vector to the *URA3* locus of the genome. To express *HSA* under the *ADH1* promoter, a *Bam*HI-*Eco*RI fragment containing the *HSA* cDNA was rendered blunt by DNA *Pol*I (Klenow fragment) and inserted into the *Sma*I site of pAD4Δ (1).

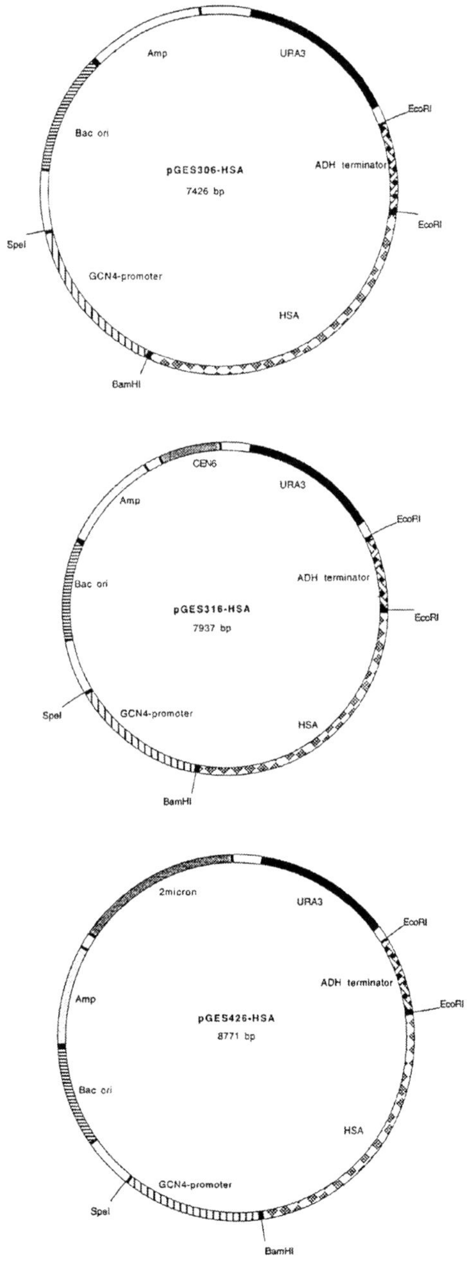

Figure 2. The basic pGES family containing the *HSA* cDNA. Note that pGES306-HSA is an integrative vector, pGES316-HSA is a centromeric vector and pGES426-HSA is a multicopy vector. All vectors shown contain the long *GCN4* promoter (see text for details).

Western Blot Analysis

Protein extracts were prepared as previously described (8). From each extract, 30 µg were used for western blot analysis. The SDS-PAGE gels contained 10% acrylamide. Protein blotting was performed with a semi-dry blotter (LKB, Sweden) for 2 h under constant current (mA = 0.8× gel area in cm^2). Polyclonal anti-HSA antibodies (RAHu/albumin) were purchased from Nordic Immunology (Tilburg, The Netherlands) and used in a dilution of 1:50 000. Peroxidase-conjugated goat anti-rabbit antibodies were purchased from Jackson ImmunoResearch Laboratories (West Grove, PA, USA) and used at the dilution of 1:25 000. Purified HSA (Sigma, St. Louis, MO, USA) was used as a standard positive control. RNA preparations and primer extension analysis were performed as previously described (6) using the following primers: HSA-extension (5′-ACCTCACTCTTGTGTGCATCTC-3′) and GCN4-extension (5′-ATAATTCGCTAGTGAAACTGATGGGC-3′).

Primer extension products were separated on 6% acrylamide 7 M urea gel.

RESULTS AND DISCUSSION

A 1067 bp Fragment of the *GCN4* Upstream Sequences is Required for Efficient Expression of a Heterologous Gene

To test whether regulatory elements of *GCN4* could be used to regulate the expression of a heterologous gene, we cloned a DNA fragment that resides upstream from the *GCN4* coding sequence. Initially, we cloned (from a genomic DNA prepared from the SP1 strain) a fragment (-706 to -1, designated short promoter; see Materials and Methods and Figure 1) previously shown to be sufficient for regulated expression of Gcn4 and of a *GCN4*/β-galactosidase (β-gal) chimeric gene (containing 53 amino acids of Gcn4 fused to β-gal) (13,14). The capability of this 706 bp fragment alone (containing no *GCN4* coding sequences) to control the transcription and translation of another gene has not been reported. We constructed a *GCN4-S-HSA* (S stands for short promoter; see Materials and Methods) cassette that was introduced into a multicopy yeast plasmid as well as into centromeric and integrated plasmids. The terminator of the *ADH1* gene was inserted downstream to *GCN4*-S-*HSA*. The resulting plasmids were introduced into yeast cells that were subsequently grown under various conditions, including amino acid starvation and 3-AT treatment.

Protein extracts were prepared from the cultures, and the expression of HSA was monitored by western blot analysis. Unexpectedly, in all cases, the expression of HSA was not detectable (data not shown). Therefore, we further cloned two longer fragments of *GCN4* upstream sequences, one containing the sequence from -904 to -1 and another containing the sequence from -1067 to -1 (designated middle-sized and long promoters, respectively; see Figure 1 and Materials and Methods). When the *HSA* gene was ligated to these fragments (plasmids pGES-M-HSA and pGES-L-HSA) and introduced into yeast cells, a high expression level of HSA was detected (Figure 3). Note that an additional band of 45 kDa cross-reacted with anti-HSA antibodies. This band, which is a result of proteolytic activity of the Yap3 protease (19), was observed also by others that expressed HSA in yeast (19,26).

In all constructs, HSA expression was barely detectable before induction with 3-AT, which suggests that expression from pGES plasmids is tightly controlled. Expression from constructs containing the long promoter was usually higher than that obtained with the middle-sized promoter (note, for example, the level of expression after 10 h induction; see Figure 3). All further experiments were carried out with the long promoter.

Expression of HSA Under *GCN4* Regulatory Elements Is Induced by 3-AT

We tested whether the cloned *GCN4* elements regulate HSA expression in a pattern similar to that of Gcn4 regulation. Yeast cells (strain BJ2168) that harbor pGES426-L-HSA were grown to logarithmic phase on SD(-uracil). Then, cells were collected and resuspended in SD(-uracil, -histidine) medium supplemented with 20 mM 3-AT. The samples were removed from the culture at indicated time points, and the expression of HSA was measured. Two hours after the addition of 3-AT, the expression of HSA was easily detected. The expression reached its maximum level between 6 and 8 h after induction (Figure 4A). These results (Figure 3 and Figure 4A) show that a vector based on *GCN4* upstream elements is useful for the inducible expression of foreign genes.

To test whether the level of expression could be finely controlled, we grew the BJ2168/pGES426-L-HSA cells to logarithmic phase, collected the cells, divided them into fractions and resuspended each fraction in medium supplemented with a different 3-AT concentration. The level of HSA expressed was proportional to the concentration of 3-AT provided up to 40 mM. Higher concentrations of 3-AT reduced the expression level of HSA (Figure 4B).

Expression of HSA from pGES Vectors Is Regulated at the Translational Level

To verify that the induction of HSA expression occurred at the translational level, we measured the levels of *HSA* mRNA and protein before and after induction with 3-AT (Figure 5). Protein extracts were tested by western blot analysis using anti-HSA antibodies (Figure 5A), while RNA preparations

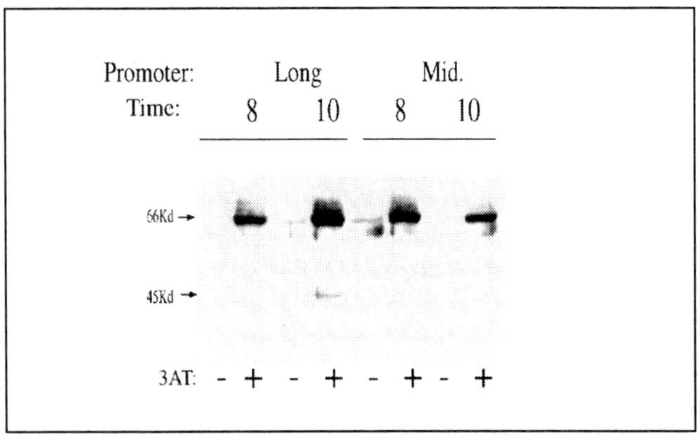

Figure 3. Induction and expression of HSA by pGES vectors containing either the long *GCN4* promoter (Long) or the middle-sized promoter (Mid.). Western blot analysis using anti-HSA antibodies was performed on protein extracts prepared from cultures (BJ2168 cells) harboring either pGES306-L-HSA (Long) or p306-M-HSA (Mid.) 8 and 10 h after addition of 3-AT.

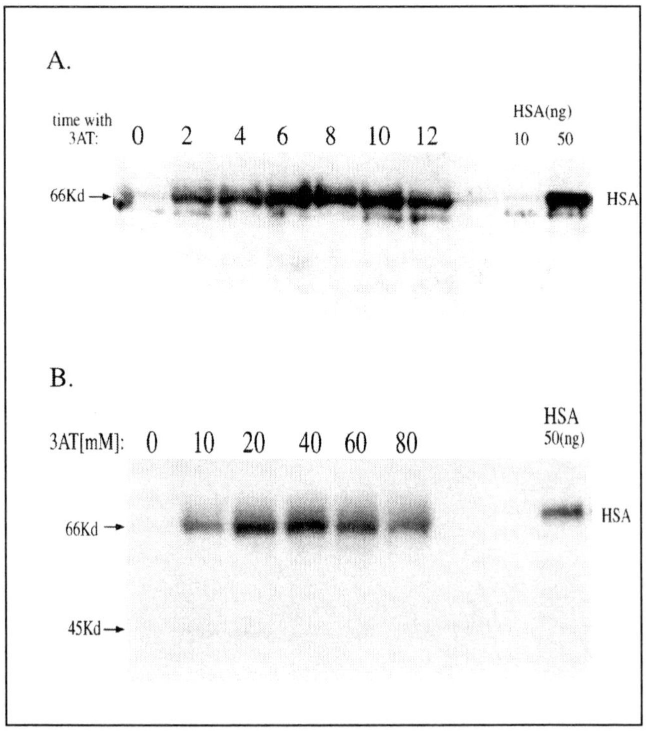

Figure 4. Time course and dose response analysis of induction of HSA expression. (A) A culture of BJ2168 cells harboring pGES426-L-HSA was grown to logarithmic phase and then supplemented with 20 mM of 3-AT. Samples were removed at indicated time points and subjected to western blot analysis. (B) A similar culture was grown to logarithmic phase and then divided into 6 fractions. Each fraction was supplemented with the indicated concentration of 3-AT. Cultures were further grown for 7 h, collected and analyzed by western blotting. Purified HSA was loaded on the right lanes of each gel.

were subjected to primer extension analysis. The extended primers were separated on sequencing gels, which allowed accurate detection of each mRNA initiation site (Figure 5, B and C). As determined by Hinnebusch (14), the transcription of *GCN4* initiates at three major sites (+1, +18 and +30) and at

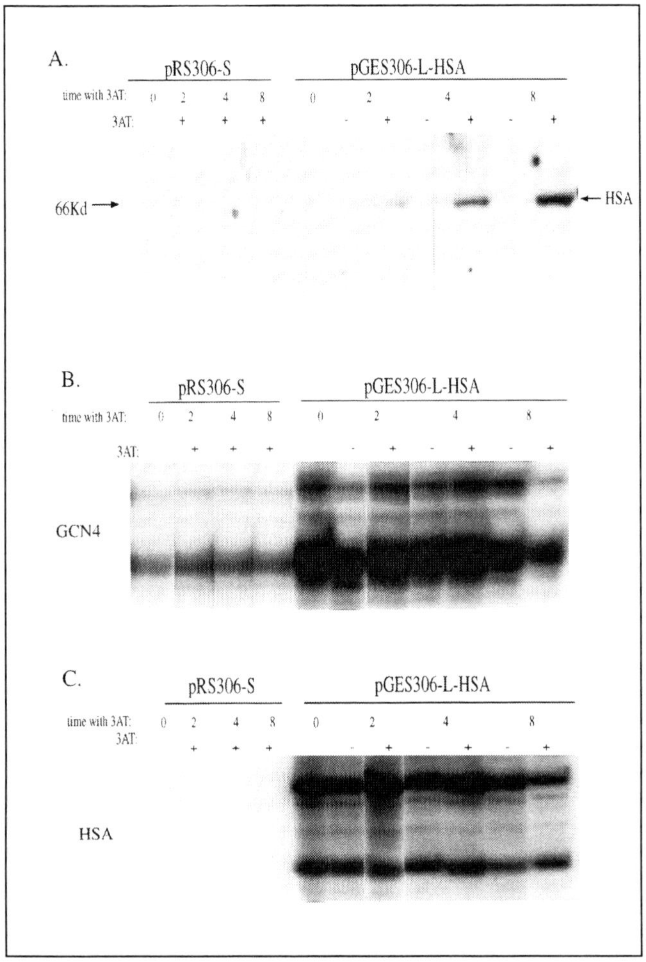

Figure 5. Expression of HSA from pGES vectors is induced at the translational level. A culture harboring an integrated copy of pGES306-L-HSA and a culture harboring an integrated copy of the control plasmid pRS306-S (a similar vector but with no *HSA* cDNA) were grown to logarithmic phase when 3-AT was added. At the indicated time points, the samples were removed. Protein extracts and RNAs were prepared from each sample and analyzed by western blot and primer extension analysis, respectively. (A) Western blot analysis. (B) Primer extension analysis using *GCN4* primer. (C) Primer extension analysis using *HSA* primer.

several minor sites, which gives rise to several mRNA species. Figure 5C shows that, as in the case of *GCN4* mRNA, the *HSA* mRNA is expressed at a similar level before and after 3-AT is provided. Most initiation sites used by RNA polymerase to transcribe the *GCN4*/*HSA* mRNA are identical to those found in *GCN4* mRNAs (compare Figure 5, B and C). Curiously, the level of *GCN4* mRNA in cells that express HSA is high (Figure 4B). The reason for this phenomenon is unknown to us. In contrast to the constitutive and uniform expression of the *HSA* mRNA, the HSA protein was barely detectable before induction with 3-AT, and its expression increased in time following induction. We conclude that 3-AT-dependent induction of HSA expression occurs at the translational level.

GCN4-Based Vector Induces Efficient Expression as an Integrated Copy

The transcription rate of *GCN4* seems to be high under most growth conditions in various strains and mutants (5,14). Accordingly, mRNA levels of *HSA* expressed from pGES plasmids do not change significantly under different growth conditions (Figure 5 and data not shown). We assume therefore that the efficient expression from the pGES426-L-HSA stems mainly from the efficient translation rate and less from the high copy number of the plasmid. If this is the case, we should be able to use the *GCN4*/*HSA* cassettes as integrated copies. Cells carrying an integrated vector could be grown on YPD medium to high cell densities and produce high yields of HSA.

To test this possibility, we have inserted the *GCN4*-L-*HSA* fragment into an integrative vector and thereby obtained the pGES306-L-HSA plasmid (see Materials and Methods). This construct was integrated into the yeast genome, and the resulting clones were grown on rich nonselective media to a high cell concentration (5×10^8 cells/mL) before the addition of 3-AT. Under these conditions, the expression of HSA was still efficiently induced (Figure 6). The level of HSA obtained per cell was just slightly less than that obtained using the multicopy pGES426-L-HSA vector. However, as the culture grown on YPD reached higher cell concentration, the quantity of intracellular HSA protein obtained (about 0.5 mg/L) was about two times higher of that obtained with the multicopy plasmid pGES426-L-HSA. Notably, all experiments described in this study were performed on cultures grown in laboratory flasks, conditions that do not support maximal yields.

To compare the pGES vector to commonly used vectors, we have ligated the cDNA that encodes *HSA* downstream

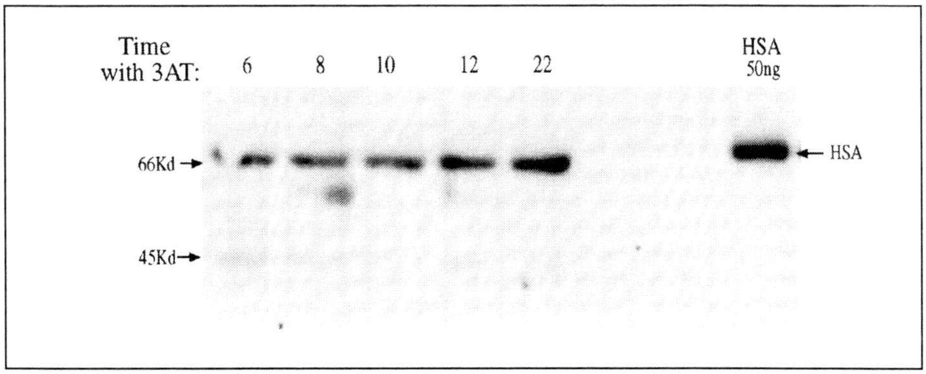

Figure 6. pGES system is inducible in highly concentrated cultures. A culture harboring an integrated copy of pGES306-L-HSA was grown on YPD medium to $A_{600} = 3.3$ when the medium was replaced with SD(-histidine) supplemented with 40 mM 3-AT. Samples were removed and analyzed by Western blot analysis at the indicated time points after induction. Purified HSA was loaded on the right lane of the gel.

from the *ADH1* promoter in the vector pAD4Δ (1). Cultures that harbor either pADH-HSA or pGES426-L-HSA were grown in parallel, and the expression of HSA was compared. In the case of pGES426-L-HSA, expression was compared after induction. The levels of HSA expression obtained from both plasmids were similar. Using commercial purified HSA as a standard, we calculated that about 250 µg/L of intracellular HSA were obtained in this experiment, about half of the amount obtained using the integrative version (pGES306-L-HSA) on rich medium. We conclude that the pGES system could be more efficient than currently available yeast expression vectors.

Although pGES plasmids already seem quite efficient, they could be regarded as archetypes that could be further manipulated and improved. For example, it might be possible to increase the mRNA level of *HSA* through a combination of a strong constitutive promoter (of another gene) with the 5′ untranslated region (UTR) of *GCN4*. Such a chimera might be more efficient than pGES and still be translationally regulated. The regulation of *GCN4* translation is well studied, and mutants are available in which the *GCN4* translation rate is constitutively high and not regulated (*gcd* mutants) (11,14,17). Also, *GCN4* translation is induced by the yeast Ras/cAMP cascade and is high in RAS^{val19} and *bcy1* strains (5). The pGES plasmids could be used in *gcd* strains and in $RAS2^{val19}$ and *bcy1* strains as constitutive vectors. These ideas are only a few of the many ways in which the pGES vectors could be used to obtain large quantities of proteins for research and industry.

ACKNOWLEDGMENTS

We wish to thank Prof. Sergei Braun for critically reading the manuscript.

REFERENCES

1. **Ballester, R., T. Michaeli, K. Ferguson, H.-P. Xu, F. McCormick and M. Wigler.** 1989. Genetic analysis of mammalian GAP expressed in yeast. Cell *59*:681-686.
2. **Chritianson, T.W., R.S. Sikorski, M. Dante, J.H. Shero and P. Hieter.** 1992. Multifunctional yeast high-copy-number shuttle vectors. Gene *110*:119-122.
3. **De Baetselier-Van Broekhoven, A.** 1994. Yeast as source for therapeutic and diagnostic proteins, p. 431-447. *In* Y. Murooka and T. Imanaka (Eds.), Recombinant Microbes For Industrial and Agricultural Applications. Marcel Dekker, New York.
4. **Dever, T.E., L. Feng, R.C. Wek, A.M. Cigan, T.F. Donahue and A.G. Hinnebusch.** 1992. Phosphorylation of initiation factor 2α by protein kinase GCN2 mediates gene-specific translational control of GCN4 in yeast. Cell *68*:585-596.
5. **Engelberg, D., C. Klein, H. Martinetto, K. Struhl and M. Karin.** 1994. The UV response involving the Ras signaling pathway and AP-1 transcription factors is conserved between yeast and mammals. Cell *77*:381-390.
6. **Engelberg, D., E. Zandi, C.S. Parker and M. Karin.** 1994. The yeast and mammalian Ras pathways control transcription of heat shock genes independently of heat shock transcription factor. Mol. Cell. Biol. *14*:4929-4937.
7. **Gao, B., S.C. Flores, S.K. Bose and J.M. McCord.** 1996. A novel *Escherichia coli* vector for oxygen-inducible high-level expression of foreign genes. Gene *176*:269-272.
8. **Garreau, H., J.H. Camonis, C. Guitton and M. Jacquet.** 1990. The *Saccharomyces cerevisiae* CDC25 gene product is a 180 kDa polypeptide and is associated with a membrane fraction. FEBS Lett. *269*:53-59.
9. **Gold, L.** 1990. Expression of heterologous proteins in *Escherichia coli*. Methods Enzymol. *185*:11-14.
10. **Harashima, S.** 1994. Heterologous protein production by yeast host-vector systems, p. 137-158. *In* Y. Murooka and T. Imanaka (Eds.), Recombinant Microbes for Industrial and Agricultural Applications. Marcel Dekker, New York.
11. **Harashima, S. and A. Hinnebusch.** 1986. Multiple GCD genes required for repression of GCN4, a transcriptional activator of amino acid biosynthetic genes in *Saccharomyces cerevisiae*. Mol. Cell. Biol. *6*:3990-3998.
12. **Hill, D.E., I.A. Hope, J.P. Macke and K. Struhl.** 1986. Saturation mutagenesis of the yeast his3 regulatory site: requirements for transcriptional induction and for binding by GCN4 activator protein. Science *234*:451-457.
13. **Hinnebusch, A.G.** 1984. Evidence for translational regulation of the activator of general amino acid control in yeast. Proc. Natl. Acad. Sci. USA *81*:6442-6446.
14. **Hinnebusch, A.G.** 1985. A hierarchy of trans-acting factors modulates translation of an activator of amino acid biosynthetic genes in *Saccharomyces cerevisiae*. Mol. Cell. Biol. *5*:2349-2360.
15. **Hinnebusch, A.G.** 1988. Mechanisms of gene regulation in the general control of amino acid biosynthesis in *Saccharomyces cerevisiae*. Microbiol. Rev. *52*:248-273.
16. **Hinnebusch, A.G.** 1992. General and pathway-specific regulatory mechanisms controlling the synthesis of amino acid biosynthetic enzymes in *Saccharomyces cerevisiae*, p. 319-414. *In* J.R. Broach, E.W. Jones and J.R. Pringle (Eds.), The Molecular and Cellular Biology of the Yeast Saccharomyces. Vol. 2: Gene Expression. CSH Laboratory Press, Cold Spring Harbor, NY.
17. **Hinnebusch, A.G.** 1997. Translational regulation of yeast GCN4. J. Biol. Chem. *272*:21661-21664.
18. **Kataoka, T., S. Powers, C. McGill, O. Fasano, J. Strathern, J. Broach and M. Wigler.** 1984. Genetic analysis of yeast RAS1 and RAS2 genes. Cell *37*:437-445.
19. **Kerry-Williams, S.M., S.C. Gilbert, L.R. Evans and D.J. Ballance.** 1998. Disruption of the *Saccharomyces cerevisiae* YAP3 gene reduces the proteolytic degradation of secreted recombinant human albumin. Yeast *14*:161-169.
20. **Mattanovich, D., R. Weik, S. Thim, W. Kramer, K. Bayer and H. Katinger.** 1996. Optimization of recombinant gene expression in *Escherichia coli*. Ann. NY Acad. Sci. USA *782*:182-190.
21. **Romanos, M.A., C.A. Scorer and J.J. Clare.** 1992. Foreign gene expression in yeast: a review. Yeast *8*:423-488.
22. **Rose, A.B. and J.R. Broach.** 1990. Propagation and expression of cloned genes in yeast: 2-µm circle-based vectors. Methods Enzymol. *185*:234-279.
23. **Schena, M., D. Picard and K.R. Yamamoto.** 1991. Vectors for constitutive and inducible gene expression in yeast. Methods Enzymol. *194*:389-398.
24. **Schneider, J.C. and L. Guarente.** 1991. Vectors for expression of cloned genes in yeast: regulation, overproduction, and underproduction. Methods Enzymol. *194*:373-388.
25. **Sikorski, R.S. and P. Hieter.** 1989. A system of shuttle vectors and yeast host strains designed for efficient manipulation of DNA in *Saccharomyces cerevisiae*. Genetics *122*:19-27.
26. **Sleep, D., G.P. Belfield and A.R. Goodey.** 1990. The secretion of human serum albumin from the yeast *Saccharomyces cerevisiae* using five different leader sequence. Biotechnology *8*:42-46.
27. **Walsh, G. and D.R. Headon.** 1994. Protein Biotechnology. John Wiley & Sons, West Sussex, England.
28. **Yasueda, H. and H. Matsui.** 1994. Overproduction of heterologous proteins in *Escherichia coli*, p. 49-70. *In* Y. Murooka and T. Imanaka (Eds.), Recombinant Microbes For Industrial and Agricultural Applications. Marcel Dekker, New York.
29. **Zubenko, G.S., A.P. Mitchell and E.W. Jones.** 1980. Mapping of the proteinase B structural gene PRB1, in *Saccharomyces cerevisiae* and identification of nonsense alleles within the locus. Genetics *96*:137-146.

Update to: *GCN4*-Based Expression System (pGES): Translationally Regulated Yeast Expression Vectors

Avraham Laban and David Engelberg[1]

Jexys (Jerusalem Expression Systems) Ltd. and [1]*The Hebrew University of Jerusalem, Jerusalem, Israel*

IMPROVED VECTORS OF THE GCN4-BASED EXPRESSION SYSTEM (pGES)

Here we describe a series of vectors, which are new members of the basic *GCN4*-based expression system (pGES vectors). As we previously suggested (2), pGES vectors could be further improved in many ways. The series presented here is the first example of such improvements. The new plasmids were constructed in order to achieve two goals. The first goal is to facilitate the use of pGES vectors (making them user-friendly) by improving several aspects: (*i*) convenient cloning, (*ii*) widening the possibility to introduce them to a wide range of yeast strains, (*iii*) providing options for easy detection and purification of expressed proteins, and (*iv*) to facilitate secretion of expressed protein. This goal was achieved through constructing pGES vectors with five different selectable markers (*URA3*, *LEU2*, *TRP1*, *HIS3* and *ADE2*) and a multicloning site comprising a large variety of cloning options (Figure 1). In addition, many of the plasmids were equipped with the sequence encoding the leader peptide of the yeast mating factor α. This peptide supports secretion of the protein. Finally, several tags were added including the sequence encoding polyhistidine (facilitating purification on Ni^{2+}-agarose beads) and the sequence encoding the cellulose binding domain (CBD) that facilitates purification by binding to cellulose (plasmids not shown). The second goal is to increase levels of expression and to have better control over expression. Several strategies are being taken towards this goal. As translation rate from pGES vectors is very high (2), one approach we take is aimed at increasing their transcription rate too. For this purpose, we constructed a vector in which the strong inducible promoter of the *GAL1* gene was ligated upstream to the 5′ untranslated reigon (UTR) of the *GCN4* gene (pGAL-GES; Figure 2A). Transcription from this vector is induced by galactose, and translation is induced by addition of 3-aminotriazole. Thus, efficient transcription and efficient translation synergize for achieving high yields of proteins.

Another approach towards improving expression is to increase translation rate. As reported previously (3), translation from the 5′ UTR of the *GCN4* gene is dramatically elevated (about 8-fold) if the 4 upstream open reading frames (uORFs) are mutated. We have, therefore, mutated those ORFs in some of the pGES plasmids (Figure 2B). Translation from these vectors is very efficient but constitutive. Namely, our control over expression was lost. To obtain a vector with efficient translation, and yet inducible, we inserted mutations in the 4uORFs of the vector pGAL-GES (Figure 2C). Expression from this plasmid is induced by galactose, and the mRNA obtained is constitutively translated with high efficiency.

The effort to make the pGES system more efficient is not restricted to improving the vectors. As we previously suggested (2), the use of the pGES vector in appropriate strains and will provide higher protein levels. Toward this goal, we are producing strains in which we systematically disrupt genes of the *GCN* and *GCD* families. As these genes encode inhibitors of *GCN4* translation (1) disruption of some of them

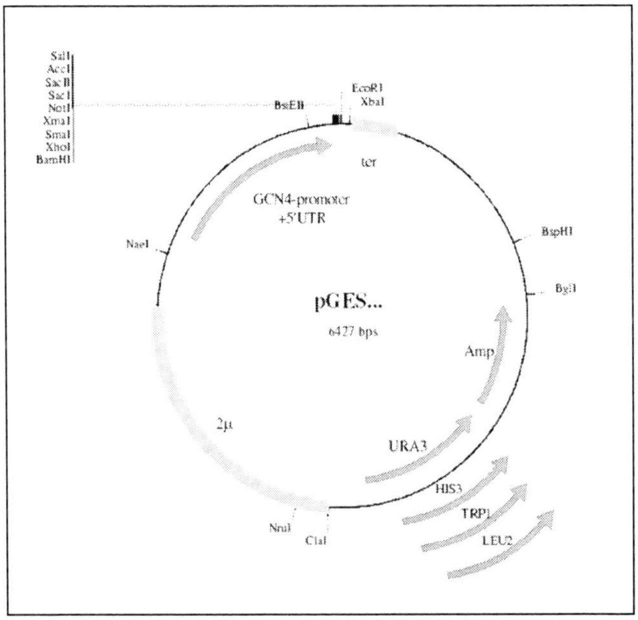

Figure 1. A schematic archetype of improved user-friendly pGES vectors. Restriction sites in the multicloning site are shown as well as optional selectable markers. Note that the plasmid shown contains the 2μ element. Equivalent plasmids without an origin of replication (for chromosomal integration) are also available.

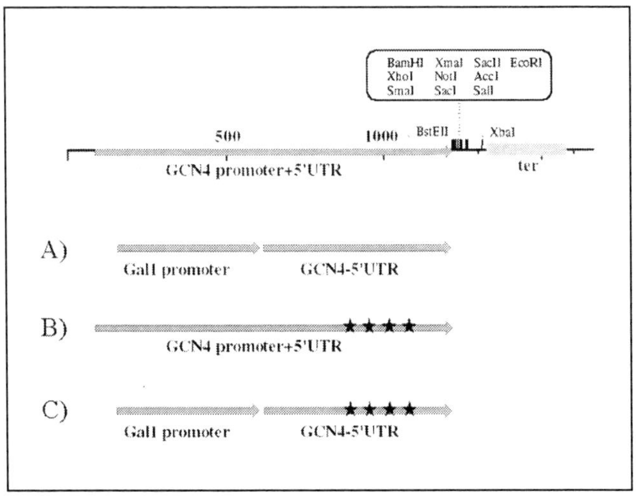

Figure 2. New strategies for improved expression level of pGES vectors. Upper panel shows the basic expression driving elements of pGES vectors, the *GCN4* promoter and 5′ UTR. Shown below are modifications of the basic elements. (A) The *GAL1* promoter fused to the *GCN4* 5′ UTR. (B) The 4 upstream ORFs of the *GCN4* 5′ UTR are eliminated (removal of an ORF is shown by a star). (C) The promoter of *GAL1* is fused to a mutated 5′ UTR of *GCN4*.

result in dramatic increase of *GCN4* translation (1).

The ultimate goal is to identify the optimal combination of a vector and a strain that gives the highest quantities of protein produced per volume of culture. Combining vectors and strains, we believe that the *GCN4*-based expression system will make a most efficient eukaryotic factory for the production of recombinant proteins.

REFERENCES

1. **Harashima, S., E.M. Hanning and A.G. Hinnebusch.** 1987. Interactions between positive and negative regulators of GCN4 controlling gene expression and entry into the yeast cell cycle. Genetics *117*:409-419.
2. **Mimran, A., I. Marbach and D. Engelberg.** 2000. *GCN4*-based expression system (pGES): translationally regulated yeast expression vectors. BioTechniques *28*:552-560.
3. **Mueller, P.P. and A.G. Hinnebusch.** 1986. Multiple upstream AUG codons mediate translational control of GCN4. Cell *45*:201-207.

CHAPTER 30

Eukaryotic Conditional Expression System

T. Sourisseau, Y. Le Dréan, G. Salbert, F. Flamant[1] and D. Michel

CNRS, Université de Rennes 1, Rennes, [1]CNRS-ENS de Lyon, Lyon, France

(Reprinted from BioTechniques 27:106-110, 1999)

ABSTRACT

Existing conditional expression systems can be classified in two major categories that are based either on the induction or on the derepression of transcription. The system described here combines both mechanisms, since a unique transcription factor can be shifted from a repression to a stimulation activity by simply changing its ligand. The resulting advantage of this system is the complete absence of basal expression before active induction. The principle of this method is based on the unexpected ability of the chimeric protein containing the DNA-binding domain of the yeast Gal4 transcription factor fused to the COOH half of the estradiol receptor (GalER), to act as a repressor when bound to the drug 4OH-tamoxifen, in the context of a previously described optimized Gal4-responsive promoter. The efficacy of this system has been assessed in transient expression assays using the chloramphenicol acetyl transferase (CAT), and in situ, through the activity of a Gal4 responsive β-galactosidase gene.

INTRODUCTION

Since no protein is completely inert when ectopically expressed, controlling transgene expression is a major challenge in biotechnology. The tight regulation of transgenes is a prerequisite for producing proteins of medical or economical interest that display some toxic or cytostatic properties for cells and organisms. In a more fundamental point of view, conditional expression systems will become a tool of choice, complementary to gene inactivation, to gain insight into the function of the unknown genes provided by systematic sequencing strategies. Two symmetrical approaches have been used to develop effector-driven expression systems: (*i*) the induction method, using ligand-dependent transcription machineries and (*ii*) the derepression method, using repressors that can be shifted in a ligand-dependent manner into conformations unable to bind DNA. These latter tools more often derive from prokaryotic gene repressors and their corresponding operator DNA-binding sites, as the tetracyclin system (4) and the lactose operator-repressor system (10).

The major limitation of the available inducible expression systems is the presence of a basal expression level in absence of active induction. This problem is likely a result of the spontaneous activity of the minimal promoters used in transgene constructions and by virtually any DNA fragment, due to the presence of cryptic binding sites for known or unknown transcription factors. This drawback is particularly obvious in the induction systems using *cis*-elements recognized by endogenous transcription factors like heat-shock elements, nuclear receptor target sites or heavy-metal responsive promoters.

The present method is based on the use of a chimeric transcription factor containing the DNA-binding domain from the yeast Gal4 transcription factor (Gal-DBD), which is known to have no equivalent in higher eukaryotes, thus avoiding undesirable interferences with endogenous transcription regulators. In the present case, Gal-DBD is fused to the ligand-dependent transcription activation domain from the human estradiol receptor (ER)-LBD. The resulting chimeric transcription factor, named GalER, has long been shown to be capable of enhancing the activity of a synthetic Gal4-responsive promoter after the addition of estrogen (9). The chimeric GalER factor presents two favorable features: (*i*) the presence of the activation domain from the carboxy-terminal end of the ER (AF2), which is well defined and clearly estradiol-dependent and, in turn, (*ii*) the absence of the NH_2-terminal domain of the ER (AF1), that displays some ligand-independent activity, which is unfavorable for the present purpose. However, the interest of such an estrogen-inducible system was greatly hindered by: (*i*) the presence of endogenous estrogens in body fluids or in the serums incorporated into cell culture media and (*ii*) by the spontaneous basal activity of Gal4-responsive promoters. To circumvent this problem, a synthetic Gal4-reponsive promoter has been constructed, with extremely low basal activity (2). This optimized construct contains four palindromic Gal4-binding sites, linked to TATA and CCAAT boxes. However, we observed that this promoter still presents a significant level of activity in the absence of induction. Here, we present an additional improvement of this GalER system, allowing in the same time, the repression of the basal expression activity and a stimulation of greater amplitude.

Cloning and Expression Vectors for Gene Function Analysis
Edited by Quinn Lu and Michael P. Weiner

MATERIALS AND METHODS

Materials

4-hydroxy-tamoxifen (4OH-tamoxifen) and estradiol were from Sigma (L'isle d'Abeau Chesnes, St. Quentin Fallavier, France). The reporter plasmid GalREnlslacZ was created by inserting a blunted fragment encoding the nuclear β-galactosidase (β-gal) (1) into the *Eco*RV site of GC vector, which carries the Gal4-responsive promoter (2). The Gal-chloramphenicol acetyl transferase (CAT) reporter plasmid is the pCAT-4 construct described previously (2) and was kindly provided by Dr. Meinrad Busslinger (Research Institute of Molecular Pathology, Vienna, Austria). The GalER expression plasmid, described by Webster et al. (9), was kindly provided by Prof. Pierre Chambon (IGBMC Strasbourg, France).

Cell Culture

CHO-CH3S cells were maintained at 37°C in 5% CO_2 atmosphere, in F12-Dulbecco's modified Eagle medium (1:1) supplemented with 10% dextran charcoal-treated fetal bovine serum (FBS) (Life Technologies, Gaithersburg, MD, USA). Phenol Red® was omitted because of its weak estrogenic activity. Medium was added with antibiotic/antimycotic solution composed of penicillin (500 U/µL), streptomycin (500 µg/µL) and amphotericin (1.25 µg/µL).

Transfections

Transfections were performed using a modified calcium phosphate precipitate method (2). Cells were seeded in 6-well tissue culture plates at the density of 10^5 cells/well one day before transfection. One hour before adding calcium phosphate/DNA co-precipitate, cells were provided with fresh medium either with or without 4OH- tamoxifen (1 µM). Twenty-four hours later, precipitate was removed by phosphate-buffered saline (PBS) washing and replaced by fresh medium containing either 4OH-tamoxifen or estradiol (1 µM each). Cells were then grown for transient expression during 24 h and then submitted to either protein extraction for CAT assays or in situ *Lac*Z detection.

CAT Assays

Cells transfected with Gal-CAT and GalER, along with cytomegalovirus (CMV) β-gal as an internal control of transfection, were harvested after 48 h transient expression by scraping in chilled PBS containing 5% glycerol and centrifuged. Cell pellets were resuspended in 130 µL of Tris-HCl, pH 7.8, containing 0.5% Triton® X-100 and then frozen and thawed twice to improve lysis. Cellular debris were then eliminated by centrifugation, and 98-µL aliquots of the supernatants were submitted to a 10-min incubation at 65°C to remove potential CAT inhibitors. Enzymatic activity was determinated using the following protocol: 98 µL of cell extracts were put in presence of 20 µL chloramphenicol, 5 mM in Tris-HCl (250 mM), pH 7.8, 0.5 µL acetyl-CoA (both from Boehringer Mannheim GmbH, Mannheim, Germany) (4.54 µM), 0.4 µL [^{3}H]acetyl-CoA (200 µCi/µM; ICN Biomedicals, Orsay, France) and 19.1 µL HCl (75 µM). After a 3-h incubation at 37°C, reactions were stopped by adding 1 mL of 5 M urea. β-gal activities were determinated using the uncolored substrate of this enzyme, namely the ortho-nitrophenyl-β-D-galactopyranoside (ONPG; Boehringer Mannheim GmbH) as follows: 30 µL of lysate were incubated with 330 µg ONPG (1.1 µg/µL) in sodium phosphate buffer (pH 7.5). Reaction products were then assayed by spectrophotometric measurement at 450 nm. The relative CAT activities were expressed as CAT/β-gal ratios. These ratios were compared to determine the fold activations of the system.

In Situ Detection of *Lac*Z-Positive Cells

Cells were initially fixed by a 15-min incubation in 1% glutaraldehyde solution at room temperature. After washing in PBS, they were then incubated at 37°C for 90 min in a coloration solution containing 5-bromo-4-chloro-3-indyl-β-D-galactoside (X-gal; Life Technologies) as a colorless substrate of β-gal. After removing the coloration solution, cells were covered with 70% glycerol and observed by light microscopy.

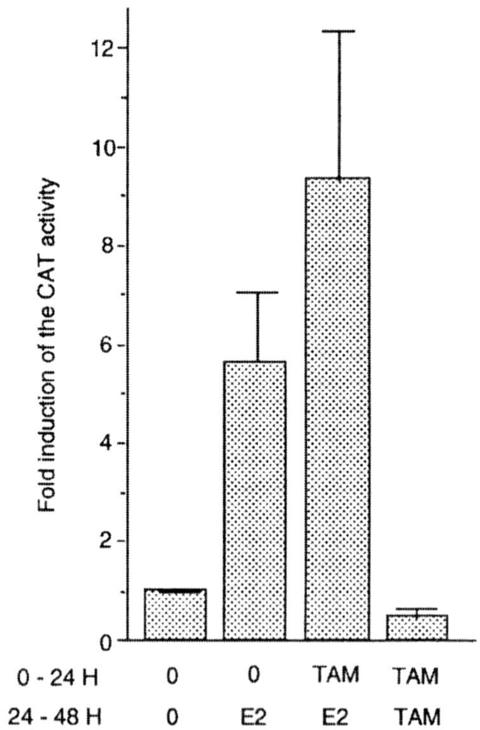

Figure 1. CHO-CH3S cells (1.5 × 10^5 cells/well) were co-transfected with the Gal-CAT reporter plasmid and a GalER expression vector, along with CMVβ-gal as an internal control of transfection. Cultures were treated with or without estradiol (E2) or 4OH-tamoxifen (TAM) during 48 h, or successively with TAM and then E2, as indicated under the histogram. Cells lysates were then tested for CAT activity as described in Materials and Methods. Results are expressed as -fold induction of the CAT activity relative to cells treated with the solvent alone (ethanol) (means ±SD; n = 3).

RESULTS

The synthetic Gal4-responsive promoter with extremely low basal activity (2) was used to direct the expression of a modified *LacZ* coding region containing the nuclear localization signal from the simian virus 40 (SV40) large T antigen (GalRE-nlsLacZ). The nuclear concentration of *LacZ* gene product further facilitates the detection of cells expressing *LacZ* at low level. The Gal4-responsive promoter used in this construction was also to be activated by the chimeric transcription factor GalER in presence of estradiol and to display a low spontaneous activity (2). However, when introduced into Chinese hamster ovary (CHO) cells, we observed a significant activity even in absence of GalER, as demonstrated for instance, by in situ *LacZ* staining after introduction of the GalRE-nlsLacZ plasmid (Figure 2A, column 1). Such a background expression is likely to prohibit the use of this system for transgenes that can affect cell viability. When attempting to lower this basal activity, we found that in presence of GalER, 4OH-tamoxifen, a low-priced compound known for its ability to act as an anti-estrogen by preventing estradiol-dependent transcription stimulation by ER (8), can significantly repress the level of background transcription. In CAT assays, done in presence of GalER, the sustained presence of 4OH-tamoxifen in the culture medium decreases, by more than half, the basal CAT activity (Figure 1, columns 1 and 3). In addition, preincubation of cells with 4OH-tamoxifen before estrogen addition, leads to an unexpected enhancement of the stimulating action by estrogen. Combined with its lessening effect on the basal expression, this enhancing activity allows a dramatic increase of the amplitude of induction by estrogen. Indeed, estrogen addition at 24 h leads to a 5.6-fold activation of the Gal4-responsive promoter (Figure 1, columns 1 and 2), while a 4OH-tamoxifen pretreatment from 0–24 h leads to a 19.4-fold induction (Figure 1, compare columns 3 and 4). In fact, this effect of 4OH-tamoxifen, which is very useful in the context of this system, could have been expected as a result of the previous observation that estradiol stimulation of the ER can be enhanced by previous addition of an anti-estrogen (5).

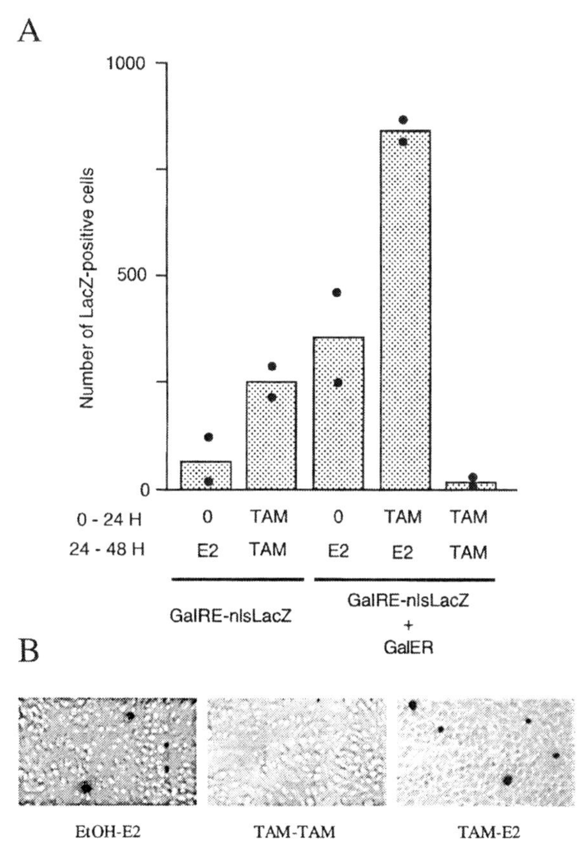

Figure 2. Effects of successive treatments by estradiol and/or 4OH-tamoxifen on the GalER-mediated activity of GalRE-nlsLacZ. (A) Number of *LacZ*-positive cells obtained after transfection with the GalRE-nlsLacZ reporter plasmid alone or co-transfected with an expression vector of GalER by the calcium phosphate method. 4OH-tamoxifen (TAM) was or was not added together with the calcium phospate DNA precipitate. Twenty-four hours later, the precipitate was washed and replaced by a fresh medium containing 1 µM TAM or estradiol (E2). This figure is obtained from two independent experiments (dots). (B) Representative regions of dishes co-transfected with GalRE-nlsLacZ and GalER and *LacZ*-stained 48 h after transfection. Cells were treated with solvent from 0–24 h and then E2 from 24–48 h (left panel), with TAM from 0–48 h (middle) or with TAM from 0–24 h and then with E2 (right).

We have also tested the validity of the system through the in situ β-gal detection assay. This alternative method was also used to determine if GalER is involved in mediating the effects of estradiol and 4OH-tamoxifen. As shown in Figure 2, the induction by estradiol of the number of blue cells is estradiol receptor-dependent, considering the 4-fold increase of *LacZ*-positive cells induced by the presence of GalER (Figure 2A, compare columns 1 and 3). One must notice, however, that the *LacZ* staining is more intense upon stimulation by estradiol (data not shown), which might reflect a certain stimulating effect of estradiol. By itself, 4OH-tamoxifen seems to have an estrogen receptor-independent stimulating effect on transgene expression (Figure 2A, column 2). In presence of GalER, 4OH-tamoxifen appears capable of strongly inhibiting the basal expression of the system (Figure 2A, column 5). This 4OH-tamoxifen-induced repression by GalER is powerful because very few *LacZ*-positive cells are detected after 4OH-tamoxifen pretreatment. For only 2 blue cells observed in one experiment in the presence of 4OH-tamoxifen, more than 800 were observed in an identical dish after replacement of 4OH-tamoxifen by estrogen. Moreover, considering that GalER and the *LacZ* genes are carried by two different plasmids transfected in an equimolar ratio, one cannot exclude that these few cells are precisely those containing only the GalRE-nlsLacZ reporter plasmid. This drastic repressive effect of 4OH-tamoxifen cannot result from a mere negative effect of this compound on transfection efficiency or on cell viability, since it is clearly dependent on the presence of GalER (Figure 2A, compare columns 2 and 5). In addition, the effect of 4OH-tamoxifen is also fully reversible because it does not forbid the action of a subsequent estradiol treatment, as demonstrated by the high inducibility by estradiol after 4OH-tamoxifen withdrawal. Indeed, as previously observed in CAT assays, the stimulating action of estradiol appears of higher amplitude if cells containing both GalER and GalRE-nlsLacZ are pretreated with 4OH-tamoxifen. 4OH-tamoxifen pretreatment enhances the subsequent stimulating effect of estradiol, leading to a 2.4-fold increase of the number of *LacZ*-positive cells (Figure 2A, com-

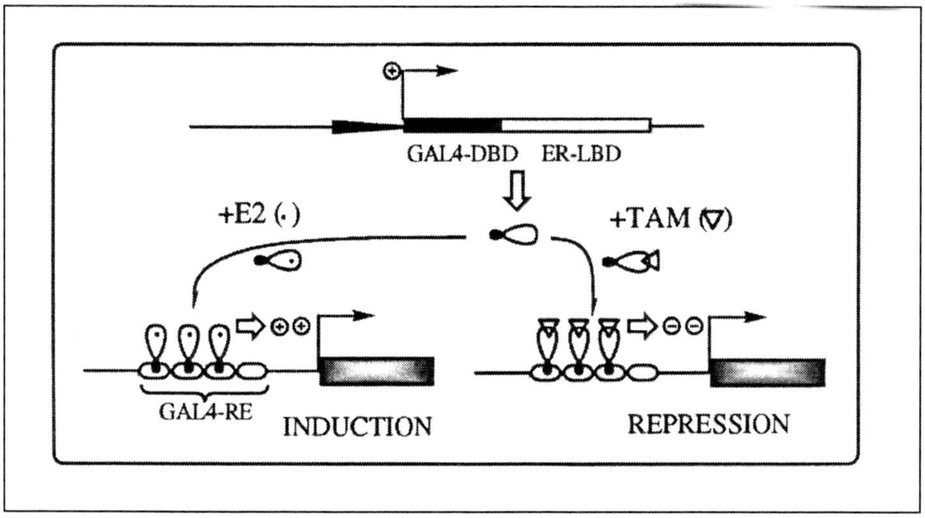

Figure 3. Schematic representation of the GalER-tamoxifen system. The same chimeric transcription factor, GalER, has the dual capacity of stimulating or inhibiting the transcription activity of the Gal4-responsive promoter, accordingly, as the ligand added is estradiol or tamoxifen, respectively.

pare columns 3 and 4). The conditions used for Figure 2A, column 4, appear optimal for using GalER. Thus, as observed through the two techniques, CAT assays and X-gal staining, the use of tamoxifen greatly improves the use of the GalER/Gal4-RE system at two levels: (*i*) by reducing the background expression and (*ii*) by enhancing the induction step.

DISCUSSION

We show that the conditional expression system, based on the Gal4-responsive promoter constructed by Braselmann et al. (2), can be strikingly improved by using the drug 4OH-tamoxifen. Fortuitously, we found that in addition to the enhanced inducibility by estradiol, 4OH-tamoxifen strongly decreases the basal activity of the Gal4-responsive promoter. This effect is clear in CAT assays and particularly obvious when using the in situ β-gal staining. The apparently more stringent regulation revealed by the latter method is probably inherent in the requirement for a threshold basal concentration of β-gal to be detectable. By itself, the repressive activity of 4OH-tamoxifen mediated by GalER, might appear surprising because contrary to other nuclear receptors, as for example the unliganded thyroid hormone (11), reports on a repressive activity of the estradiol receptor are very few. Interestingly, it has been described that the AF2 domain of the estrogen receptor, present in GalER, can acquire a repressive activity through interaction with C/EBP (7). This finding is of particular interest considering that the Gal4-responsive promoter used in this system does contain a typical C/EBP target CCAAT box. It has also recently been shown that tamoxifen can promote the binding of co-repressors to the ER (6). Elucidating the molecular mechanism underlying the 4OH-tamoxifen-induced repression would be of fundamental interest. In the context of this work, this feature makes it possible to shift from a tight repression to a tight induction of the expression by using a single transcription factor (see Figure 3 for a synthetical scheme of the system). While only one out of the two activities of induction and repression is exploited at the same time in the other conditional expression systems, they are reunited in the present one, as a result of the dual activity of 4OH-tamoxifen. Taken together, the absence of Gal4-RE-binding factors and Gal4 consensus recognition sites in higher eukaryotes, the lack of cellular toxicity of 4OH-tamoxifen and of estradiol and the strong repressive activity of 4OH-tamoxifen-bound GalER make this system particularly useful for finely regulating transgene expression. As an ultimate test, we plan to use this system for controlling the expression of toxic transgenes. In conclusion, the improved system presented here allows the reconciliation of most requirements for an ideal inducible system; specifically, a strong activity upon induction and, more importantly, a low activity before induction.

ACKNOWLEDGMENTS

This work was funded by Education Nationale (DRED), Centre National de la Recherche Scientifique (CNRS) and supported by the Association pour la Recherche contre le Cancer (ARC).

REFERENCES

1. **Bonnerot, C., D. Rocancourt, P. Briand, G. Grimber and J.F. Nicolas.** 1987. A beta-galactosidase hybrid protein targeted to nuclei as a marker for developmental studies. Proc. Natl. Acad. Sci. USA *84*:6795-6799.
2. **Braselmann, S., P. Graninger and M. Busslinger.** 1993. A selective transcriptional induction system for mammalian cells based on Gal4-estrogen receptor fusion proteins. Proc. Natl. Acad. Sci. USA *90*:1657-1661.
3. **Chen, C. and H. Okayama.** 1987. High-efficiency transformation of mammalian cells by plasmid DNA. Mol. Cell Biol. *7*:2745-2752.
4. **Gossen, M. and H. Bujard.** 1992. Tight control of gene expression in mammalian cells by tetracycline-responsive promoters. Proc. Natl. Acad. Sci. USA *89*:5547-5551.
5. **Lascombe, I., P. Mougin, C. Vuillermoz, G.L. Adessi and M. Jouvenot.** 1995. Gene transfer into subcultured endometrial cells using lipofection. BioTechniques *20*:88-91.
6. **Lavinsky, R.M., K. Jepsen, T. Heinzel, J. Torchia, T.M. Mullen, R. Schiff, A. Del-Rio, M. Ricote et al.** 1998. Diverse signaling pathways modulate nuclear receptor recruitment of N-CoR and SMRT complexes. Proc. Natl. Acad. Sci. USA *95*:2920-2925.
7. **Stein, B. and M.X. Xang.** 1995. Repression of the interleukin-6 promoter by estrogen receptor is mediated by NF-κB and C/EBPβ. Mol. Cell Biol. *15*:4971-4979.
8. **Sutherland, R.L., L.C. Murphy, M. San Foo, M.D. Green, A.M. Whybourne and Z.S. Krozowski.** 1980. High-affinity anti-estrogen binding site distinct from the estrogen receptor. Nature *288*:273-275.

9. **Webster, N.J., S. Green, J.R. Jin and P. Chambon.** 1988. The hormone-binding domains of the estrogen and glucocorticoid receptors contain an inducible transcription activation function. Cell *54*:199-207.

10. **Wu, J.D., H.C. Hsueh, W.T. Huang, H.S. Liu, H.W. Leung, Y.R. Ho, M.T. Lin and M.D. Lai.** 1997. The inducible lactose operator-repressor system is functional in the whole animal. DNA Cell Biol. *16*:17-22.

11. **Zhang, J., I. Zamir and M.A. Lazar.** 1997. Differential recognition of liganded and unliganded thyroid hormone receptor by retinoid X receptor regulates transcriptional repression. Mol. Cell Biol. *17*:6887-6897.

Update to: Eukaryotic Conditional Expression System

T. Sourisseau, Y. Le Dréan, G. Salbert, F. Flamant[1] and D. Michel

CNRS, Université de Rennes 1, Rennes, [1]CNRS-ENS de Lyon, Lyon, France

The gene expression system described in this paper was primarily intended to work in the context of transfected plasmids. The use of episomic vectors, such as plasmids or artificial chromosomes, offers certain advantages over genomic integration, in that they allow position-independent and copy number-dependent transgene expression. Conversely, the main drawback of episomic vectors, regardless of the "fold-induction" parameter, is generally their high level of basal expression, which makes them unsuitable for many applications using somehow toxic transgenes. The GalRE/GalER system that we proposed allows to reduce the level of basal expression, thanks to an active repression mediated by 4OH-tamoxifen-liganded GalER and by using a DNA-binding factor (GalER) and a DNA target sequence (GalRE: 5'-CGGAGGACAGTCCTC-CG-3') that are absent from higher eukaryotes, thus preventing undesired stimulation by endogenous DNA-binding factors. Indeed, the Gal4 transcription factor combines two determinant advantages: *(i)* though of eukaryotic origin (yeast), Gal4 does not have any equivalent in higher eukaryotes, as well as are the prokaryotic molecules often selected in other eukaryotic expression systems, such as T7 RNA polymerase or tetracyclin repressors; and *(ii)* contrary to prokaryotic molecules, Gal4 possesses a nuclear targeting signal, which is functional in all eukaryotic cells.

In our hands, the GalRE/GalER molecular tool works well in the conditions described in the article when using charcoal-treated serum. We must, however, indicate that we observed a decrease of the repressive activity of 4OH-tamoxifen when using untreated serum still containing liposoluble factors as steroids. This condition may appear surprising, considering the purported anti-estrogenic action of 4OH-tamoxifen, and may impede the use of this tool for in vivo purposes. However, in its present state, this system already provides an interesting alternative to other inducible eukaryotic expression vectors for cultured cells.

The (GalRE)4-CCAAT-TATA promoter used in this study was dissected from the plasmid named pCAT-4 described in Braselman et al. (see Reference 2 in original article) and was obtained from Dr. Meinrad Busslinger (Research Institute of Molecular Pathology, Vienna, Austria). The GalRE-LacZ construct, containing this promoter in front of the *LacZ* reporter gene, is available from Dr. Frederic Flamant (Laboratoire de Biologie Moléculaire et Cellulaire, Lyon, France). The GalER construct, described in Webster et al. (see Reference 9 in original article), was kindly provided by Pr. Pierre Chambon (Institut de Génétique et de Biologie Moléculaire et Cellulaire, Illkirch, France).

CHAPTER 31

Lac/Tet Dual-Inducible System Functions in Mammalian Cell Lines

H.-S. Liu, C.-H. Lee, C.-F. Lee, I.-J. Su and T.-Y. Chang

National Cheng Kung University, Tainan, Taiwan, ROC

(Reprinted from BioTechniques 24:624-632, 1998)

ABSTRACT

The Escherichia coli *Lac repressor (Lac system) and tetracycline responsive promoter (Tet system) systems have been used individually to regulate gene expression at the cellular as well as at the organismal levels. In this study, these two systems were combined (designated Lac/Tet dual-inducible system) to regulate two inducible genes simultaneously in a single cell. The isopropyl-β-D-thiogalactopyranoside (IPTG) and tetracycline (used for the operation of the Lac and the Tet systems) were non-cytotoxic to the cells when added together into the cells at around the optimal concentrations (IPTG: ≤5 mM; tetracycline: <1.5 μg). The rate and efficiency of induction and repression of two inducible genes regulated by the Lac/Tet dual-inducible system were similar to the results obtained when one inducible gene is regulated by one inducible system in a single cell. The Lac/Tet dual-inducible system could function in many cell lines, which was demonstrated by regulating the expression of β-galactosidase and luciferase reporter genes in five tumor cell lines by transient transfection analysis. The feasibility of introducing a second inducible system into an already established inducible cell line was confirmed. Finally, we showed that the Lac/Tet dual-inducible system functions at translational and at functional levels in a stable cell line named 7-4-b, which contains the Ha-*ras *and* bcl-2 *inducible genes. In conclusion, this study extends the application of prokaryotic inducible systems from the regulation of a single gene to two genes and helps clarify the relationship between two genes and the effects of two genes on the cells.*

INTRODUCTION

The concept of using *Escherichia coli* regulatory systems in mammalian cells has numerous attractive features, including having the potential for imposing very stringent regulation and reducing the likelihood that prokaryotic regulatory proteins will interact with host DNA sequences that are similar to target sequences in a biologically meaningful fashion. Moreover, the bacterial regulatory elements can function and regulate gene expression in mammalian cells with less restriction of promoter specificity and low side effects to the cells (5–7,12,14,17). We have demonstrated that the bacterial lactose repressor system (Lac system) can regulate the expression of a reporter gene by the binding of *lac* repressor (Lac R) to a *lac* operator sequence (Lac O), which is located between the TATA box and the transcription starting site of the gene in mammalian cells and mice. The reporter gene can be specifically activated by addition of the lactose analogue, isopropyl-β-D-thiogalactopyranoside (IPTG), which binds to the repressor and disables it from binding to the operator sequence by conformational change (12,13).

The tetracycline-dependent expression system (Tet system) reported by Gossen and Bujard (6) is based on two regulatory elements derived from the tetracycline (Tc)-resistance operon of the *E. coli* Tn10 transposon. They are the tet repressor protein (Tet R) and the tet operator sequence (Tet O) to which Tet R binds (6). The gene to be expressed is cloned into the "response" plasmid, which contains the $P_{hCMV\text{-}1}$ promoter and the subsequent gene of interest. $P_{hCMV\text{-}1}$ is a compound promoter consisting of the tet-responsive element (TRE), which contains seven copies of Tet O, and the minimal immediate early promoter of cytomegalovirus (P_{minCMV}). The second key component of the system is a "regulator" plasmid, which expresses a hybrid protein known as the tet-controlled transcriptional activator (tTA). tTA is encoded by pTet-off and is a fusion of the wild-type Tet R to the VP16 activation domain of herpes simplex virus. tTA binds and thereby activates transcription in the absence of Tc. Thus, as Tc is added to the culture medium, transcription is turned off (3,6).

Comparatively, the Tet system shows faster responses and higher induction efficiency than the Lac system (6). However, in the Tet expression system, the restrictedly used CMV minimal promoter has cell specificity and is not inducible in some of the cells (1). On the contrary, the Lac system is widely accepted by many cell lines because the cell-type specific pro-

Cloning and Expression Vectors for Gene Function Analysis
Edited by Quinn Lu and Michael P. Weiner

Table 1. Cytotoxicity of IPTG Plus Tetracycline to HA22T Hepatoma Cells

IPTG (mM)	2.5	2.5	2.5	2.5	2.5	5	5	5	5
Tetracycline (µg)	1	1.25	1.5	2	2.5	1	1.25	1.5	2
24 h[b]	**0[a]**	**0**	**8.23**	**8.69**	**8.9**	**0**	**0.69**	**8.11**	21.32
48 h	**0**	**4.51**	**7.42**	**4.57**	**8.69**	**0**	**6.28**	**10**	11.67
72 h	**7.05**	12.86	13.35	19.10	19.89	**4.95**	17.86	16.39	19.61

[a]Numbers indicate the percentage of cell death analyzed by the microassay analysis (see Materials and Methods).
[b]Hours post-treatment.
Bold numbers represent the cell death less than 10%.

moter can be used (12). Regardless of the variations in expression features of these two systems, they are among the best tools to study transient and temporary effects of gene function in cells and in animals (4,8,9,15,17). To further extend the application of the inducible system in gene functional study, we combined these two systems into a dual-inducible system and observed the repression and induction of two reporter genes in individual cells. Our data demonstrate the feasibility of operating the dual-inducible system, which regulates two genes at gene expression and at functional levels in many cell lines.

MATERIALS AND METHODS

Cells and Cell Culture

The cell lines of 212, 7-4, 7-4-2 and 7-4-b, derived from NIH/3T3 cells, were established in our laboratory (11,13). J82, TCCSUP, T1 and T2 are bladder carcinoma cell lines. The first two were from ATCC (Rockville, MD, USA), and the last two were established by Dr. C.C. Tzeng (16). HA22T (a hepatoma cell line) was from Dr. L.M. Chang (18). All the cells were maintained in Dulbecco's modified Eagle medium (DMEM) supplemented with 10% fetal bovine serum (FBS) (both from Life Technologies, Gaithersburg, MD, USA).

Plasmids

Plasmids pSVlacOZ, pSVlacOras, pTREbcl-2 and pHβlacINLSneo were constructed in our laboratory (12,13). Plasmids pTREluc and pTet-off were purchased from CLONTECH Laboratories (Palo Alto, CA, USA) (3).

DNA Transfection

The cells (1×10^5) in a 6-well tray were co-transfected with various plasmids (5 µg/dish) by the LIPOFECTIN® method (Life Technologies) for 5 h. The cells were then cultured in DMEM containing 10% FBS in either the presence or the absence of IPTG and/or tetracycline for 24–48 h. The cells were harvested, and lysates were assayed to determine gene expression activity.

Microassay for Cytotoxicity

The microassay for the cytotoxicity of IPTG and tetracycline was performed as described by Ito (10). Briefly, cells were seeded in the 96-well, flat-bottom microplates (Nalge Nunc International, Kamstrapvej, Denmark) at appropriate concentrations (1.5×10^4 cells/100 µL/well) and incubated at 37°C in a CO_2 incubator for 4 h for cell attachment. The medium was then replaced with either fresh medium or medium containing appropriate concentrations of tested agents (100 µL/well). Eight wells were used for each agent concentration. The cells were incubated for 1–3 days in the presence of IPTG plus tetracycline. At the end of the incubation, the medium was removed, and the cells in the wells were then fixed with 10% formalin (50 µL/well) for 10 min, following which they were stained with 0.05% crystal violet (30 µL/well) for 30 min. The stained microplates were rinsed with tap water and then air-dried. For the determination of the percent inhibition, colorimetry of the dye from cells was carried out. Dye was eluted by the addition of 50% ethanol/0.1% acetic acid (150 µL/well) and quantified immediately by optical density (OD) measured at 590 nm (OD_{590}) with a Model MR5000 Microplate Reader (Dynatech Laboratories, Chantilly, VA, USA). The percent inhibition was calculated according to the following equation:

$$\%\ \text{Inhibition} = 100 \times \frac{OD_{590}\,(\text{sample}) - OD_{590}\,(\text{medium control})}{OD_{590}\,(\text{cell control}) - OD_{590}\,(\text{medium control})}$$

Luciferase and β-galactosidase Activity Assay

The luciferase and β-galactosidase (β-gal) activities were determined by a Dual-Light™ Luciferase and β-galactosidase Reporter Gene Assay System (Tropix, Bedford, MA, USA) (14).

Western Blot Analysis

Fifteen micrograms of total protein from the cells were denatured in sodium dodecyl sulfate (SDS) lysis buffer and loaded onto duplicate 12% SDS polyacrylamide gels. After electrophoresis at 100 V for 3 h, one gel was stained with Coomassie® Brilliant Blue (Sigma Chemical, St. Louis, MO, USA) as a quantitative control, and the other was transferred to a polyvinylidine difluoride membrane (Stratagene, La Jolla, CA, USA) and blocked with 5% skim milk in phosphate-buffered saline Tween® (PBST) solution containing 0.2% Tween-20, pH 7.5 at 4°C overnight. After washing with

Table 2. Lac/Tet Dual-Inducible System Functions in Five Human Tumor Cell Lines

Cell Line	Induction Dual[a]	Induction Single[b]	Luciferase (RLU/µg)[d] Repressed	Luciferase (RLU/µg)[d] Induced	Fold[c] Increase	β-gal (RLU/µg)[d] Repressed	β-gal (RLU/µg)[d] Induced	Fold[c] Increase
T1	+		2490	234773	**94**	604	2416	**4**
T1		+	5599	244131	**44**	1074	6055	**5.6**
T2	+		2252	487816	**216**	366	1142	**3.2**
T2		+	5996	744972	**124**	590	5888	**10**
J82	+		41	1748	**43**	270	1431	**5.1**
J82		+	2268	52844	**23**	348	1119	**3**
TCCSUP	+		109	10704	**98**	446	2676	**6**
TCCSUP		+	69	31153	**451**	414	975	**2.4**
HA22T	+		923	310296	**336**	938	4461	**4.8**
HA22T		+	1502	869791	**579**	3152	32278	**10**

pTREluc (2 µg), pTet-off (1 µg), pSVlacOZ (2 µg) and pHβlacINLSneo (1 µg) were co-transfected into the tumor cells. The cell extracts were analyzed at 48 h post-transfection. This experiment was repeated three times.

[a]Both luciferase and β-gal genes were activated simultaneously.

[b]Only luciferase or β-gal gene was activated.

[c]The fold increase is a comparison between repressed and induced levels of expression.

[d]Relative luciferase unit per microgram cell extract.

PBST and PBS buffer, the membrane was incubated with the anti-human BCL-2 monoclonal antibody (PharMingen, San Diego, CA, USA) or anti-v-H-RAS antibody (Oncogene Science, San Diego, CA, USA) at 37°C for 1 h and then incubated with horseradish peroxidase-conjugated goat anti-mouse IgG (Jackson ImmunoResearch Laboratories, West Grove, PA, USA) at 25°C for 1 h. The membrane was exposed to X-ray film (Fuji, Tokyo, Japan) for 10 min after the ECL™ Detection Reagents (Amersham, Arlington Heights, IL, USA) were added (2).

RESULTS AND DISCUSSION

No Evident Cytotoxicity was Detected in Tumor Cells Treated with IPTG and Tetracycline in 48 h

Agents required for operating Lac and Tet systems are IPTG and tetracycline (Sigma Chemical). Their optimal concentrations are 2.5 mM and 1 µg, respectively (4,6). It has been reported that no side effects were detected while the agents were used to treat cells separately. However, to our knowledge, whether or not the agents will become toxic when they are added together has never been reported. Table 1 shows the cytotoxicity (% of inhibition) of IPTG and tetracycline to HA22T cells (a hepatoma cell line). Our data show that IPTG and tetracycline can be added together into the cells to operate the Lac/Tet dual-inducible systems when the concentration of IPTG is 5 mM or less and that of tetracycline is 1.5 µg or less. However, IPTG and tetracycline co-existing longer than 48 h may cause some cytotoxicity to the cells.

The Kinetics of the Induction and Repression of Luciferase and β-gal Transgene Expression in NIH/3T3 Cells under the Regulation of Lac/Tet Dual-Inducible System

To investigate the regulation efficiency of each individual transgene expression under the regulation of the Lac/Tet dual-inducible system by the manipulation of IPTG (5 mM) and tetracycline (1.25 µg), we transiently transfected pTREluc, pTet-off, pSVlacOZ and pHβlacINLSneo plasmids into NIH/3T3 cells by the LIPOFECTIN method. The first two plasmids are for the Tet system, and the remaining two are for the Lac system. Figure 1A shows that in the presence of IPTG and in the absence of tetracycline, both β-gal and luciferase genes were activated (4× and 40×, respectively) and reached the peak at 24–28 h post-treatment. Figure 1B shows that in the presence of both IPTG and tetracycline, β-gal expression was activated (5.3×); however, luciferase gene expression decreased (1/29×) because of the removal of Tet-controlled transactivator by tetracycline. Figure 1C shows that in the absence of both IPTG and tetracycline, β-gal gene expression dropped because of the repression of lac repressor (1/3×), and luciferase gene expression increased because of the activation of Tet-controlled transactivator (35×). Figure 1D shows that in the absence of IPTG and presence of tetracycline, the expression of both β-gal and luciferase decreased (1/3× and 1/25×). Relatively, the increase for luciferase gene expression was 35-fold to 40-fold, and, for β-gal, gene expression was 4-fold to 5.3-fold. These results demonstrate the capability of regulating two genes simultaneously in a single cell.

A Lac/Tet Dual-Inducible System Functions in the 7-4-b Cell Line, which Contains Ha-*ras* Plus *bcl*-2 Inducible Genes and Constitutively Expresses Lac Repressor and Tet-Controlled Transactivator Genes

We previously demonstrated that the Ha-*ras* inducible 7-4 cells underwent apoptosis, while the activated Ha-*ras* gene was overexpressed by IPTG induction under serum-deprived condition. This apoptosis was demonstrated by DNA fragmentation (Figure 2) and could be prevented by *bcl*-2 overexpression, which was demonstrated by a 7-4 derivative of 7-4-2 cells containing another constitutively expressed *bcl*-2 transgene (11). We further transfected a Tet-controlled transactivator and its regulated *bcl*-2 genes into the 7-4 cell line and established a stable cell line 7-4-b. These cells contain both the Ha-*ras* oncogene regulated by the Lac system and the *bcl*-2 gene regulated by the Tet system. Our data show that while *bcl*-2 and Ha-*ras* transgenes were induced simultaneously, Ha-*ras* increased 25-fold, and *bcl*-2 increased 15-fold, which was confirmed by western blot analysis (Figure 3, lane 7). While the transgenes were induced individually, Ha-*ras* and *bcl*-2 genes were induced 10-fold and 25-fold, respectively (Figure 3, lanes 4 and 5). Cell death was determined by evaluating the activity of β-gal gene, which was transiently transfected into the analyzed cells. Our data demonstrate that while only Ha-*ras* was overexpressed, cells went into apoptosis (Figure 2, lanes 4 and 9; Figure 3, lane 5), and this kind of cell death could be reduced by *bcl*-2 overexpression (Figure 2, lanes 3 and 7; Figure 3, lane 7). Cell death drops dramatically when the cells express *bcl*-2 only or no transgene is expressed (Figure 2, lanes 1, 2, 6 and 8; Figure 3, lanes 3, 4 and 6). This result is consistent with our previous observation (11). The phenomena of inducing and preventing apoptosis was demonstrated by overexpression of Ha-*ras* alone in 7-4 cells and Ha-*ras* plus *bcl*-2 genes in 7-4-2 cells. The same phenomena reappeared in 7-4-b cells when both of the transgenes were introduced and under the regulation of the Lac/Tet dual-inducible system. This study demonstrated that the dual bacterial Lac/Tet-inducible system could control the expression of two transgenes either individually or simultaneously in a stable cell line at translational and functional levels.

Taken together, through the manipulation of IPTG and tetracycline, two inducible genes can be activated or suppressed simultaneously or separately in a single cell. This was demonstrated by the regulation of β-gal and luciferase reporter gene activation in the transient transfected NIH/3T3 cells and by the regulation of Ha-*ras* and *bcl*-2 gene activation in the stable cell line 7-4-b, indicating the wide application of this system for manipulating diverse genes. Moreover, when two inducible genes were regulated simultaneously by the Lac/Tet regulatory system, neither synergism nor interference of the gene expression was detected in the cells.

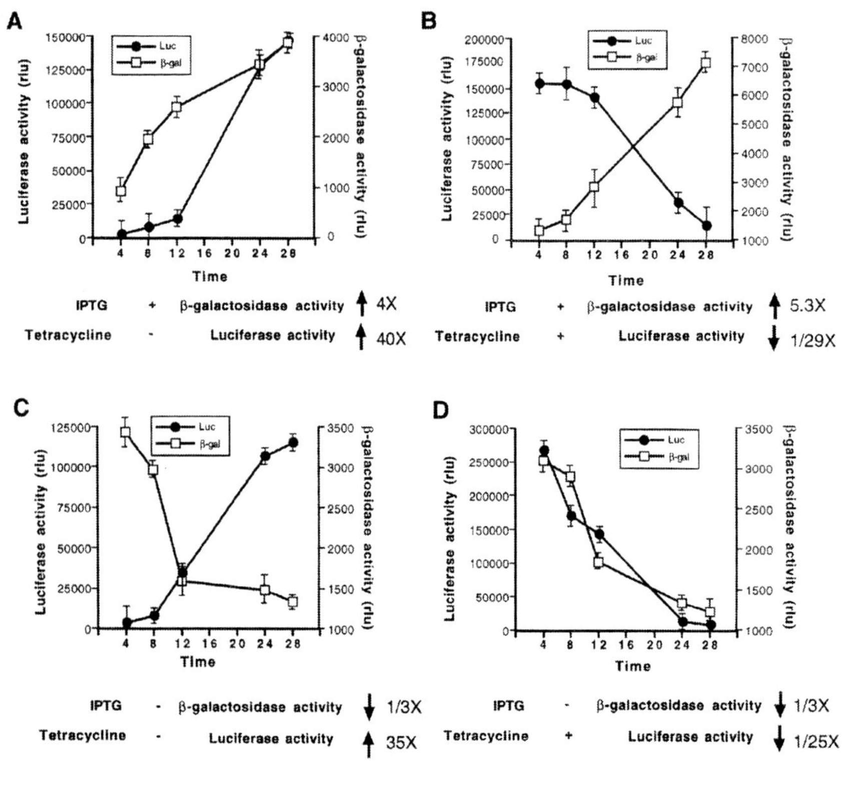

Figure 1. The kinetics of luciferase and β-gal gene expression under the regulation of the Lac/Tet dual-inducible system in NIH/3T3 cells. NIH/3T3 cells (1 × 10^5/well) in the 6-well tray were transfected with pTREluc (2 μg), pTet-off (1 μg), pSVlacOZ (2 μg) and pHβlacINLS-neo (1 μg). The cells were harvested at 48 h post-transfection. The protein was extracted and analyzed by the Dual-Light analysis. The administration of IPTG (5 mM) and tetracycline (1.25 μg), and the corresponding gene activation fold increase/decrease of gene expression are indicated at the bottom of each diagram.

Table 3. Lac/Tet Dual-Inducible System Functions in 7-4 Cells

DNA[a] Added	Induction		Luciferase (RLU/μg)		Fold[d] Increase	β-gal (RLU/μg)		Fold[d] Increase
	Dual[b]	Single[c]	Repressed	Induced		Repressed	Induced	
1-1-3	+		5258	286901	**54.6**	2085	6753	**3.2**
1-1-3		+	5258	111620	**21.2**	2085	12003	**2.3**
1-2-2	+		2097	140784	**67.1**	839	1956	**2.3**
1-2-2		+	2097	111189	**53**	839	3567	**4.3**
2-1-2	+		1146	323390	**282.2**	627	5733	**9.1**
2-1-2		+	1146	104790	**91.4**	627	7324	**11.7**

1. pTREluc, pTet-off and pSVlacOZ were co-transfected into 7-4 cells. This experiment was repeated three times.

[a]DNA added (μg) → pTet-off: pTREluc : pSVlacOZ

[b]Both luciferase and β-gal genes were activated simultaneously.

[c]Only luciferase or β-gal gene was activated.

[d]The fold increase is a comparison between repressed and induced levels of expression.

2. No significant correlation was found between molar ratio of the transgenes and fold increase of the reporter gene expression.

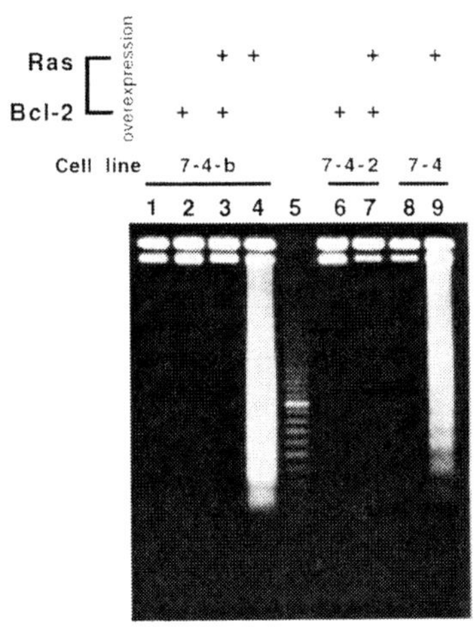

Figure 2. DNA fragmentation in 7-4, 7-4-2 and 7-4-b cells was affected by overexpression of Ha-*ras* and *bcl*-2 transgenes, which were regulated by the Lac and Lac/Tet dual-inducible systems. Cells (1×10^6/100-mm plate) were maintained under 0.2% serum-deprived condition in the presence or absence of IPTG (5 mM) and/or tetracycline (1 μg) for 48 h. Lanes 1–4 are 7-4-b cells. Lane 1: no transgene was overexpressed. Lane 2: only *bcl*-2 was overexpressed. Lane 3: both *bcl*-2 and Ha-*ras* were overexpressed. Lane 4: only Ha-*ras* was overexpressed, and DNA fragmentation was detected. Lane 5 is 100-bp DNA marker. Lanes 6 and 7 are 7-4-2 cells containing inducible Ha-*ras* and constitutively expressed *bcl*-2 gene, respectively. Lane 6: only *bcl*-2 was overexpressed. Lane 7: both *bcl*-2 and Ha-*ras* were overexpressed. Lanes 8 and 9 are 7-4 cells containing inducible Ha-*ras* transgene. Lane 8: no transgene was overexpressed. Lane 9: only Ha-*ras* was overexpressed, and DNA fragmentation was detected.

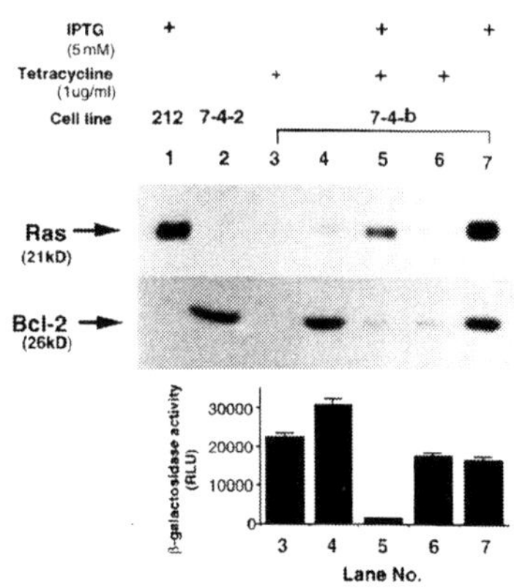

Figure 3. Selective activation of Ha-*ras* and *bcl*-2 transgenes using the dual-inducible system was demonstrated at translational and at functional levels by the administration of IPTG and tetracycline in 7-4-b cells. The experimental procedure is the same as described in Figure 2. Ha-*ras* and *bcl*-2 gene expressions at translational level under the regulation of IPTG and tetracycline were detected by western blot analysis using anti-RAS and anti-BCL-2 monoclonal antibodies. The plasmid pCMVlacZ (CLONTECH) was transiently transfected into 7-4-b cells, and β-gal activity representing cell survival ratio was measured. Lane 1: 212 cell, containing a Ha-*ras* inducible gene, is used as a control of Ras. Lane 2: 7-4-2 cell, constitutively expressing *bcl*-2, is used as a control of Bcl-2. Lanes 3–7 are 7-4-b cells. Lanes 3 and 6 are duplicates of 7-4-b cells without transgene expression. Lane 4: the cells express *bcl*-2 only. Lane 5: the cells express Ha-*ras* transgene only. Lane 7: the cells express both Ha-*ras* and *bcl*-2.

Lac/Tet Dual-Inducible System Functions in Five Tumor Cell Lines Expressing Luciferase and β-gal Reporter Genes by Transient Transfection Analysis

In an attempt to test whether the *E. coli* Lac/Tet dual-inducible system functions in widely used mammalian cell lines, the luciferase gene under tetracycline regulation was co-transfected with its activator Tet R gene (pTet-off), and the β-gal gene under lactose analogue IPTG regulation was co-transfected with its repressor Lac R gene (pHβlacINLS*neo*) by transient transfection analysis. A total of five cell lines were tested (T1, T2, J82, TCCSUP and HA22T). Our data show that both luciferase and β-gal genes are inducible. When both genes were activated simultaneously, the increase for luciferase induction ranged between 43-fold and 336-fold, and that for β-gal induction ranged between 3.2-fold and 6-fold. Moreover, when only one gene was activated at a time (Table 2), the increase for luciferase ranged between 23-fold and 579-fold, and that for β-gal ranged between 2.4-fold and 10-fold. The significant variation of the repressed expression and fold increase for luciferase or β-gal in the five cell lines tested was possibly affected by the specificity of the cells and promoters and transfection efficiency. Nevertheless, we demonstrate that two transgenes can be regulated simultaneously by the Lac/Tet dual-inducible system in many cell lines, indicating the wide application capability of this system. In this study, the promoter used for the Tet-inducible gene is the minimal CMV, and, for the Lac-inducible gene, the simian virus 40 (SV40) promoter is used. Both Tet- and Lac-inducible systems are commercially available.

A Dual-Inducible System was Established in the 7-4 Cells that Already Contained a Constitutively Expressed Lac Repressor Gene by Transiently Introducing the Second Tet System and β-gal Gene

We introduced the plasmids of the inducible luciferase gene (pTREluc), Tet transactivator gene (pTet-off) and β-gal reporter gene (pSVlacOZ) by transient transfection into the 7-4 cell line. In 7-4 cells, the Lac repressor gene is constitutively expressed, therefore can regulate the expression of β-gal transgene. Table 3 demonstrates that constitutively expressed Lac repressor plus transiently expressed Tet-controlled transactivator could activate β-gal and luciferase transgenes in 7-4 cells. Table 3 lists their induction folds. Briefly, when both genes are activated, the fold increase for luciferase gene ranged between 54.6 and 282.2, and, for the β-gal gene, it ranged between 3.2 and 9.1. Moreover, when only one gene was activated at a time, the fold increase for the luciferase gene was between 21.2 and 91.4, and, for the β-gal, it was between 2.3 and 11.7. This study demonstrates the feasibility of introducing a second inducible system into an already established inducible cell line.

In our study of the Lac/Tet dual-inducible system, the maximal increase for the Lac system is around 12-fold, and for the Tet system it is around 459-fold. There are two possible reasons causing this tremendous difference in induction. First, the activation of the reporter gene by the Tet system is through an activator (HSV VP16), but that by the Lac system is through a repressor (4,6). Second, the parameters such as half-life of the reporter genes, sensitivity of analyzing reporter gene expression and the affinity of IPTG to the Lac repressor and of tetracycline to the Tet-controlled activator are all involved. Although the inductive efficiency of the Lac system as compared with that of the Tet system is quite low, it is still usable for the regulation of many genes at the cellular level and at the organismal level (i.e., in transgenic animals and plants) (6,8,9,15,17). This is also true in our dual-inducible regulatory study. In conclusion, this study demonstrated the feasibility of regulating two genes simultaneously in individual cells using a dual-inducible system that is especially useful for the study on the interaction between two genes and on the effects of two genes expressed simultaneously or separately in cells.

ACKNOWLEDGMENTS

This work was supported by grants from National Health Research Institutes (DD01-86IX-MG601P) and National Science Council (NSC86-2314-B006-050), Taiwan, ROC. Address correspondence to Dr. Hsiao-Sheng Liu, Department of Microbiology & Immunology, Medical College, National Cheng Kung University, Tainan, Taiwan, ROC. Internet: a713@mail.ncku.edu.tw

REFERENCES

1. **Ackland-Berglund, C.E. and D.A. Leib.** 1995. Efficacy of tetracycline-controlled gene expression is influenced by cell type. BioTechniques *18*:196-200.
2. **Chang, M.Y., S.J. Won and H.S. Liu.** 1997. A ribozyme specifically suppresses transformation and tumorigenicity of Ha-ras oncogene transformed NIH/3T3 cell lines. J. Cancer Res. Clin. Oncol. *123*:91-99.
3. **Diehl, P.** 1996. Tet-off and Tet-on gene expression systems and cell lines. Clontechniques *XI*:2-5.
4. **DuCoeur, L., D.L. Wyborski and J.M. Short.** 1993. Control of gene expression in eukaryotic cells using the lac repressor system. Strategies *5*:70-72.
5. **Gossen, M., A.L. Bonin, S. Freundlieb and H. Bujard.** 1994. Inducible gene expression systems for higher eukaryotic cells. Curr. Opin. Biotechnol. *5*:516-520.
6. **Gossen, M. and H. Bujard.** 1992. Tight control of gene expression in mammalian cells by tetracycline-responsive promoters. Proc. Natl. Acad. Sci. USA *89*:5547-5551.
7. **Hoffmann, A., M. Villalba, L. Journot and D. Spengler.** 1997. A novel tetracycline-dependent expression vector with low basal expression and potent regulatory properties in various mammalian cell lines. Nucleic Acids Res. *25*:1078-1079.
8. **Howe, J.R., B.V. Skryabin, S.M. Belcher, C.A. Zerillo and C. Schmauss.** 1995. The responsiveness of a tetracycline-sensative expression system differs in different cell lines. J. Biol. Chem. *270*:14168-14174.
9. **Iida, A., S.T. Chen, T. Friedmann and J.K. Yee.** 1996. Inducible gene expression by retrovirus-mediated transfer of a modified tetracycline-regulated system. J. Virol. *70*:6054-6059.
10. **Ito, M.** 1984. Microassay for studying anticellular effects of human interferons. J. Interferon Res. *4*:603-608.
11. **Liu, H.S., C.Y. Chen, C.H. Lee and Y.I. Chou.** Selective induction of oncogenic Ha-ras induced apoptosis in NIH/3T3 cells. Br. J. Cancer (In press).
12. **Liu, H.S., E.S. Feliciano and P.J. Stambrook.** 1989. Cytochemical observation of regulated bacterial-galactosidase gene expression in mammalian cells. Proc. Natl. Acad. Sci. USA *86*:9951-9955.
13. **Liu, H.S., H.J. Scrable, D.B. Viliaret, M.A. Lieberman and P.J. Stambrook.** 1992. Control of Ha-*ras* and the transformed phenotype in mammalian cells by the *E. coli* regulatory elements. Cancer Res. *52*:983-989.
14. **Martin, C.S., P.A. Wight, A. Dobretsova and I. Bronstein.** 1996. Dual luminescence-based reporter gene assay for luciferase and β-galactosidase. BioTechniques *21*:520-524.

15. **Shockett, P., M. Difilippantonio, N. Hellmanand and D.G. Schatz.** 1995. A modified tetracycline-regulated system provides autoregulatory, inducible gene expression in cultured cells and transgenic mice. Proc. Natl. Acad. Sci. USA *92*:6522-6526.
16. **Tzeng, C.C., H.S. Liu, C. Li, Y.T. Jin, R.M. Chen, W.H. Yang and J.S. Lin.** 1996. Characterization of two urothelium cancer cell lines derived from a blackfoot disease endemic area in Taiwan. Anticancer Res. *16*:1797-1804.
17. **Wu, J.D., H.C. Hsueh, W.T. Huang, H.S. Liu, H.W.C. Leung, Y.R. Ho, M.T. Lin and M.D. Lai.** 1997. The inducible lactose operator-repressor system is functional in the whole animal. DNA Cell Biol. *16*:17-22.
18. **Wuu, K.D., L. Kang, Y.M. Su, S. Wang-Wuu, C.P. Hu and L.M. Chang.** 1987. Karyotypic characterization of an established human hepatoma cell line HA22T/VGH. Cancer Genet. Cytogenet. 26:279-286.

Update to: Lac/Tet Dual-Inducible System Functions in Mammalian Cell Lines

Hsia-Sheng Liu

National Cheng Kung University, Tainan, Taiwan, ROC

A plasmid pTet-Lac-Hyg containing a lac repressor gene (lacI), a tetracycline activator gene (tTA) and a hygromycine resistant gene has been constructed, which can regulate two inducible genes in either transiently or stably transfected mammalian cells. We demonstrate that the lac repressor and tTa were expressed efficiently from pTet-Lac-Hyg and could regulate their target genes in various mammalian cell lines. Most importantly, this unique plasmid is very useful for studying the transient effects of two genes on the cell either individually or simultaneously, as well as the relationship between these two inducible genes. In conclusion, pTet-Lac-Hyg simplifies the transfection procedure and extends the application of prokaryotic inducible systems from the regulation of a single gene to two genes. It will be valuable for the functional study of multiple novel genes discovered from the human genomic projects.

CHAPTER 32

Cloning Trap for Signal Peptide Sequences

Siew Pheng Lim and Alfredo Garzino-Demo[1]

Institute of Molecular and Cell Biology, Singapore; [1]University of Maryland, Baltimore, MD, USA

(Reprinted from BioTechniques 28:124-130, 2000)

ABSTRACT

Novel secreted and/or type I transmembrane proteins containing N-terminal signal sequences have been successfully cloned using the signal sequence trapping (SST) method. Often this involves random cloning of short 5′ cDNA terminal ends into an epitope-tagged expression vector and the detection of expressed recombinant proteins on the cell surfaces of transfected cells with an antibody to the tagged epitope. Here, we report a novel cloning system for the detection of secreted proteins also using SST. In this method, we used the human immunodeficiency virus (HIV-1) p24 as the epitope for tagging. To test the system, two constructs were created. The 5′ terminal end of a human beta-chemokine (which was regulated upon activation, expressed by normal T cells and presumably secreted [RANTES]) and the 5′ end of a human CD4 receptor were cloned upstream of and in-frame with the p24 cDNA. Secreted p24 was detectable in the culture media two days after transfection of either DNA construct into the human cell lines, HeLa and 293T. When the chimeric p24 expression constructs were transfected at a ratio of 1:100 to the vector pcDNA3.1(+), p24 could still be detected in cell supernatants. The use of a secreted viral antigen like HIV-1 p24 (or of any noncellular protein) as a marker in SST cloning approaches is likely to be advantageous because it reduces the background noise in detection and also renders this system suitable for high-throughput screening.

INTRODUCTION

Signal peptide sequences are required for proper orientation of the amino terminal of mature secreted proteins and type I membrane proteins inside the endoplasmic reticulum and exocytotic vesicles. The signal sequence trap (SST) (6) takes advantage of the presence of these signal sequences for the selective cloning of new members of these proteins without the use of biological functional assays. Several strategies have been reported (3,8,10) and usually use an expression vector containing a membrane-anchored protein, such as the IL-2α receptor (8,10) or CD4 (3), devoid of its signal sequence. The 5′ terminal ends of cDNAs, approximately 400 bp in length, are randomly cloned into these expression vectors. Inserts with signal sequences cloned in-frame with the correct orientation resulted in expression of fusion proteins on plasma membranes of transfected cells. They were detected by flow cytometry with specific antibodies to the membrane-anchored proteins used in the vector (3,8,10).

Using this method, large-scale screening of clones from an SST cDNA library may be possible (8). Usually, methods based on flow cytometry involve transfecting a large number of clones into target cells, followed by repeated sorting for positive cells that express the surface tag. The plasmid is subsequently rescued from the sorted cells and characterized. We have devised a simpler and alternative method for SST. This involves the cloning of 5′ terminal cDNA ends upstream to the human immunodeficiency virus (HIV-1) p24 cDNA in an expression vector. Chimeric proteins bearing 5′ terminal signal sequences cloned in-frame to p24 will be secreted into the cell culture of transfected cells and can be measured with an ELISA kit for the HIV-1 p24 antigen. In this way, large pools of recombinant clones may be easily subjected to high-throughput screening in a 96-well plate format for novel secreted proteins.

MATERIALS AND METHODS

Cell Lines

The human epitheloid cervical carcinoma cell line, HeLa, and the human embryonic kidney cell line, 293, bearing the large T antigen from SV40 (293T) were purchased from ATCC (Manassas, VA, USA) and cultured at 37°C in 5% CO_2 in EMEM containing 2 mM L-glutamine, 1.5 g/L sodium bicarbonate, 0.1 mM nonessential amino acids, 1 mM sodium pyruvate and 10% fetal bovine serum.

Cloning and Expression Vectors for Gene Function Analysis
Edited by Quinn Lu and Michael P. Weiner

Plasmids and Constructs

The HIV-1 p24 cDNA was amplified by PCR (94°C for 1 min, 62°C for 1 min and 68°C for 1 min) from the plasmid pHXB2gpt (9) with P24FOR, 5′-GGCCTCGAGCCTATAGTGCAGAACATCCAG-3′ and P24REV, 5′-GGCTCTAGATTATTGTGACGAGGGGTCGTT-3′. The HXB2 $Pr41^{gag}$ protein was amplified with P17FOR, 5′-GGCCTCGAGAGAATGGGTGCGAGAGCGTC-3′ and P24REV. PCR products were digested with *Xho*I and *Xba*I and cloned into the pcDNA3.1(+) vector from Invitrogen (Carlsbad, CA, USA) to generate pCMV-P24 and pCMV-P17P24, respectively. The 5′ fusion of the human RANTES and CD4 signal sequences to the p24 DNA was carried out using a two-step "megaprimer" method for site-directed mutagenesis (6). The 5′ terminal region of the human RANTES cDNA (7) (nucleotides 27–95) was amplified using primers RANFOR, 5′-GCGAATTCACCATGAAGGTCTCCGCGG-3′ and RANP24REV, 5′-GTTCTGCACTATAGGTGGGGAGGCAGATGCA-3′ (p24 nucleo-tides are underlined) at 94°C for 1 min, 52°C for 1 min and 68°C for 1 min with a RANTES expression vector as template.

The resultant PCR product and pCMV-P24 were used as templates in a second round of PCR (94°C for 1 min, 62°C for 1 min and 68°C for 1 min) with the primers RANFOR and P24REV. To generate the chimeric CD4-p24 product, the 5′ terminal region of the human CD4 cDNA (4) (nucleotides 76–160) was similarly amplified with CD4FOR, 5′-GCGAATTCACAATGAACCGGGGAGTCCC-3′ and CD4-P24REV, 5′-GTTCTGCACTATAGGTTTCTTTCCCTGAGTGG-3′ (p24 nu- cleotides are underlined) using an expression plasmid for a human CD4 as template. The PCR product and pCMV-P24 were used as templates in a second round of PCR with primers CD4FOR and P24REV. The 1.18 kb PCR products were digested with *Eco*RI and *Xba*I and cloned into pcDNA3.1(+) to generate pCMV-RANSSP24 and pCMV-CD4SSP24, respectively.

DNA Sequencing

DNA sequencing on all constructs created was carried out in the Biopolymer Sequencing Facility at the University of Maryland at Baltimore, using the *Taq* DyeDeoxy™ Terminator Cycle Sequencing Kit and the automated Model 377 DNA Sequencer from PE Biosystems (Foster City, CA, USA).

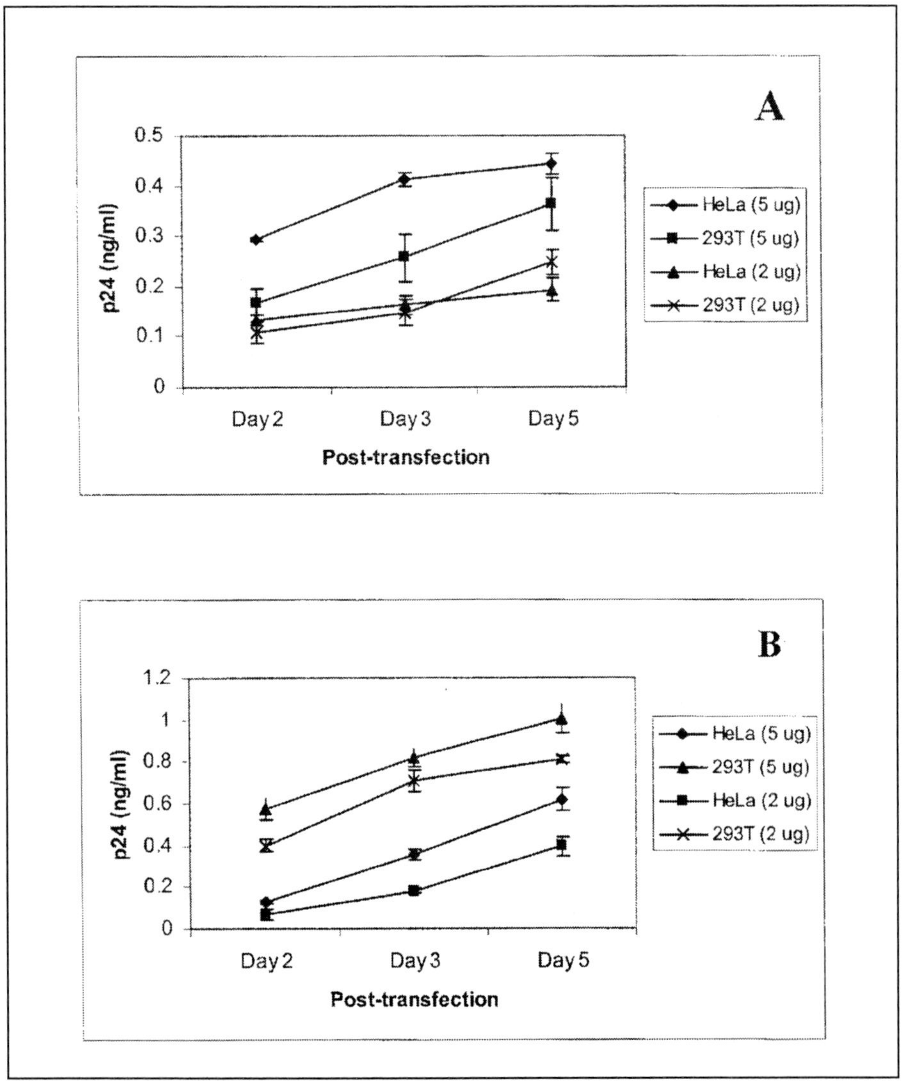

Figure 1. Analysis of p24 levels in culture media of 293T and HeLa cell lines five days posttransfection with 2 and 5 µg of pCMV-RANSSP24 (A) or pCMV-CD4SSP24 (B). Transfections were carried out with SuperFect as described in Materials and Methods. Cells were kept in culture at 37°C in 5 mL of complete EMEM. Results shown are the average of two readings. Error bars correspond to the standard error of the mean value.

Cell Transfections

Transfection experiments were performed using SuperFect™ Transfection Reagent from Qiagen (Valencia, CA, USA), according to the manufacturer's protocol. Briefly, 1.5×10^5 HeLa or 293T cells were plated into 6-well tissue culture plates and exposed to a mixture of 5 µg of plasmid and 75 µg SuperFect in 150 µL serum-free EMEM for 3 h at 37°C and 5% CO_2. After this step, the cells were washed and added with fresh complete media. At the indicated times, an aliquot of cell media was harvested for p24 ELISA analysis. Where less than 5 µg of DNA were used for transfection, the amounts of SuperFect and serum-free media used were scaled down according to the manufacturer's recommendations.

Flow Cytometry

The 293T cells were harvested two days posttransfection, washed in PBS and resuspended at 1×10^6 cells/mL. Then, 0.5 mL of cells were incubated with an anti-human CD4 antibody or an isotype-matched control from Sigma (St. Louis, MO, USA) for 30 min at 4°C, washed and re-incubated with a goat anti-mouse IgG antibody conjugated with phycoerythrin (Sigma). Cells were washed as before and analyzed on a FACScalibur™ Flow Cytometer (Becton Dickinson, San Jose, CA, USA). Live cells were gated, and a total of 20 000 events were collected per analysis.

p24 ELISA

p24 ELISAs were performed at the core facility at the Institute of Human Virology (University of Maryland at Baltimore) using a p24 ELISA Kit™ from R & D Systems (Minneapolis, MN, USA), following the manufacturer's instructions. In all experiments, standards were simultaneously assayed over a range from 6.25–400 pg/mL.

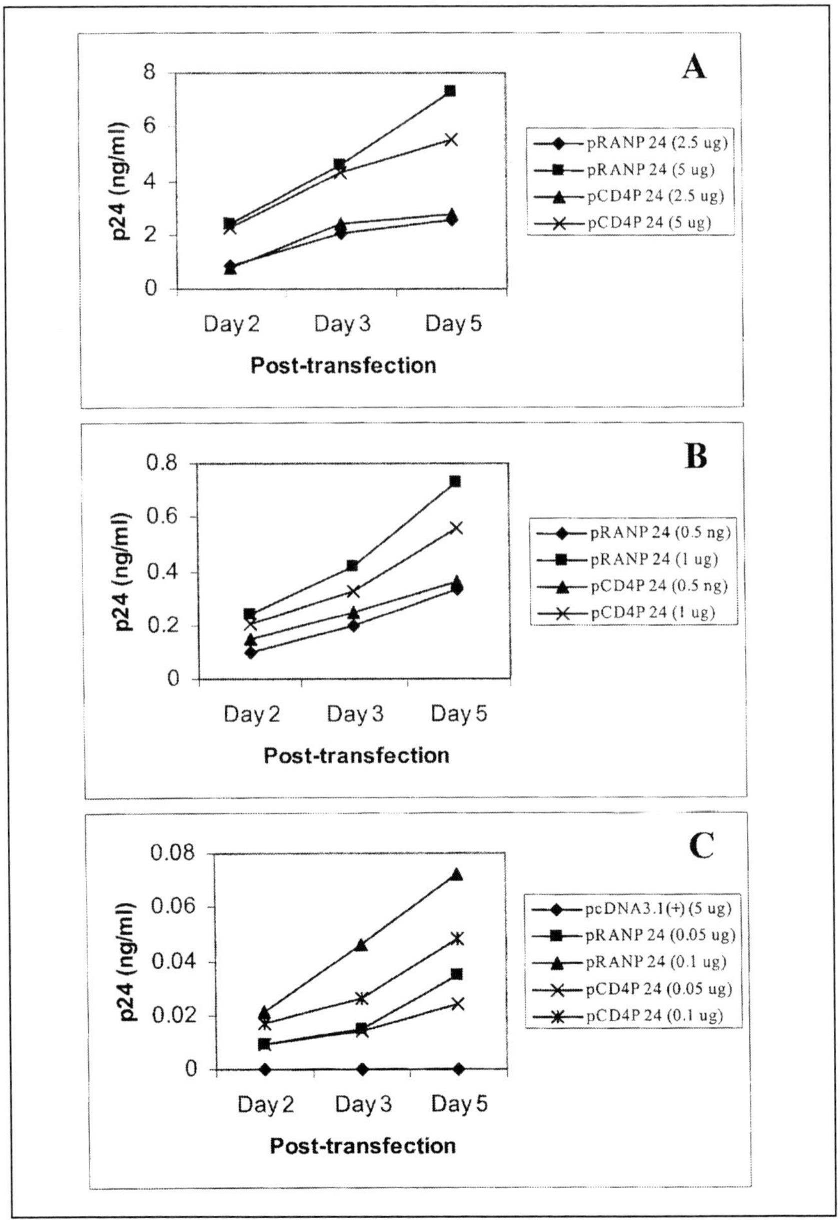

Figure 2. Analysis of p24 levels in culture media of 293T cells five days posttransfection in a titration experiment with decreasing amounts of pCMV-RANSSP24 or pCMV-CD4SSP24. Transfections were carried out with SuperFect as described in Materials and Methods with a total of 5 µg DNA. Chimeric p24 expression constructs and pcDNA3.1(+) were transfected at a ratio of 5:0 and 1:1 (A), 1:5 and 1:10 (B) and 1:50, 1:100 and 0:5 (C). Cells were kept in culture at 37°C in 2.5 mL of complete EMEM. Results shown are the average of two readings. Error bars correspond to the standard error of the mean value.

RESULTS AND DISCUSSION

Release of P24 Antigen in Tissue Culture Supernatants When Coupled With Heterologous Signal Peptides

The constructs pCMV-P24, pCMV-P17P24, pCMV-RANSSP24 and pCMV-CD4SSP24, expressing HIV-1 p24, p41, p24 with a RANTES or a CD4 signal peptide, respectively, and the vector pCDNA3.1(+) were separately transfected into 1×10^5 HeLa and 293T cell lines using SuperFect. At days 2, 3 and 5 posttransfection, 500 µL of culture media were removed and assayed for secreted p24 antigen using a commercial ELISA kit. As expected, media harvested from cells transfected pCMV-P24 did not contain detectable levels of p24 antigen (data not shown), because the p24 cDNA in this construct does not encode any signal peptide at its 5′ end to facilitate its extracellular release. Interestingly, transfection of the construct pCMV-P17P24 also did not give detectable levels of p24 antigen in the cell media (data not shown). This was unexpected as it has been previously demonstrated that the first 10 amino acids of the $Pr55^{gag}$ polyprotein alone was sufficient to provide a secretory signal for release of viral particles (1,5). Nevertheless, work by Gheysen et al. indicates that in the absence of the HIV protease or of other gag components such as the C terminal peptide, p16, the $Pr41^{gag}$ polyprotein may not be efficiently released into the cell media (2) and may explain why we failed to observe secretion of the HIV-1 gag protein from our pCMV-P17P24 construct. However, chimeric p24 antigens were readily detected in both HeLa and 293T cells after transfection with pCMV-RANSSP24 (Figure 1A) and pCMV-CD4SSP24 (Figure 1B). The p24 levels present in the cell media also increased in a dose-dependent manner with 2 and 5 µg of plasmid DNA transfected into the cell lines (Figure 1). In addition, culture media harvested from all transfectants exhibited a steady increase in the levels of p24 antigen from days 2–5 (Figure 1).

When the amount of transfected plasmid DNA was decreased to 0.5 µg, p24 was still detectable in tissue culture supernatants of both HeLa and 293T cell lines (data not shown). It should be noted that we did not alter the codon usage of the p24, which is not optimal for expression in mammalian cells. We anticipate that optimizing codon usage should result in higher gene expression and further improve the detection of proteins fused with p24.

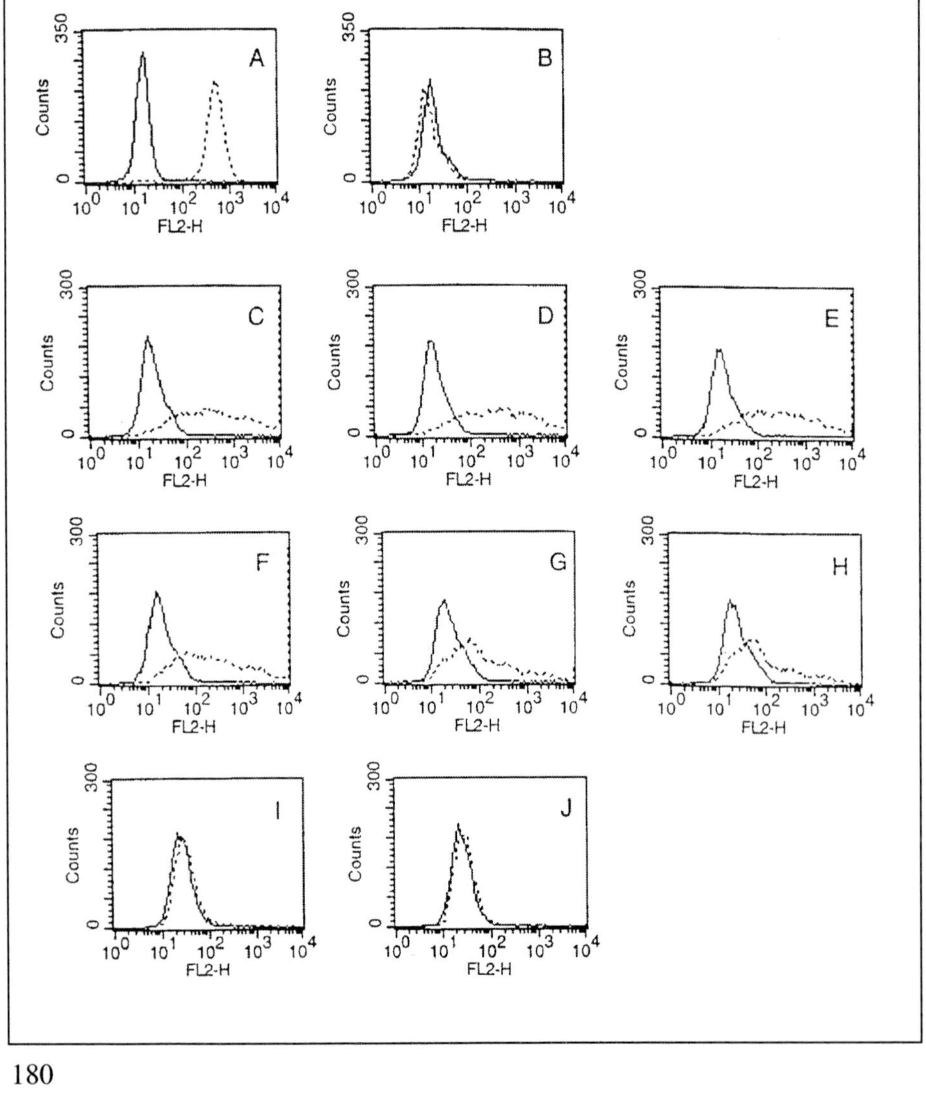

Figure 3. Detection of surface CD4 receptors on transfected 293T cells by flow cytometry. Transfections were carried out as described in Materials and Methods with a total of 5 µg DNA, and cells were analyzed two days later. The human CD4 expression construct and pcDNA3.1(+) were transfected at a ratio of 0:5 (B), 5:0 (C), 1:1 (D), 1:5 (E), 1:10 (F), 1:50 (G), 1:100 (H), 1:200 (I) and 1:500 (J). The human T-cell line CEMSS was used as a positive control (A). In each panel, staining with an isotype-matched control is denoted by a solid line, and staining with anti-CD4 antibody is denoted by a hatched line.

High Sensitivity of the p24-Based Detection System is Highly Sensitive

To determine the sensitivity of our cloning system, we carried out a series of dilution experiments in which pCMV-SSRANP24 or pCMV-CD4P24 was transfected together with the vector pcDNA3.1 (+) into 293T cells. Ratios ranged from 5:0–1:100, with the total amount of DNA transfected kept at 5 µg in all cases. Transfection of 5 µg of pcDNA3.1(+) did not produce any detectable p24 antigens (Figure 2C). As we observed previously, transfection of 5 µg of chimeric p24 constructs (or a ratio of 5:0) produced high levels of p24 antigens in the culture media (Figure 2A). Transfection of the p24 expression plasmids and pcDNA3.1(+) at a ratio of 1:1 readily yielded detectable levels of p24 (Figure 2A). When the ratios of the chimeric p24 expression constructs to pcDNA3.1(+) were progressively decreased to 1:5, 1:10, 1:50 and 1:100 (equivalent to 0.5, 0.1 and 0.05 µg of p24 expression plasmids respectively), p24 antigens continued to be released into the culture supernatants, in a dose-dependent fashion (Figure 2, B and C). In all cases, release of p24 in the tissue culture supernatant increased with time (Figure 2, A–C). The cut-off for detection occurred at the dilution ratio of 1:100, as further lower dilution ratios such as 1:150 and 1:200 did not produce detectable p24 signals (data not shown).

To compare the sensitivity of our assay with the conventional method by FACS analysis, we co-transfected 293T cells with a human CD4 expression construct and pcDNA3.1(+) at similar ratios as above. Two days after transfection, cells were harvested and analyzed for the expression of CD4 receptors by flow cytometry. Cells transfected with pcDNA3.1(+) alone did not express CD4 markers (Figure 3, panel B). Positive staining for CD4 receptors was observed when cells were transfected at ratios of 5:0–1:100 (Figure 3, panels C–H). Transfection at ratios lower than 1:100 also yielded marginal levels of surface CD4 markers that were barely detectable by flow cytometry (Figure 3, panels I and J). Thus, the sensitivity of the signal sequence trapping method we describe here is at least comparable to the conventional method by flow cytometry.

Methods to isolate secreted proteins based on the SST have been described (3,6,8,10), but they have some disadvantages: *(i)* membrane-bound markers are used, and their detection is based on flow cytometry-based techniques, which is time consuming and expensive; and *(ii)* cellular proteins are used as markers, thus raising the possibility of a background signal caused by cellular expression of the wild-type marker. By using our method of detection, the process of screening large libraries can be made simpler, quicker and less labor intensive because the secreted viral antigen can be detected by a standard, commercially available ELISA test.

The use of a viral antigen is less likely to produce a background noise that could interfere with detection. Although it may be possible to use transfected cells in ELISAs, the experimental procedures for recovery of viable cells will be more difficult and may produce a higher background. Our method for signal peptide trapping using the HIV-1 p24 antigen is highly sensitive. As we have demonstrated, as little as 0.05 µg of a positive expression construct or when the construct is present at a ratio 1:100 in a mixed pool readily yields a detectable signal two days after transfection. We expect that optimizing the codon usage of the viral protein marker will further enhance the sensitivity of detection.

ACKNOWLEDGMENTS

The authors thank R.C. Gallo, A.L. DeVico, M.S. Reitz, Jr., G.K. Lewis and R.K. Lewis for valuable suggestions and discussions. We acknowledge support from the Institute of Human Virology (University of Maryland at Baltimore, Baltimore, MD, USA) and the National Science and Technology Board (Singapore).

REFERENCES

1. **Garnier, L., L. Ratner, B. Rovinski, S.X. Cao and J.W. Willis.** 1998. Particle size determinants in the human immunodeficiency virus type 1 gag protein. J. Virol. *72*:4667-4677.
2. **Gheysen, D., E. Jacobs, F. de Foresta, C. Thiriart, M. Francotte, D. Thines and M. De Wilde.** 1989. Assembly and release of HIV-1 precursor Pr55^{gag} virus-like particles from recombinant baculovirus-infected insect cells. Cell *59*:103-112.
3. **Imai, T., T. Yoshida, M. Baba, M. Nishimura, M. Kakizaki and O. Yoshie.** 1996. Molecular cloning of a novel T-cell directed CC chemokine expressed in thymus by signal sequence trap using Epstein-Barr virus vector. J. Biol. Chem. *271*:21514-21521.
4. **Maddon, P.J., D.R. Littman, M. Godfrey, D.E. Maddon, L. Chess and R. Axel.** 1985. The isolation and nucleotide sequence of a cDNA encoding the T cell surface protein T4: A new member of the immunoglobulin gene family. Cell *42*:93-104.
5. **Parent, L.J., C.B. Wilson, M.D. Resh and J.W. Wills.** 1996. Evidence for a second function of the MA sequence in the rous sarcoma virus gag protein. J. Virol. *70*:1016-1026.
6. **Sarker, G. and S.S. Somer.** 1990. The "megaprimer" method of site-directed mutagenesis. BioTechniques *8*:404-407.
7. **Schall, T.J., J. Jongstra, B.J. Dyer, J. Jorgensen, C. Clayberger, M.M. Davis and A.M. Krensky.** 1988. A DNA sequence encoding the mature human RANTES protein sequence. J. Immunol. *141*:1018-1025.
8. **Shirozu, M., H. Tada, K. Tashiro, N.D. Lopez, M. Nazarea, T. Hamada, T. Sato, T. Nakano and T. Honjo.** 1996. Characterization of novel secreted and membrane proteins isolated by the signal peptide trap method. Genomics. *37*:273-280.
9. **Starcich, B., L. Ratner, S.F. Josephs, T. Okamoto, R., C. Gallo and F. Wong-Staahl.** 1985. Complete nucleotide sequence of the AIDS virus, HTLV-III. Nature *313*:277-284.
10. **Tashiro, K., H. Tada, R. Heilker, M. Shirozu, T. Nakano and T. Honjo.** 1993. Signal sequence trap: a cloning strategy for secreted proteins and type I membrane proteins. Science *261*:600-603.

Update to: Cloning Trap for Signal Peptide Sequences

Siew Pheng Lim and Alfredo Garzino-Demo[1]

Institute of Molecular and Cell Biology, Singapore and [1]University of Maryland, Baltimore, MD, USA

Functional cloning of proteins containing signal peptide signals still holds importance in the era of high-throughput genome sequencing, as “signatures” of signal peptides may not be evident to computer (“in silico”) analysis (2,4); further, genomic sequencing of the vast majority of organisms is not likely to become a routine operation at least for the next few years. We have proposed a novel detection system, based on the detection of the human immunodeficiency virus (HIV)-p24 antigen by a standard immunoassay. We engineered two constructs (Figure 1), containing the signal peptides from CD4 and RANTES, to prove the concept that such an assay is suitable to detect signal peptides.

A further interesting improvement of signal peptide trap cloning systems (referred to as signal-exon trap) has been published recently (5). This method is based on genomic DNA cloning, rather than on cDNA, and it is a modification of exon trapping/amplification (1,3,6), thus allowing for the identification of signal peptides in genomic DNA. One advantage of this system is that genomic DNA is inherently normalized, so that the problem that may occur with cDNA-based cloning, i.e., missing underrepresented RNA species, is avoided. In addition, this method potentially allows for the cloning of any signal peptide-containing sequence, while cDNA-based methods are cell- or tissue-specific. Therefore, while cDNA cloning may still remain the method of choice for identifying factors that are expressed at high levels by a given cell population or tissue, the signal-exon trap system can be useful to identify a broader spectrum of signal peptide-containing elements. One potential disadvantage of this system, however, is that it will not identify genes that do not contain introns (6). The authors used this cloning system to clone 3 signal peptide-containing clones in a 6.25 MB fragment of genomic DNA. Their detection system is based on an assay for enzymatic activity of a cytplasmatically inactive, secretory pathway-specific form of CD26, a dipeptidyl-peptidase normally anchored on the cell membrane. We envision that the use of a noncellular detection system, i.e., the release of p24 may further improve the detection specificity of this system.

REFERENCES

1. **Buckler, A.J., D.D. Chang, S.L. Graw, J.D. Brook, D.A. Haber, P.A. Sharp and D.E. Housman.** 1991. Exon amplification: a strategy to isolate mammalian genes based on RNA splicing. Proc. Natl. Acad. Sci. USA *88*:4005-4009.
2. **Diehn, M., M.B. Eisen, D. Botstein and P.O. Brown.** 2000. Large-scale identification of secreted and membrane-associated gene products using DNA microarrays. Nat. Genet. *25*:58-62.
3. **Duyk, G.M., S.W. Kim, R.M. Myers and D.R. Cox.** 1990. Exon trapping: a genetic screen to identify candidate transcribed sequences in cloned mammalian DNA. Proc. Natl. Acad. Sci. USA *87*:8995-8999.
4. **Ladunga, I.** 2000. Large-scale predictions of secretory proteins from mammalian genomic and EST sequences. Opin. Biotechnol. *11*:13-18.
5. **Peterfy, M., T. Gyuris and L. Takacs.** 2000. Signal-exon trap: a novel method for the identification of signal sequences from genomic DNA. Nucleic Acids Res. *28*:E26.
6. **Smith, C.W., J.G. Patton and B. Nadal-Ginard.** 1989. Alternative splicing in the control of gene expression. Annu. Rev. Genet. *23*:527-577.

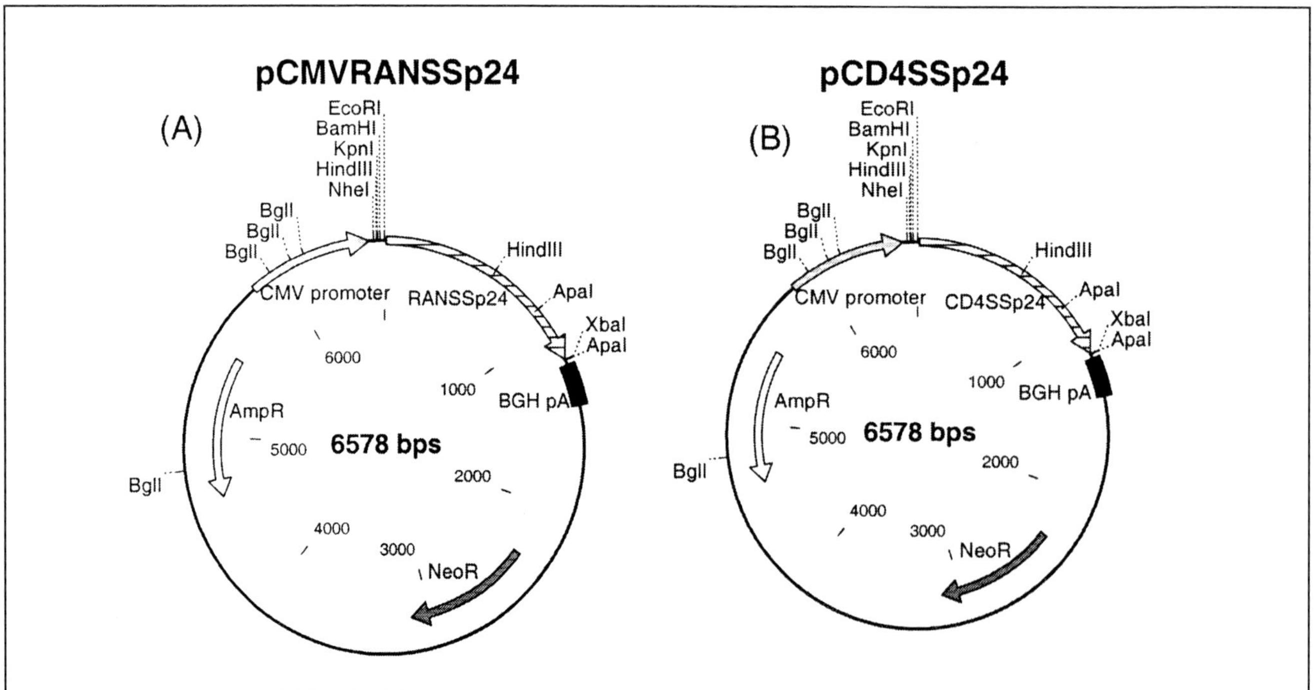

Figure 1. Maps of constructs encoding HIV-p24, cloned downstream from signal peptides from human RANTES (A) or CD4 (B). Both constructs were derived from pCDNA3.1+ (Invitrogen).

CHAPTER 33

Vectors to Target Protein Domains to Different Cellular Compartments

Garabet Toby[1,2], Susan F. Law[1] and Erica A. Golemis[1]

[1]Fox Chase Cancer Center, Philadelphia, PA, USA and [2]Mediterranean Agronomic Institute of Chania, Crete, Greece

(Reprinted from BioTechniques 24:637-640, 1998)

ABSTRACT

The pcDNA3 mammalian expression vector uses the cytomegalovirus (CMV) promoter to express proteins cloned into an adjacent polylinker. We have modified this vector to create three different targeting constructs: Go to Plasma Membrane (pGTM), Go to Nucleus (pGTN) and Go to Nucleus, Activate Transcription (pGTNAT). pGTM expresses a protein as a fusion to a myristic acid attachment signal, which targets proteins to the cell membranes. pGTN expresses an inserted protein as a fusion to a nuclear localization sequence (NLS), targeting it to the nucleus. pGTNAT incorporates an NLS (directing proteins to the nuclear compartment), an acid blob (a transcriptional activation domain) and a hemagglutinin epitope tag, creating a 107-amino acid fusion domain. We then cloned green fluorescent protein (GFP) into the three novel vectors and pcDNA3, and we transfected HeLa cells to test the new targeting constructs. Immunofluorescence analysis showed that the GFP protein localizes to the nucleus when over-expressed in pGTN and pGTNAT, localizes to the plasma membrane and perinuclear membrane when in pGTM and is ubiquitous through the cell in pcDNA3. We anticipate that these new vectors will prove very useful in future expression studies to examine the function of particular proteins when they are localized to specific cellular compartments.

INTRODUCTION

One of the approaches commonly followed in characterizing a novel protein is to over-express the intact protein, or protein truncations, from an expression vector and then establish interactions between particular domains of this protein and its cellular partners. One problem with such an approach is that the dissection of a protein into constituent domains may deprive some domains of particular signals that formerly directed the full-length protein to specific subcellular compartments. With the protein domains removed from their normal cellular compartment, and hence sequestered from their cellular partners, it becomes problematic to assess any interaction, even between molecules that normally associate with high affinity. To overcome this localization problem, we have modified the widely used mammalian expression vector pcDNA3 (Invitrogen, Carlsbad, CA, USA) to create three different targeting vectors that direct proteins or their domains to particular cellular compartments.

pcDNA3 utilizes a cytomegalovirus (CMV) promoter in conjunction with the bovine growth hormone (BGH) polyadenylation transcription termination signal sequences to allow high-level mRNA expression of inserted genes. It also encompasses a neomycin-resistance gene (neo^R) to allow the selection of stable mammalian cell lines. Here we describe three novel versions of pcDNA3: Go to Membrane (pGTM), Go to Nucleus (pGTN) and Go to Nucleus, Activate Transcription (pGTNAT). pGTM expresses a protein as a fusion to a myristic acid attachment signal (MAS) (8), which targets proteins to the cell periphery. pGTN expresses an inserted protein as a fusion to a nuclear localization sequence (NLS) targeting it to the nucleus. pGTNAT incorporates an NLS as with pGTN, but also includes an acid blob (AB) (a transcriptional activation domain) and a hemagglutinin epitope tag; it is suitable for use in mammalian two-hybrid strategies (10).

As a test of function, we cloned green fluorescent protein (GFP) of the jelly fish *Aequorea victoria* (2,7) into the three new vectors and the parent vector pcDNA3, and transfected HeLa cells. Immunofluorescence analysis showed that the GFP localizes to the nucleus when over-expressed in pGTN and pGTNAT, localizes to the cell plasma membrane and perinuclear membrane when in pGTM and is ubiquitous throughout the cell in pcDNA3. We anticipate these vectors will facilitate study of protein interactions at specific subcellular locations.

Cloning and Expression Vectors for Gene Function Analysis
Edited by Quinn Lu and Michael P. Weiner

MATERIALS AND METHODS

Preparation of pGTM

To construct pGTM, we designed two primers (EG436 and EG437). When annealed, these primers possessed the sequences in Table 1, which encodes the MAS (8) MGSSKSK with overhangs complementary to *Hin*dIII (5′) and *Eco*RI (3′). The annealed oligonucleotides were gel-purified and ligated into *Hin*dIII-*Eco*RI-restricted pcDNA3 and transformed into *E. coli*. The MAS sequence was confirmed by automated sequencing. Figure 1A shows the domain structure and cloning sites of pGTM.

Preparation of pGTNAT

To construct pGTNAT, pJG4-5 (4) was used as a source of targeting and transcriptional activation. pJG4-5 contains a 320-bp fusion cassette incorporating an NLS from simian virus 40 (SV40) (5), an AB that activates transcription (6) and a hemagglutinin tag (HA-tag). The gel-purified fusion cassette was ligated into *Hin*dIII-*Eco*RI-restricted pcDNA3 and transformed into *E. coli*. The sequence of the fusion cassette insert was confirmed by automated sequencing. Figure 1B shows shows the domain structure and cloning sites of pGTNAT.

Preparation of pGTN

To construct pGTN, we designed two primers (EG377 and EG398) that possess the sequence 5′-CTGGCTTATC-GAAATTAAT-3′ and 5′-GGGGAATTCACCAGCTACGTTT-C-3′, encoding the SV40 large T derived NLS (PPKKKRK-VA) flanked by *Hin*dIII and *Eco*RI restriction sites. We performed a polymerase chain reaction (PCR) using pGTNAT as template. The amplified product was gel-purified and ligated into restricted *Sma*I pBluescript® (Stratagene, La Jolla, CA,

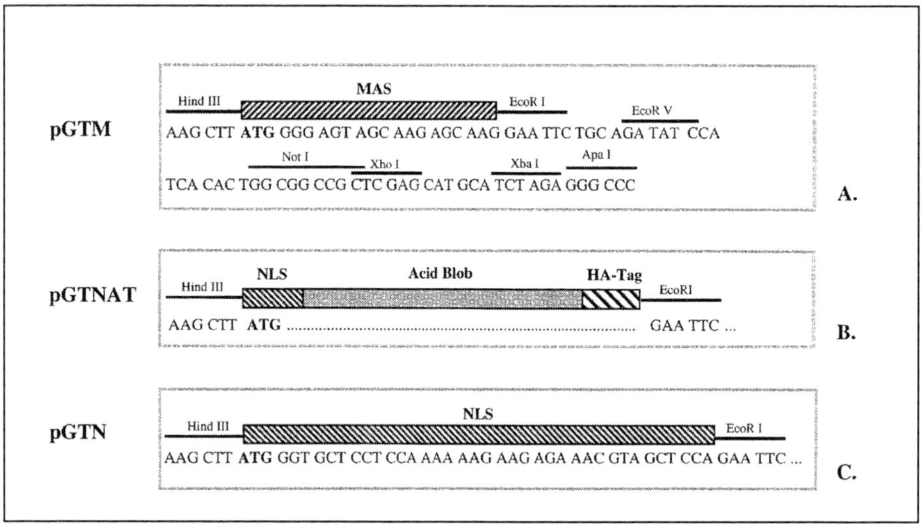

Figure 1. (A) Polylinker region of pGTM showing the reading frame of the MAS in relation to available restriction sites. (B) Polylinker region of pGTNAT with the reading frame of the NLS, AB and HA-Tag. (B) Polylinker region of pGTN with the coding frame of the NLS shown relative to the cloning sites.

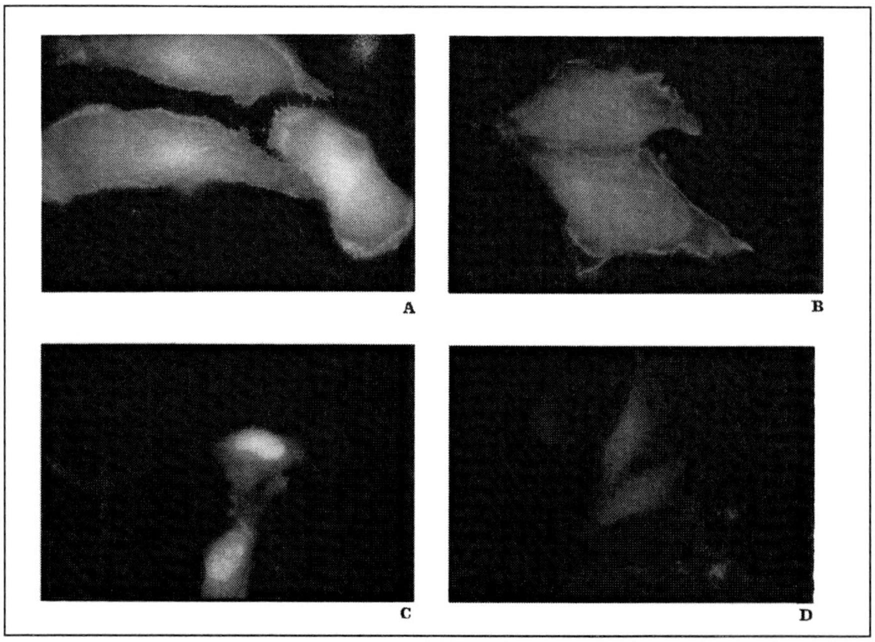

Figure 2. HeLa cells seen under fluorescence microscope using a 40× oil-immersion objective after 14-h recovery post-transfection. With pGTM-GFP, cells show fluorescence at the cell membrane site (A and B); pGTN-GFP cells show heavy nuclear fluorescence (C); and pcDNA3-GFP cells show weaker staining ubiquitous throughout the cell (D).

Table 1. EG436 and EG437 Primer Sequences

5′-AG CTT ATG GGG AGT AGC AAG AGC AAG G-3′
3′-A TAC CCC TCA TCG TTC TCG TTC CTT AA-5′
M G S S K S K

USA), from which the NLS was then excised and ligated into *Hin*dIII-*Eco*RI-restricted pcDNA3. The sequence of the NLS was confirmed by automated sequencing. Figure 1C shows the domain structure and cloning sites of pGTN.

Cloning the GFP into the Targeting Vectors and pcDNA3

To test the new vectors, GFP was used as a reporter. The complete GFP coding sequence was excised from pUC19 using an *Xba*I-*Apa*I digestion. The purified product was inserted into pGTM, pGTN, pGTNAT and pcDNA3 similarly cleaved with *Xba*I and *Apa*I as cloning sites. Ligations were transformed into DH5α^{TM} cells (Life Technologies, Gaithersburg, MD, USA), then plasmids were isolated using a Midi Prep Kit (Qiagen, Chatsworth, CA, USA). DNA sequencing was performed for each vector to check for the insertion of GFP in-frame.

Transfection and Immunofluorescence Analysis

Cells were split 24 h before transfection onto coverslips. Transfection of cells was carried with LIPOFECTAMINE™ Reagent (Life Technologies) using 1 μg of the plasmid and 6 μL of LIPOFECTAMINE in 200 μL OPTI-MEM® I Reduced Serum Media (Life Technologies). Cells were incubated for 6 h with the transfection mixture and then supplemented with 2 mL Dulbecco's modified Eagle medium (DMEM). Cells were then incubated for an additional 12 h before immunofluorescence analysis; coverslips were washed twice with 1× phosphate-buffered saline (PBS) then fixed with 3.5% paraformaldehyde for 7 min. Following that, cells were washed with 1× KB (0.01 M Tris, pH 7.5, 0.15 M NaCl and 0.1% bovine serum albumin [BSA]) for 5 min and then were observed microscopically using a 40× oil-immersion objective.

RESULTS AND DISCUSSION

Transfection of HeLa cells with GFP in the three targeting vectors and the parent plasmid pcDNA3 followed by immunofluorescence was used to test the function of the three novel targeting vectors.

Because the CMV promoter used in pcDNA3 is so potent, proteins are expressed at high levels intracellularly. With standard incubation post-transfection (24–48 h), GFP expressed from the parental and targeting vectors is abundant ubiquitously throughout the cell, with obvious artifactual aggregates of proteins. This high over-expression is linked to substantial cellular toxicity, as we noted lysed or abnormally shaped cells on plates at those latter times. In contrast, we determined that at earlier time points following transfection (12–15 h), protein levels were lower, and little if any toxicity was noted.

When a plasmid over-expressing GFP from pGTM was transfected into HeLa cells, immunofluorescence analysis at an optimal period post-transfection (ca. 14 h) showed a bright fluorescence concentrated at the cell membranes (Figure 2, A and B). Bright fluorescence was observed both at the cell periphery, reflecting localization to the plasma membrane and also at the nuclear envelope. Cytoplasmic staining was relatively low in comparison. Transfection of pGTN (Figure 2C) or pGTNAT (not shown) into HeLa cells resulted in intense nuclear staining, with a considerable reduction in general cytoplasmic staining (Figure 2C). In contrast, following transfection of GFP expressed from the parent vector pcDNA3, the protein was found to be ubiquitous throughout the cell, with no specific localization to any subcellular compartment (Figure 2D). In summary, these data indicate that GFP localizes to the cell membranes when expressed from pGTM, to the nucleus when expressed from pGTN or pGTNAT and is ubiquitous in the cell when expressed from pcDNA3.

A major problem in studying proteins and protein domains is the localization of the truncated domains to subcellular compartments that do not match the normal location of the protein. This is generally addressed by over-expressing proteins to high levels, so that some population reaches the intended location; the disadvantage of this approach is that high-level expression can have cellular toxicity and may lead to nonspecific interactions. We have overcome this problem by constructing three mammalian targeting vectors (pGTM, pGTN and pGTNAT) to direct the expressed protein to the cell membranes or to the nucleus. We note that in addition to their function in targeting proteins, some other aspects make these vectors useful in mammalian cells studies. pGTNAT in particular might be useful in mammalian two-hybrid applications (10).

In addition, the testing of the novel vectors resulted in three independent constructs: pGTM-GFP, pGTN-GFP and pGTNAT-GFP, which incorporate a reporter gene for protein expression, GFP, under the control of the CMV promoter. These can be used as protein localization markers facilitating analysis of subcellular compartments by immunofluorescence. GFP presents some advantages over other markers (such as β-galactosidase and luciferase) in that it does not require the addition of exogeneous substrates or cofactors, but is detected by simply irradiating the cell with near UV or blue light, making it very easy to monitor protein localization within the cell.

Finally, we anticipate that other mammalian vectors will be constructed to direct proteins and protein domains to other locations in the cell by the use of different targeting signals. Candidates for such signals might be mitochondrial (1), focal adhesion targeting (9), Golgi complex targeting (3) and others. This approach might also be modified to include the use of weaker or inducible promoters to lessen the toxic effects of over-expression as well as to allow a tighter control of compartment localization.

ACKNOWLEDGMENTS

These studies were supported by National Cancer Institute/NIH Grant Nos. R29-CA63366 and RO1 CA70841 to E.G. and core funds Grant No. CA-06927 to Fox Chase Cancer Center; and by American Cancer Society Grant No. CB-

121 to E.G. We are grateful to Anita Cywinski for help in sequencing. Address correspondence to Dr. Erica A. Golemis, Fox Chase Cancer Center, 7701 Burholme Ave, Philadelphia, PA 19111, USA. Internet: ea_golemis@ fccc. edu

REFERENCES

1. **Borgese, N., D. Aggujaro, P. Carrera, G. Pietrini and M. Bassetti.** 1996. A role for N-Myristoylation in protein targeting: NADH-cytochrome b5 reductase requires myristic acid for association with the outer mitochondrial but not ER membranes. J. Cell Biol. *135*:1501-1513.
2. **Chalfie, M., T. Yuan, G. Euskirchen, W. Ward and D. Prasher.** 1994. Green fluorescent protein as a marker for gene expression. Science *263*:802-805.
3. **Di Paolo, G., R. Lutjens, V. Pellier, S.A. Atimpson, M.H. Beuchat, S. Catsicas and G. Grenningloh.** 1997. Targeting of SCG10 to the area of the Golgi complex is mediated by its NH2-terminal region. J. Biol. Chem. *272*:5175-5182.
4. **Gyuris, J., E. Golemis, H. Chertkov and R. Brent.** 1993. Cdi1, a human G1 and S phase protein phosphatase that associates with Cdk2. Cell *75*:791-803.
5. **Hanover, J.** 1992. The nuclear pore: at the cross roads. FASEB J. *6*:2288-2295.
6. **Ma, J. and M. Ptashne.** 1987. A new class of yeast transcriptional activators. Cell *51*:113-119.
7. **Niwa, H., S. Inouye, T. Hirano, T. Matsuno, S. Kojima, M. Kubota, M. Ohashi and F.I. Tsuji.** 1996. Chemical nature of the light emitter of the *Aequorea* green fluorescent protein. Proc. Natl. Acad. Sci. USA *93*:13617-13622.
8. **Resh, M.D.** 1994. Myristylation and palmytylation of src family members: the fats of the matter. Cell *76*:411-413.
9. **Tachibana, K., T. Sato, N. D'Avirro and C. Morimoto.** 1995. Direct association of pp125FAK with paxillin, the focal adhesion-targeting mechanism of pp125FAK. J. Exp. Med. *182*:1089-1099.
10. **Ying, L., A. Batalao, H. Zhou and L. Zhu.** 1997. Mammalian two-hybrid system: a complementary approach to the yeast two-hybrid system. BioTechniques *22*:350-352.

Update to: Vectors to Target Protein Domains to Different Cellular Compartments

Garabet Toby

Fox Chase Cancer Center, Philadelphia, PA, USA

We would like to make the following additions to our original article.

Corrections

First we wish to make a correction to the sequence of the nuclear localization signal (NLS) in pGTN (See Figure 1 in original paper). The correct NLS DNA sequence and corresponding coded amino acids used to construct pGTN is: AAGCTTATGGGTGCTCCTCCAAAAAAGAAGAGAAACGTAGCTGGTGAATTC encoding the peptide sequence KLMGAPPKKKRNVAGEF, replacing AAGCTTATGGGTGCTCCTCCAAAAAAGAAGAGAAACGTAGCTCCAGAATTC and KLMGAPPKKKRKVAPEF, respectively. It should be noted that, in spite of these amino acid alterations, the NLS functions effectively in targeting fusion proteins to the nucleus, as reported by a number of investigators that have used the pGTN plasmid.

Additional References

A number of additional papers related to protein targeting are listed below. These describe additional domains/motifs that have been established to target fused proteins to different cell compartments. These include targeting to the endoplasmic reticulum (3), the plasma membrane (4), the chloroplast (6), focal adhesions (2), the peroxisomal matrix (1) and the mitochondria (5).

Commercially Available Targeting Vectors

Finally, it should be noted that a number of subcellular compartment targeting vectors are now available commercially. Invitrogen (Carlsbad, CA, USA) offers a series of vectors that target proteins to the mitochondria, nucleus and cytoplasm. Clontech Laboratories (Palo Alto, CA, USA) also offers a selection of targeting vectors (actin filaments, microtubules, endoplasmic reticulum, Golgi apparatus, mitochondria, cell membrane and nucleus) and additionally offers variants of these vectors that include expression of green fluorescent protein (GFP) reporters as fusions to the different targeting domains. Vectors from both companies feature different epitope tags and multiple cloning sites among other characteristics. Detailed information related to these vectors can be viewed by accessing (**www.invitrogen.com or www.clontech.com**).

REFERENCES

1. **Dingwall, C. and R.A. Laskey.** 1991. Nuclear targeting sequences—a consensus? Trends Biochem. Sci. *16*:478-481.
2. **Hettema, E.H., B. Distel and H.F. Tabak.** 1999. Import of proteins into peroxisomes. Biochim. Biophys. Acta. *1451*:17-34.
3. **Pelham, H.R.** 1996. The dynamic organisation of the secretory pathway. Cell Struct. Funct. *21*:413-419.
4. **Resh, M.D.** 1994. Myristylation and palmitylation of Src family members: the fats of the matter. Cell *76*:411-413.
5. **Rizzuto, R., A.W. Simpson, M. Brini and T. Pozzan.** 1992. Rapid changes of mitochondrial Ca2+ revealed by specifically targeting recombinant aequorin. Nature *358*:325-327.
6. **Settles, A.M. and R. Martienssen.** 1998. Old and new pathways of protein export in chloroplasts and bacteria. Trends Cell Biol. *8*:494-501.

CHAPTER 34

Green Fluorescent Protein Labeling of Cytoskeletal Structures—Novel Targeting Approach Based on Leucine Zippers

B.-Z. Katz, D. Krylov[1], S.-i. Aota, M. Olive[1], C. Vinson[1] and K.M. Yamada

National Institute of Dental Research and [1]National Cancer Institute, NIH, Bethesda, MD, USA

(Reprinted from BioTechniques 25:298-304, 1998)

ABSTRACT

Green fluorescent protein (GFP) is a valuable marker for intracellular protein localization. However, the fusion of GFP with structural proteins can alter their properties, resulting in a loss of fusion protein localization, decreased GFP fluorescence or both. We describe a novel targeting approach based on noncovalent heterodimerization of GFP and cytoplasmic structural proteins. The formation of structural protein/GFP complexes was mediated by modified leucine zipper protein spacers designed to form high-affinity heterodimers. The complexes localized accurately to specific sites within cells, providing selective fluorescence labeling of subcellular structures such as microfilaments or focal contacts.

INTRODUCTION

Studies of the dynamics of specific cytoplasmic components within living cells are becoming practical by using green fluorescent protein (GFP) as a molecular marker. A variety of cytoplasmic proteins (e.g., enzymes and structural molecules) have been successfully fused to GFP, creating chimeric proteins that are utilized as valuable tools to study cytoplasmic processes by direct visualization of specific proteins (5,14). Nevertheless, anecdotal experience indicates that certain GFP-tagged proteins fail to show fluorescence in patterns that match the endogenous protein.

Detailed mutational analysis revealed that the structure of GFP is extremely sensitive to molecular or biochemical modifications (7,13). For example, deletion of the first amino acid (methionine) or the last seven amino acids (aa) result in total loss of fluorescence activity (7). Similar deleterious effects can result from other structural modifications or biochemical conditions (13,17). Moreover, unlike short peptide epitopes that serve as molecular tags (e.g., Flag® or hemagglutinin epitope) GFP is relatively large (238 aa). For these reasons, direct fusion between GFP and the protein being examined can alter the function of GFP, its fusion partner or both. Indeed, the expression of the GFP/actin fusion protein in *Dictyostelium* results in dominant-negative effects on cell spreading, cytokinesis and motility (1). Here, we observed a nonspecific localization pattern of several fusion proteins of GFP with cytoskeletal molecules (see Results).

To provide an alternative approach to overcoming this and related problems resulting from covalent fusions of GFP with other proteins, we have developed a general methodology based on leucine zippers as linkers and spacers to form a modular, noncovalent GFP-protein complex. In this system, GFP is targeted to specific molecules and thereby to specific sites in cells using modified dimerization domains of a leucine zipper protein, the vitellogenin gene binding protein (VBP) (11). The leucine zipper domain of VBP has been modified to generate leucine zipper pairs that preferentially heterodimerize. The use of leucine zipper structures to mediate heterodimerization has several advantages. First, leucine zipper domains can form a specific and stable, yet noncovalent bond (11), thereby providing linkages that are stable but less likely to transmit structural distortions from one protein component to another in the absence of a direct linear covalent link. In addition, the dimeric coiled-coil structure formed by the leucine zipper provides a spacer between the partners of the complex. A wide range of leucine zipper domains with different intermolecular specificities and dimerization affinities has been designed (11). Because these dimerization domains are short and their sequences lack common restriction enzyme sites, polymerase chain reaction (PCR)-based construction of custom adapters for incorporating full-length cDNA molecules into an expression plasmid is simple and requires sequencing of only a short DNA segment to confirm accuracy of the PCR step.

Cloning and Expression Vectors for Gene Function Analysis
Edited by Quinn Lu and Michael P. Weiner

In this study, we fused an acidic leucine zipper dimerization domain to the C terminus of GFP and its basic counterpart to the N-terminal head domain of α-actinin. Co-expression of these components in the cytoplasm resulted in specific targeting of GFP to cytoskeletal structures, especially to actin-containing stress fibers.To establish the generality of this approach, we also examined a leucine zipper/vinculin fusion protein. Vinculin targeted GFP to distinct sites, the "focal contacts" involved in cellular adhesion.

MATERIALS AND METHODS

Generation of Leucine Zipper Fusion Proteins

Expression vectors were constructed using standard molecular biology methods (15). Briefly, leucine zipper $E{\cdot}E_{34}$ coding cDNA was amplified from template plasmids (11) by PCR. The cDNA coding for the α-actinin N terminus (nucleotides [nt] 1–861 of human non-muscle α-actinin coding region; GenBank® Accession No. g178051) was prepared from RNA of human foreskin fibroblasts by reverse-transcription (RT) followed by PCR using the RiboClone® System (Promega, Madison, WI, USA). The fragments were then ligated into the vector pRc/CMV containing the Flag epitope tag (Scientific Imaging System [Eastman Kodak], New Haven, CT, USA) in-frame, upstream to the insert. The resulting plasmid was sequenced and confirmed to encode the Flag epitope followed by a modified dimerization leucine zipper domain and α-actinin N terminus, and was designated pFlag/$E{\cdot}E_{34}$/α-actinin.

Full-length chicken vinculin cDNA with 5′ *Hin*dIII and a 3′ *Xba*I sites (kindly provided by B. Geiger, The Weizmann Institute of Science, Israel) was ligated into *Hin*dIII/*Xba*I restricted pRc/CMV vector containing the Flag epitope and the $E{\cdot}E_{34}$ form of VBP leucine zipper domain described above. The resulting plasmid was designated pFlag/$E{\cdot}E_{34}$/Vinculin.

For GFP expression, we used the vector pCA1 containing a cytamegalovirus (CMV) promoter followed by GFP cDNA [with the S65T mutation (10)] and *Hin*dIII and *Xba*I restriction sites. Modified leucine zipper dimerization domains of VBP were amplified by PCR from template plasmids (11), restricted by *Hin*dIII/*Xba*I and ligated into the GFP vector. The resulting plasmid was designated pGFP/$R{\cdot}R_{34}$. Detailed descriptions of oligonucleotides used to generate the PCR products are available upon request.

Western Blotting Analysis of Flag Epitope

Adherent human foreskin fibroblasts (kindly provided by Susan S. Yamada, National Institute of Dental Research, NIH) were transfected with plasmid DNAs by electroporation as described in Reference 12. After 48 h, the cells were lysed with RIPA buffer as previously described (6), and 10% sodium dodecyl sulfate polyacrylamide gel electrophoresis (SDS-PAGE) was performed, followed by Western blotting analysis with anti-Flag Monoclonal Antibody M2 (International Biotechnologies, New Haven, CT, USA) (6).

Indirect Immunofluorescence

Human foreskin fibroblasts on coverslips were fixed with 4% formaldehyde with 5% sucrose in phosphate-buffered

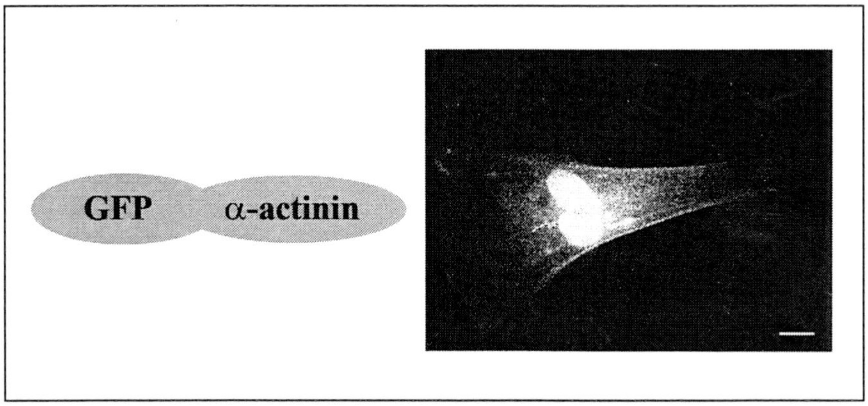

Figure 1. Expression of GFP/α-actinin head fusion protein in human foreskin fibroblasts. GFP is localized nonspecifically in the nucleus and weakly in the cytoplasm. Scale bar = 6 μm.

heptad		1	2	3	4
$R{\cdot}R_{34}$	ITI	RAAFLEK	ENTALRT	RVAELRK	RVGRCRNI
coiled coil		g a b c d e f	g a b c d e f	g a b c d e f	g a b c d e f g
$E{\cdot}E_{34}$	ITI	RAAFLEK	ENTALRT	EVAELEK	EVGRCENI
coiled coil		g'a'b'c'd'e'f'	g'a'b'c'd'e'f'	g'a'b'c'd'e'f'	g'a'b'c'd'e'f' g'

Figure 2. Leucine zipper domains designed for high-affinity heterodimerization. Amino acid sequences and 4-heptad organization of the leucine zipper domains used in this study. The underlined amino acids indicate the substitutions made to generate the $E.E_{34}$ and the $R.R_{34}$ forms. Below the VBP aa sequence (capital letters) is the nomenclature for coiled coil positions (lower case letters). Amino acids at the g positions of the $R.R_{34}$ form interact with the following e′ position of the $E.E_{34}$ form, as previously described (11).

saline (PBS) for 10 min and permeabilized with 0.5% Triton® X-100 in PBS for 10 min. Immunofluorescence staining was performed as previously described (12). In some cases, the cells were examined by confocal microscopy (Model TCS 4D; Leica, Deerfield, IL, USA).Optical sections were performed at 250-nm intervals in the Z plane.

RESULTS

GFP was fused to the globular head domain of α-actinin (the N terminus including aa 1–252) with the aim of labeling actin-containing cytoskeletal structures to study their dynamics. The head domain of α-actinin contains two actin-binding sites that mediate the α-actinin/actin association (2). However, transfection of fibroblasts with the GFP/α-actinin, N-terminus fusion protein resulted in a nonspecific pattern of localization with prominent nuclear localization, which failed to match the pattern of actin microfilaments (Figure 1). The lack of GFP fusion localization and its low fluorescence were not a result of negative effects on the actin cytoskeleton (not shown). Other GFP fusions with the cell-cell adherence junction proteins α-catenin and plakoglobin also resulted in low fluorescence intensity and lack of localization when expressed in epithelial cells (data not shown). We therefore generated noncovalent complexes of GFP with cytoskeletal components based on heterodimerization mediated by modified leucine zippers.

Generation of Leucine Zipper Fusion Proteins

A pair of heterodimerizing leucine zipper domains, 31 aa in length, was selected to mediate intermolecular GFP targeting. The mutated E•E_{34} and the R•R_{34} forms of the VBP leucine zipper dimerization domain (11) fulfilled requirements for high affinity and minimal self-interaction, and Figure 2 shows their amino acid sequences and heptad organization. This pair of leucine zipper modules has a low affinity for homodimerization ($K_d = 8.1 \times 10^{-4}$ M, with a melting temperature [T_m] of 21.6°C for the E•E_{34} form, and $K_d = 3.9 \times 10^{-5}$ M with a T_m of 27.1°C for the R•R_{34} form, ensuring a lack of dimerization under routine cell culture conditions of 37°C) (11). The E•E_{34} and the R•R_{34} leucine zipper domains form heterodimers with a high affinity of $K_d = 7 \times 10^{-9}$ M and a T_m of 51.5°C (11).

The E•E_{34} form was fused to the N terminus of α-actinin or of vinculin, and the R•R_{34} form was fused to the C terminus of GFP. These orientations of fusions were chosen to reduce possible steric interference with the noncovalent leucine zipper bond by the fusion proteins (see below). GFP alone without any fusion protein was used as a routine control in the following experiments.

Expression and Localization of Leucine Zipper Fusion Proteins in Fibroblasts

The plasmid containing the Flag/E•E_{34}/α-actinin cDNA was transfected into human foreskin fibroblasts. Forty-eight hours after transfection, total protein extracts (50 µg from transfected compared to nontransfected control cells) were analyzed by Western blotting with anti-Flag monoclonal antibody (MAb). As shown in Figure 3, cells transfected with the E•E_{34}/α-actinin plasmid expressed an approximately 35-kDa protein (the predicted molecular size of the fusion protein from the amino acid sequence) containing the Flag epitope. No additional bands were detected, indicating that the fusion protein was not cleaved by proteases. Immunofluorescence staining of the transfected cells with anti-Flag MAb showed localization of the fusion protein both along the actin stress fibers and in the cortex of the cells (Figure 4A). No morphological changes or modifications of the actin cytoskeleton were observed in the transfected cells compared to the nontransfected cells according to phase-contrast microscopy, and staining of the cells with rhodamine-labeled phalloidin to stain F-actin filaments (not shown).

GFP/R•R_{34} was expressed in human foreskin fibroblasts. Forty-eight hours following transfection, the cells were fixed, but not permeabilized, and were examined by confocal microscopy. Figure 4B shows an optical section 1 µm above the ventral surface of the cell. GFP/R•R_{34} was distributed relatively diffusely both in the cytoplasm and nucleus, as previously reported for the distribution of GFP alone (with no fusion) (4). Similar patterns of expression were observed for other optical sections throughout the cells (data not shown). GFP without fusion to any other protein had a similar diffuse distribution pattern (not shown). The transfection efficiency with GFP/R•R_{34} was approximately 30%, and no morphological changes were observed compared to nontransfected cells.

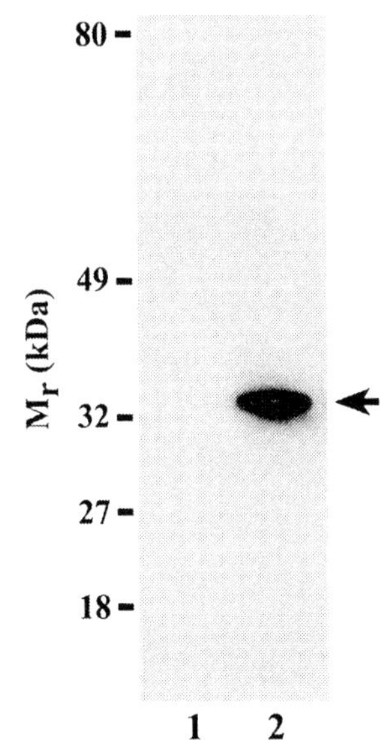

Figure 3. Expression of leucine zipper/α-actinin N-terminus fusion protein after transfection. Western immunoblotting with anti-Flag MAb to characterize expression of the Flag/E•E_{34}/α-actinin N-terminus fusion protein in human foreskin fibroblasts. Lane 1: protein extract from nontransfected cells. Lane 2: protein extract from cells transfected with pFlag/E•E_{34}/α-actinin. Arrow indicates the position of the Flag/E•E_{34}/α-actinin head domain recombinant protein.

Leucine Zipper-Mediated Targeting of GFP to Actin Subcellular Structures

To test the full system of leucine zipper-mediated linkages to GFP, we co-transfected the plasmids coding for GFP/R•R_{34} and Flag/E•E_{34}/α-actinin into human foreskin fibroblasts. The distribution of GFP in the transfected cells was analyzed using confocal microscopy 48 h after transfection. About 30% of the cells were positive for expression and most expressed both proteins. A series of optical sections along the Z-axis was performed; Figure 4C presents two of them. An optical section 1 µm above the ventral surface (bottom) of the cell shows a peripheral, cortical pattern of distribution of fluorescence, with relatively little GFP localized at the center of the cell (Figure 4C). An optical section 0.25 µm above the ventral surface of the cell shows a filamentous pattern typical for actin stress fibers that is parallel to the bottom of the cell (Figure 4C). No changes in cellular morphology or actin organization were observed in the transfected cells compared to the non-transfected cells. In later studies, we examined GFP heterodimer localization in living transfected cells for 8 h using time-lapse video fluorescence microscopy (exposures at 10-min intervals, 3 s/exposure). GFP targeted by α-actinin continued to label the actin cytoskeleton for the entire period, even as the cell underwent movements (not shown).

Leucine Zipper-Mediated Targeting of GFP to Focal Contacts by Vinculin

To examine the general utility of this approach, i.e., labeling other proteins that target other sites in cells, we generated GFP plus vinculin complexes. Vinculin is one of the major structural components of the well-studied cell-matrix adhesion complexes termed focal contacts (9). We used a leucine zipper/vinculin fusion protein to attempt to direct the localization of GFP to the focal contacts of human foreskin fibroblasts.

Human foreskin fibroblasts transfected with pFlag/E•E_{34}/vinculin expressed a single 125-kDa protein, the expected molecular size of the E•E_{34}/vinculin fusion protein (not shown). When pFlag/E•E_{34}/vinculin and pGFP/R•R_{34} were co-transfected into human foreskin fibroblasts, GFP became localized into the focal contacts, and did not co-localize with actin microfilaments (Figure 4D). Additional studies showed that the leucine zipper bond was maintained after cells were detached and kept in suspension, and appropriate targeting of several different leucine zipper fusion proteins into focal adhesions was accomplished following cell attachment to fibronectin (unpublished).

DISCUSSION

GFP is a 37-kDa protein that serves as part of the bioluminescence mechanism of *Aequorea victoria* (16). During the short period of time since its discovery, many improvements have been introduced in the molecular structure of GFP to produce better emission intensities, e.g., point mutations as well as codon changes to enhance translation in mammalian cells (3,8,18). Such modifications provided improved GFP variants that are currently being used to study intracellular dynamic processes. However, changes in GFP structure can result in decreases or loss of its ability to fluoresce. Fusion between GFP and a molecular partner can also result in an alteration in the localization and/or activity of the protein being examined. A GFP-actin fusion protein has dominant-negative effects on cell spreading, cytokinesis and motility in *Dictyostelium* (1). Another striking example was the aberrant localization of the GFP/α-actinin N-terminus fusion protein in this study, as well as α-catenin and plakoglobin constructs. To provide an alternative approach, we developed a general-purpose modular system using high-affinity, noncovalent linkers and spacers based on the leucine zipper. The coding

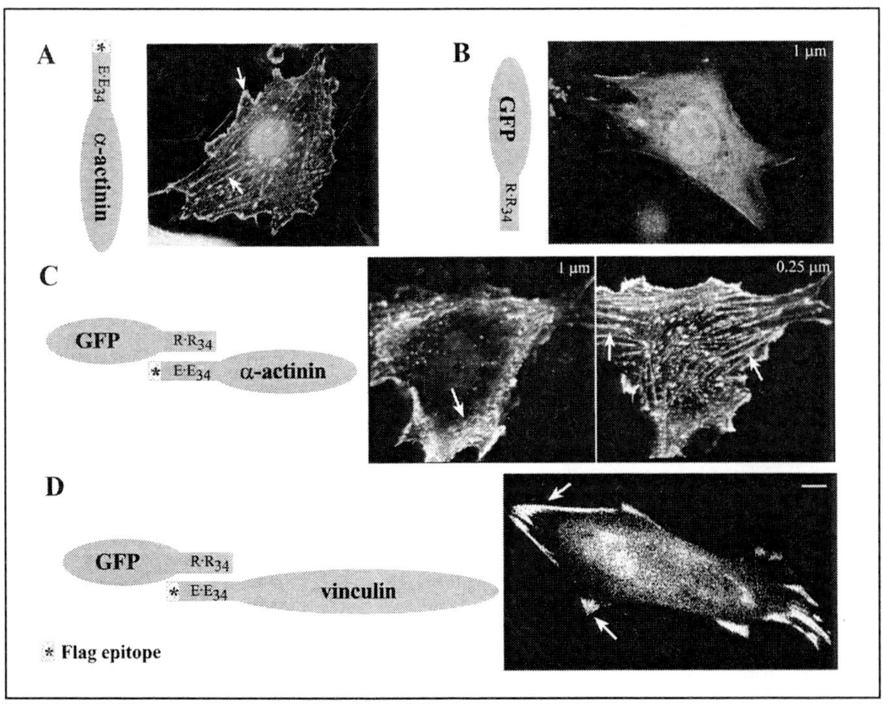

Figure 4. Localization of GFP with a cytoskeletal protein using leucine zipper domains in human foreskin fibroblasts. (A) Flag/E.E_{34}/α-actinin. Immunofluoresence staining for Flag in cells transfected with pFlag/E.E_{34}/α-actinin containing the Flag epitope. Arrows indicate cortical staining and localization along stress fibers. (B) GFP/R.R_{34}. A confocal optical section, 1 µm from the ventral surface (bottom) of a non-permeabilized cell transfected with pGFP/R.R_{34}. Relatively diffuse GFP fluorescence was observed both in the cytoplasm and in the nucleus. (C) GFP/R.R_{34} and Flag/E.E_{34}/α-actinin. Confocal optical sections, 0.25 and 1 µm from the ventral surface of a non-permeabilized cell co-transfected with pGFP/R.R_{34} and pFlag/E.E_{34}/α-actinin are shown. Arrow indicates the localization of GFP fluorescence to the cortex of the cell (the 1-µm section). In the 0.25-µm section, arrows indicate the localization of the GFP fluorescence along stress fibers oriented parallel to the ventral surface of the cell. (D) Localization of GFP in a cell co-transfected with pGFP/R.R_{34} and pFlag/E.E_{34}/vinculin. Arrows indicate immunofluorescence labeling of focal contacts. Scale bar = 6 µm.

sequence of leucine zippers is short (128 nt), and it does not contain any internal restriction sites for commonly used enzymes. These properties make it convenient to fuse leucine zippers to various proteins by simple PCR-mediated fusion or ligation of PCR products, as used in this study.

One theoretical concern is that the leucine zipper constructs, even though synthetic, might dimerize with endogenous leucine zippers to cause mislocalization of the expressed constructs or disruption of normal cellular functions. Consequently, each component of the complex was evaluated. The GFP/zipper was examined for its localization and fluorescence properties, and then the leucine zipper/α-actinin fusion was examined for localization (using the Flag tag). No evidence for mislocalization or deleterious effects on cell behavior or actin cytoskeleton were observed. Following these analyses, both leucine zipper chimeric proteins were co-transfected into cells, and GFP localization of the complexes was determined. In this study, we have demonstrated, to our knowledge for the first time, a GFP-labeled system able to localize along actin filaments in mammalian cells.

We then utilized another cytoskeletal protein, vinculin, to demonstrate leucine zipper-mediated targeting of GFP to this protein. The GFP/vinculin complex localized to a different site, the focal contacts involved in cell adhesion. These studies with GFP/vinculin show that the system can be used for other proteins, suggesting the general utility of this system for analyzing subcellular protein localization to various sites. Replacing the GFP with enzymes could facilitate their intracellular targeting to specific sites so that enzymatic modulation of specific sites within the cytoplasm can be possible using this same general approach.

ACKNOWLEDGMENTS

We are grateful to Dr. Sankar Adhya for facilitating this collaboration and to Dr. Roger Tsien for providing the green fluorescent protein construct. Flag and anti-Flag are registered trademarks of Immunex Corporation.

REFERENCES

1. **Aizawa, H., M. Sameshima and I. Yahara.** 1997. A green fluorescent protein-actin fusion protein dominantly inhibits cytokinesis, cell spreading, and locomotion in *Dictyostelium*. Cell Struct. Funct. *22*:335-345.
2. **Blanchard, A., V. Ohanian and D. Critchley.** 1989. The structure and function of α-actinin. J. Muscle Res. Cell Motil. *10*:280-289.
3. **Crameri, A., E. Whitehom, E. Tate and W. Stemmer.** 1996. Improved green fluorescent protein by molecular evolution using DNA shuffling. Nature Biotechnol. *14*:315-319.
4. **Cubitt, A.B., R. Heim, S.R. Adams, A.E. Boyd, L.A. Gross and R.Y. Tsien.** 1995. Understanding, improving and using green fluorescent proteins. Trends Biochem. Sci. *20*:448-455.
5. **DeGiorgi, F., M. Brini, C. Bastianutto, R. Marsault, M. Montero, P. Pizzo, R. Ross and R. Rizzuto.** 1996. Targeting aequorin and green fluorescent protein to intracellular organelles. Gene *173*:113-117.
6. **De Nichilo, M.O. and K.M. Yamada.** 1996. Integrin αvβ5-dependent serine phosphorylation of paxillin in cultured human macrophages adherent to vitronectin. J. Biol. Chem. *271*:11016-11022.
7. **Dopf, J. and T. Horiagon.** 1996. Deletion mapping of the *Aequorea victoria* green fluorescent protein. Gene *173*:39-44.
8. **Ehrig, T., D.J. O'Kane and F.G. Prendergast.** 1995. Green-fluorescent protein mutants with altered fluorescence excitation spectra. FEBS Lett. *367*:163-166.
9. **Geiger, B., K.T. Tokuyasu, A.H. Dutton and S.J. Singer.** 1980. Vinculin, an intracellular protein localized at specialized sites where microfilament bundles terminate at cell membranes. Proc. Natl. Acad. Sci. USA *77*:4127-4131.
10. **Heim, R., D.C. Prasher and R.Y. Tsien.** 1994. Wavelength mutations and post-translational autoxidation of green fluorescent protein. Proc. Natl. Acad. Sci. USA *91*:12501-12504.
11. **Krylov, D., I. Mikhailenko and C. Vinson.** 1994. A thermodynamic scale for leucine zipper stability and dimerization specificity: e and g interhelical interactions. EMBO J. *13*:2849-2861.
12. **LaFlamme, S.E., S.K. Akiyama and K.M. Yamada.** 1992. Regulation of fibronectin receptor distribution. J. Cell Biol. *117*:437-447.
13. **Lim, C., K. Kimata, M. Oka, K. Nomaguchi and K. Kohno.** 1995. Thermosensitivity of green fluorescent protein fluorescence utilized to reveal novel nuclear-like compartments in a mutant nucleoporin NSP1. J. Biochem. *118*:13-17.
14. **Rizzuto, R., M. Brini, F. DeGiorgi, R. Rossi, R. Heim, R.Y. Tsien and T. Pozzan.** 1996. Double labeling of subcellular structures with organelle-targeted GFP mutants in vivo. Curr. Biol. *6*:183-188.
15. **Sambrook, J., E.F. Fritch and T. Maniatis.** 1989. Molecular Cloning: A Laboratory Manual, 2nd ed., CSH Laboratory Press. Cold Spring Harbor, NY.
16. **Ward, W.** 1979. *In* K. Smith (Ed.), Photochemical and Photobiological Reviews, p. 1-57. Plenum Press, New York.
17. **Ward, W.W. and S.H. Bokman.** 1982. Reversible denaturation of Aequorea green-fluorescent protein: physical separation and characterization of the renatured protein. Biochemistry *21*:4535-4540.
18. **Youvan, D.C. and M.E. Michel-Beyerle.** 1996. Structure and fluorescence mechanism of GFP. Nature Biotechnol. *14*:1219-1220.

Update to: Green Fluorescent Protein Labeling of Cytoskeletal Structures—Novel Targeting Approach Based on Leucine Zippers

Ben-Zion Katz[1,3], Dmitry Krylov[2], Charles Vinson[2], and Kenneth M. Yamada[1]

[1]National Institute of Dental and Craniofacial Research, [2]National Cancer Institute, National Institutes of Health, Bethesda, MD, USA and [3]The Hematology Institute, Tel-Aviv Medical Center, Tel-Aviv, Israel

NOVEL APPLICATION FOR PRODUCING SITE-SPECIFIC MODIFICATIONS OF SUBCELLULAR STRUCTURES: ENZYMATIC MODULATION OF FOCAL CONTACTS BY LEUCINE ZIPPER-MEDIATED TARGETING

Targeted Modulation of Subcellular Structures

Since publication of our original paper describing leucine zipper chimeras with green fluorescent protein (GFP), we have developed a novel extension of its approaches to achieve site-specific targeting of a candidate enzyme to a selected cellular structure. We focused on a method to selectively target a well-characterized class of subcellular structures that mediate strong cell adhesion to substrates termed focal contacts or focal adhesions. Previously, subcellular structures such as focal contacts (FC) were studied by two general approaches: pharmacological or molecular manipulation. The pharmacological approach involves drugs or chemical modifying agents (e.g., inhibitors of tyrosine or serine kinases, tyrosine phosphatases, etc.). Such studies demonstrated that tyrosine (as well as serine/threonine) phosphorylation may regulate the formation of FC (1,2). Moreover, the recruitment of some cytoskeletal and signaling molecules to experimentally-induced adhesion sites using fibronectin-coated beads was inhibited by tyrosine phosphorylation inhibitors such as herbimycin A or genistein (11). Pervanadate, a potent inhibitor of phosphotyrosine phosphatases, can induce an increase in total protein tyrosine phosphorylation in cells and results in the accumulation of cytoskeletal components into FC (1). Nevertheless, such drugs may have toxic side effects, and most importantly, can have global effects on cellular physiology. They may consequently affect structures and functions via indirect mechanisms. For example, pervanadate treatment of 3T3 cells results in transient assembly of cell-cell adherence junctions, in addition to FC assembly (10).

The second major previous approach has been molecular modulation of single cytoplasmic components. This strategy has included site-directed mutagenesis and deletions in FC components to generate activated or dominant-inhibitory molecules (e.g., FAK, Src, etc.) or deleting or modulating function by gene targeting (5–7,13). The molecular modulation approach has been used extensively to study the role of Src and FAK in the function of FC (5–7,13).

Specific Targeting of Subcellular Sites

We describe a novel subcellular targeting approach based on the formation of complexes linked by leucine zippers that incorporate a targeting element and an enzymatically active component (in this case a kinase: c-Src). Leucine zippers that were designed for specific heterodimer formation were used to mediate formation of a bimolecular complex (8.9). Previous studies have established that leucine zipper chimeras can be used to form artificial dimers, albeit extracellularly. Dimerization of the bacterial aspartate receptor and of the IL-3 receptor could be induced by GCN4 and c-fos/c-jun leucine zipper domains, respectively, and resulted in the activation of signal transduction pathways (3,12).

Our previous study indicated that a molecule that lacks any subcellular localization signal (e.g., nonfused GFP) could be directed to specific cytoplasmic structures using a targeting system based on leucine zippers (8). We speculated that the acidic leucine zipper dimerization domain could be fused to the C terminus of any enzyme inserted in place of the GFP, while the basic zipper counterpart could be fused to the N terminus of structural molecules resident in FC (e.g., α-actinin or vinculin). Because co-expression of these components in the cytoplasm resulted in specific targeting of GFP to cytoskeletal structures, especially to stress fibers and FC (8), we examined whether an enzyme substituted for GFP could be targeted to FC to produce enzymatic modulation of function.

Design of an Enzymatic Targeting System for Focal Contacts

We used a targeting system based on high-affinity, noncovalent leucine-zipper complexes between the cytoskeletal molecule vinculin as a targeting element and c-Src kinase as the active component. The effects of localized modification of tyrosine phosphorylation by Src on the structure and function of cell-matrix adhesion sites were evaluated. As in the previous paper, this system placed the high-affinity $E{\cdot}E_{34}$ (acidic) and $R{\cdot}R_{34}$ (basic) forms of leucine zippers as a spacer between the two molecules in order to minimize structural distortions from the chimeric partner and to permit testing of each component of the system separately as biological controls.

Vinculin was selected to serve as the targeting component of the system because of its lack of any known enzymatic activity and its specific subcellular distribution to FC in fibroblasts (4). As shown in Figure 1, the $E{\cdot}E_{34}$ form of leucine zipper vitellogenin gene binding protein (VBP) was fused to the N terminus of c-Src. The fusion protein also contained a FLAG®-epitope tag (Scientific Imaging [Eastman

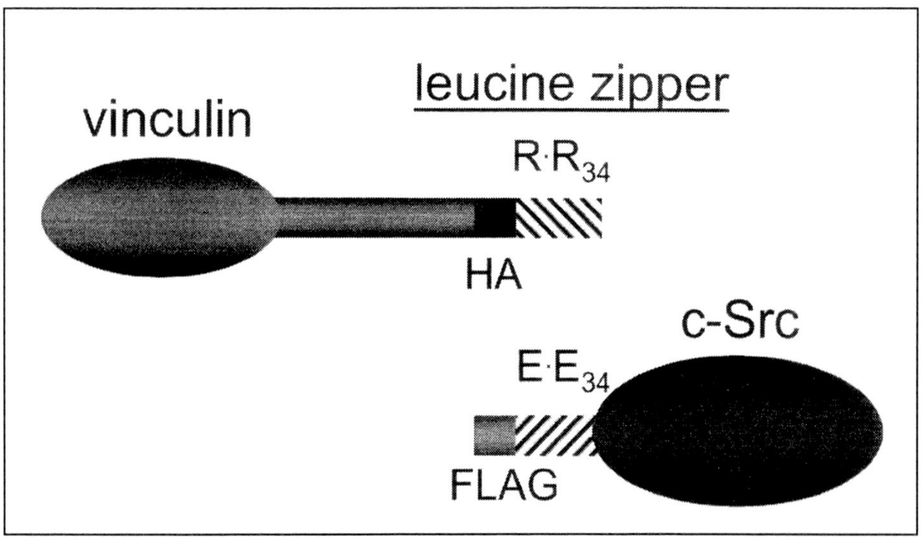

Figure 1. Relative orientations of leucine zipper domains and the different fusion partners of each of the constructs. In the $E.E_{34}$ form, the amino acids in the e and g positions are glutamic acid, while in the $R.R_{34}$ form, the e and g positions contain arginine residues (9). HA, hemagglutinin epitope; F, FLAG epitope.

Kodak], New Haven, CT, USA) at its N terminus. The R·R_{34} form of leucine zipper VBP was fused to the C terminus of vinculin. This fusion protein contained a hemagglutinin-epitope tag (Figure 1). The fusions were designed to involve the two opposing termini of the leucine zipper components in order to reduce possible interference with the leucine zipper bond (Figure 1).

Leucine Zippers Target c-Src via Vinculin to Focal Contacts

Human foreskin fibroblasts were cotransfected with vinculin/R·R_{34} and E·E_{34}/c-Src. Forty-eight hours after transfection, the cells were fixed and permeabilized; then double immunofluorescence staining was performed with anti-HA

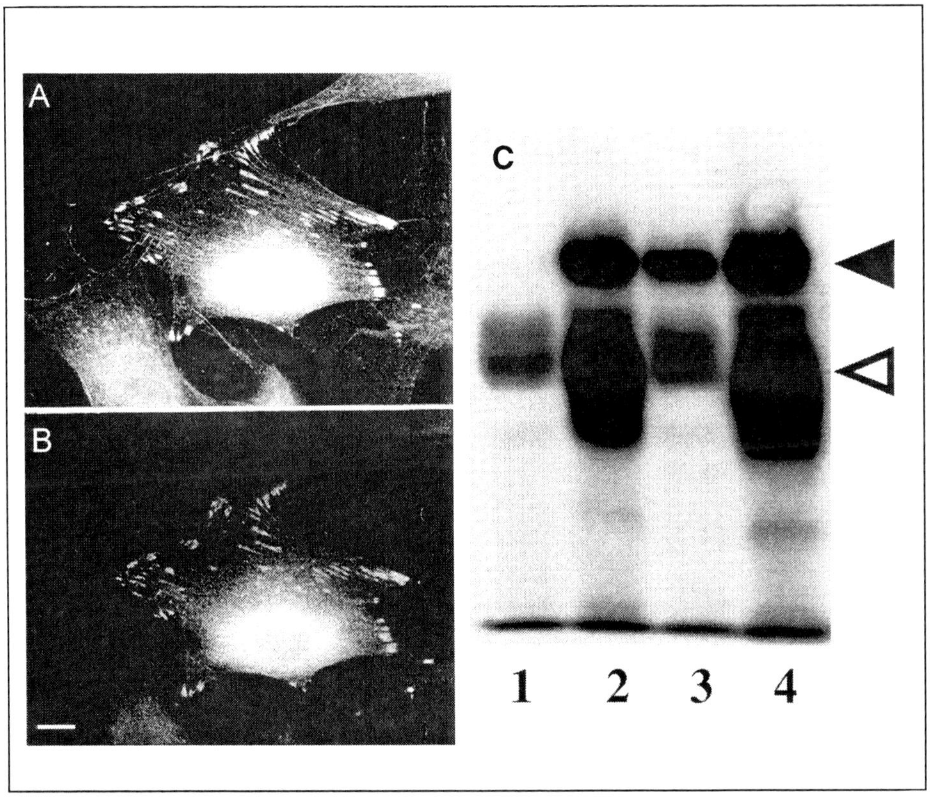

Figure 2. Localization and physical association of the leucine zipper fusion proteins. (A and B) Human foreskin fibroblasts were transfected with E.E_{34}/c-Src and vinculin/R.R_{34} to evaluate efficiency of Src targeting. Forty-eight hours after transfection the cells were fixed, permeabilized and double-stained for FLAG (A) and HA (B). Bar = 10 µm. (C) Co-immunoprecipitation of the E.E_{34}/c-Src and vinculin/R.R_{34} fusion proteins. Cells were cotransfected with E.E_{34}/c-Src and vinculin/R.R_{34} (lanes 3 and 4) or with vinculin and E.E_{34}/c-Src (lanes 1 and 2). Immunoprecipitation was performed 48 h following transfection using rabbit anti-c-Src polyclonal antibody (lanes 2 and 4) or anti-vinculin mAb (lanes 1 and 3) from aliquots containing equal total protein concentrations. The precipitates were analyzed by western blotting with rabbit anti-c-Src polyclonal antibody. Empty and filled arrowheads indicate the position of endogenous and the E.E_{34}/c-Src fusion protein, respectively.

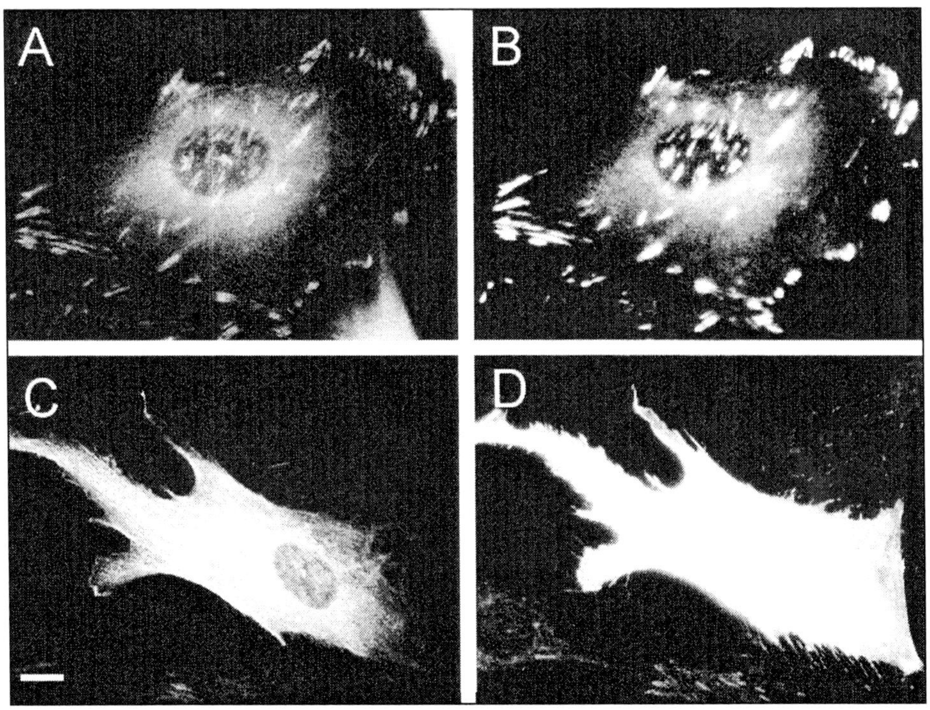

Figure 3. Localized c-Src elevates FC tyrosine phosphorylation. Cells were transfected with either E.E_{34}/c-Src and vinculin/R.R_{34} (A and B) or with E.E_{34}/c-Src and vinculin (C and D). Forty-eight hours later, cells were fixed and permeabilized. Double immunofluorescence was performed with rabbit anti-c-Src polyclonal antibody, followed by fluorescein isothiocyanate (FITC)-labeled goat anti-rabbit polyclonal antibody (A and C) or mouse anti-pTyr mAb, followed by rhodamine-labeled goat anti-mouse polyclonal antibody (B and D). Bar = 10 µm.

and anti-FLAG monoclonal antibodies. As expected, vinculin/R·R_{34} localized to the FC of the fibroblasts as demonstrated by staining of the transfected cells with anti-HA mAb (Figure 2B). As shown in Figure 2A, FLAG immunofluorescence staining for the E·E_{34}/c-Src fusion protein also localized to FC, where the vinculin/R·R_{34} was present.

To confirm that vinculin/R·R_{34} and E·E_{34}/c-Src were physically associated, immuoprecipitations for vinculin or c-Src were performed from lysates prepared from cells transfected with both vinculin/R·R_{34} and E·E_{34}/c-Src, or from cells transfected with E·E_{34}/c-Src and vinculin (which did not contain the leucine zipper R·R_{34} domain). The precipitates were resolved by sodium dodecyl sulfate-polyacrylamide gel electrophoresis (SDS-PAGE) (10%) and subjected to western blotting with anti-Src antibody. As shown in Figure 2C, lane 3, Src was co-precipitated with vinculin in cells expressing both vinculin/R·R_{34} and E·E_{34}/c-Src. However, in cells transfected with E·E_{34}/c-Src and vinculin that did not contain the leucine zipper domain, these molecules did not co-precipitate (Figure 2C, lane 1). Immunoprecipitation of c-Src from these cells confirmed that equal quantities of E·E_{34}/c-Src were expressed in both (Figure 2C, lanes 2 and 4). The co-precipitation experiment was performed in high-stringency RIPA buffer (1% Triton X-100, 1% sodium deoxycholate, 0.1% SDS, 150 mM NaCl, 50 mM Tris, pH 8.0, and 2 mM PMSF), indicating that the leucine zipper-mediated c-Src/vinculin bond was of high affinity. These data demonstrate that leucine zipper dimerization domains can mediate the targeting of c-Src to FC by high-affinity physical association with the structural protein vinculin.

Local Modulation of Tyrosine Phosphorylation of Focal Contacts

The effect of c-Src targeting on the levels of tyrosine phosphorylation in FC was examined. Cells cultured on coverslips for 48 h following transfection with both Src and targeting leucine zipper fusion proteins expressed Src localized in FC (Figure 3B). As shown in Figure 3, A and B, these cells also displayed a high level of tyrosine phosphorylation that was strikingly confined to the FC, and it colocalized with Src. In contrast, both the distribution of E·E_{34}/c-Src and protein tyrosine phosphorylation were diffuse throughout the cell when expressed without the targeting element, as evaluated by immunofluorescence staining for Src (Figure 3C) and antiphosphotyrosine antibody (Figure 3D). The patterns of the E·E_{34}/c-Src fusion protein localization and tyrosine phosphorylation in the control-transfected cells were similar to that of nonmodified c-Src (not shown).

Conclusions

In this supplement, we extended the original concept of leucine zippers chimeras to develop a novel intracellular targeting strategy that can locally modulate tyrosine phosphorylation of focal contacts. This novel approach can generate site-specific modification of a subcellular structure using a noncovalent chimera with targeting and enzymatic domains. To our knowledge, this represents the first time that an individual cytoplasmic adhesive/signaling structure has been selectively targeted and modified by enzymatic means. Targeting of c-Src to focal contacts via a leucine zipper-vinculin system resulted in highly selective protein tyrosine phosphorylation of focal contacts, whereas diffusely expressed c-Src produced widespread intracellular tyrosine phosphorylation.

This overall approach can provide a novel tool for dissecting function by specific enzymatic modifications of individual subcellular structures in living cells.

REFERENCES

1. **Ayalon, O. and B. Geiger.** 1997. Cyclic changes in the organization of cell adhesions and the associated cytoskeleton, induced by stimulation of tyrosine phosphorylation in bovine aortic endothelial cells. J. Cell Sci. *110*:547-556.
2. **Burridge, K., M. Chrzanowska-Wodnicka and C.-L. Zhong.** 1997. Focal adhesion assembly. Trends Cell Biol. *7*:342-347.
3. **Cochran, A.G. and P.S. Kim.** 1996. Imitation of *Escherichia coli* aspartate receptor signaling in engineered dimers of the cytoplasmic domain. Science *271*:1113-1116.
4. **Geiger, B., K.T. Tokuyasu, A.H. Dutton and S.J. Singer.** 1980. Vinculin, an intracellular protein localized at specialized sites where microfilament bundles terminate at cell membranes. Proc. Natl. Acad. Sci. USA *77*:4127-4131.
5. **Gilmore, A.P. and L.H. Romer.** 1996. Inhibition of focal adhesion kinase (FAK) signaling in focal adhesions decreases cell motility and proliferation. Mol. Biol. Cell *7*:1209-1224.
6. **Ilic, D., Y. Furuta, S. Kanazawa, N. Takeda, K. Sobue, N. Nakatsuji, S. Nomura, J. Fujimoto, M. Okada and T. Yamamoto.** 1995. Reduced cell motility and enhanced focal adhesion contact formation in cells from FAK-deficient mice. Nature *377*:539-544.
7. **Kaplan, K.B., K.B. Bibbins, J.R. Swedlow, M. Arnaud, D.O. Morgan and H.E. Varmus.** 1994. Association of the amino-terminal half of c-Src with focal adhesions alters their properties and is regulated by phosphorylation of tyrosine 527. EMBO J. *13*:4745-4756.
8. **Katz, B.-Z., D. Krylov, S. Aota, M. Olive, C. Vinson and K.M. Yamada.** 1998. Green fluorescent protein labeling of cytoskeletal structures—novel targeting approach based on leucine zippers. BioTechniques *25*:298-304.
9. **Krylov, D., I. Mikhailenko and C. Vinson.** 1994. A thermodynamic scale for leucine zipper stability and dimerization specificity: e and g interhelical interactions. EMBO J. *13*:2849-2861.
10. **Michalides, R., T. Volberg and B. Geiger.** 1994. Augmentation of adherens junction formation in mesenchymal cells by co-expression of N-CAM or short-term stimulation of tyrosine-phosphorylation. Cell Adhes. Commun. *2*:481-490.
11. **Miyamoto, S., H. Teramoto, O.A. Coso, J.S. Gutkind, P.D. Burbelo, S.K. Akiyama and K.M. Yamada.** 1995. Integrin function: molecular hierarchies of cytoskeletal and signaling molecules. J. Cell Biol. *131*:791-805.
12. **Patel, N., J.M. Herrman, J.C. Timans and R.A. Kastelein.** 1996. Functional replacement of cytokine receptor extracellular domains by leucine zippers. J. Biol. Chem. *271*:30386-30391.
13. **Thomas, S.M., P. Soriano and A. Imamoto.** 1995. Specific and redundant roles of Src and Fyn in organizing the cytoskeleton. Nature *376*:267-271.

CHAPTER 35

Establishment of Stable Cell Lines Expressing Potentially Toxic Proteins by Tetracycline-Regulated and Epitope-Tagging Methods

Shwu-Yuan Wu and Cheng-Ming Chiang

University of Illinois at Urbana-Champaign, Urbana, IL, USA

(Reprinted from BioTechniques 21:718-725, 1996)

ABSTRACT

A tetracycline-regulated expression system is combined with the FLAG®-epitope tagging method for conditional expression of potentially toxic proteins in mammalian cells. This strategy allows a controlled expression of exogenous gene products and also provides a unique way of protein purification. Two mammalian expression plasmids containing the FLAG sequence and flanking multiple cloning sites were created for conditional protein expression. The cDNAs encoding human basal transcription factors TBP, $TAF_{II}55$ and the p62 subunit of TFIIH were individually cloned into these vectors and introduced into a HeLa-derived cell line that constitutively expresses a tetracycline-regulated transactivator (tTA). The established clonal human cell lines express FLAG-tagged basal transcription factors in a manner modulated by the amount of tetracycline in the growth medium. In the absence of tetracycline, tTA binds to the DNA recognition sites of the expression plasmid and induces the expression of tagged proteins. When tetracycline is added back to the growth medium, the induced protein starts to decay. This provides us with an estimation of the in vivo half-lives of TBP and $TAF_{II}55$, which were assessed to be less than 20 and 6 h, respectively, in HeLa cells. The level of induced proteins in the absence of tetracycline could be further enhanced by including the antibiotic G418 to presumably boost the production of tTA which in turn activates the expression of tagged proteins.

INTRODUCTION

The biological function of proteins is often investigated by transient overexpression of a particular gene product in cultured cells. Transient transfection offers a quick and simple assay to evaluate the potential activity of proteins, such as transcription factors and cell-cycle regulatory proteins. However, due to the variable proportion of cells taking up the DNA and to variations in culture conditions, transient transfection is sometimes difficult to repeat, and frequently cannot provide enough transfected cells for biochemical analysis. In our experience, successful purification of a multisubunit protein complex is unlikely to be achieved by simply overexpressing a single protein subunit by transient transfection. This problem is further complicated by the stoichiometric ratio among all other subunits in a multiprotein complex. To overcome these difficulties, we and others have used retrovirus-mediated gene transfer techniques to establish stable cell lines that constitutively express epitope-tagged proteins for purification of multiprotein complexes, such as the basal transcription factor TFIID (1,13). The introduced epitope tag also provides a convenient site for affinity purification by monoclonal antibody-conjugated beads. This purification strategy has been successfully applied to many other protein complexes (1; S.-Y. Wu and C.-M. Chiang, unpublished data). Nevertheless, attempts to establish stable cell lines that express other transcription factors by a retroviral approach have failed, partly due to the potential cytotoxicity of these proteins.

Conditional expression of a foreign gene product is invaluable in assessing the biological function of a protein during specific stages of development, cellular growth and division cycles. Conversely, inhibition of endogenous protein expression can also reveal the effect(s) of the protein on cellular activities. However, it has been difficult to produce effective regulated expression for studies in vertebrate cells. Recently, a tetracycline-regulated expression system (9) has been successfully applied to mammalian cells for controlled expression of an ectopic gene product. This inducible system relies on a bacterial tetracycline repressor (TcR) that binds to DNA recogni-

Cloning and Expression Vectors for Gene Function Analysis
Edited by Quinn Lu and Michael P. Weiner

tion sites only in the absence of tetracycline. In the presence of the antibiotic, ligand-induced conformational change of the TcR DNA-binding domain prevents it from binding to the cognate DNA sequence. This "on" and "off" regulation offers a precise control for protein expression in mammalian cells because tetracycline does not change the overall metabolism in mammalian cells, as do other inducers such as heat shock, heavy metals and hormones (8). This approach has also been used to create transgenic mice for conditional expression of specific gene products (7). For this system to work in mammalian cells, the DNA-binding domain of TcR was linked to the acidic activation domain of VP16. This tetracycline-dependent transactivator (tTA), i.e., TcR-VP16 fusion protein, acts as a transcriptional activator when it binds to specific DNA sites located upstream of eukaryotic promoters used to drive the expression of inserted genes. A HeLa-derived cell line (HtTA-1) that constitutively expresses tTA was established for induced expression of foreign gene products (9).

Overexpression of a crucial transcription factor in cells may disturb cellular metabolism, resulting in cell death. This might be due to cytotoxicity exerted by unregulated expression of ectopic gene products. Since we could not establish stable cell lines expressing several epitope-tagged transcription factors using retrovirus-mediated gene transfer, we have resorted to the tetracycline-regulated expression system that can turn gene expression "on" or "off" depending on the availability of tetracycline (9). In this report, we describe an adaptation of the tetracycline-regulated expression system, in combination with FLAG® epitope-tagging (2), to establish stable cell lines that conditionally express potentially toxic proteins which can be easily purified for biochemical analysis.

MATERIALS AND METHODS

Construction of Tetracycline-Regulated Plasmids Expressing FLAG-Tagged Basal Transcription Factors

Two oligonucleotides, 5′-CATGGACTACAAAGACGATGACGATAAACA-3′ and 5′-TATGTTTATCGTCATCGTCTTTGTAGTC-3′, were individually synthesized and annealed to form a double-stranded oligonucleotide. This double-stranded oligonucleotide containing the FLAG epitope sequence with flanking *Nco*I and *Nde*I sites was used to replace the original FLAG sequence that also contains a kinase phosphorylation site in both pFLAG(S)-7 (2) and pFLAG(AS)-7 (1) to create, respectively, pFLAG°(S)-7 and pFLAG°(AS)-7 (Figure 1A). All these FLAG-containing cloning plasmids have a Kozak consensus sequence (10) inserted between the *Bgl*II and *Nco*I sites to facilitate expression in mammalian cells. The *Eco*RI-*Bam*HI fragment of pFLAG°(S)-7 was then cloned into pUHD10-3 (11) at the corresponding sites to generate pTetCMV-F°(S) (Figure 1B). Similarly, the *Cla*I-*Xba*I fragment of pFLAG°(AS)-7 was cloned into pUHD10-3 between the *Eco*RI and *Xba*I sites after Klenow filling-in at the *Cla*I and *Eco*RI sites (which regenerates the *Eco*RI site) to make pTetCMV-F°(AS) (Figure 1B). For conditional expression of FLAG-tagged human TBP, the *Nde*I-*Bam*HI fragment of TBP was transferred from pF:TBP(S)-7 (2) to pTetCMV-F°(S) to create pTetCMV-F°:hTBP. To make tetracycline-regulated human $TAF_{II}55$ expression plasmids, the *Nde*I-*Sac*I fragment of pF:55(ORF)-11d (3) and the *Sac*I-*Eco*RI fragment of 7-21 (3) were cloned into pFLAG°(AS)-7 between the *Nde*I and *Eco*RI sites to create pF°:55-7, which contains $TAF_{II}55$ cDNA from the initiation codon to the end of the 3′ untranslated region. The *Nco*I and partial *Xba*I digestion fragments of pF°:55-7 were then cloned into pTetCMV-F°(AS) between the *Nco*I and *Xba*I sites to generate pTetCMV-F°:h$TAF_{II}55$ and pTetCMV-F°:h$TAF_{II}55$/A^+. The former plasmid has a short 3′ untranslated region ending at the first *Xba*I site of $TAF_{II}55$, while the latter plasmid contains the entire 3′ untranslated region, including the poly(A) signal, of $TAF_{II}55$ and terminates at the *Xba*I site of the vector. The p62 cDNA of human TFIIH was isolated from pET11a/p62 (6) as N-terminal *Nde*I (introduced at the first initiation codon)-*Eco*RV and C-terminal *Eco*RV-*Bam*HI fragments, and cloned into pFLAG(S)-7 and pF:TBP-11d (2), after removing the TBP insert, between the *Nde*I and *Bam*HI sites to generate pF:62(H)-7 and pF:62(H)-11d, respectively. The entire FLAG-tagged p62 cDNA was then cleaved from pF: 62(H)-7 by *Bam*HI and partial *Eco*RI digestion, and cloned into pUHD10-3 between the *Eco*RI and *Bam*HI sites to create pTetCMV-F:62(H) for conditional expression of the p62 subunit of human TFIIH in mammalian cells.

Establishment of Tetracycline-Regulated Clonal Human Cell Lines

To establish stable cell lines that conditionally express epitope-tagged TBP, $TAF_{II}55$ and p62, we cotransfected 10 µg of each *Pvu*I-linearized pTetCMV-F°:hTBP, pTetCMV-F°:h$TAF_{II}55$, pTetCMV-F°:h$TAF_{II}55$/A^+ or pTetCMV-F:62(H) with 50 µg of sheared calf thymus DNA and with 0.5 µg of *Sac*I-linearized pREP4 (Invitrogen, San Diego, CA, USA) that contains the hygromycin-resistant gene into HtTA-1 (9), a HeLa-derived cell line that constitutively expresses a tetracycline-controlled *tet*R-VP16 transactivator (tTA). Transfection was performed by electroporation (Bio-Rad, Hercules, CA, USA) at 960 µF and 200 V as previously described (4). After electroporation, the cells were left at room temperature for 10 min and then transferred to a 15-mL centrifuge tube containing 10 mL growth medium (Dulbecco's modified Eagle medium [DMEM] plus 10% fetal bovine serum [FBS]). Following spinning at 1000 rpm using a Sorvall® RC-3B Plus centrifuge with an H6000A rotor (Du Pont, Wilmington, DE, USA) for 5 min to remove debris, the cells were plated onto five 100-mm plates and left in a 37°C, 5% CO_2 incubator for two days before initiating drug selection (400 µg/mL of G418, 200 µg/mL of hygromycin B and 2 µg/mL of tetracycline). G418 sulfate (Geneticin®; Life Technologies, Gaithersburg, MD, USA) and tetracycline (Sigma Chemical, St. Louis, MO, USA) were dissolved in water as 1 mg/mL stock solution, whereas hygromycin B (Boehringer Mannheim, Indianapolis, IN, USA) was purchased as 20 mg/mL solution. These antibiotics were added to the growth medium before use. The medium was changed every 3 or 4 days and selection was continued for approximately 3 weeks until drug-resistant colonies were clearly visible. Twelve colonies were then picked up and expanded into cell lines. To identify positive clones responding to tetracycline, cells were plated onto two 60-mm dishes in selection medium, with or without tetracycline, for 48 to 72 h. Subconfluent cells were

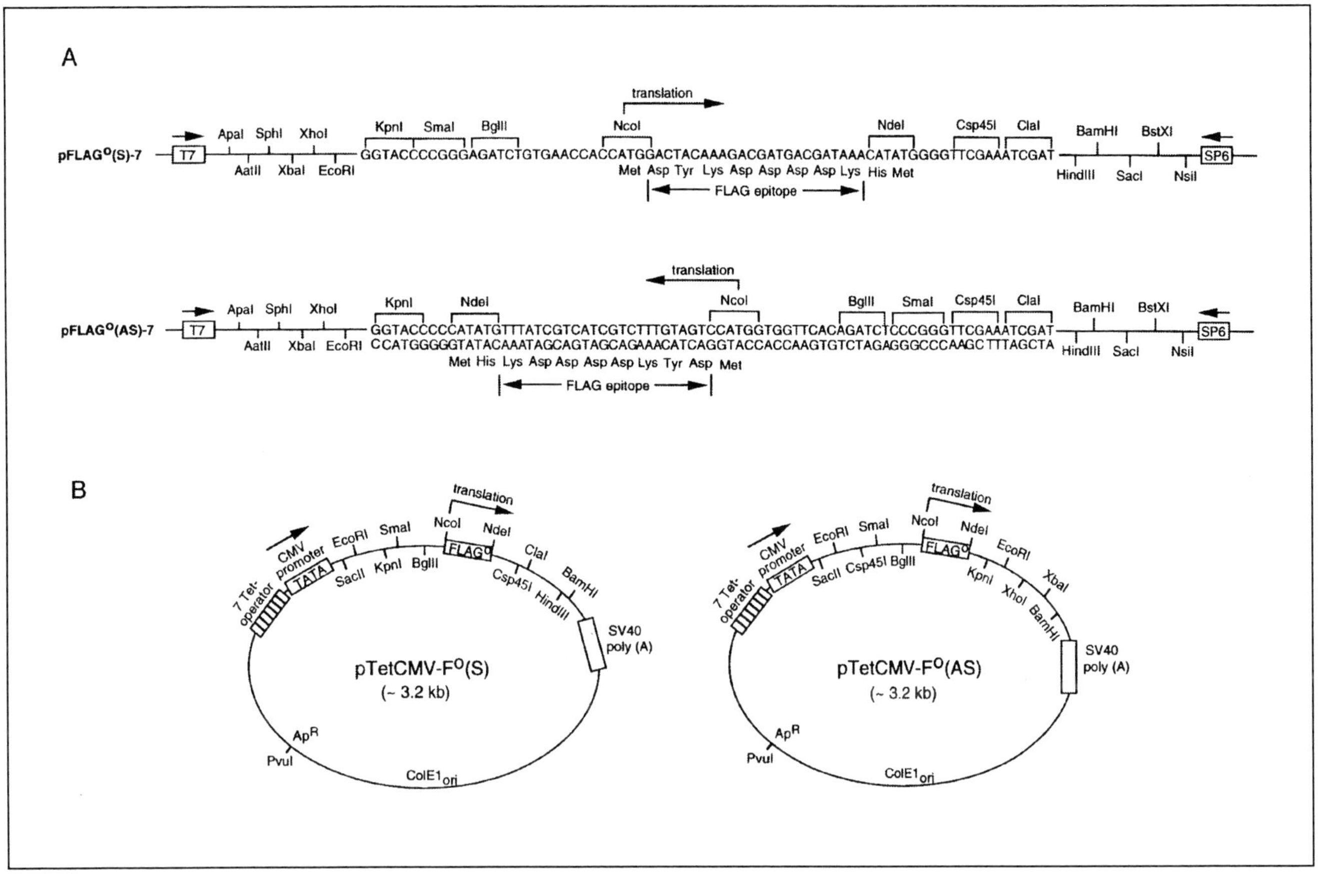

Figure 1. Construction of FLAG-tagged plasmids for inducible protein expression. (A) pFLAG°(S)-7 and pFLAG°(AS)-7. The backbones of both FLAG cloning plasmids were derived from pGEM®7-Zf(+) (Promega, Madison, WI, USA). Small arrows above the T7 and SP6 promoters indicate the direction of transcription, whereas the extended arrows over the *Nco*I sites specify the direction of translation from the FLAG sequence. (B) Inducible FLAG-tagged mammalian expression plasmids. The backbones of both pTetCMV-F°(S) and pTetCMV-F°(AS) were derived from pUHD10-3 (11). Unique enzyme cutting sites surrounding the FLAG sequence are *Sac*II, *Eco*RI, *Bgl*II, *Nco*I, *Nde*I, *Cla*I and *Bam*HI in pTetCMV-F°(S), and *Sac*II, *Bgl*II, *Nco*I, *Nde*I, *Xba*I and *Bam*HI in pTetCMV-F°(AS). Arrows above the CMV promoter and the FLAG sequence indicate the directions of transcription and translation, respectively.

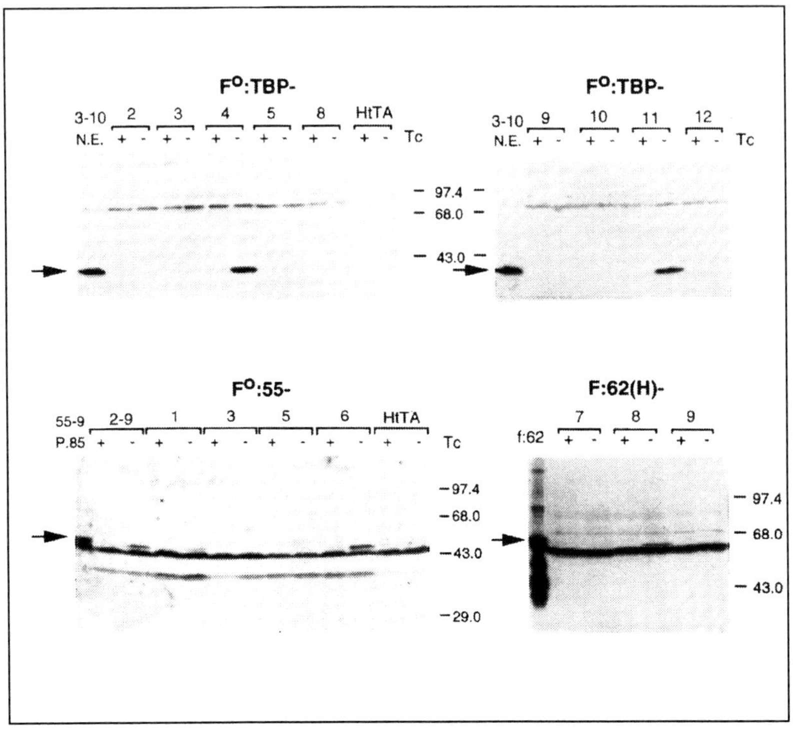

Figure 2. Establishment of stable cell lines expressing inducible FLAG-tagged human TBP, $TAF_{II}55$ and the p62 subunit of human TFIIH. Identification of tetracycline-regulated drug-resistant cell lines was performed by western blotting. Numbers above the lanes indicate individual cellular isolates. The 2-9 cell line contains the longer version of the $TAF_{II}55$ cDNA (pTetCMV-F°:h$TAF_{II}55/A^+$), whereas the rest of the $TAF_{II}55$ cell lines express the shorter version of the $TAF_{II}55$ cDNA (pTetCMV-F°:h$TAF_{II}55$). The presence (+) or absence (-) of tetracycline (Tc) for each isolate, as well as for parental cells (HtTA), is also indicated. The nuclear extract (N.E.) of 3-10 (1), the P11 0.85 M KCl fraction (P.85) of 55-9 (3) and FLAG-tagged p62 (f:62) purified from bacteria were used as positive controls for FLAG-tagged TBP, FLAG-tagged $TAF_{II}55$ and f:62 (denoted by arrows), respectively. For TBP cell lines, anti-FLAG M2 monoclonal antibodies were used at 1000-fold dilution. For $TAF_{II}55$ and p62 cell lines, anti-$TAF_{II}55$ (3) and anti-p62 polyclonal antibodies were applied at the same dilution. The protein bands directly underneath FLAG-tagged $TAF_{II}55$ and FLAG-tagged p62 are endogenous $TAF_{II}55$ and p62, respectively. The rest of the protein bands present in all the samples are nonspecific, cross-interacting species also detected by individual antibodies. Protein signals were detected by ECL (Amersham) according to the manufacturer's instruction. Molecular weights (in kDa) of protein size markers are indicated at the sides.

then lysed in 250 µL of sodium dodecyl sulfate-polyacrylamide gel electrophoresis (SDS-PAGE) sample buffer and used for western blotting.

Western Blotting

Proteins were transferred from either 10% (for TBP and $TAF_{II}55$) or 8% (for p62) SDS-PAGE gels to nitrocellulose filters at 0.8 mÅ/cm^2 for an hour by a NovaBlot™ semi-dry blotting apparatus (Pharmacia Biotech, Piscataway, NJ, USA). The anti-FLAG M2 monoclonal antibody (Scientific Imaging Systems [Eastman Kodak], New Haven, CT, USA), anti-$TAF_{II}55$ (3) and anti-p62 polyclonal antibodies (obtained from H. Xiao, Rockefeller University, New York) were diluted at 1:1000 fold in 1× TBST (10 mM Tris-HCl, pH 8.0, 150 mM NaCl, 0.05% Tween® 20) containing 5% nonfat milk. These primary antibodies were incubated, individually, with filters at room temperature for an hour. After brief rinsing with distilled water (two times) and extensive washing with 1× TBST (four times, 5 min each wash), the filters were incubated with 2000-fold dilutions of horseradish peroxidase-conjugated secondary antibodies (Amersham, Arlington Heights, IL, USA) in 1× TBST plus 5% non-fat milk at room temperature for 20 min. The filters were extensively washed again as described above. Signals were then detected by ECL™ (Amersham) according to the manufacturer's instructions.

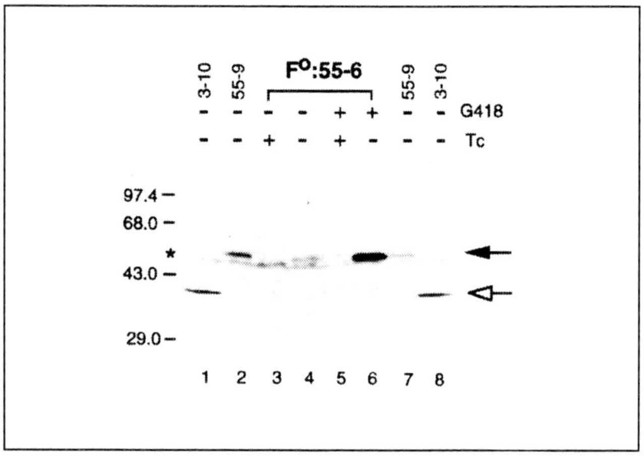

Figure 3. G418 enhances tetracycline-regulated gene expression. Western blotting was performed as described in Materials and Methods with the anti-$TAF_{II}55$ antiserum No. 3095 (3), which also contains antibodies against the FLAG-epitope tag. F°:55-6 cells in a 100-mm plate were pre-selected in G418 (400 µg/mL), in the presence of tetracycline (1 µg/mL), for two weeks before cells in a subconfluent (ca. 80%–90% confluent) 100-mm plate were split to twenty-four 60-mm plates for protein induction. At that time, G418 was no longer included in the medium. Cell lysates were collected after 72 h and used for western blotting. 3-10 (1) and 55-9 (3) cells, which were established by retrovirus-mediated gene transfer, express FLAG-tagged TBP (open arrow) and FLAG-tagged $TAF_{II}55$ (closed arrow), respectively. The asterisk (*) indicates the position of endogenous $TAF_{II}55$. Protein size markers (in kDa) are listed on the left. An UltroScan™ XL laser densitometer (Model 2222-020; Pharmacia Biotech) was used to quantitate the levels of the individual protein bands on the X-ray film.

Cell Maintenance

HtTA-1 cells were grown in DMEM supplemented with 10% FBS either with or without G418 (0.4 mg/mL of active components). All clonal human cell lines expressing FLAG-tagged proteins were usually maintained in DMEM plus 10% FBS with 1 µg/mL of tetracycline, unless otherwise specified in individual experiments.

Protein Purification

The p62 bacterial expression plasmid, pF:62(H)-11d, was introduced into a bacteria strain BL21(DE3)pLysS (12). Purification of bacterially expressed FLAG-tagged p62 (f:62) was conducted as previously described for FLAG-tagged TBP, without rifampicin (2). Nuclear extracts of the 3-10 human cell, a HeLa-derived cell line that constitutively expresses FLAG-tagged human TBP (1), were prepared according to the published procedure (5). The 0.85 M KCl fraction of the phosphocellulose P11 (Whatman, Hillsboro, OR, USA) ion-exchange column was isolated from nuclear extracts of 55-9, a HeLa-derived cell line that constitutively expresses human FLAG-tagged $TAF_{II}55$ (3).

RESULTS

Characterization of Tetracycline-Regulated Clonal Human Cell Lines

To establish clonal human cell lines that conditionally express FLAG-tagged basal transcription factors, including human TBP, the $TAF_{II}55$ subunit of human TFIID and the p62 subunit of human TFIIH, we have modified a pre-existing

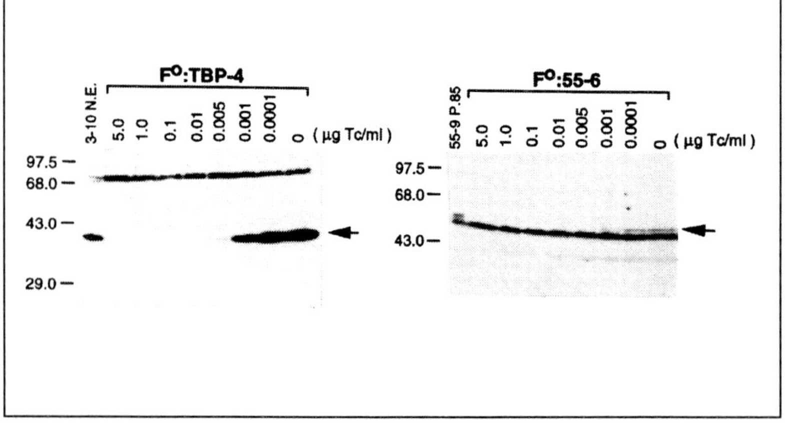

Figure 4. Induced protein expression is modulated by the amount of tetracycline in the growth medium. Western blotting was conducted with either anti-FLAG M2 monoclonal antibodies (for F°:TBP-4 cells) or anti-$TAF_{II}55$ polyclonal antibodies (for F°:55-6 cells). F°:TBP-4 and F°:55-6 cells were maintained in 100-mm plates in the presence of tetracycline (1 µg/mL). For the experiment, cells in a subconfluent (ca. 80%–90% confluent) 100-mm plate were split to twenty-four 60-mm plates with various concentrations of tetracycline in the growth medium. Cell lysates were then collected after 72 h. The nuclear extract (N.E.) of 3-10 (1) and the P11 0.85 M KCl fraction (P.85) of 55-9 (3) were used as positive controls for FLAG-tagged TBP and FLAG-tagged $TAF_{II}55$, respectively. Arrows point to the positions of FLAG-tagged proteins. The protein band detected below FLAG-tagged $TAF_{II}55$ is endogenous $TAF_{II}55$. Protein size markers (in kDa) are indicated on the left of each gel.

tetracycline expression plasmid to include a FLAG-epitope tag (Figure 1), thereby facilitating later purification of these proteins for biochemical analysis. This simpler version of the FLAG-epitope tag (F°) does not contain a heart muscle kinase phosphorylation site initially introduced in our FLAG expression vectors (1,2). To enhance the transfection efficiency, we employed electroporation to deliver cloned genes into HtTA-1 cells that express a tetracycline-regulated transactivator (9). From twelve drug-resistant colonies of each screening, we obtained two cell lines that expressed FLAG-tagged TBP (f°:TBP), three cell lines that expressed FLAG-tagged $TAF_{II}55$ (f°:$TAF_{II}55$) and only one cell line that expressed FLAG-tagged p62 (f:62) in the absence of tetracycline (Figure 2). The FLAG epitope tag (F°) at the N-termini of f°:TBP (molecular weight around 40 kDa) and f°:$TAF_{II}55$ does not include the HMK site as introduced in f:62. Amounts of f°:TBP and f°:$TAF_{II}55$ in these tetracycline-regulated cell lysates are comparable to the levels of FLAG-tagged TBP (f:TBP) detected in the nuclear extract of 3-10 (1) and FLAG-tagged human $TAF_{II}55$ (f:$TAF_{II}55$) present in the P11 0.85 M KCl fraction of 55-9 (3). The f:TBP and f:$TAF_{II}55$ proteins constitutively produced in 3-10 and 55-9 cells, which were both established by retrovirus-mediated gene transfer, contain an HMK site linked to the FLAG epitope tag.

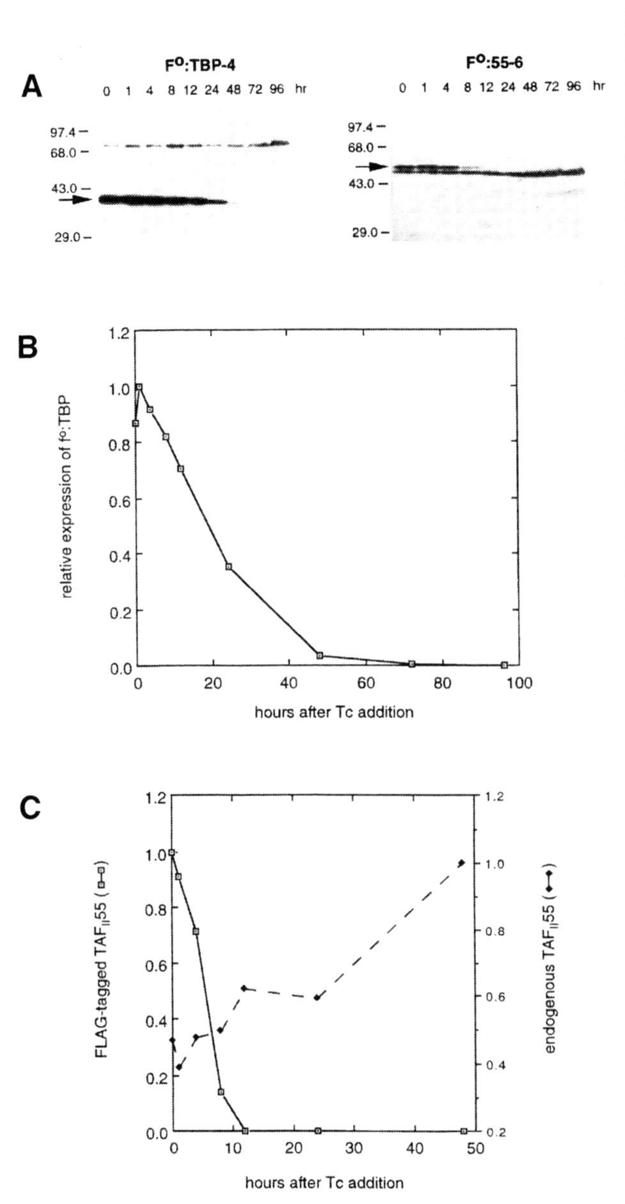

Figure 5. Protein decay following addition of tetracycline to growth medium. (A) Western blotting. F°:TBP-4 and F°:55-6 cells were maintained in 100-mm plates without tetracycline for 3 days. Subconfluent cells were then split from a 100-mm plate to twenty-four 60-mm plates in the presence of tetracycline (1 μg/mL). Cell lysates were collected at different time points as indicated and used for western blotting with either anti-FLAG M2 monoclonal antibodies (for F°:TBP-4) or anti-$TAF_{II}55$ (for F°:55-6) polyclonal antibodies as primary antibodies. Arrows indicate the positions of FLAG-tagged proteins. Protein size markers are listed on the left of the individual gels. Quantitation of FLAG-tagged TBP (B) and $TAF_{II}55$ (C) signals on X-ray film were performed using a Model 2222-020 UltraScan XL laser densitometer. Numbers on the sides of the graphs (B and C) correspond to the relative ratios of protein expression as scanned by the densitometer. The dashed line (C) indicates the level of endogenous $TAF_{II}55$.

Preselection of Clonal Cell Lines with G418 Enhances Tetracycline-Regulated Gene Expression

To compare the levels of protein expression achieved between the tetracycline-regulated expression system and the retroviral system, we conducted western analysis using cell lysates harvested from 3-10, 55-9 and F°:55-6 cells. Since the amount of the tetracycline-regulated transactivator (tTA) in F°:55-6 cells could affect the level of tagged protein expression and a G418 selection plasmid was used initially for cotransfection with a tTA expression plasmid to establish HtTA-1 cells (9), we assumed that inclusion of G418 in the growth medium would probably enhance the expression of tetracycline-regulated plasmids. Interestingly, the level of f°:$TAF_{II}55$ expression in tetracycline-regulated F°:55-6 cells was indeed augmented approximately 10-fold after preselection of F°:55-6 cells in the presence of G418 for 2 weeks (Figure 3, compare lane 4 with lane 6). Moreover, the relative ratio between f°:$TAF_{II}55$ and endogenous $TAF_{II}55$ was dramatically increased from 1 (lane 4), without preselection with G418, to more than 100-fold (lane 6) after G418 selection. Induction of f°:$TAF_{II}55$ expression was specific, as no tagged proteins were detected in the presence of tetracycline (lanes 3–6), which prevents tTA from binding to tetracycline-regulated plasmids. The levels of tagged $TAF_{II}55$ expression were comparable between the tetracycline-regulated system (without G418) and the retroviral system (lanes 2, 4 and 7). The anti-$TAF_{II}55$ antisera (3), raised in rabbits against FLAG-tagged $TAF_{II}55$, also contained antibodies that recognized the FLAG epitope tag. The f:TBP signals detected in 3-10 cells (lanes 1 and 8), therefore, served as controls for protein transfer on either side of the gel.

Modulation of the Protein Expression by the Amount of Tetracycline

The tetracycline-regulated expression system also offers a unique method to modulate the level of protein expression (9). Depending on the amount of tetracycline included in the growth medium, the levels of f°:TBP and f°:$TAF_{II}55$ could be varied over three orders of magnitude (Figure 4). Induced expression of FLAG-tagged proteins was clearly detected by western blotting when the concentrations of tetracycline dropped below 1 ng/mL. The tagged proteins, if present, were barely detected when tetracycline was higher than 10 ng/mL. The dosage-responsive effects were very similar among clonal cell lines that expressed different proteins (Figure 4), suggesting that the induced protein expression is mainly determined by the DNA-binding property of tTA, regardless of the nature of the expressed protein.

Measurement of the In Vivo Protein Half-Life

By controlling the availability of tetracycline, we can turn "on" and "off" a particular gene expression at a specific time and location. To prevent a potential cytotoxic effect resulting from the overexpression of tTA and tetracycline-regulated gene products, we usually maintain clonal cells in the presence of only 1 μg/mL tetracycline. Once tetracycline is removed from the growth medium, expression of the tetracycline-regulated proteins is induced and reaches the maximum level in approximately 24 h for $TAF_{II}55$ and 48 h for TBP, and remains constant for at least 5 days (data not shown). Interestingly, when tetracycline was added back to the cells, the induced proteins started to decay (Figure 5A). The levels were reduced by approximately one-half within 20 h for TBP (Figure 5B) and within 6 h for $TAF_{II}55$ (Figure 5C). In the case of $TAF_{II}55$, the reduction of exogenous FLAG-tagged $TAF_{II}55$ was clearly compensated for by the increase in endogenous $TAF_{II}55$ (Figure 5C), indicating a tightly regulated control of the overall protein level on this critical transcription factor that mediates activator response. Tetracycline-regulated expression of FLAG-tagged TBP and FLAG-tagged $TAF_{II}55$ was specific, as no change in the protein levels of other TFIID subunits was observed in the presence or absence of tetracycline (data not shown). This provided us with an estimation of the in vivo protein half-lives inferred directly from the protein level, not relying on the amplified signals through enzymatic activities when LacZ and luciferase reporter genes were used.

DISCUSSION

The tetracycline-regulated expression system is an effective approach for conditional expression of foreign gene products in mammalian cells. This system allows the expression of potentially toxic proteins in a controlled manner, depending on the amount of tetracycline in the growth medium. Introduction of the FLAG-epitope tag into the tetracycline-regulated plasmids (Figure 1) further extends the potential of the system for biochemical analysis. The introduced FLAG tag offers several advantages. *(i)* It serves as a marker for tracking proteins before anti-protein antibodies become available. *(ii)* It is synthetic. Very few, if any, cross-reacting species are detected in bacterial, yeast, insect and mammalian cells. *(iii)* It is very hydrophilic and tends to help proteins solubilize in aqueous solution. *(iv)* Purification of FLAG-tagged proteins can be easily achieved by affinity purification and peptide elution methods (2). Proteins with the FLAG-epitope tag will be retained on the anti-FLAG M2 monoclonal antibody-conjugated agarose beads. After extensive washing off the unbound proteins, the FLAG-tagged proteins can be easily isolated by incubation with the synthetic FLAG peptide, which competes with the antibody-binding sites on the M2-agarose beads, thereby allowing recovery of FLAG-tagged proteins (and associated polypeptides) under mild conditions in a form ready for functional assays. *(v)* A heart muscle kinase (HMK) phosphorylation site linked to the FLAG sequence in our original FLAG expression vectors (2) offers a unique way to label proteins with radioisotope ^{32}P after purification. The labeled proteins can be used for interaction studies, expression library screening, protein tracking and so forth. However, the HMK site may also complicate the issues for cell cycle studies, such as monitoring protein phosphorylation changes through cell cycle progression. Therefore, a simpler version of the FLAG tag sequence was used in our current studies for the inducible expression of TBP and $TAF_{II}55$.

Our previous attempts to establish stable cell lines that constitutively express some critical transcription factors, such as the p62 subunit of human TFIIH, were not successful. This is presumably due to the cytotoxicity of these transcription factors on cellular metabolism when overexpressed. Using the combined tetracycline-regulated and FLAG-epitope tagging methods, we are able to conditionally express the FLAG-tagged p62 (f:62) subunit of human TFIIH in HeLa cells (Figure 2). The establishment of the f:62-expressing cell line F:62(H)-8 (Figure 2) by the tetracycline-regulated expression system is a major breakthrough in studies of transcriptional regulation, as it is now possible to purify sufficient amounts of homogeneous human TFIIH for biochemical analysis.

Expression of tetracycline-regulated proteins was greatly enhanced by preselection of clonal cells with G418 (Figure 3). Since high levels of tTA expression may be toxic to cells, the amount of transactivators present in HtTA-1 and its derived cell lines may become limiting through multiple passages. This reduction presumably protects cells from cytotoxicity exerted by overexpression of tTA. Interestingly, we were able to further enhance the expression of the tagged proteins at least another tenfold by preselection with G418, which was employed as a selection marker initially to cotransfect with a tTA expression plasmid for the establishment of HtTA-1 cells. Since very few, if any, cells died during the 2-week G418 selection, we tend to believe that G418 actually increases the level of tTA expression in individual cells. The possibility that cells lose the tTA gene is less likely.

By modulating the amount of tetracycline in clonal human cell lines, we are able to quantitate the decline in levels of human TBP and $TAF_{II}55$ (Figure 5). Although this method includes a summation of the decay rates of protein and mRNA, it provides a measure of the rate at which the levels of these transcription factors can be regulated. To our knowledge, this is the first description of the in vivo half-lives of basal transcription factors that are required for transcription of most eukaryotic genes by RNA polymerase II.

ACKNOWLEDGMENTS

We are grateful to H. Bujard for providing pUHD10-3 and for permission to obtain HtTA-1 cells from M. Henry and F. Solomon, whose kindness in circulating the cell line is appreciated. We also thank H. Xiao for furnishing pET11a/p62 and anti-p62 antisera, and D. Shapiro and members of the Chiang laboratory, particularly Samuel Hou, Eric Kershnar and Joseph Lasky, for comments on the manuscript. C.-M.C. is a

Pew Scholar in the Biomedical Sciences. This research is in part aided by Grant No. 96-41 from the American Cancer Society, Illinois Division, Inc.

REFERENCES

1. **Chiang, C.-M., H. Ge, Z. Wang, A. Hoffmann and R.G. Roeder.** 1993. Unique TATA-binding protein-containing complexes and cofactors involved in transcription by RNA polymerases II and III. EMBO J. *12*:2749-2762.
2. **Chiang, C.-M. and R.G. Roeder.** 1993. Expression and purification of general transcription factors by FLAG epitope-tagging and peptide elution. Pept. Res. *6*:62-64.
3. **Chiang, C.-M. and R.G. Roeder.** 1995. Cloning of an intrinsic human TFIID subunit that interacts with multiple transcriptional activators. Science *267*:531-536.
4. **Chiang, C.-M., M. Ustav, A. Stenlund, T.F. Ho, T.R. Broker and L.T. Chow.** 1992. Viral E1 and E2 proteins support replication of homologous and heterologous papillomaviral origins. Proc. Natl. Acad. Sci. USA *89*:5799-5803.
5. **Dignam, J.D., R.M. Lebovitz and R.G. Roeder.** 1983. Accurate transcription initiation by RNA polymerase II in a soluble extract from isolated mammalian nuclei. Nucleic Acids Res. *11*:1475-1489.
6. **Fischer, L., M. Gerard, C. Chalut, Y. Lutz, S. Humbert, M. Kanno, P. Chambon and J.-M. Egly.** 1992. Cloning of the 62-kilodalton component of basic transcription factor BTF2. Science *257*:1392-1395.
7. **Furth, P.A., L.S. Onge, H. Böger, P. Gruss, M. Gossen, A. Kistner, H. Bujard and L. Hennighausen.** 1994. Temporal control of gene expression in transgenic mice by a tetracycline-responsive promoter. Proc. Natl. Acad. Sci. USA *91*:9302-9306.
8. **Gossen, M., A.L. Bonin and H. Bujard.** 1993. Control of gene activity in higher eukaryotic cells by prokaryotic regulatory elements. Trends Biochem. Sci. *18*:471-475.
9. **Gossen, M. and H. Bujard.** 1992. Tight control of gene expression in mammalian cells by tetracycline-responsive promoters. Proc. Natl. Acad. Sci. USA *89*:5547-5551.
10. **Kozak, M.** 1986. Point mutations define a sequence flanking the AUG initiator codon that modulates translation by eukaryotic ribosomes. Cell *44*:283-292.
11. **Resnitzky, D., M. Gossen, H. Bujard and S.I. Reed.** 1994. Acceleration of the G1/S phase transition by expression of cyclins D1 and E with an inducible system. Mol. Cell. Biol. *14*:1669-1679.
12. **Studier, F.W., A.H. Rosenberg, J.J. Dunn and J.W. Dubendorff.** 1990. Use of T7 RNA polymerase to direct expression of cloned genes. Methods Enzymol. *185*:60-89.
13. **Zhou, Q., P.M. Lieberman, T.G. Boyer and A.J. Berk.** 1992. Holo-TFIID supports transcriptional stimulation by diverse activators and from a TATA-less promoter. Genes Dev. *6*:1964-1974.

Update to: Establishment of Stable Cell Lines Expressing Potentially Toxic Proteins by Tetracycline-Regulated and Epitope-Tagging Methods

Samuel Y. Hou, Shwu-Yuan Wu and Cheng-Ming Chiang

Case Western Reserve University, Cleveland, OH, USA

The establishment of stable cell lines under the control of the tetracycline-regulated expression system with FLAG®-epitope (Scientific Imaging [Eastman Kodak], New Haven, CT, USA) tagging has provided us with a unique opportunity for biochemical purification of potentially toxic protein complexes from mammalian cells. Purification of multiprotein complexes, such as human TFIIH and RNA polymerase II, by conventional chromatography has been a tedious task, often requiring many chromatography steps, and the overall yield after purification is generally low. With the establishment of stable cell lines that conditionally express a FLAG-tagged subunit of a multiprotein complex, we have purified several multisubunit protein complexes, including human TFIIH and RNA polymerase II, via one-step immunoaffinity purification (4). The introduction of the FLAG-epitope tag not only simplifies the purification scheme but also enhances the overall yield. This experimentation is the first documentation for purification of multiprotein complexes from an inducible expression system in mammalian cells.

Although FLAG-epitope tagging in a conditional expression system provides a unique method for tracing and purification of multisubunit protein complexes, a high level of protein induction in mammalian suspension culture may sometimes be difficult to achieve, likely due to low expression of the tetracycline-regulated transactivator (tTA) within the cell or due to residual amounts of tetracycline in the induction medium. To ensure high protein induction, we follow four general guidelines. First, preselect cells with antibiotic (G418) at a small culture (approximately 100 mL) in the presence of tetracycline for at least 3 days before further expansion in the absence of antibiotic (G418). Presumably, this will boost the expression of tTA in the cell, thereby enhancing the level of transactivation from the tetracycline-regulated promoter. Second, wash cells at least 3 times with 1× phosphate-buffered saline (PBS) to remove tetracycline prior to induction. Third, use 250-mL conical centrifuge tubes instead of 1-L flat bottom bottles to spin down cells when washing with 1× PBS. We have found that conical centrifuge tubes allow for better cell adhesion, thus medium and PBS washes can be completely removed after each successive centrifugation without losing cell volume. The 1-L flat bottom centrifuge bottles, although convenient to use, cannot completely remove original medium; the minor amount of residual tetracycline will prohibit the induction. Fourth, use tetracycline-free serum as growth supplement when protein induction is desired. We have discovered that some batches of calf serum contain endogenous tetracycline, likely resulting from the use of tetracycline in cattle feed to protect animals from bacterial infection.

Once a high level of protein induction has been achieved, the cells are then separated into three fractions: nuclear extract, cytoplasmic S100 and nuclear pellet (2). To successfully isolate multisubunit complexes, we first perform Western blotting to determine which cellular fraction our protein resides in. Once determined, careful consideration is then given to the conditions for immunoaffinity purification of epitope-tagged protein complexes from the particular cellular fraction. We not only consider wash and elution variables but also the source of material to be purified (i.e., nuclear extract, S100 or nuclear pellet). The three main components of the wash steps in our immunoaffinity purification buffers are salt (KCl), detergent (Nonidet® P-40 [NP40]) and chaotropic agent (urea). An example of varying these conditions, as shown in Figure 1, is given in the purification of different multisubunit complexes from hRPB9-3, a HeLa-derived cell line that conditionally expresses the FLAG-tagged RPB9 subunit of human RNA polymerase II (7). Depending on the washing conditions, either core RNA polymerase II or an RNA polymerase II holoenzyme complex, which contains core RNA polymerase II and other associated polypeptides, including a subset of general transcription factors, chromatin remodeling factors and other cellular proteins, can be purified from the cytoplasmic S100 fraction derived from the hRPB9-3 cell line (4,7-9). When a low salt (i.e., 100 mM KCl) buffer is used for washing during immunoaffinity purification, a transcriptionally active RNA polymerase II holoenzyme complex is purified (7,9). If we increase the salt (850 mM KCl) and also include 1.0 M urea in the wash, only core RNA polymerase II is purified (4,8).

The elution conditions also play an important role in immunoaffinity purification. Ideally, we like to store proteins in low salt-containing buffers which allow us to better manipulate the salt conditions for later functional assays. But when comparing the elution conditions, we find that some proteins, such as the FLAG-tagged human papillomavirus E2 protein, elute better in 300 mM KCl-containing buffer (3), while the majority of other proteins elute well in 100 mM KCl-containing buffer. Inclusion of a minor amount (0.1%–0.01%) of detergents, such as NP40 or Triton® X-100, in the elution buffer prevents protein/bead aggregation, allowing the immunoaffinity beads to rotate better during the elution step. The caveat in using detergent, however, is that the activity of some proteins (or complexes) may be sensitive to high concentrations of detergent. Luckily, our purified protein complexes, including human TFIIH, RNA polymerase II, and RNA polymerase II holoenzyme, are fully functional in both basal and activated transcription (4,7-9). Finally, we must consider the source of material to be purified (i.e., nuclear extract, cytoplasmic S100, or chromosomal nuclear pellet). Purification of RNA polymerase II can be achieved from all three fractions although slightly different results are obtained from each fraction. For example, both core RNA polymerase II and RNA polymerase II holoenzyme can be purified from either nuclear extract or cytoplasmic S100 fractions, depending on the washing conditions (see Figure 1). These purified protein complexes contain only the IIA form, which comprises only the nonphosphorylated carboxy-terminal domain (CTD) in the largest subunit (RPB1), of RNA polymerase II.

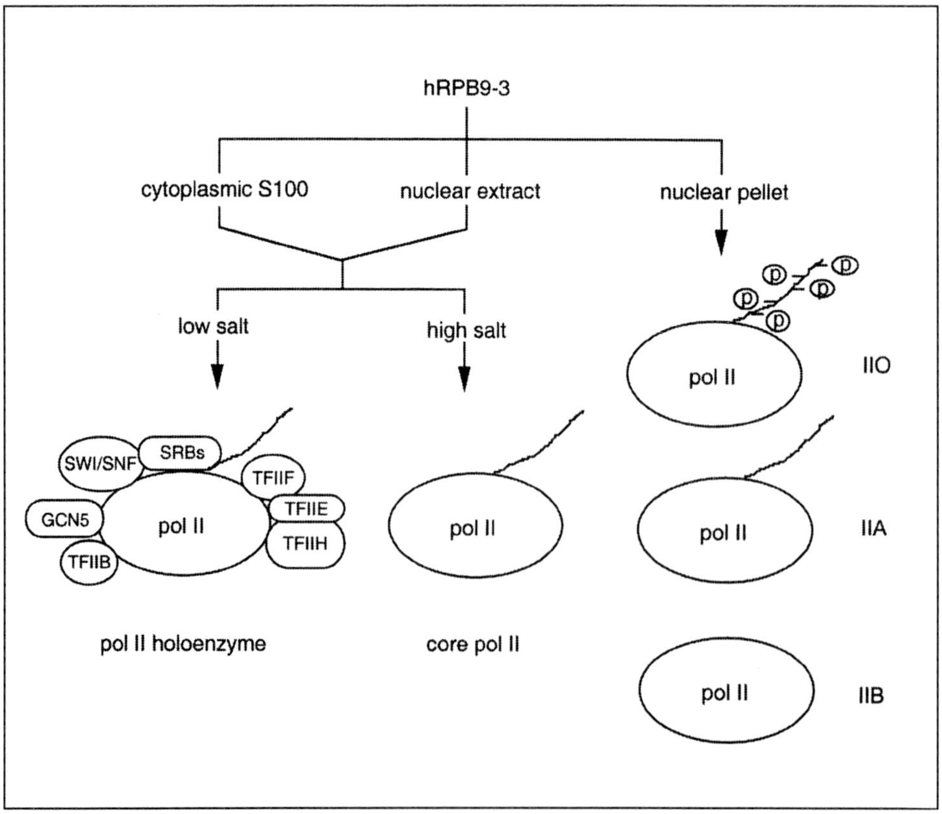

Figure 1. Purification of human RNA polymerase II complexes. The hRPB9-3 cells (7), derived from human HeLa cells that conditionally express FLAG-tagged RPB9 (a subunit of human RNA polymerase II), are first fractionated into cytoplasmic S100, nuclear extract and nuclear pellet. Immunoaffinity purification was then performed with S100 or nuclear extract under either low salt (100 mM KCl-containing buffer) or high salt (850 mM KCl- and 1.0 M urea-containing buffer) wash conditions, which results in the purification of human RNA polymerase II (pol II) holoenzyme and core pol II, respectively. The pol II holoenzyme complex purified from S100 or nuclear extract contains core pol II, a subset of general transcription factors (TFIIB, TFIIE, TFIIF and TFIIH), SRBs, histone acetyltransferase GCN5 and chromatin remodeling factor SWI/SNF (7,9). Additional forms of pol II complexes with different phosphorylation status on the largest subunit (RPB1) of pol II can also be purified from the nuclear pellet under the high salt wash condition. The IIO form of pol II contains the highly phosphorylated carboxy-terminal domain (CTD), indicated by a tail, in RPB1, whereas the IIA and IIB forms of pol II have nonphosphorylated or truncated CTD in RPB1, respectively.

Interestingly, while RNA polymerase II could be efficiently purified from the S100 and nuclear extracts of the hRPB9–3 cells, the majority of RNA polymerase II complexes still remain in the chromosomal fraction (i.e., nuclear pellets). Purification from this fraction yielded multiple forms of RNA polymerase II, including not only the IIA form, but also the IIO form that contains the highly phosphorylated CTD and the IIB form which has the CTD deleted in RPB1. Thus different forms of multisubunit protein complexes can be selectively purified from the same cell, depending on the conditions and the material used for protein purification.

Although isolation of complexes may be achieved by one-step immunoaffinity purification, additional chromatography steps may be included to further enhance protein purification or to separate different forms of tagged proteins into various fractions. For example, fractionation of the hRBP9-3 S100 over a P11 phosphocellulose ion-exchange column prior to the immunoaffinity purification step not only removes several contaminant proteins, but also eliminates the free form of FLAG-tagged RPB9 protein copurified with the RNA polymerase II complexes (4).

Albeit conditional expression of a FLAG-tagged subunit of a multiprotein complex in an established cell line offers a unique way to trace and purify the protein complexes, screening of the clonal cell lines that conditionally express the tagged protein has been a tedious process. Usually only 5%–10% of cell clones are positive in expressing the tagged protein. One way to overcome this problem is to combine the tetracycline-regulated expression system with retrovirus-mediated gene transfer (5), as the retroviral vector provides site-specific integration between two viral long-terminal repeats (LTRs) and thereby will not disrupt the expression cassette located between the two LTRs. In fact, essentially all cell clones turn out to be positive in expressing the tagged protein, although the expression level varies among individual isolates (1). The pBPSTR1 plasmid, created by Reeves and colleagues, is a self-contained tetracycline-regulated retroviral expression vector containing both the tTA gene and also the tetracycline-regulated response elements that control the expression of exogenous gene products located between the 5′ and 3′ LTR within the vector (5). This vector has been used to carry the FLAG-tag sequence to facilitate the tracing and purification of multiprotein complexes as described above (our unpublished data). Another advantage is that this pBPSTR1-derived FLAG-containing vector can be introduced into an amphotropic packaging cell line. The viruses harvested are then used to express the tagged protein in any mammalian cell lines.

Another problem we have noticed in the tetracycline-regulated expression system is that without constant drug selection, conditional expression of the tagged protein may decrease over prolonged passages. This phenomenon may reflect transcriptional silencing due to packaging of the expression cassette into heterochromatin, when an ectopic gene is integrated into chromosome in the stable cell lines. To overcome the effect of heterochromatin gene silencing, we will introduce chicken ß-globin insulator (or boundary) elements, which have been shown to protect the expression cassette from surrounding sequence effects (6), into the FLAG-tagged tetracycline-regulated expression vector. We believe that, once a stable cell line is established, the insulator elements will prolong the expression of the tagged gene and, therefore, prevent silencing that is often seen when the cultured cells are not maintained under constant drug selection.

So far, we have effectively combined the tetracycline-regulated expression system with FLAG-epitope tagging to establish stable cell lines which conditionally express potentially cytotoxic proteins. From these cell lines, we are able to easily purify multisubunit protein complexes such as human TFIIH and RNA polymerase II. By expanding our current approach to include retrovirus-mediated gene transfer and boundary elements, we hope to establish stable cell lines that conditionally express ectopic genes of interest at high levels of induction in a wide variety of host cells. The development of this methodology will greatly enhance our biochemical characterization of protein complexes isolated from a variety of mammalian cell types.

ACKNOWLEDGMENTS

C.-M.C. is a Pew Scholar in the Biomedical Sciences and a Mt. Sinai Health Care Foundation Scholar. Our research is supported by the American Cancer Society Research Project Grant No. RPG-97-135-01-MBC and by National Institutes of Health Grant Nos. CA81017 and GM59643.

REFERENCES

1. **Chiang, C.-M., H. Ge, Z. Wang, A. Hoffmann and R.G. Roeder.** 1993. Unique TATA-binding protein-containing complexes and cofactors involved in transcription by RNA polymerases II and III. EMBO J. *12*:2749-2762.

2. **Dignam, J.D., R.M. Lebovitz and R.G. Roeder.** 1983. Accurate transcription initiation by RNA polymerase II in a soluble extract from isolated mammalian nuclei. Nucleic Acids Res. *11*:1475-1489.

3. **Hou, S.Y., S.-Y. Wu, T. Zhou, M.C. Thomas and C.-M. Chiang.** 2000. Alleviation of human papillomavirus E2-mediated transcriptional repression via formation of a TATA binding protein (or TFIID)-TFIIB-RNA polymerase II-TFIIF preinitiation complex. Mol. Cell. Biol. *20*:113-125.

4. **Kershnar, E., S.-Y. Wu and C.-M. Chiang.** 1998. Immunoaffinity purification and functional characterization of human transcription factor IIH and RNA polymerase II from clonal cell lines that conditionally express epitope-tagged subunits of the multiprotein complexes. J. Biol. Chem. *273*:34444-34453.

5. **Paulus, W., I. Baur, F.M. Boyce, X.O. Breakefield and S.A. Reeves.** 1996. Self-contained, tetracycline-regulated retroviral vector system for gene delivery to mammalian cells. J. Virol. *70*:62-67.

6. **Pikaart, M.J., F. Recillas-Targa and G. Felsenfeld.** 1998. Loss of transcriptional activity of a transgene is accompanied by DNA methylation and histone deacetylation and is prevented by insulators. Genes Dev. *12*:2852-2862.

7. **Wu, S.-Y. and C.-M. Chiang.** 1998. Properties of PC4 and an RNA polymerase II complex in directing activated and basal transcription in vitro. J. Biol. Chem. *273*:12492-12498.

8. **Wu, S.-Y., E. Kershnar and C.-M. Chiang.** 1998. TAF_{II}-independent activation mediated by human TBP in the presence of the positive cofactor PC4. EMBO J. *17*:4478-4490.

9. **Wu, S.-Y., M.C. Thomas, S.Y. Hou, V. Likhite and C.-M. Chiang.** 1999. Isolation of mouse TFIID and functional characterization of TBP and TFIID in mediating estrogen receptor and chromatin transcription. J. Biol. Chem. *274*:23480-23490.

CHAPTER 36

Highly Efficient Eukaryotic Gene Expression Vectors for Peptide Secretion

Te-Hua Tearina Chu, Idali Martinez, Paul Olson and Ralph Dornburg

University of Medicine and Dentistry of New Jersey, Piscataway, NJ, USA

(Reprinted from BioTechniques 18:890-896, 1995)

ABSTRACT

Recently, we constructed a series of highly efficient universal eukaryotic gene expression vectors (Sheay et al., BioTechniques 15:856-862, 1993). Such vectors contain a viral promoter and enhancer followed by the adenovirus tripartite leader sequence, a multiple cloning site for the insertion of the gene of interest and a polyadenylation sequence. To enable expression of peptides to be secreted into the tissue culture medium or to be incorporated into the cell membrane, several modifications have been introduced into such vectors: stop codons in all three reading frames were inserted at the end of the multiple cloning site and a DNA sequence coding for a signal peptide for transport through the endoplasmatic reticulum (ER) was introduced downstream of the adenovirus tripartite leader sequence followed by two unique restriction enzyme recognition sites. A protocol is described that allows fast and easy cloning of peptide-coding regions, i.e., PCR products, for expression and secretion. The transport of a genetically engineered chimeric transmembrane protein connected to this ER leader sequence was as efficient as that of the original protein from which the ER sequence has been derived. These universal vectors can also be used for the easy construction of any chimeric transmembrane or secretion proteins.

INTRODUCTION

Expression of eukaryotic genes in bacteria has become a powerful method to produce proteins in large quantities for many different purposes. However, proteins secreted from eukaryotic cells are usually glycosylated and/or contain other posttranslational modifications essential for function, solubility and/or stability. Bacteria are unable to introduce such modifications. For this reason, it is difficult or impossible to manufacture such proteins in bacteria. Many such proteins can only be produced in tissue culture cells (9,15,20).

Eukaryotic proteins which are secreted from the cell or which are inserted into the cell membrane contain a hydrophobic stretch of amino acids at the amino terminus. This hydrophobic leader sequence interacts with a protein-ribonucleic acid complex in the endoplasmatic reticulum (ER) to initiate transport into the lumen of the ER (7,14). Usually, during this transport, the hydrophobic leader is removed from the peptide by proteolytic cleavage. Next, the protein is usually glycosylated and/or subjected to further modifications. If the protein contains one or more stretches of hydrophobic transmembrane domains, it is usually retained in the membrane. If no such transmembrane domains are present, the peptide is secreted from the cell (7,14).

In modern protein engineering, truncated peptides, instead of their full-length counterpart, are often manufactured, because they are sometimes more stable and/or active and/or less immunogenic. The engineering of genes coding for amino-terminal-truncated peptides for secretion requires that a hydrophobic leader sequence is fused in-frame to the actual peptide-coding region.

We constructed universal gene expression vectors that contain a DNA region coding for a hydrophobic leader sequence, followed by two unique restriction enzyme sites and stop codons in all three reading frames. Such vectors allow direct cloning of polymerase chain reaction (PCR) products of any peptide-coding region without further manipulations. These new vectors have been derived from strong gene expression vectors that we recently constructed (17). They contain a strong viral promoter, followed by the adenovirus tripartite leader sequence that further enhances gene expression. The vectors described in this report can be used in various applications requiring secretion of proteins from eukaryotic cells in large quantities or their incorporation into the cell membrane.

Cloning and Expression Vectors for Gene Function Analysis
Edited by Quinn Lu and Michael P. Weiner

MATERIALS AND METHODS

Plasmid Constructions

Plasmids were constructed using standard procedures (13). DNAs were digested with restriction enzymes under conditions recommended by the supplier. DNA-modifying enzymes were mainly purchased from New England Biolabs (Beverly, MA, USA), Boehringer Mannheim (Indianapolis, IN, USA), Promega (Madison, WI, USA) and Stratagene (La Jolla, CA, USA). Small DNA fragments and plasmids were purified by HPLC (using a GenPak™ FAX column purchased from Waters Corporation, Milford, MA, USA [R. Dornburg, unpublished]). Plasmids were transformed into *Escherichia coli* strain DH5.

Plasmids pRD131, pRD132 and pRD133 (Figure 1) were made in two cloning steps: first, plasmid pRD113 (17) was digested with *Xho*I plus *Kpn*I, followed by the insertion of a synthetic DNA fragment containing three stop codons in the three reading frames. The synthetic DNA fragment was made as follows: two oligonucleotides (purchased from DNA International, Lake Oswego, OR, USA) were dissolved in TNE buffer (10 mM Tris-HCl, pH 8.0, 100 mM NaCl, 1 mM EDTA) at a concentration of 1 μg/mL, mixed in equimolar concentration, heated to 75°C in a heating block for 2 min, followed by slow cooling to room temperature. The resulting double-stranded oligonucleotide had sticky ends at both sides enabling cloning into the *Xho*I-, *Kpn*I-digested plasmid DNA. After cloning, the plasmid was termed pRD118 and sequenced to verify the presence of the stop codons. pRD131, pRD132 and pRD133 were made by inserting the *Sac*II/*Kpn*I fragment, isolated from pRD118 by HPLC fractionation, into *Sac*II-, *Kpn*I-digested plasmids pRD114, pTC1 and pWS4 (17), respectively.

pTC13, pTC15 and pTC16 were made in several cloning steps: A plasmid containing the envelope gene of spleen necrosis virus (SNV) was subjected to site-directed mutagenesis to introduce an *Hpa*I site downstream of the sequence coding for the hydrophobic leader region. The resulting plasmid (termed pSNV-env-mA) was digested with *Hpa*I. A synthetic linker containing *Nru*I, *Stu*I and *Afl*II sites (made as described above) was inserted. The resulting plasmid was termed pRD130 and was sequenced to confirm the correct linker insertion. pRD130 was digested with *Sac*II (located 69 bp upstream of the ATG codon of the hydrophobic leader) plus *Afl*II. The corresponding 210-bp fragment was isolated by HPLC on a GenPak FAX column and cloned into pRD131, pRD132 and pRD133 digested with *Sac*II plus *Afl*II. The resulting plasmids were termed pTC13, pTC15 and pTC16, respectively (Figure 2).

Plasmid pRD134 contains the complete wild-type envelope gene of SNV cloned in pRD114 (17). Plasmid pIM16 is identical to pRD134, except that it contains an *Hpa*I site immediately downstream of the hydrophobic leader sequence (as described above). The creation of the *Hpa*I site by site-directed mutagenesis resulted in two changes of the amino acid sequence (gly--val; tyr--lys; I. Martinez and R. Dornburg, unpublished). Plasmid pTC21 has been described recently (3). It contains a single-chain anti-DNP antibody gene fused in-frame to the carboxy-terminal half of the SNV envelope gene. The hydrophobic leader sequence is that of the immunoglobulin heavy chain gene (3). Plasmids pTC23 and pTC24 have been made following the strategy outlined in Figure 3. The coding region of the single-chain antibody (scA) gene B6.2 [made for expression in *E. coli* (2,18) kindly provided by ENZON, Piscataway, NJ, USA] has been amplified by PCR using primers that added an *Eco*47 III site at the 5′ end of the coding region and an *Nru*I site at the 3′ end. The PCR product was cloned into the *Sma*I site of pRD15 (a vector derived from pUC19, Reference 17). After DNA sequencing to verify the absence of PCR artifacts, the modified B6.2 gene was isolated as an *Eco*47 III/*Nru*I fragment and cloned into the *Nru*I site of pTC13 to generate plasmid pTC23 (Figure 4). In plasmid pTC24, carboxy-terminal parts of the SNV envelope were fused to the B6.2 gene.

PCR and Cloning of PCR Products

PCR with primers that have an 8-bp non-hybridizing

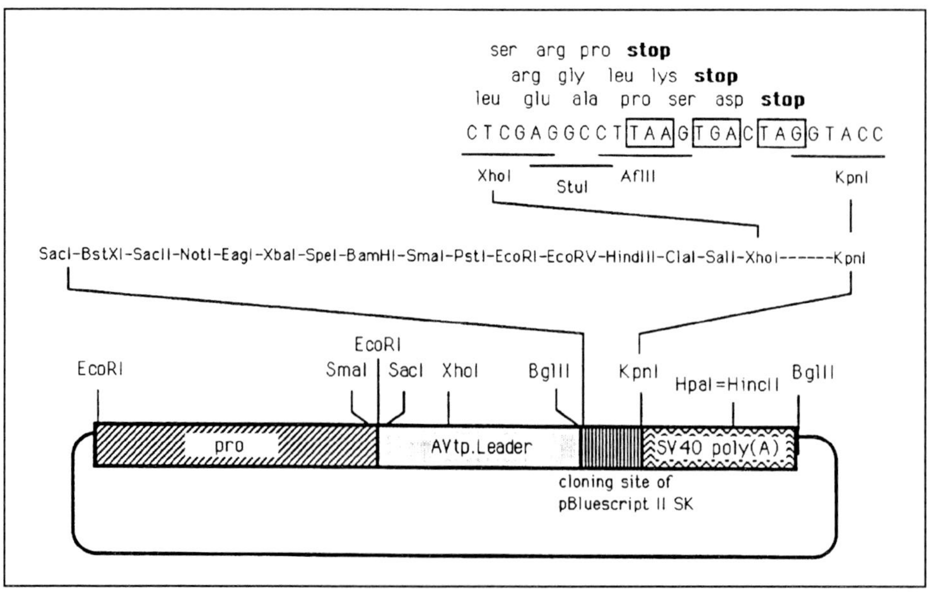

Figure 1. Universal eukaryotic gene expression plasmids. All plasmids contain the ampicillin resistance gene, an origin of replication and a multiple cloning site (referred to here as a polylinker). A linker containing three stop codons has been inserted into gene expression plasmids pRD114, pTC1 and pWS4 (described in Reference 17). pro: pRD131 contains the U3 promoter and enhancer of MLV; plasmid pRD132 contains the promoter and enhancer of SNV; and plasmid pRD133 contains the promoter and enhancer of CMV. AVtl: tripartite leader sequence of adenovirus 2. SV40 poly(A): polyadenylation addition signal sequence of simian virus 40.

region at the 5′ end was performed under the following conditions: Five micrograms of plasmid DNA in water were denatured in the presence of 0.2 M NaOH and 0.1 mM EDTA for 10 min. The solution was neutralized with 5 M sodium acetate pH 4.5. The DNA was precipitated with 2.5 volumes of ethanol. The pellet was rinsed with 70% ethanol and then dried, and the DNA was dissolved in 50 µL of distilled water. One microliter was used for PCR in a 100-µL assay using $Vent_R$® DNA polymerase under the salt concentrations recommended by the supplier (New England Biolabs). 5′ Phosphorylated primers were purchased from DNA International. The DNA was cycled 10 times at 94°C (1.5 min), 50°C (2 min) and 72°C (3 min). The PCR products were phenol-chloroform-extracted three times and precipitated with ethanol. The pellet was dried and re-suspended in 50 µL of distilled water. To remove nucleotides and primer, the DNA was further purified using the GENECLEAN® kit purchased from Bio 101 (La Jolla, CA, USA). The purified DNA was blunt-ended using *E. coli* DNA polymerase I under standard conditions (13). The DNA was purified again using GENECLEAN. Finally the DNA was cloned into blunt-ended, phosphatase-treated vector DNA under standard conditions (13).

Transfections

Plasmid DNAs were transfected into D17 cells (a dog osteosarcoma cell line obtained from ATCC [Rockville, MD, USA]) using the polybrene (hexadimethrine bromide)-dimethyl sulfoxide (DMSO) method (8). Polybrene and DMSO were purchased from Sigma Chemical (St. Louis, MO, USA) and Fisher Scientific (Pittsburgh, PA, USA), respectively.

Fluorescence-Activated Cell Sorting (FACS) Analysis

Intracellular processing and transport of the chimeric envelope proteins to the cell surface were determined by FACS. In this analysis the primary antibody used was a polyclonal rabbit anti-SNV envelope, a gift from Dr. Antonito Panganiban (University of Wisconsin) (6). A goat anti-rabbit IgG conjugated with fluorescein (Pierce, Rockford, IL, USA) was used as the secondary antibody. FACS analysis was performed using 10^6 chimeric envelope expressing cells. The cells were incubated with 100 µL of anti-SNV envelope antibody (diluted 1:10 in phosphate-buffered saline [PBS] containing 0.5% sodium azide) on ice for 30 min. Cells were washed twice with 4 mL of PBS-0.5% sodium azide and resuspended in 100 µL of anti-IgG (diluted 1:200 in PBS-0.5% sodium azide). The secondary antibody was also incubated on ice for 30 min. Cells were washed once as before, re-suspended in 0.4 mL of PBS-0.5% sodium azide and analyzed by flow cytometry. The green fluorescence was measured at 520 nm. Under these experimental conditions, only proteins that appear on the cell surface are detected (16).

RESULTS AND DISCUSSION

Recently, we constructed a series of highly efficient universal eukaryotic gene expression vectors (17). All vectors contain a strong viral promoter and enhancer followed by the adenovirus tripartite leader sequence (1,11), a multiple cloning site and the polyadenylation addition site of simian virus 40 (SV40; Reference 19). They all have an identical

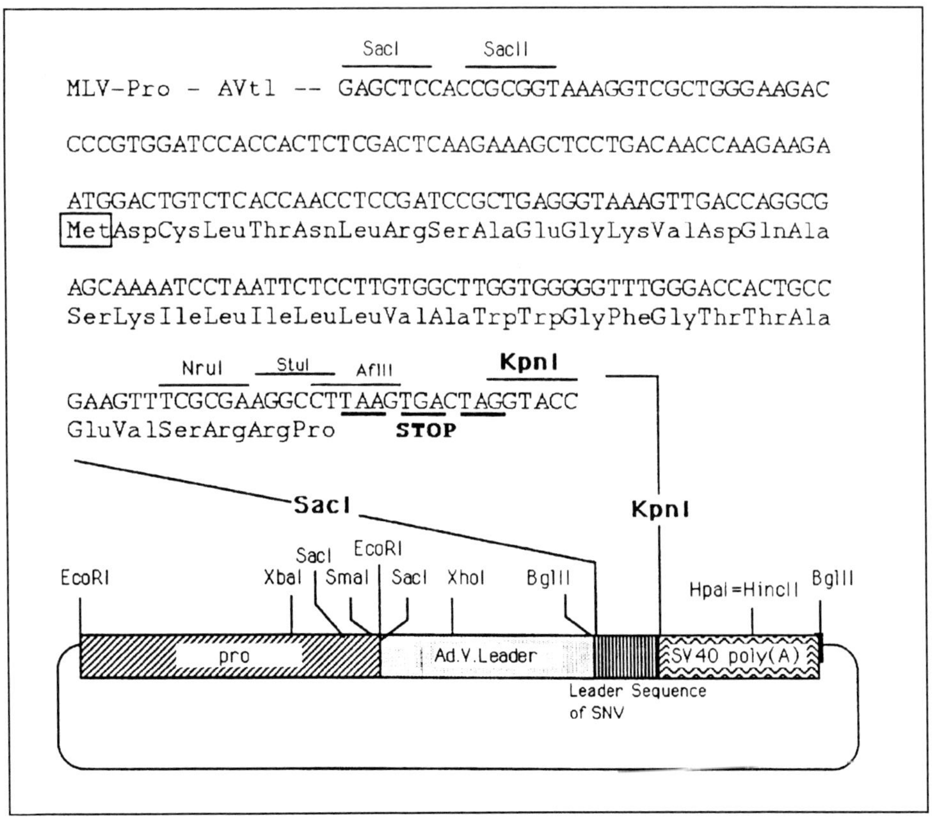

Figure 2. Universal eukaryotic gene expression vectors for secretion of peptides. pro: plasmid pTC13 contains the promoter and enhancer of MLV; plasmid pTC15 contains the promoter and enhancer of SNV; and plasmid pTC16 contains the promoter and enhancer of CMV. All vectors were derived from pRD131, pRD132, and pRD133 (Figure 1).

backbone including the ampicillin resistance gene and an origin of DNA replication (17). These constructs were further modified to obtain vectors for secretion of peptides.

In the first step, a linker was inserted that contains stop codons in all three reading frames (Figure 1). The vectors containing the stop codons were used to further develop the universal gene expression vectors for peptide secretion (Figure 2). These vectors contain the coding region for the hydrophobic leader of the envelope gene of SNV, followed by restriction enzyme recognition sites for *Nru*I and *Stu*I. *Nru*I was chosen as the first site because the first three bases of this enzyme recognition site code for the amino acid serine. Serine is a small, charged amino acid which delineates the end of the hydrophobic amino acid leader sequence and which was not expected to interfere with the folding of the peptide expressed from the DNA fragment of interest that abuts this codon after cloning (also see Figure 3).

Construction of Specific Peptide Secretion Vectors

The research in our laboratory is focused on construction of retroviral vectors displaying scA on the viral surface for cell-type-specific gene transfer (3,4). For incorporation of scAs into the viral particle, chimeric genes are constructed that contain the scA fused to the carboxy terminus of the viral envelope protein (3,4). In the past few years, several scA genes have been constructed for expression in *E. coli* (2,18). To modify such scA genes for construction of chimeric proteins for expression and membrane incorporation in eukaryotic cells, we used the protocol described in Figure 3 to modify an scA gene for expression and membrane incorporation in eukaryotic cells.

General Experimental Protocol for Cloning of PCR Products

To enable the in-frame fusion of the B6.2 peptide-coding region to the hydrophobic leader sequence, the following strategy was employed: PCR primers that contained 8-bp tails at their 5′ ends and that did not hybridize to the DNA region to be amplified (Figure 3) were designed. In the case of the 5′ primer, the first base of the region that hybridizes to the gene of interest is complementary to the first base of the first codon of interest. Likewise, the first base of the 3′ primer that hybridizes to the DNA of interest, base-pairs with the last base of the last codon. The 8-bp overhanging regions contain recognition sites for restriction enzymes that create blunt ends (for details, see Figure 3). We chose and recommend using enzymes whose recognition sites code for small amino acids with or without charge (e.g., ser, thr, gly, ala). For example, the restriction enzyme site *Eco*47 III (AGC GCT) codes for ser-ala. Further, we used primers with two different enzyme recognition sites to avoid cross-hybridization between the two primers during the PCR. Since one of the amino acids encoded by the restriction enzyme recognition site will be present in the final engineered construct (see below), the likelihood that

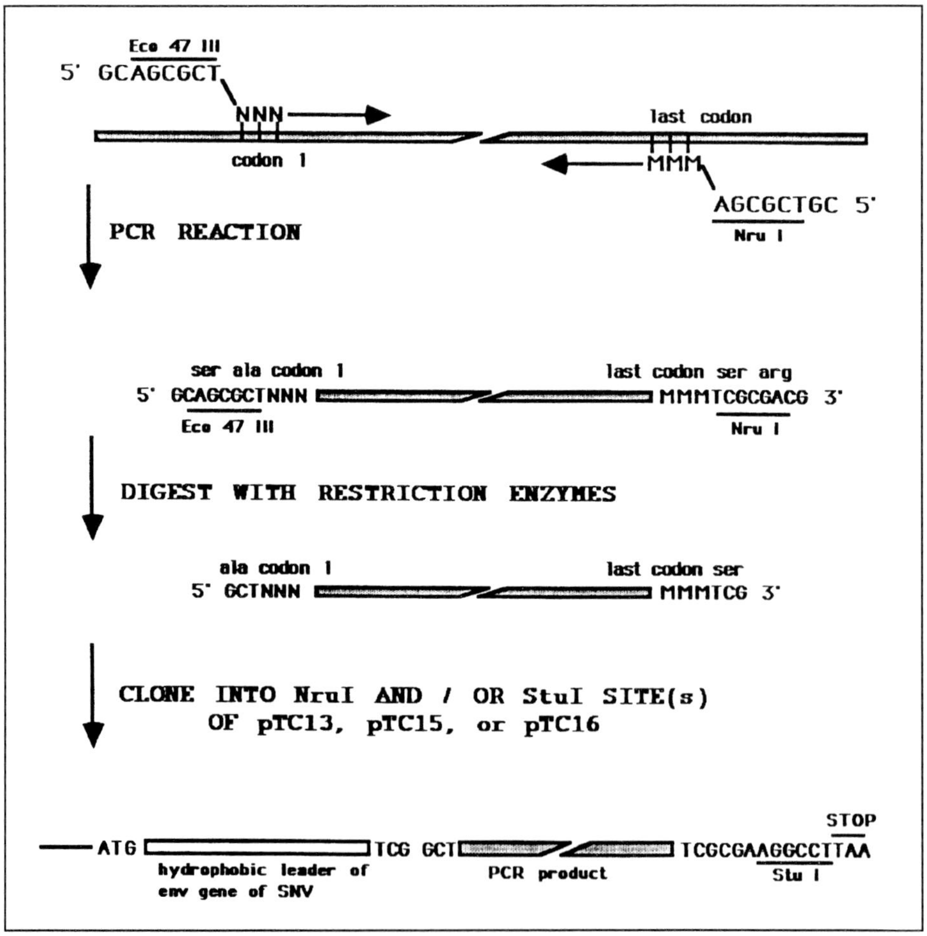

Figure 3. Strategy for cloning peptide coding regions into a peptide secretion vector. Primers are synthesized that contain an 8-bp-long region that contains a restriction enzyme recognition site and that does not hybridize to the DNA fragment of interest. NNN: codon 1 of the peptide coding region of interest; MMM: last codon of the peptide coding region of interest. For more details see text.

the insertion of such small amino acids will have an adverse effect on protein structure, function and stability is low. However, the choice of the restriction enzyme sites is dependent on the sequence of the DNA fragment that needs to be amplified by PCR. Enzyme sites should be added that do not exist within the gene of interest.

The reason for using primers with overhangs is that after PCR, depending on the enzyme and the protocol used, the resulting PCR products have frequently 1- to 2-bp deletions which would make it necessary to screen and sequence a large amount of clones to obtain one construct that contains the insert in the correct reading frame (5,10,12). The use of primers as described eliminates this problem. For the engineering of chimeric proteins, we always follow the general strategy of using only restriction enzymes that generate blunt ends. Furthermore, we generate the corresponding sites in such a way that the enzymes always cut between two codons. This facilitates the maintenance of the reading frame when constructing chimeric protein-coding sequences.

After PCR amplification, the complete amplified gene was cloned into the SmaI site of pUC19 by blunt-end ligation followed by blue/white colony screening and selection (13). The PCR product was sequenced to verify the absence of PCR artifacts. We found that PCR products sometimes contain a single base pair deletion close to the PCR primer. Such a deletion would cause a shift of the reading frame and render the PCR product useless. Thus, sequencing of the ends appears to be essential. Next, the plasmid was digested with the enzymes that cut within the PCR primers. This digest removed five terminal bases of the original PCR product (Figure 3). The resulting DNA fragment, of which the first base corresponds to the first base of a codon, was cloned into the *Nru*I site of the secretion vector pTC13.

The resulting plasmid (termed pTC23) was sequenced again to verify the correct reading frame. Unfortunately, we found that restriction enzyme digestion with *Nru*I, followed by ligation with another fragment, can result in products that contain a one- or two-bp deletion of the remaining nucleotides of the *Nru*I site; depending on the source of the enzyme and lot number, frequently only one out of three to four constructs contained the complete sequence as expected (data not shown). We assume that such deletions stem from low levels of exonuclease contaminations of the enzyme supplied by the manufacturer rather than from an unfaithful function of the enzyme. A similar phenomenon has been observed in our laboratory using the enzyme *Sma*I. This problem has not been observed with any other restriction enzyme that generates blunt ends (e.g., *Sca*I, *Eco*47 III, *Msc*I, *Hpa*I, *Stu*I). Thus, to avoid this problem with *Nru*I, the second restriction enzyme site (*Stu*I) may be used if expression of a single peptide is wanted. In any case, we recommend to always sequence the fusion product to verify the reading frame. Plasmid pTC23 was further modified by fusing carboxy-terminal parts of the envelope gene of SNV to the scA to generate plasmid pTC24 (Figure 4). Again, DNA sequencing confirmed the in-frame fusion.

FACS Analysis

To test whether the hydrophobic leader sequence could initiate transport of the chimeric protein through the ER to the cell surface, pTC24 was transfected into D17 cells. Membrane incorporation was compared with that of the SNV envelope protein (construct pRD134) from which the hydrophobic leader has been derived. All protein-coding sequences are

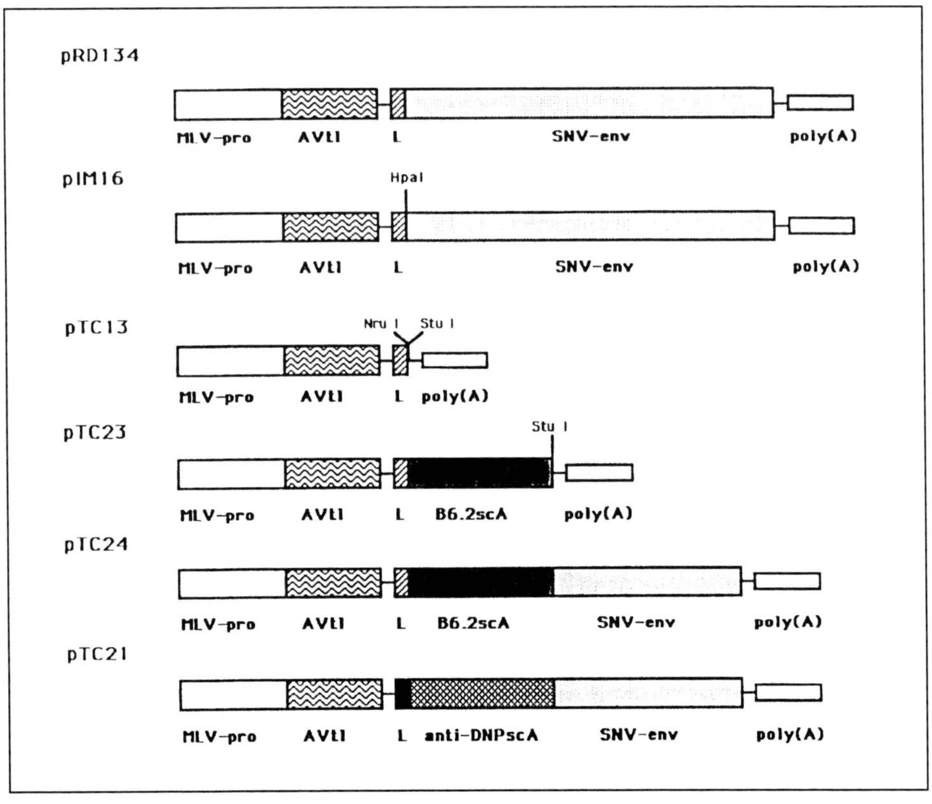

Figure 4. Gene expression vectors expressing SNV envelope (env) gene or scA-env fusion proteins. pRD134 contains the complete envelope gene of SNV. pIM16 is identical to pRD134, except an *Hpa*I site downstream of the hydrophobic leader sequence (L). pTC23 contains a PCR product of the scA gene B6.2 (B6.2scA) inserted into the *Nru*I site of pTC13 (also see Figures 2 and 3). In pTC24, the B6.2scA gene is fused in-frame to carboxy-terminal regions of the SNV envelope gene (SNV-env). pTC21 is similar to pTC24. It contains an anti-DNP antibody with its original ER leader fused to carboxy-terminal sequences of the SNV envelope gene. All vectors contain identical plasmid backbones (derived from pUC19, not shown) that abut the sequences shown. MLV-pro: promoter and enhancer of MLV; AVtl: tripartite leader sequence of adenovirus 2; poly(A): polyadenylation recognition site of SV40.

inserted into the same vector backbone. Thus, efficiency of transcription is the same in all constructs. We had compared SNV envelope expression from pRD134 with that from various other constructs present in different retroviral packaging lines. We found that envelope expression in pRD134 was by far the highest among all cell lines and constructs tested. For example, envelope gene expression from the Rous sarcoma virus (RSV) promoter was hardly detectable by FACS analysis (data not shown).

We also compared membrane incorporation of the chimeric protein with that of pIM16 (envelope gene containing the HpaI site) and pTC21. Plasmid pTC21 is a construct similar to pTC24. However, it contains the original hydrophobic leader sequence of the heavy chain gene of an antibody directed against the hapten dinitrophenyl (DNP; Reference 3). Please note that all plasmids are identical except for the protein-coding regions of interest (Figure 4).

Membrane incorporation of all proteins was measured by FACS analysis using a polyclonal antibody directed against the envelope portion of the chimeric construct. We found that the introduction of the *Hpa*I site, which causes two mutations in the amino acid sequence, did not abolish processing and transport of the SNV envelope protein to the cell surface (Figure 5). These data indicate that the mutagenesis did not have an adverse effect on processing of the hydrophobic leader sequence.

The chimeric scA-envelope proteins derived from pTC24 were transported to the cell surface with an efficiency similar to that of the wild-type envelope protein (Figure 5); the shift (mean value) of fluorescent cells was about 75% of that of cells expressing the wild-type envelope. The slightly higher fluorescence of cells expressing the wild-type envelope most probably results from the presence of additional epitopes recognized by the first antibody and which are missing in the chimeric construct. This conclusion is based on the observation that a similar construct (pTC21) that had the original antibody ER sequence gave a shift of fluorescent cells similar to that of pTC24. These data indicate that the SNV envelope's hydrophobic leader sequence can initiate the transport of a heterologous peptide through the ER as efficiently as in connection with its original peptide.

The biological function of the chimeric scA-envelope protein has been tested in a variety of experiments. We found that such proteins were incorporated into retroviral vector particles; they were able to bind antigen and enabled the cell-type-specific infection of selected target cells (4).

For experimental reasons that are outlined elsewhere (4), all gene expression vectors used in this study were derived from pTC13 (Figure 1), which contains the murine leukemia virus (MLV) promoter and enhancer (17). We have constructed similar vectors containing the cytomegalovirus (CMV) or SNV promoter and enhancer (Figure 2). Previously, we showed that the CMV promoter is about threefold stronger in D17 cells than the MLV promoter. Thus, based on our earlier studies (17), we would expect that the other vectors (pTC15 and pTC16) work as well or even better than pTC13, depending on the cell type being used. Furthermore, we also made a series of other constructs (termed pRD127, pRD128 and pRD129) which are identical to pTC13, pTC15 and pTC16, respectively, except that they contain the hydrophobic leader sequence of the MLV envelope gene. However, the function of these vectors has not been tested at this point.

In summary, we constructed a series of universal eukaryotic gene expression vectors for the expression and ER transport of any peptide-coding region. We show that the addition of the coding region of the SNV envelope hydrophobic leader to an unrelated peptide-coding region results in a gene product that is efficiently transported through the ER. Further, the presence of two unique cloning sites downstream of the hydrophobic leader coding region allows the easy construction of chimeric proteins. For example, the second site can be used to add a TAG region if affinity purification of the secret-

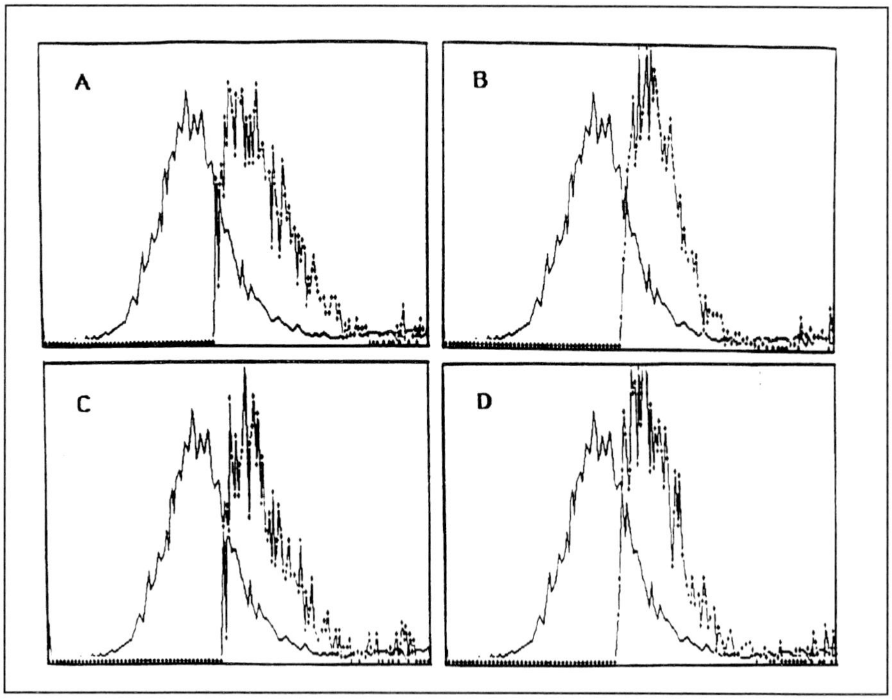

Figure 5. FACS analysis of D17 cells stable-transfected with plasmids pRD134 (A), pIM16 (B), pTC21 (C) and pTC24 (D). All samples were compared with plain D17 cells treated with the same antibodies under identical conditions (solid line). The degree of the shift of the intensity of the fluorescence, indicated by the dashed lines (after background subtraction), indicates the amount of incorporation of the protein expressed from above vectors into the cell membrane.

ed protein is desired. Thus, the universal cloning sites and easy cloning strategy as well as the high expression potential make these vectors attractive for a large variety of other applications and purposes.

ACKNOWLEDGMENTS

We thank Dr. Antonito Panganiban of the University of Wisconsin, Madison, Wisconsin, for the anti-SNV envelope antibody and for supplying us with the complete DNA sequence of the SNV genome. We also thank ENZON for the B6.2 scA gene. This research was supported by grants of the New Jersey Commission on Cancer Research, the March of Dimes Birth Defect Foundation and the UMDNJ Foundation.

REFERENCES

1. **Berkner, K.L.** 1988. Development of adenovirus vectors for the expression of heterologous genes. BioTechniques *6*:616-624.
2. **Bird, R.E., K.D. Hardman, J.W. Jacobson, S. Johnson, B.M. Kaufman, S.-M. Lee, T. Lee, S.H. Pope, G.S. Riordan and M. Whitlow.** 1988. Single-chain antigen-binding proteins. Science *242*:423-426.
3. **Chu, T.-H.T., I. Martinez, W. Sheay and R. Dornburg.** 1994. Cell targeting with retroviral vector particles containing antibody-envelope fusion proteins. Gene Therapy *1*:292-299.
4. **Chu, T.-H.T. and R. Dornburg.** Retroviral vector particles displaying the antigen-binding site of an antibody enable cell-type-specific gene transfer. J. Virol. (In press).
5. **Crowe, J.S., H.J. Cooper, M.A. Smith, M.J. Sims, D. Parker and D. Gewert.** 1991. Improved cloning efficiency of polymerase chain reaction (PCR) products after proteinase K digestion. Nucleic Acids Res. *19*:184.
6. **Delwart, E. and A.T. Panganiban.** 1989. Role of reticuloendotheliosis virus envelope glycoprotein in superinfection interference. J. Virol. *63*:273-280.
7. **Hershey, J.W.B.** 1991. Translational control in mammalian cells. Annu. Rev. Biochem. *60*:717-755.
8. **Kawai, S. and M. Nishizawa.** 1984. New procedure for DNA transfection with polycation and dimethyl sulfoxide. Mol. Cell. Biol. *4*:1172-1174.
9. **Kriegler, M.** 1990. Gene transfer and expression. W.H. Freeman and Company, New York.
10. **Liu, Z. and L.M. Schwarz.** 1992. An efficient method for blunt-end ligation of PCR products. BioTechniques *12*:28-30.
11. **Logan, J. and T. Shenk.** 1984. Adenovirus tripartite leader sequence enhances translation of mRNAs late after infection. Proc. Natl. Acad. Sci. USA *81*:3655-3659.
12. **Lohff, C.J. and K.B. Caese.** 1992. PCR using a thermostable polymerase with 3′ to 5′ exonuclease activity generates blunt end products suitable for direct cloning. Nucleic Acids Res. *20*:144.
13. **Maniatis, T., E.F. Fritsch and J. Sambrook.** 1982. Molecular Cloning: A Laboratory Manual. Cold Spring Harbor Laboratory, Cold Spring Harbor, NY.
14. **Merrick, W.C.** 1992. Mechanism and regulation of eucaryotic protein synthesis. Microbiol. Rev. *56*:291-315.
15. **Miller, J. and M. Calos.** 1987. Gene transfer vectors for mammalian cells. Cold Spring Harbor Press, Cold Spring Harbor, NY.
16. **Shapiro, H.M.** 1983. Multistation multiparameter flow cytometry: A critical review and rationale. Cytometry 3:227-243.
17. **Sheay, W., I. Martinez, T.-H.T. Chu, S. Nelson, S. Bhatia and R. Dornburg.** 1993. Downstream insertion of the adenovirus leader sequence enhances gene expression in universal eukaryotic vectors. BioTechniques *15*:856-862.
18. **Whitlow, M. and D. Filpula.** 1991. Single-chain Fv proteins and their fusion proteins. Methods: Companion Methods Enzymol 2:1-9.
19. **Wickens, M. and P. Stephenson.** 1984. Role of the conserved AAUAAA sequence. Four AAUAAA point mutants prevent messenger RNA 3′ end formation. Science *226*:1045-1051.
20. **Yan, S.C.B., W. Grinnell and F. Wold.** 1989. Post-translational modifications of proteins: some problems left to solve. Trends Biol. Sci. *14*:264.

CHAPTER 37

Bicistronic Vector for the Creation of Stable Mammalian Cell Lines that Predisposes All Antibiotic-Resistant Cells to Express Recombinant Protein

S. Rees, J. Coote, J. Stables, S. Goodson, S. Harris and M.G. Lee

Glaxo Research and Development, Ware, Herts and Greenford, Middlesex, UK

(Reprinted from BioTechniques 20:102-110, 1996)

ABSTRACT

We have developed an improved vector for the stable expression of recombinant protein in mammalian cells. In this vector, designated pCIN, both the recombinant cDNA and the neomycin phosphotransferase selection marker are transcribed from a single promoter element. To facilitate translation of the second open reading frame, the encephalomyocarditis virus internal ribosome entry site has been inserted into the expression cassette immediately before the start codon of this sequence. We report the use of this vector to generate stable cell lines expressing the human 5-HT$_{1D\alpha}$ serotonin receptor and show that following transfection and clonal selection, all ten cell lines characterized express similar and high levels of receptor (1.5–11.9 pmol receptor/mg protein). Use of pCIN should permit the rapid and efficient production of stable mammalian cell lines for the characterization of recombinant protein, as this vector appears to predispose all transfected cells to express such protein.

INTRODUCTION

The expression of recombinant protein in mammalian cell lines is a fundamental technique in modern biology. Such studies have been used to characterize many proteins, including members of the superfamily of 7-transmembrane domain, guanine-nucleotide binding protein (G protein) coupled receptors (6). The stable expression of such receptors has facilitated the identification of selective ligands (21), the understanding of receptor/G protein interaction and the identification of second messenger signaling mechanisms (6,20,23). Indeed these studies have revealed several features of this class of protein not predicted by traditional pharmacological analysis. Many receptors over-expressed in cell lines show constitutive activity, that is, the ability to signal in the absence of ligand. This observation has led to the identification of a class of compounds, termed inverse agonists, which are able to lower this basal level of activity (3,14). However, the study of these and other proteins expressed in stable cell lines is hampered by the technical problems associated with their development.

Standard methods for the creation of a stable cell line involve transfection into a host cell of two expression cassettes: the first for expression of recombinant protein and a second for expression of an antibiotic resistance gene, carried on either two plasmids (co-transfection) (23) or on a single "all-in-one" vector (5). However, following transfection and clonal selection for stable integrants, the proportion of antibiotic resistant clones that also express recombinant protein is rarely more than 30%. This is a result of deletion or inactivation of the expression cassette for the production of recombinant protein during plasmid integration, or in the case of co-transfection, due to integration of only the plasmid carrying the antibiotic selection marker. Furthermore, the level of expression is impossible to predict, is often low and will often decrease during time in culture, as there is no positive selection to maintain the transcriptional activity of the site of vector integration.

The standard model of protein translation that applies to the vast majority of eukaryotic and prokaryotic mRNAs involves ribosome entry at the 5′ cap structure followed by scanning of the mRNA in a 5′ to 3′ direction until the initiation codon is reached with the subsequent initiation of protein translation (13). This model of translation does not appear to apply to the picornaviridae family of viruses, which includes encephalomyocarditis virus (EMCV), foot-and-mouth disease virus and poliovirus. These viruses possess a single-stranded RNA genome of positive polarity containing a 600–1200 nucleotide (nt) 5′ untranslated region (UTR) followed by a single open reading frame (ORF) encoding the viral polyprotein. Gene expression is unusual in that the ribosome is able to recognize and bind directly to a 450–600-nt secondary structure element located within the 5′ UTR termed an internal ribosome entry site (IRES) (8,9). The 5′ UTR contains numerous AUG initiation codons; however, following ribo-

Cloning and Expression Vectors for Gene Function Analysis
Edited by Quinn Lu and Michael P. Weiner

some entry at the IRES, a number of these are bypassed to allow translation initiation at the authentic AUG (AUG 11 or AUG 12 for EMCV) (8). IRES elements have been removed from their viral setting and linked to unrelated genes to produce polycistronic RNAs. Several reports describe their application in retroviral mediated gene transfer, in the creation of transgenic mice and for the expression of multiple ORFs from a single transcript in mammalian cells (10,11,15).

To facilitate the creation of stable cell lines, we have developed a bicistronic expression vector that predisposes every transfected cell to express recombinant protein and at apparently high levels. This is achieved through inclusion of the EMCV IRES (8) into the expression cassette to permit the translation of two ORFs, the recombinant protein and the neomycin phosphotransferase (NPT II) antibiotic resistance marker, from a single messenger RNA. We demonstrate the use of this vector to express the human 5-$HT_{1D\alpha}$ serotonin receptor in HEK 293 cells.

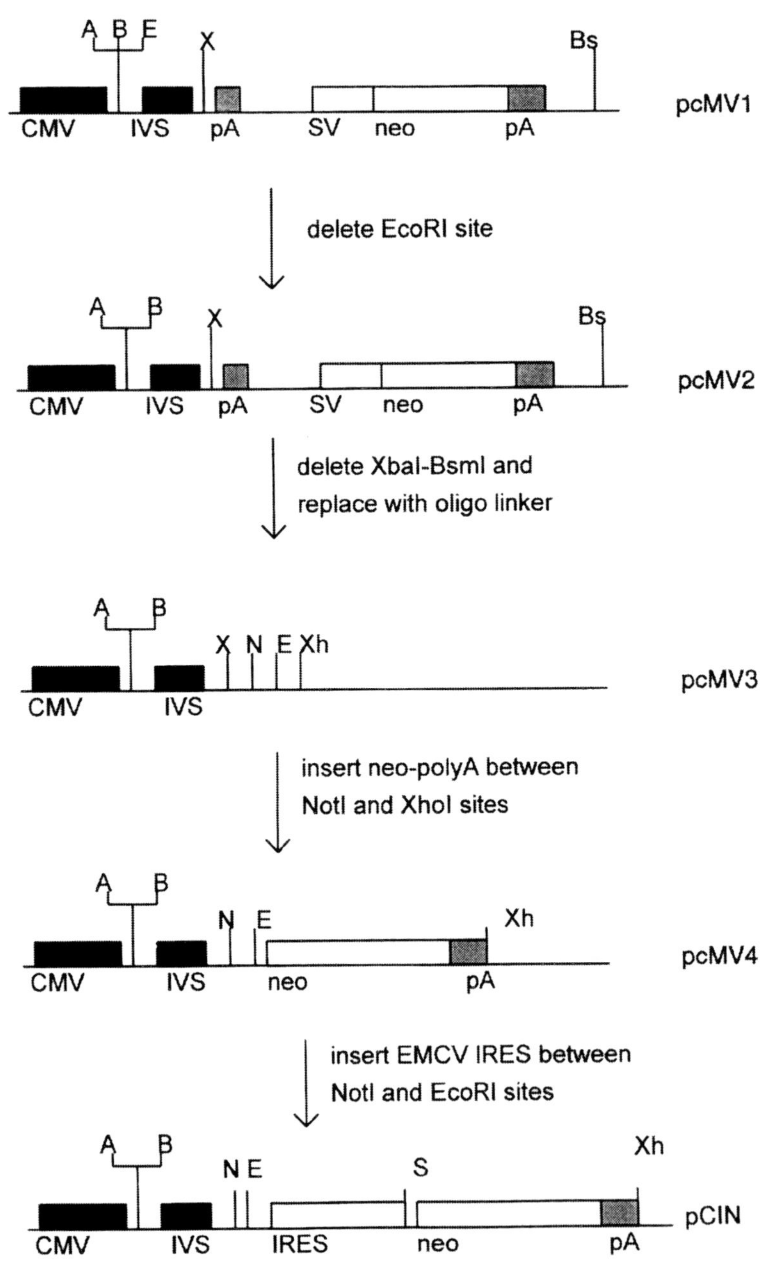

Figure 1. Construction of the expression vector pCIN. A = *Asp*718, B = *Bam*HI, Bs = *Bsm*I, E = *Eco*RI, N = *Not*I, S = *Sma*I, X = *Xba*I, Xh = *Xho*I; CMV = cytomegalovirus intermediate early promoter, IVS = intron, pA = polyadenylation sequences, SV = SV40 early promoter/origin of replication, neo = neomycin phosphotransferase cDNA and IRES = internal ribosome entry site.

MATERIALS AND METHODS

Plasmid Constructions

All DNA manipulations were performed using standard methods unless otherwise described (17).

The construction of pCIN is summarized in Figure 1. A synthetic intron (7) was cloned from pOG44 (Stratagene Ltd, Cambridge, UK) between the *Eco*RI and *Sal*I sites of pcDNA3 (Invitrogen, San Diego, CA, USA) to generate pcMV1, followed by deletion of the *Eco*RI site in this vector by digestion with this enzyme and filling the resulting sticky ends with Klenow DNA Polymerase to create pcMV2. The NPT II expression cassette in pcMV2 was excised by digestion with *Xba*I and *Bsm*I with simultaneous insertion of a synthetic linker created following annealing of the oligonucleotides 5′-CTAGGGCGGCCGCGTACGAATTCGATACTCGAGTG-3′ and 5′-CTCGAGTATCGAATTCGTACGCGGCCGCC-3′ encoding the restriction sites *Xba*I, *Not*I, *Eco*RI and *Xho*I, to generate pcMV3. Following this, the NPT II cDNA and the bovine growth hormone polyadenylation sequence were excised from the plasmid pPol2sneobpA (18) by digestion with *Eco*RI and *Xho*I and inserted between the *Eco*RI and *Xho*I sites of pcMV3 to create pcMV4. Finally, the EMCV IRES was amplified using the polymerase chain reaction (PCR) from pIRESβ-geo (Dr. P. Mountford, University of Edinburgh, UK) with the primers 5′-GATCGCGGCCGCACTAGAGGAATTCCGCCCCTCTCC-3′, incorporating *Not*I and *Eco*RI sites (underlined) and 5′-CATGGATATCCCGGGTTGTGGCAAGCTTATCATCGT-3′, containing overlapping *Sma*I and *Eco*RV sites (again underlined). The PCR product was digested with *Not*I and *Eco*RV and cloned between the *Not*I and Klenow-treated *Eco*RI sites of pcMV4 to generate pCIN. The IRES was re-sequenced following PCR.

To validate these vectors, the human $5\text{-HT}_{1D\alpha}$ serotonin receptor cDNA was amplified from pCD-MA6A (4) by PCR with the primers 5′-CCAGAATTCCATGTCCCCACTGAACCAGTCAGCAG-3′ and 5′-CCGAGAATTCCTAGGAGGCCTTCCGGAAAGGG-3′ incorporating *Eco*RI restriction sites at their 5′ termini (underlined). The product was digested with this enzyme and subcloned into the *Eco*RI site of pcMV1 to generate pcMV1/$5\text{-HT}_{1D\alpha}$. The cDNA was re-sequenced following PCR. For expression studies, the $5\text{-HT}_{1D\alpha}$ cDNA was further transferred into the *Eco*RI site of pCIN to create pCIN/$5\text{-HT}_{1D\alpha}$.

A control vector without the IRES sequence was built by insertion of a *Asp*718-*Sac*II fragment from pcMV1/$5\text{-HT}_{1D\alpha}$, encoding the $5\text{-HT}_{1D\alpha}$ cDNA and part of the intron, between the *Asp*718 and *Sac*II sites of pcMV4 (recreating an intact intron) to generate pΔIRES/$5\text{-HT}_{1D\alpha}$.

Expression Studies

All cells were maintained at 37°C with 5% CO_2 and 92% humidity, in Dulbecco's modified Eagle Medium/ Ham's F-12 (1:1) media, supplemented with 10% fetal calf serum and 4 mM glutamine (all reagents from Life Technologies, Gaithersburg, MD, USA).

For transient transfection, Cos M6 cells, grown to 40% confluence in 500 cm^2 triple flasks, were transfected with 50 µg of plasmid using the DEAE dextran procedure (12). Cell extracts were prepared 48 h post-transfection. Cells were washed twice in Hank's balanced salt solution (HBSS), detached from the plate by incubation in HBSS containing 600 µM EDTA and collected by centrifugation at 200× *g* in a Heraeus Bactifuge® (Heraeus Instruments GmbH, Hanau, Germany). The cell pellet was resuspended in 1 mL of phosphate-buffered saline (PBS) containing 1 mM phenylmethyl sulphonyl fluoride, and the cells lysed by three alternate brief immersions into a dry ice/ethanol bath followed by a 37°C water bath. Cell debris were removed by centrifugation at 500× *g* for 5 min at 4°C in a microcentrifuge (Model 5415C; Eppendorf, Hamburg, Germany) and the supernatant was collected and stored at -80°C for NPT II assay.

For stable transfection, 10 µg of pCIN/$5\text{-HT}_{1D\alpha}$ were linearized with *Ssp*I and transfected into HEK 293 cells, plated at 1% confluence in a 10-cm-diameter tissue-culture dish, using the calcium phosphate procedure. Two days following transfection, the growth medium was replaced with medium supplemented with 1 mg/mL G418 (Life Technologies) (19). Following application of antibiotic selection, single colonies were picked and expanded through several passages to provide cells for homogenate preparation. Cells, grown to confluence in 175-cm^2 flasks, were detached by incubation in PBS containing 5 mM EDTA, counted and harvested by centrifugation at 4500× *g* for 35 min [SORVALL® RC-3 Centrifuge; Du Pont (UK) Ltd., Herts, UK]. The cell pellet was re-suspended in assay buffer (100 mM Tris-HCl, 1 mM EDTA and 1% ascorbic acid, pH 7.7) to yield a concentration of 10^7 cells/mL. Cells were disrupted by homogenization, and the homogenates were stored at -80°C for ligand binding studies.

NPT II Assay

The sandwich enzyme-linked immunosorbent assay (ELISA) for NPT II activity was performed according to the protocol supplied with the NPT II assay kit (CP Laboratories, Bishop's Stortford, Herts, UK), except that the final incubation of bound streptavidin alkaline phosphatase with *p*-nitrophenolphosphate was performed at room temperature. NPT II levels were determined according to reaction rate using the kinetic mode of the $\text{THERMO}_{\text{MAX}}$ microplate reader at 405

```
                 10          11              12
EMCV   AAAAACACGATGATAATATGGCCACAACCATGGGAA

                                                                    NEO
pCIN   AAAAACACGATGATAAGCTTGCCACAACCCCGGGATAATTCCTGCAGCCAATATG
                                   SmaI
```

Figure 2. Structure of the junction between the EMCV IRES and the NPT II cDNA in the expression vector pCIN. EMCV ATG 10, 11 and 12 and the neomycin phosphotransferase ATG are indicated.

nm (Molecular Devices, Menlo Park, CA, USA). Four assays were performed on each transfection.

Protein Assay

Protein concentrations were determined with bovine serum albumin as a standard (1). Six independent assays were performed on each cell extract or homogenate.

Radioligand Binding Assay

Saturation experiments were carried out under the following conditions: cell homogenate (11–41 µg protein in 750- µL assay buffer), [^{3}H]-5-carboxytryptamine (5-CT; 0.1–30 nM final concentration, 200 µL; Du Pont NEN, Boston, MA, USA) and unlabeled 5-CT or assay buffer (50 µL) were incubated at 37°C for 30 min. Reactions were terminated by rapid filtration (Brandel 48-well harvester; Semat, Luton, UK) through glass fiber filters (GF/B; Whatman) presoaked in 0.1% polyethyleneimine for 30 min. The filters were washed four times with 1 mL ice-cold wash buffer (100 mM Tris-HCl, 1 mM EDTA, pH 7.6) and placed in vials with scintillation fluid. Bound radioactivity was measured on a liquid scintillation counter (Model 2200CA; Packard, Pangbourne, UK). Under these conditions, specific [^{3}H]-5-CT binding (defined as that inhibited in the presence of 10 mM 5-CT) represented greater than 95% of total binding, while less than 10% of total ligand was bound. The K_D and B_{max} values were calculated using the curve-fitting program LIGAND (16).

RESULTS AND DISCUSSION

Plasmid Construction

To facilitate the creation of stable mammalian cell lines, we have developed a bicistronic expression vector in which both the antibiotic resistance marker and the recombinant protein are under the transcriptional control of a single promoter element. The expression cassette in pCIN consists of the human cytomegalovirus (CMV) major intermediate early promoter/enhancer sequence, followed by a multiple cloning site, a synthetic intron shown to enhance message stability (7), the EMCV IRES followed by the NPT II cDNA, for the selection of stable transformants in the presence of the antibiotic G418, and the bovine growth hormone polyadenylation signal. The remainder of the plasmid consists of the *Bsm*I to *Bgl*II region of pcDNA3 (Invitrogen) and encodes the β-lactamase gene to confer ampicillin resistance in *Escherichia coli* and the bacterial ColE1 origin of replication (Figure 1). We also constructed a control vector, pΔIRES, identical to pCIN but lacking the IRES element.

The 5′ UTR (IRES) of EMCV is 718-nt long and contains 10 AUG initiation codons scattered throughout the sequence (8). The major site of initiation for viral polyprotein synthesis is AUG 11, with some initiation at AUG 12 (8). To achieve maximal translation through the IRES, the position of AUG 11 relative to the upstream IRES must be maintained (8,9). However, during the construction of pCIN, it was decided to disable the ability of the IRES to mediate NPT II translation by displacing the start codon 44 nt downstream of AUG 11 (Figure 2). We expect this to result in the generation of cell lines likely to express high levels of recombinant protein, since in order to express sufficient NPT II to allow growth in 1 mg/mL G418, we would select for sites of high gene expression. Furthermore, to prevent inappropriate initiation of protein translation at either EMCV AUG 11 or AUG 12, both of these start codons have been destroyed (Figure 2).

Table 1. Assay of the Ability of the EMCV IRES to Permit Translation of NPT II

Vector	pg NPT II/ µg Protein
no DNA	0.0 ± 0
pΔIRES/5-HT$_{1D\alpha}$	0.0 ± 0.3
pCIN/5-HT$_{1D\alpha}$	4.0 ± 0.4
pcMV1/5-HT$_{1D\alpha}$	59.0 ± 3.8

NPT II assays were performed on cell extracts prepared following transient transfection of expression vector into Cos M6 cells. Results are expressed as the mean of 4 inde- pendent assays in pg NPT II per µg of total cellular protein.

To characterize these vectors, the human 5-HT$_{1D\alpha}$ serotonin receptor cDNA (4) was subcloned into the multiple cloning site of each to generate pCIN/5-HT$_{1D\alpha}$, pΔIRES/5-HT$_{1D\alpha}$ and pcMV1/5-HT$_{1D\alpha}$.

Transient Assay for IRES Function

To demonstrate the ability of the ribosome to enter the messenger RNA through the disabled IRES in pCIN/5-HT$_{1D\alpha}$, a transient expression study was performed. Protein extracts prepared following transfection of this plasmid, pcMV1/5-HT$_{1D\alpha}$ and pΔIRES/5-HT$_{1D\alpha}$ into Cos M6 cells were assayed for NPT II using a sandwich ELISA assay. We found significant NPT II expression in extracts prepared following transfection with pcMV1/5-HT$_{1D\alpha}$ (59 pg NPT II/µg protein; Table 1). In this vector, NPT II expression is under the transcriptional control of the simian virus 40 (SV40) early promoter. However, the plasmid also contains the SV40 origin of replication, which permits episomal replication in Cos M6 cells due to the stable expression of the SV40 large T-antigen in this cell line. Consequently, the high level of NPT II can be attributed to both the plasmid amplification seen in this expression system and the relative strength of the SV40 early promoter. As expected, no NPT II was seen in cells transfected with pΔIRES/5-HT$_{1D\alpha}$, as this vector does not contain the IRES (Table 1). However, we did observe NPT II activity in extracts prepared from cells transfected with pCIN/5-HT$_{1D\alpha}$ (4 pg/µg; Table 1). While the level of NPT II expression is only 7% of that seen when this enzyme is expressed directly from the SV40 early promoter, the disabled IRES is clearly able to permit ribosome entry to lead to translation of this ORF.

Generation of Stable Cell Lines with pCIN

pCIN/5-HT$_{1D\alpha}$ was transfected into HEK 293 cells. Following application of the antibiotic G418 (1 mg/mL), 10 colonies, arising from 10 single cells, were isolated, expanded

Table 2. Expression of the Human 5-$HT_{1D\alpha}$ Serotonin Receptor in HEK 293 Cells Transfected with pCIN/5-$HT_{1D\alpha}$

Clone	K_D (nM)	B_{max} (pmol/mg)
1	1.5	1.5
2	1.4	2.9
4	1.6	3.2
13	1.7	1.8
14	1.9	3.4
15	2.4	11.9

Saturation analysis with the high-affinity 5-HT_{1D} selective ligand [3H]-5-CT was performed on cell homogenates as described. The equilibrium dissociation constant (K_D) and receptor density (B_{max}) were determined using the curve-fitting program LIGAND. Each value is calculated from a single experiment performed in triplicate.

and characterized for expression of the 5-$HT_{1D\alpha}$ receptor by ligand binding to cell homogenates of the high affinity 5-HT_{1D} agonist radioligand [^{3}H]-5-CT (2). Unlike untransfected cells, all 10 cell lines exhibited specific binding of this ligand. Homogenates harvested from six of these clones displayed an apparently homogenous population of high-affinity, saturable, [^{3}H]-5-CT binding sites (data not shown). The receptor density (B_{max}) was between 1.5 and 11.9 pmol receptor/mg of total cellular protein, with an equilibrium dissociation constant (K_D) for the ligand 5-CT of between 1.5 and 2.4 nM consistent with binding to the 5-$HT_{1D\alpha}$ receptor (Table 2) (4,23). We have examined the stability of expression in four of these cell lines and find no loss of expression following 20 passages in culture.

Since both the 5-$HT_{1D\alpha}$ receptor and the NPT II selection marker are translated from the same bicistronic mRNA, we might expect to see a correlation between the expression level of these two proteins. However, we find no such correlation (data not presented). This observation may imply that individual members of a clonal population are not genetically and biochemically identical; each of the derived cell lines may have different intrinsic resistance to the antibiotic G418 or a differing ability to tolerate the cellular effects caused by over-expression of this receptor. However, it remains a possibility that the level of expression within an individual cell line may be regulated through alterations in the concentration of G418.

Literature reports of the level of expression of the 5-$HT_{1D\alpha}$ serotonin receptor vary from 98 fmol/mg in the human cervical carcinoma HeLa cell line (4) to 4 pmol/mg in mouse LM(tk-) fibroblasts (24) compared with levels of expression in this study of 1.5–11.9 pmol/mg. Similarly, in our laboratory we have generated stable cell lines with many 7-transmembrane receptors and rarely see expression at greater than 2 pmol/mg. Thus, we appear to have succeeded in generating a vector that predisposes transfected cells to express high levels of recombinant protein, presumably due to selection for sites of high transcriptional activity. The implication would be that low expressing cell lines are not isolated, as they produce insufficient NPT II to allow growth in 1 mg/mL G418, although selection with less antibiotic might be expected to result in the identification of clones with a lower level of expression.

There are a number of reports of the development of retroviral vectors containing drug selectable genes under the control of the EMCV IRES including the multiple drug resistance and NPT II selection markers (15,22) and a single report describing the use of such vectors for the derivation of stable mammalian cell lines (10). The vector used in the latter study, pED4 carries dihydrofolate reductase (DHFR) rather than neomycin selection and was used to express the eIF-2α transcription factor in DHFR-deficient CHO cells with the result that all 14 cell lines characterized were shown to express a full-length bicistronic mRNA. However, there is no data presented concerning the number of clones that express eIF-2α protein or the level of expression obtained with the bicistronic DHFR selection vector. In contrast, we have shown that following transfection with pCIN, every cell line characterized expresses recombinant protein and at apparently high levels. Furthermore, the use of neomycin rather than DHFR within a bicistronic mammalian expression vector should greatly increase the versatility of such vectors.

Previously we have attempted to generate stable cell lines expressing the 5-$HT_{1D\alpha}$ receptor using the vector pcDNA3 (Invitrogen) and found a single expressing clone from 19 cell lines screened (unpublished results). The ability of the bicistronic vector to ensure that all antibiotic-resistant clones express recombinant protein saves considerable time and manpower during the generation of a stable cell line, and for certain studies, may eliminate the need for clonal selection to identify expressing cells for analysis. Furthermore, the use of pCIN to generate cell lines predicted to express high levels of recombinant protein would facilitate the preparation of recombinant protein for functional analysis. In particular, the cell lines generated in this study may be used for high-throughput compound screening to identify novel ligands for the human 5-$HT_{1D\alpha}$ serotonin receptor and would allow an examination of constitutive activity and the identification of inverse agonist ligands. We have recently generated CHO/human adenosine A2a, Cos-7/human 5-HT_{5A} serotonin receptor and CHO/human β3-adrenoreceptor expressing cell lines with similar results.

ACKNOWLEDGMENTS

Address correspondence to Stephen Rees, Molecular Discovery Department, Glaxo Wellcome Research and Development, Gunnels Wood Road, Stevenage, Herts, SG1 2NY, UK (email: esr1353@glaxowellcome.co.uk).

REFERENCES

1. **Bradford, M. M.** 1976. A rapid and sensitive method for the quantitation of microgram quantities of protein utilising the principle of protein-dye binding. Anal. Biochem. *72*:248-254.
2. **Beer, M.S., J.A. Stanton, Y. Beven, N.S. Chauhan and D.N. Middlemiss.** 1992. An investigation of the 5-HT1D receptor binding affinity of 5-hydroxytryptamine, 5-carboxyamidotryptamine and sumatriptan in the central nervous system of seven species. Eur. J. Pharmacol. *213*:193-197.
3. **Chidiac, P., T.E. Hebert, M. Valiquett, M. Dennis and M. Bouvier.** 1994. Inverse agonist activity of β-adrenergic antagonists. Mol. Pharma-

col. *45*:490-499.
4. **Hamblin, M.W. and M.A. Metcalf.** 1991. Primary structure and functional characterization of a human 5-HT_{1D} type serotonin receptor. Mol. Pharmacol. *40*:143-148.
5. **Hamblin, M.W., M.A. Metcalf, R.W. McGuffin and S. Karpells.** 1992. Molecular cloning and functional characterization of a human 5-HT1B serotonin receptor: a homologue of the rat 5-HT1B receptor with 5-HT1D-like pharmacological specificity. Biochem. Biophys. Res. Commun. *184*:752-759.
6. **Hoyer, D., D.E Clarke, J.R. Fozard, P.R. Hartig, G.R. Martin, E.J. Mylecharane, P.R. Saxena and P.P.A. Humphrey.** 1994. The IUPHAR classification of receptors for 5-Hydroxytryptamine (serotonin). Pharmacol. Rev. *46*:157-204.
7. **Huang, T.F.M. and C.M. Gorman.** 1990. Intervening sequences increase efficiency of RNA 3′ processing and accumulation of cytoplasmic RNA. Nucleic Acids Res. *18*:937-947.
8. **Jackson, R.J., M.T. Howell and A. Kaminski.** 1990. The novel mechanism of initiation of picornavirus RNA translation. Trends Biochem. Sci. *15*:-477-483.
9. **Jang, S.K., H. Krausslich, M.J.H. Nicklin, G.M. Duke, A.C. Palmenberg and E. Wimmer.** 1988. A segment of the 5′ nontranslated region of encephalomyocarditis virus RNA directs internal entry of ribosomes during in vitro translation. J. Virol. *62*:2636-2643.
10. **Kaufman, R.J., M.V. Davies, L.C. Wasley and D. Michnick.** 1991. Improved vectors for stable expression of foreign genes in mammalian cells by use of the untranslated leader sequence from EMC virus. Nucleic Acids Res. *19*:4485-4490.
11. **Kim, D.G., H.M. Kang, S.K. Jang and H.S. Shin.** 1992 Construction of a bifunctional mRNA in the mouse by using the internal ribosome entry site of encephalomyocarditis virus. Mol. Cell Biol. *12*:3636-3643.
12. **Kluxen, F.W. and H. Lubbert.** 1993. Maximal expression of recombinant cDNAs in Cos cells for use in expression cloning. Anal. Biochem. *208*:352-356.
13. **Kozak, M.** 1989. The scanning model for translation: an update. J. Cell Biol. *108*:229-241.
14. **Lefkowitz, R.J., S. Cotecchia, P. Samama and T. Costa.** 1994. Constitutive activity of receptors coupled to guanine nucleotide regulatory proteins. TIPS *14*:303-307.
15. **Morgan, R.A., L. Couture, O. Elroy-Stein, J. Ragheb, B. Moss and W. French Anderson.** 1990. Retroviral vectors containing putative internal ribosome entry sites: development of a polycistronic gene transfer system and applications to human gene therapy. Nucleic Acids Res. *20*:1293-1299.
16. **Munson, P.J. and D. Rodbard.** 1980. LIGAND: a versatile computerized approach for characterization of ligand-binding systems. Anal. Biochem. *107*:220-239.
17. **Sambrook, J., E.F. Fritsch and T. Maniatis.** 1989. Molecular Cloning: A Laboratory Manual, 2nd ed. CSH Laboratory Press, Cold Spring Harbor, NY.
18. **Soriano, P., L. Montgomery, R. Geske and A. Bradley.** 1991. Targeted disruption of the c-src proto-oncogene leads to osteopetrosis in mice. Cell *64*:693-702.
19. **Southern, P.J. and P. Berg.** 1982. Transformation of mammalian cells to antibiotic resistance with a bacterial gene under the control of the SV40 early region promoter. J. Mol. Appl. Genet. *1*:327.
20. **Spengler, D., C. Waeber, C. Pantaloni, F. Holsboer, J. Bockaert, P.H. Seeberg and L. Journot.** 1993. Differential signal transduction by five splice varients of the PACAP receptor. Nature *365*:170-175.
21. **Stables, J., I.J.M. Beresford, S. Arkinstall, S.J. Ireland, D.M. Walsh, P.W. Seale, P. Ward and R.M. Hagan.** 1994. GR138676, a novel peptidic antagonist which is potent at NK3 receptors. Neuropeptides *27*:333-341.
22. **Sugimoto, Y., I. Aksentijevich, M.M. Gottesman and I. Pastan.** 1994. Efficient expression of drug-selectable genes in retroviral vectors under control of an internal ribosome entry site. Bio/Technology *12*:694-698.
23. **Van Sande, J., A. Allgeier, C. Massart, A. Czernilofsky, G. Vassart, J.E. Dummont and C. Maenhaut.** 1993. The human and dog 5-HT_{1D} receptors can both activate and inhibit adenylate cyclase in transfected cells. Eur. J. Pharmacol. *247*:177-184.
24. **Weinshank, R.L., J.M. Zgombick, M. Macchi, N. Adham, H. Lichtblau, T.A. Branchek and P. Hartig.** 1990. Human serotonin 1D receptor is encoded by a subfamily of two distinct genes: 5-$HT_{1D\alpha}$ and 5-$HT_{1D\beta}$. Mol. Pharmacol. *38*:681-688.

Update to: Bicistronic Vector for the Creation of Stable Mammalian Cell Lines that Predisposes All Antibiotic-Resistant Cells to Express Recombinant Protein

Stephen Rees

Glaxo Wellcome, Stevenage, Herts, England, UK

INTERNAL RIBOSOME ENTRY SITE EXPRESSION SYSTEMS

Introduction

Expression vectors containing internal ribosome entry site (IRES) elements are in widespread use in mammalian gene expression studies. A cursory literature search will generate several hundred recent examples of the development and use of expression vectors containing such elements. Our original application of IRES elements was to create a bicistronic expression vector to facilitate the construction of stable mammalian cell lines (25). Within the expression vector pCIN, the gene of interest and the neomycin phosphotransferase selection marker were expressed on a single messenger RNA (mRNA) with the translation of the neomycin phosphotransferase open reading frame being mediated through internal ribosome entry at the encephalomyocarditis virus (EMCV) IRES. Through the use of the vector pCIN for the creation of stable mammalian cell lines, we observed that the vast majority, and often all, of the neomycin-resistant colonies expressed the gene of interest (25). This was in contrast to previous studies using so-called all-in-one vectors, in which the gene of interest and the selection marker are expressed on different expression cassettes within the same plasmid, in which we rarely observed greater than 20% of the antibiotic-resistant colonies expressed the gene of interest. Moreover, the use of the pCIN vector appeared to allow for the identification of cell lines expressing very high levels of recombinant protein (25). Following publication, we have distributed the pCIN vector and other vectors listed in Figure 1, to over 200 laboratories worldwide, where it has been used for the stable expression of many different proteins in a wide range of cell types.

In recent years, IRES elements have been incorporated into bicistronic and tricistronic plasmid and viral expression systems, and such vectors have been used to facilitate gene expression in both mammalian cell lines and transgenic animals. While the majority of such studies utilize the EMCV IRES, such elements from other viruses have been adapted for use in recombinant expression studies. This includes such elements from hepatitis C virus (29), foot and mouth disease virus (FMDV) (5) and poliovirus (10). While this article does not attempt a comprehensive review of a now extensive literature, I have tried to provide the reader with a number of examples of the application and advantages of the use of IRES-containing expression systems.

The Progeny of pCIN

In the years since the publication of the pCIN expression vector, we have developed a range of IRES-containing expression vectors for the stable expression of recombinant proteins in mammalian cells (Figure 1). Vectors have been developed carrying the neomycin, hygromycin, puromycin and zeocin selection markers and with expanded multiple cloning sites to facilitate subcloning. These vectors have been used extensively in our laboratory for the stable expression of many G-protein-coupled receptors including the human ORL-1 receptor (2) and the human corticotrophin releasing hormone receptor (21). In addition, we have successfully used these vectors for the stable expression of heterotrimeric G-protein alpha subunits (28), reporter proteins such as aequorin (28), ion channels (15), integrins, transporters and other proteins. Using each of these vectors, we routinely observe that a high proportion, and often all, of the antibiotic-resistant clones obtained following transfection express the protein of interest (26). These vectors have been used with success in a number of cell lines including Chinese hamster ovary (CHO), human embryonic kidney (HEK) 293, HeLa, NIH3T3, Swiss 3T3, K562, ND7/23, SK-N-MC, SK-N-SH and A549 cells. To date, these vectors have worked in all cell lines tested. A particular application of IRES-containing vectors appears to be the stable expression of protein in cell lines that have been difficult to transfect for the expression of proteins that have proven to be difficult to express. The original pCIN and pCIH vectors, now designated pIRES-neo and pIRES-hyg, are available from Clontech Laboratories (Palo Alto, CA, USA). Also available from Clontech are IRES-expression vectors containing the puromycin or bleomycin selection markers.

Bicistronic Plasmid Expression Systems

A number of other groups have developed similar bicistronic expression plasmids for the stable expression of recombinant protein. Hobbs et al. (11) have developed a vector pEFIRES-P in which the gene of interest and the puromycin selection marker are expressed as a single transcript from the human polypeptide chain elongation factor 1a promoter. The authors report the use of this vector to generate stable cell lines expressing human pyruvate carboxylase and show that following clonal selection all 20 puromycin-resistant cell lines obtained expressed very high levels of functional pyruvate carboxylase (11). In a second example, a bicistronic expression plasmid in which the glutamine synthetase selection marker is translated following internal ribosome entry through the EMCV IRES has been developed. This vector has been used to stably express a soluble form of the interleukin-4 (IL-4) receptor alpha chain in CHO-K1 cells. Within the glutamine synthetase expression system, plasmid amplification is achieved following treatment of transfected cells with the glutamine synthetase inhibitor methionine sulphoximine. Following transfection and amplification, it was demonstrated that the IRES-containing vector was superior to a conventional plasmid vector in terms of both levels of protein expression and amplification efficiency (23).

In addition to the use of bicistronic expression plasmids for the rapid generation of stable mammalian cell lines, such vectors have been developed for a number of further laboratory applications. There are many reports that describe the creation and use of bicistronic vectors in which one of the color variants of the *Aequorea victorea* green fluorescent protein (GFP) is included as the second cistron. Following transfection with such a vector, transfected cells can be identified according to whether they express the intrinsically fluorescent GFP. Such vectors have been used for the enrichment of cell populations expressing the gene of interest by fluorescence-activated cell sorting (FACS) (22). In such studies, the levels of expression of the gene of interest over time can be simply monitored by examining the fluorescence intensity of the cell culture by FACS. We have also used GFP-containing bicistronic vectors to identify cells for patch clamp electrophysiology analysis. In this application, a bicistronic vector containing an ion channel and GFP is transfected into the cell line. Transfected cells are identified by fluorescence microscopy prior to patch clamp studies. The use of the bicistronic vector enables the simple identification of transfected cells for further study. Bicistronic vectors containing GFP have also been used in inducible expression systems (20). Vectors have been developed based upon the tetracycline expression system such that inducible expression of the gene of interest can be monitored according to the fluorescence intensity of the cell population. A number of bicistronic expression plasmids containing different color variants of GFP can also be purchased from Clontech.

The bicistronic expression plasmid pIRES-CD4t has been developed for both FACS and magnetic cell sorting of transfected cells to derive cell populations enriched for the expression of the transfected gene. Within this vector, the gene of interest is coexpressed together with a truncated CD4 marker (9). Transfected cells are identified according to the presence of the CD4 marker on the cell surface. In a further application, a bicistronic vector, pPV, has been developed to facilitate antibody generation as a consequence of plasmid immunization (4). The use of this vector allows the transcription of a single mRNA carrying the epitope of interest, together with a control epitope, human interleukin-5 (IL-5), with which to monitor the generation of an immune response. Following direct intramuscular injection of the plasmid, the immunological response can be monitored according to the generation of antibodies to the IL-5 control epitope. The authors present the use of this vector to successfully generate antibodies to human interleukin-1b. In other studies, a bicistronic vector has been developed to mediate intracellular peptide production (6).

Bicistronic vectors have been used to simultaneously express two proteins of interest. We have expressed the P2X2 and P2X3 subunits of an ATP-gated cation channel in a vector in which translation of the P2X3 subunit was mediated through the EMCV IRES (15). While we have used this approach successfully in both stable and transient expression studies, the use of such a vector for the generation of a stable cell line requires that the antibiotic selection marker be expressed on a second cotransfected plasmid. As a consequence, this approach results in a low percentage of antibiotic-resistant cell lines expressing the proteins of interest.

There are few studies characterizing the parameters influencing translational efficiency in IRES-based bicistronic expression vectors. We have made a number of preliminary observations that indicate that translation of the first cistron on the mRNA, mediated through standard ribosome scanning, is more efficient than translation of the second, or indeed a subsequent cistron, through internal ribosome entry. As expected, the presence of additional start codons or sequences with the capacity to form hairpins immediately downstream of the 3′ border of the IRES strongly reduces the translation efficiency of a downstream cistron. This has been demonstrated for the FMDV IRES (5), and we have made preliminary observations indicating that the same is true at the EMCV IRES. In this study, it was shown that an increase of distance between the FMDV IRES and the start codon of the downstream cistron did not have an effect on translation efficiency (5).

We have observed using the EMCV IRES that we generate cell lines with vastly different levels of expression dependent upon the position of the neomycin phosphotransferase start codon with respect to the upstream IRES element (26). Within EMCV, the major site of initiation of the viral polyprotein is AUG11 (13). In the original pCIN vector, and most of the IRES vectors listed in Figure 1, AUG11 was deleted, and the start codon of the antibiotic selection marker was displaced 35 bp downstream (25). The high levels of expression obtained with pCIN are believed to be due to the selection of genome integration sites that permit high levels of gene expression when using a vector in which the antibiotic selection marker is expressed less efficiently. We have since generated two novel vectors, designated pCIN5 and pCIN9 (Figure 1), in which the neomycin phosphotransferase initiation codon is in the precise position with respect to the upstream IRES element, as EMCV AUG11. These vectors were constructed in order to maximize the translation efficiency, and hence level of expression, of the antibiotic selection marker. As the level of expression of the selection marker would be higher, we rationalized that the use of pCIN5 or pCIN9 would allow the selection for sites of plasmid integration of lower transcriptional efficiency. The use of these vectors would thus allow the isolation of antibiotic-resistant cell lines expressing lower levels of recombinant protein than had been obtained through the use of pCIN. In unpublished studies, we have demonstrated this to be the case. Stable expression of the 5-$HT_{1D\alpha}$ serotonin receptor in pCIN in HEK 293 cells generated clonal cell lines with receptor expression levels of between 1.5 and 11.9 pmol/mg (25). When these studies were repeated with the same receptor expressed in the same cell line in pCIN5, we obtained a number of clonal isolates with levels of expression of between 0.02 and 0.11 pmol/mg. We have since observed this phenomena with several proteins and routinely use pCIN5 or pCIN9 in instances where we do not require over-expression of the transfected gene.

Bicistronic Viral Expression Systems

There are in excess of 30 reports describing the development and characterization of polycistronic viral expression systems. The majority of these reports describe the use of such vectors to simultaneously express two proteins. For example, Qiao and coworkers created a recombinant adenoviral vector to co-express the p40 and p35 subunits of human IL-12 in human tumor cell lines (24). Viral bicistronic expression systems have been used to transfect immortalized cell lines in culture, primary cell lines and following injection a number of different cell types in vivo. Bicistronic viral expression vectors have been used for the efficient transfection of cell lines that are refractory to transfection with more traditional vector systems. For example, a bicistronic retroviral expression vector has been used to overcome the low efficiency of retroviral transfection into immature T cells (12). As examples of the use of these elements in viral expression systems, I shall describe their application to retroviral, adenoviral and human immunodeficiency virus (HIV)-based expression systems.

A number of bicistronic retroviral vectors have been described. Julius and coworkers have developed a series of such vectors, designated Q vectors, for applications in which high viral titers and high level stable or transient gene expres-

Vector	Sites
Neomycin selection	
pCIN	BamHI - BstXI - intron - NotI - EcoRI
pCIN3	BamHI - EcoRI - NotI - EcoRV - BstXI
pCIN4	EcoRV - NotI - EcoRI - BamHI - BstXI
pCIN8	BamHI - BstXI - intron - - NsiI - NotI - NheI - SgrAI - AgeI - AscI - EcoRI
Optimized Neomycin Expression	
pCIN5	BamHI - BstXI - intron - NotI - EcoRI
pCIN9	BamHI - BstXI - intron - NsiI - NotI - NheI - SgrAI - AgeI - AscI - EcoRI
Hygromycin selection	
pCIH	BamHI - BstXI - intron - NotI
pCIH5	SmaI - NotI - NheI - AgeI - EcoRV - BamHI
pCIH6	BamHI - EcoRV - AgeI - NheI - NotI - SmaI
Zeocin selection	
pCIZ	EcoRV - NotI - EcoRI - BamHI - BstXI
Puromicin selection	
pCIP4	EcoRV - NotI - EcoRI - BamHI - BstXI

Figure 1. Bicistronic mammalian expression vectors derived from pCIN. All vectors contain the pCIN backbone (ColE1 origin of replication and the ampicillin resistance gene for selection in *E. coli*). All vectors have the following basic structure: CMV - MCS - intron - EMCV IRES - resistance marker - polyA. CMV, cytomegalovirus intermediate early promoter; EMCV IRES, encephalomyocarditis virus internal ribosome entry site; MCS, multiple cloning site; pA, bovine growth hormone polyadenylation sequence.

sion is required (14). These vectors contain a 3′ long terminal repeat (LTR) of the self-inactivating class, a 5′ LTR that is a hybrid of the cytomegalovirus (CMV) enhancer and the mouse sarcoma virus promoter, an internal CMV promoter to drive expression of the transduced gene, the neomycin phosphotransferase selection marker, an expanded multiple cloning site and the EMCV IRES element. The vector backbone also contains the simian virus 40 (SV40) origin of replication to promote high copy number replication in cell lines expressing the SV40 large T antigen. As a second example, a single transcript tricistronic retroviral expression vector was developed in which translation of both the second and third cistrons was mediated through the EMCV IRES (19). This vector was characterized using the multidrug resistance, GFP and neomycin phosphotransferase genes as the first, second and third cistrons. The vector was transduced into NIH 3T3 cells, and all three cistrons were expressed.

In a further example, a bicistronic retroviral expression vector was developed to enable the sustained and high level expression of apolipoprotein A-1 (apo A-1) and lethicin-cholesterol acyl transferase (LCAT) in transduced cell lines (7). Two retroviral vectors were generated, one containing LCAT and the neomycin phosphotransfersase resistance marker, and the second containing apo A-1, LCAT and the neomycin phosphotransferase gene. Both the bi- and tricistronic vectors efficiently co-expressed all cistrons when transfected into mouse C2C12 muscle cells or human HEK 293 cells. After 30 days under antibiotic selection, the retroviral vector sequences were retained by the host cells, whereas those of a conventional plasmid vector were lost. Furthermore, transfected C2C12 cells retained the ability for heterologous expression of LCAT and apo A-1 after differentiation into myotubes. In further studies, a bicistronic murine stem cell-based retroviral vector was developed containing the amphotropic receptor (Pit 2) and GFP (18). The use of this vector to transduce bone marrow stromal cells (MSCs) resulted in a gene transfer efficiency of greater than 80% as determined by GFP fluorescence. In this study, both genes within the bicistronic vector were expressed for at least 6 months. The authors speculate that this vector could be used for the development of gene therapy protocols.

IRES elements have been incorporated into adenovirus and adenoassociated virus (AAV) vector systems. Such vectors have been used in both cell culture experiments and in vivo. As an example of the development and use of such vectors, a bicistronic AAV vector was developed in which brain-derived neurotrophic factor (BDNF) and GFP were expressed off the neuron-specific enolase (NSE) promoter. This vector was injected into the rat substantia nigra pars compacta to investigate the influence of BDNF on nigrostriatal dopaminergoc activity (16). Through the use of an IRES element, a reporter gene, such as GFP, can be included in the vector to act as a marker of both the site and the level of expression of the transfected gene in vivo. Within this study, transgene expression persisted for at least 9 months, during which time expression was monitored using the GFP marker.

As a final example of the use of IRES elements in viral vectors, a bicistronic HIV-1-based vector was generated to express a selection marker together with an endoplasmic recticulum-targeted intrabody directed against a subunit of the interleukin-2 (IL-2) receptor (27). The use of this vector allowed the high level expression of the intrabody in approximately 90% of transduced cells. In the T cell line Kit225 and in primary human T cells stably transduced with the intrabody vector, cell surface expression of IL-2a was reduced to low or undetectable levels, and responsiveness to IL-2 was reduced 10-fold. The authors describe the use of this vector system to transfect a number of primary cell types.

Bicistronic Expression in Transgenic Mice

IRES elements have found widespread use in transgenic studies for both gene addition and gene deletion experiments. The inclusion of an IRES/reporter gene cassette, using reporter genes, such as GFP, β-galactosidase (β-gal), or β-geo within a gene addition construct, allows the rapid detection of the site and time of expression of the transgene by simple detection of the reporter protein. Similarly, the inclusion of an IRES/reporter gene cassette within a gene deletion construct allows for the detection of the temporal, site and level of expression of the deleted gene.

One of the first applications of IRES elements was in gene trap experiments in transgenic mice. A construct was developed containing the EMCV IRES upstream of a fusion protein between β-gal and neomycin phosphotransferase (β-geo). As the β-geo reporter/selection marker contains no upstream promoter element, the generation of a neomycin-resistant embryonic stem cell is only possible should the vector integrate into an expressed gene. The use of this vector for gene trap experiments revealed a 15-fold increase in the number of gene trap events in comparison to a conventional vector (3). The characterization of β-gal expression both in embryonic stem cells in culture and, following the use of such cells to create transgenic animals, in vivo suggests that the IRES vector is able to capture a wide range of genes expressed in a variety of tissues and at different developmental stages. It also appears to allow for the trapping of genes expressed at very low levels in embryonic stem cells. In a further application, an exchangeable gene trap vector has been developed in which IRES elements have been used in combination with the Cre/loxP recombination system (1).

Tricistronic Expression

IRES elements have been used in both plasmid and viral expression vectors for the expression of three cistrons from a single mRNA. Within such vectors, there are examples of the use of different or the same IRES element to mediate translation of the second and third cistrons. We and others have developed tricistronic IRES plasmid vectors to permit the rapid generation of stable cell lines expressing three proteins. For example, we have expressed the human Adenosine $A_{2\alpha}$ receptor, the G-protein alpha subunit $G\alpha_S$ and the neomycin phosphotransferase selection marker in CHO cells on a single mRNA with $G\alpha_S$ translation mediated through the FMDV IRES and neomycin phosphotransferase translation through the EMCV IRES. This enabled the creation of stable cell lines over-expressing both the receptor and the G-protein following a single transfection. In such cells, the presence of additional G-protein allowed adenosine receptor agonists to mediate a greater stimulation of the effector enzyme adenylate cyclase compared with cells expressing the receptor alone at a similar receptor density. In a second example, we have co-expressed the P2X2 and P2X3 subunits of an ATP-gated cation channel in both HEK 293 and CHO cells. In this study, a vector was created in which translation of both the P2X3 ion channel and neomycin phosphotransferase was mediated through separate EMCV IRES elements. Following transfection, a high proportion of transfectants expressed the P2X2 and P2X3 heteromeric ion channel (15). In further studies, tricistronic vectors have been generated expressing the blue, green and yellow variants of GFP for the demonstration of triple label detection on the flow cytometer (31). Tricistronic retroviral vectors have also been described (for example see Reference 7). As a final example, Fussenegger and coworkers (8) have developed a tricistronic plasmid vector family, named pTRIDENT, for the simultaneous expression of three genes in mammalian cells. This vector family utilizes the EMCV and poliovirus IRES elements to mediate translation of the second and third cistrons on the mRNA. The vectors have been engineered to contain 3 multiple cloning sites, containing a total of 18 unique restriction endonuclease cleavage sites, to facilitate construction of the tricistronic vector (8).

The creation of a tricistronic expression vector is technically complex due to the limited number of restriction endonuclease sites available for cloning. The construction of such vectors can also be complicated if it is necessary to create optimal junctions between the IRES element and the downstream cistron in order to maximize translation efficiency. An alternative method for the creation of stable cell lines expressing two proteins following a single transfection is to transfect cells with two IRES expression vectors carrying different selection markers, for example pCIN and pCIH (Figure 1). We routinely use this method for the generation of cell lines expressing two proteins. Following selection in both antibiotics, we observe that a large number of doubly resistant colonies express the protein of interest. The functional metabotropic GABA(B) receptor is a heteromer of the $GABA(B)_{R1}$ and $GABA(B)_{R2}$ gene products (30). We found that the most effective way to reconstitute the functional heterodimeric receptor in stable cell lines was to simultaneously cotransfect cells with the $GABA(B)_{R1}$ receptor in pCIN and the $GABA(B)_{R2}$ receptor in pCIH. Following transfection and the application of both neomycin and hygromycin, greater than 90% of cell lines exhibited functional responses to GABA receptor agonists. Using a similar approach, we have reconstituted the functional calcitonin receptor in mammalian cell lines following co-expression of the calcitonin receptor-like receptor and the accessory protein RAMP1 (17).

Concluding Remarks

IRES-containing expression vectors are now in widespread use for recombinant gene expression both in vitro and in vivo. The use of such vectors has greatly simplified a number of experimental procedures including the identification of transfected cells for subsequent experimental analysis, the transfection of both immortalized and primary mammalian cell lines and the generation of transgenic animals. There is increasing evidence that vectors containing such elements may also have application for human gene therapy. In recent years, the phenomena of internal ribosome entry has been observed not just in viral genomes, but also in eukaryotic genomes. The isolation of IRES-like elements from other genomes may, in the coming years, enable the use of IRES-containing expression vectors in other expression systems including yeast, baculovirus and *Xenopus* oocytes.

REFERENCES

1. **Araki, K., T. Imaizumi, T. Sekimoto, K. Yoshinobu, J. Yoshimuta, M. Akizuki, K. Miura, M. Araki and K. Yamamura.** 1999. Exchangeable gene trap using the Cre/mutated lox system. Cell. Mol. Biol. *45*:737-750.
2. **Bevan, N., S. Scott, P.E. Shaw, M.G. Lee, F.H. Marshall and S. Rees.** 1998. Nociceptin activates Elk-1 and Sap1a following expression of the ORL-1 receptor in Chinese hamster ovary cells. Neuroreport *9*:2703-2708.
3. **Bonaldo, P., K. Chowdhury, A. Stoykova, M. Torres and P. Gruss.** 1998. Efficient gene trap screening for novel developmental genes using IRES b-geo vector and in vitro preselection. Exp. Cell Res. *244*:125-136.
4. **Clarke, N.J., P. Hissey, K. Buchan and S. Harris.** 1997. pPV: a novel IRES-containing vector to facilitate plasmid immunisation and antibody response characterisation. Immunotech. *3*:145-153.
5. **de Quinto, S.L. and E. Martinez-Salas.** 1998. Parameters influencing translational efficiency in aphthovirus IRES based bicistronic expression vectors. Gene *217*:51-56.
6. **Duch, M., A. Tolstrup, I. Dalum, T. Jespersen, S. Mouritsen and F.S. Pedersen.** 1999. Functional testing of a bicistronic retroviral vector for intracellular peptide production. BioTechniques *26*:1032-1034.
7. **Fan, L., J.S. Owen and G. Dickson.** 1999. Construction and characterisation of polycistronic retrovirus vectors for sustained and high level co-expression of apolipoprotein A-1 and lethicin-cholesterol acyltransferase. Atherosclerosis *147*:139-145.
8. **Fussenegger, M., X. Mazur and J.E. Bailey.** 1998. pTRIDENT, a novel vector family for tricistronic gene expression in mammalian cells. Biotech. Bioeng. *57*:1-10.
9. **Gaines, P. and D.M. Wojchowski.** 1999. pIRES-CD4t, a dicistronic expression vector for MACS- or FACS-based selection of transfected cells. BioTechniques *26*:683-688.
10. **Hildinger, M., B. Fehse, S. Hegewisch-Becker, J. John, J.R. Rafferty, W. Ostertag and C. Baum.** 1998. Dominant selection of haematopoietic progenitor cells with retroviral MDR1 coexpression vectors. Hum. Gene Ther. *9*:33-42.
11. **Hobbs, S., S. Jitrapakdee and J.C. Wallace.** 1998. Development of a bicistronic vector derived by the human polypeptide chain elongation factor 1a promoter for the creation of stable mammalian cell lines that express very high levels of recombinant proteins. Biochem. Biophys. Res. Comm. *252*:368-372.
12. **Hozumi, K., R. Ohtsuka, D. Suzuki, K. Ando, M. Ito, T. Nishimura, M. Merkenschlager and S. Habu.** 2000. Establishment of efficient reaggregation culture system for gene transfection into immature T cells by retroviral vectors. Immunol. Lett. *71*:61-66.
13. **Jackson, R.J., M.T. Howell and A. Kaminski.** 1990. The novel mechanism of initiation of picornavirus RNA translation. Trends. Biochem. Sci. *15*:477-483.
14. **Julius, M.A., Q. Yan, Z. Zheng and J. Kitajewski.** 2000. Q vectors, bicistronic retroviral vectors for gene transfer. BioTechniques *28*:702-708.
15. **Kawashima, E., D. Estoppey, C. Virginio, D. Fahmi, S. Rees, A. Surprenant and R.A. North.** 1998. A novel and efficient method for the stable expression of heteromeric ion channels in mammalian cells. Receptors Channels *5*:53-60.
16. **Klein, R.L., M.H. Lewis, N. Muzyczka and E.M. Meyer.** 1999. Prevention of 6-hydroxydopamine induced rotational behaviour by BDNF somatic gene transfer. Brain Res. *847*:314-320.
17. **Main, M.J., J. Brown, S. Brown, N.J. Fraser and S.M. Foord.** 1998. The CGRP receptor can couple via pertussis toxin sensitive and insensitive G-proteins. FEBS Lett. *441*:6-10.
18. **Marx, J.C., J.A. Allay, D.A. Persons, S.A. Nooner, P.W. Hargrove, P.F. Kelly, E.F. Vanin and E.M. Horwitz.** 1999. High-efficiency transduction and long term gene expression with a murine stem cell retroviral vector encoding the green fluorescent protein in human marrow stromal cells. Hum. Gene Ther. *10*:1163-1173.
19. **Metz, M.Z., A. Piitchler, K. Kuchler and S.E. Kane.** 1998. Construction and characterisation of single-transcript tricistronic retroviral vectors using two internal ribosome entry sites. Somat. Cell Mol. Genet. *24*:53-69.

20. **Mosser, D.D., A.W. Caron, L. Bourget, P. Jolicoeur and B. Massie.** 1997. Use of a dicistronic expression cassette encoding the green fluorescent protein for the screening and selection of cells expressing inducible gene products. BioTechniques *22*:150-161.
21. **Parsons, S.W.J., S.A. Rhodes, H.E. Connor, S. Rees, J. Brown and H. Giles.** 2000. Use of a dual firefly and Renilla luciferase reporter gene assay to simultaneously determine drug selectivity at human corticotrophin releasing hormone 1 and 2 receptors. Anal. Biochem. *281*:187-192.
22. **Persons, D.A., J.A. Allay, E.R. Allay, R.A. Ashmun, D. Orlic, S.M. Jane, J.M. Cunningham and A.W. Nienhuis.** 1999. Enforced expression of the GATA-2 transcription factor blocks normal haematopoiesis. Blood *93*:488-499.
23. **Pu, H., L.M. Cashion, P.J. Kretschmer and Z. Liu.** 1998. Rapid establishment of high producing cell lines using dicistronic vectors with glutamine synthetase as the selection marker. Mol. Biotechnol. *10*:17-25.
24. **Qiao, J., S.H. Chen, K.B. Pham-Nguyen, J. Mandeli and S.L. Wu.** 1999. Construction and characterisation of a recombinant adenoviral vector expressing human interleukin-12. Cancer Gene Ther. *6*:473-479.
25. **Rees, S., J. Coote, J. Stables, S. Goodson, S. Harris and M.G. Lee.** 1996. Bicistronic vector for the creation of stable mammalian cell lines that predisposes all antibiotic resistant cells to express recombinant protein. BioTechniques *20*:102-110.
26. **Rhodes, A.D., N. Bevan, K. Patel, M. Lee and S. Rees.** 1998. Mammalian expression of transmembrane receptors for pharmaceutical applications. Biochem. Soc. Trans. *26*:699-704.
27. **Richardson, J.H., W. Hofmann, J.G. Sodroski and W.A. Marasco.** 1998. Intrabody-mediated knock-out of the high affinity IL-2 receptor in primary human T cells using a bicistronic lentivirus vector. Gene Ther. *5*:635-644.
28. **Stables, J., A. Green, F. Marshall, N. Fraser, E. Knight, M. Sautel, G. Milligan, M. Lee and S. Rees.** 1997. A bioluminescent assay for agonist activity at potentially any G-protein coupled receptor. Anal. Biochem. *252*:115-126.
29. **Urabe, M., Y. Hasumi, Y. Ogasawara, T. Matsushita, N. Kamoshita, A. Nomoto, P. Colosi, G.J. Kurtzman, K. Tobati and K. Ozawa.** 1997. A novel dicistronic AAV vector using a short IRES segment derived from hepatitis C virus genome. Gene *200*:157-162.
30. **White, J.H., A. Wise, M.J. Main, A. Green, N.J. Fraser, G.H. Disney, A.A. Barnes, P. Emson, S.M. Foord and F.H. Marshall.** 1998. Heterodimerisation is required for the formation of a functional GABA(B) receptor. Nature *396*:679-682.
31. **Zhu, J., M.L. Musco and M.J. Grace.** 1999. Three colour flow cytometry analysis of tricistronic expression of eBFP, eGFP, and eYFP using EMCV-IRES linkages. Cytometry *37*:51-59.

CHAPTER 38

Novel Cloning Method for Recombinant Adenovirus Construction in *Escherichia coli*

David W. Souza and Donna Armentano

Genzyme Corporation, Framingham, MA, USA

(Reprinted from BioTechniques 26:502-508, 1999)

ABSTRACT

pAd$_{vantage}$ is a rapid cloning system for generating recombinant adenoviruses. The system is based on manipulating the full-length adenovirus genome as a stable plasmid in E. coli *using intron-encoded endonucleases. These intron-encoded endonucleases cut their recognition sequences, which range from 15–39 bp, with high specificity. Their unusual long homing sequence makes them rare-cutting and ideal for use as cloning sites. We report how transgenes can easily be cloned directly into the E1 region of an adenoviral plasmid, followed by transfection into a mammalian packaging cell line, to produce homogeneous recombinant viruses without the need for plaque purification.*

INTRODUCTION

Adenoviruses are double-stranded linear DNA viruses with a 36-kb genome. Several features of adenovirus have made them a popular choice as a transgene delivery vehicle for therapeutic applications. They have a broad host range, can easily be propagated and purified (4) and have a reasonable cloning capacity. Up to 7.5 kb of heterologous sequence can substitute for the E3 and E1 regions of adenovirus using complementation in 293 cells (5). Deletions in E4 together with a complementing E4/E1 cell line permits the inclusion of larger inserts up to 10 kb (6).

The most common approach in generating adenovirus recombinants involves in vivo homologous, intramolecular recombination between overlapping Ad sequences in plasmids and restriction-cleaved Ad viral DNA (10). However, problems with this approach include high background due to contaminating parental virus, which necessitates repeated rounds of plaque purification, and the low efficiency of in vivo homologous recombination. Other recombinant adenoviral systems have been developed to overcome some of these technical challenges. Several current methods involve generating recombinant adenoviruses utilizing the rec system in *Escherichia coli*. (1,6). The method described in this report utilizes the homing sequence of the intron-encoded endonuclease I-*Ceu* I as a cloning site for direct transgene insertion into a viral plasmid.

Intron-encoded endonucleases are enzymes that are encoded within a group I intron, which mediates the intron's transposition (9). The intron-encoded endonucleases have an unusual long homing sequence ranging from 15–39 bp, which makes them rare-cutting and ideal to use as cloning sites. The intron encoding I-*Ceu* I is called CeLSU.5 and is found in the large subunit rRNA gene of *Chlamydomonas eugametos* (7,8,11). The encoded protein I-*Ceu* I has a 26-bp homing site 5′-TAACTATAACGGTCCTAAGGTAGCGA-3′ in which we report to be a very good cloning site for the generation of recombinant adenoviruses.

MATERIALS AND METHODS

Plasmid Constructions

The shuttle plasmid pSV-ICEU I was constructed within a pBluescript® SK(+) (Stratagene, La Jolla CA, USA) backbone. The cytomegalovirus (CMV) promoter, multiple cloning sites and the bovine growth hormone (BGH) poly(A) were polymerase chain reaction (PCR)-amplified from pcDNA3 (Invitrogen, Carlsbad, CA, USA) using the primers: 5′SV-, 5′-TGCACGCGTAACTATAACGGTCCTAAGGTAGCGATCGATTGACATTGAT TATTGACTAGTTATTAATAGTAATC-AATTACG-3′ and 3′SV-, 5′-TGCACGCGTCGCTACCTTAGGACCGTTATAGTTAGCTAGCTCCCCAGCATGCCTGCTATTGTCTTCCCAATCCTCCCCCTT-3′ with Vent® DNA Polymerase (New England Biolabs, Beverly MA, USA) at cycle parameters 1× (4 min at 94°C, 2 min at 45°C and 5 min at 72°C) and 25× (1 min at 94°C, 2 min at 55°C and 5 min at 72°C). A 1.2-kb PCR product was purified from a 1% agarose gel slice by a QIAEX II™ Gel Extraction Kit (Qiagen, Chatsworth, CA, USA). The purified 1.2-kb PCR product was

Cloning and Expression Vectors for Gene Function Analysis
Edited by Quinn Lu and Michael P. Weiner

digested with *Mlu*I and cloned into *Bss*HII-digested pBluescript SK(+) to generate pSV-ICEU I. The PCR-amplified portion was sequenced on a Model 377 Sequencer (PE Biosystems, Foster City, CA, USA) to verify sequence integrity. The plasmid pSV2-ICEU I (Figure 1B) is derived from pSV-ICEU I and contains a mutation (TACGTA to TAC<u>CA</u>A) that eliminates the *Sna*BI site in the CMV promoter. This was accomplished by replacing a *Kpn*I-*Nde*I fragment from pSV-ICEU I with a *Kpn*I-*Nde*I fragment from the plasmid called pCMV (Gary White, Genzyme, unpublished), which carries the mutated *Sna*BI site within the CMV promoter.

The viral vector $pAd_{vantage}$ was constructed in a multi-step manner, as depicted in Figure 1A. First, pAd2-ICEU I was generated from its parent vector pAd (D. Souza, unpublished). pAd is a pBr322-backbone-based vector, which contains Ad2 sequence nucleotides (nt) 1–357 and nt 3328–10685. A polylinker containing unique restriction sites supplants the Ad2 E1 region. An oligonucleotide containing the I-*Ceu* I recognition site was cloned into the *Spe*I and *Mlu*I within pAd polylinker to generate pAd2-ICEU I.

Second, a *Sna*BI site was created just outside the inverted terminal repeat (ITR) in pAd2-ICEU I by PCR. The *Sna*BI recognition site was included within the 5′ PCR oligonucleotide that was used to amplify the ITR. The ITR PCR product was cloned into the unique *Eco*RI and *Spe*I sites of pAd-ICEU I, replacing the former ITR to create pAd2-ICEU I-*Sna*BI.

Third, the vector pAdE4/E3-2.9 (Cathy Sookdeo, Genzyme, unpublished), which contains Ad2 sequence from nt 23428 to 3′ ITR with a deletion in E3 from nt 27971–30937,

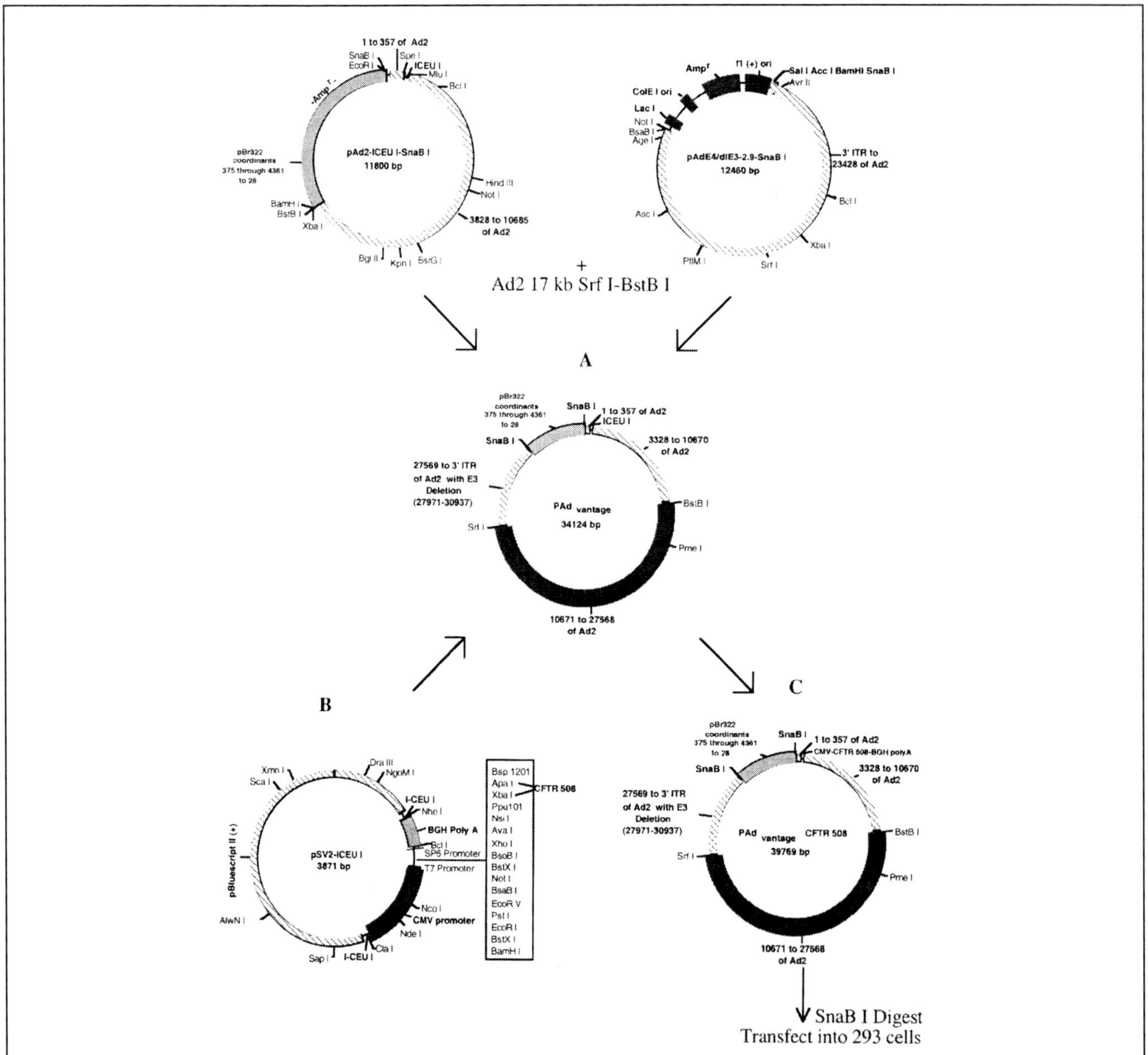

Figure 1. Overview of the $pAd_{vantage}$ system. (A) Adenoviral cloning vector $pAd_{vantage}$. (B) Shuttle vector pSV2-ICEU I. (C) Recombinant adenoviral vector harboring the *CFTR* gene with the ΔF508 mutation.

was modified to create a *Sna*BI site flanking the 3′ ITR by PCR. The ITR was amplified with a 5′ oligonucleotide that introduced the *Sna*BI site. The ITR product was cloned into the unique *Sal*I and *Avr*II sites of pAdE4/E3-2.9, replacing the existing ITR to create pAdE4/E3-2.9 *Sna*BI.

A triple ligation of *Bam*HI-*Bst*BI cut pAd2-ICEU I- *Sna*BI vector with a 5.4-kb *Bam*HI-*Srf*I fragment from pAdE4/E3-2.9 *Sna*BI and a 17-kb *Bst*BI-*Srf*I fragment from Ad2 was executed to generate $pAd_{vantage}$. The ligation was transformed into DH5™ competent cells (Life Technologies, Gaithersburg, MD, USA). Putative recombinant clones were screened by standard mini-prep analysis.

[The $pAd_{vantage}$ vector, the shuttle vector and any precursor vectors will be made available to all interested readers after they fill out a standard material and transfer agreement.]

Preparation of $pAd_{vantage}$ for Cloning

Five micrograms of $pAd_{vantage}$ were digested with a total of 30 U (30 μL) I-*Ceu* I (New England Biolabs) in 500 μL 1× I-*Ceu* I buffer at 37°C over 3 h. Ten units of I-*Ceu* I were added every hour to achieve a total of 30 U. Twenty units (1 μL) alkaline phosphatase (Boehringer Mannheim GmbH, Mannheim, Germany) were then added, and the reaction was incubated at 37°C for an additional 20 min. The DNA was extracted twice with equal volumes of (25:24:1, vol/vol) phenol:chloroform: isoamyl alcohol (Life Technologies) and then precipitated by adding 1/16 vol of 5 M NaCl and 2 vol of 100% ethanol. The DNA pellet was resuspended in 20 μL 1× TE (10 mM Tris-HCl, pH 8.0, 1 mM EDTA). Two microliters (0.25 μg) of vector were used for cloning with threefold molar excess of insert in a total volume of 50 μL of 1× T4 ligation buffer and 1 μL high-concentration T4 ligase (New England Biolabs). The reaction was incubated at room temperature for a minimum of 1 h.

Cell Culture Medium and Reagents

The 293 cell line was purchased from ATCC (Rockville, MD, USA) and was propagated in Dulbecco's modified Eagle medium (DMEM; high glucose) with 10% fetal bovine serum (FBS) and 1% penicillin/streptomycin (Life Technologies).

Viral Packaging in 293 Cells

Twenty micrograms of plasmid DNA were digested with 5 μL (20 U) *Sna*BI in 100 μL 1× NEB No. 4 Buffer (New England Biolabs) at 37°C for 1 h. The DNA was precipitated with 6 μL 5 M NaCl and 200 μL ethanol and collected by microcentrifugation at 14000 rpm in a Model 5415C Centrifuge (Eppendorf, Madison, WI, USA) for 10 min. The 293 cells seeded in a 10-cm plate at 2×10^6 cells per plate were transfected with *Sna*BI-digested DNA using the MBS Mammalian Transfection Kit (Stratagene). Plaques were picked 10 days post-transfection by 10-mm cloning cylinders in a total volume of 200 μL DMEM plus 10% FBS per plaque.

Isolation of Viral DNA

Six-well plates were seeded with the 293 cells at a density of 2.5×10^5 cells per well. The cells were infected the next day with 100 μL putative recombinant viral plaque. When cytopathic effect (CPE) was visible (generally about 3–5 days post-infection), cells and supernatant were harvested. The supernatant was saved as a source of virus, while the cells were pelleted and resuspended in 650 μL Hirt lysis buffer (10 mM Tris-HCl, pH 7.5, 10 mM EDTA, 0.6% sodium dodecyl sulfate [SDS]). One hundred and sixty three microliters of 5 M NaCl were then added, and samples were incubated at -20°C for 10 min. Cellular debris was pelleted at room temperature by microcentrifugation at 14000 rpm as above for 15 min. The lysate was transferred, leaving cellular debris behind, and was incubated at 37°C for 1 h with proteinase K (Life Technologies) at a final concentration of 200 μg/mL. Samples were extracted once with an equal volume of (25:24:1, vol/vol) phenol:chloroform:isoamyl alcohol, and the DNA was precipitated with the addition of 2 vol of 100% ethanol. DNA pellets were resuspended in 50 μL of 1× TE, and 5 μL were used for restriction analysis.

Note: For 10-cm plates, cells were seeded at a density of 1×10^6 per plate, and the reagent amounts were doubled.

RESULTS AND DISCUSSION

Design of Cloning Vectors for Recombinant Adenovirus Construction

The adenoviral vector and shuttle vector were constructed in a stepwise manner (Figure 1, A and B) as described in Materials and Methods. This quick cloning system involves three steps. (*i*) The transgene is cloned into the shuttle vector pSV2-ICEU I. (*ii*) It is then subcloned from pSV2-ICEU I by I-*Ceu* I digestion into the I-*Ceu* I site of the viral vector $pAd_{vantage}$. (*iii*) The $pAd_{vantage}$-based recombinant adenoviral vector is cleaved with *Sna*BI to expose the inverted terminal repeats and is then transfected into 293 cells for generation of virus. Viral plaques are visible within seven to ten days. The total time to generate a recombinant virus is about three weeks.

Generation of Recombinant Adenovirus Ad2 CMV CFTR ΔF508 ΔE3-2.9

To test the efficiency of the system, a cystic fibrosis transmembrane conductance regulator (CFTR) ΔF508 recombinant adenovirus was generated. Since the *CFTR* gene is 4.5 kb in size, it was chosen as a practical virus to make, demonstrating the flexibility of the system to accommodate large inserts. A *CFTR* gene containing the ΔF508 mutation was cloned into the *Xba*I-*Apa*I sites of pSV2-ICEU I as a 4.5-kb *Spe*I-*Apa*I fragment from pCMVCFTR 936c ΔF508 (D. Souza, unpublished). The gene was then subcloned into $pAd_{vantage}$ to generate $pAd_{vantage}$-CFTR 508 (Figure 1C). One out of sixteen putative $pAd_{vantage}$-CFTR 508 recombinant colonies screened were positive for transgene insertion, resulting in a 6% cloning efficiency. The clone appeared to be stable in *E. coli*, as no rearrangements were detected by restriction analysis (data not shown). The low efficiency for this cloning was a result of either a poorly dephosphorylated vector or uncut vector background. The method for preparing the viral vector has been improved since this construct was made. To date, based on five independent experiments, we achieve a 53% cloning efficiency. The method for preparing the viral vector stated in Materials and Methods is current.

DNA from 1 L of a $pAd_{vantage}$- CFTR 508 culture was isolated using the QIAfilter™ Plasmid Mega Kit and QIAGEN®-tip 2500 (Qiagen). Twenty micrograms of DNA were digested with *Sna*BI and then transfected into 293 cells. At day seven, one plaque was picked to infect a single well from a 6-well plate. When most of the cells had rounded and detached from the plate (see Materials and Methods), cells were harvested for Hirt analysis.

Since the virus is cloned, it is not necessary to pick isolated plaques. The whole viral population should be homogeneous, and, therefore, the entire plate can be harvested. However, we have found that we obtain confirmation of viral identity and virus expansion quicker if the first visible plaque is picked and used to infect a small population of cells (6-well seeded at 2.5 $\times$ 10^5 cells per well) rather than wait for the entire plate to develop CPE.

Structure of Ad2CMVCFTR ΔF508 ΔE3-2.9 was confirmed by restriction analysis of Hirt DNA (Figure 2). Lanes 2 and 3 represent parental virus Ad2ΔE1ΔE3-2.9 ($pAd_{vantage}$) and Ad2CMVCFTR ΔF508 ΔE3-2.9 Hirt DNA digested with *Bam*HI, respectively. As observed, fragments at 11.4, 7.8, 6.2 and 4.7 kb were detected for Ad2ΔE1ΔE3-2.9, and additional bands characteristic of CFTR at 2.5, 1.5 and 1.1 kb were detected for Ad2CMVCFTR ΔF508 ΔE3-2.9. Lanes 4 and 5 represent parental virus Ad2ΔE1ΔE3-2.9 and Ad2CMVCFTR ΔF508 ΔE3-2.9 Hirt DNA digested with *Psh*AI, respectively. As observed, fragments at 27 and 3.0 kb were detected for Ad2ΔE1ΔE3-2.9, while fragments at 27 and 8.7 kb were detected for Ad2CMVCFTR ΔF508 ΔE3-2.9. Based on the *Bam*HI and *Psh*AI diagnostic digestions, the CFTR recombinant virus generated appears to contain the correct genomic structure and is free from parental virus contamination.

Expression of Recombinant Adenovirus Ad2CMVCFTR ΔF508 ΔE3-2.9

We next wanted to demonstrate transgene expression from Ad2CMVCFTR ΔF508ΔE3-2.9 generated with the $pAd_{vantage}$ system. For this purpose, 293 cells were infected with Ad2-CMVCFTRΔF508ΔE3-2.9 or Ad2 CFTR-16 at a multiplicity of infection (MOI) of 10. Sodium butyrate was added to the media at a final concentration of 5 mM to enhance transcription and thus facilitate detection of ΔF508 CFTR. Cells were harvested 72 h post-infection to determine CFTR production by in vitro phosphorylation assay as previously described (2). Lanes 3 and 4 in Figure 3 indicate the characteristic band C observed for the mature form of CFTR, while lanes 1 and 2 show only band B form, which is consistent with the ΔF508 mutation (3).

Second Generation $pAd_{vantage}$ Vectors

The system is limited to expression cassettes that do not contain a *Sna*BI. At the time the system was designed, we were concerned with the affect that DNA bases that flank the viral ITR's had on viral DNA replication. Upon consulting the literature, we decided that the first generation viral vector should contain a restriction endonuclease recognition site that leaves the minimum number of bases that flank the viral ITR's after enzyme excision of viral DNA from plasmid backbone to minimize possible inhibition of viral DNA replication. Subsequent viral vectors would address the tolerance of additional flanking DNA bases on viral DNA replication.

We are currently testing whether an 18-bp recognition site for the restriction encoded endonuclease I-*Ppo* I would prove

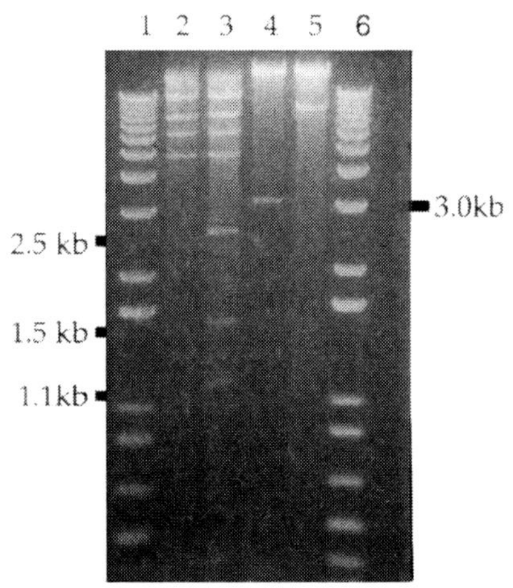

Figure 2. Restriction analysis of recombinant Ad2CMVCFTR ΔF508 ΔE3-2.9 Hirt DNA. Lane 1 is 1-kb plus marker (1 µg; Life Technologies) Lane 2 is Ad2 ΔE1ΔE3-2.9 Hirt DNA *Bam*HI-digested. Lane 3 is Ad2CMVCFTR ΔF508 ΔE3-2.9 Hirt DNA *Bam*HI-digested. Lane 4 is Ad2 ΔE1ΔE3-2.9 Hirt DNA *Psh*AI-digested. Lane 5 is Ad2CMVCFTR ΔF508 ΔE3-2.9 Hirt DNA *Psh*AI-digested. Lane 6 is 1-kb plus marker (1 µg; Life Technologies).

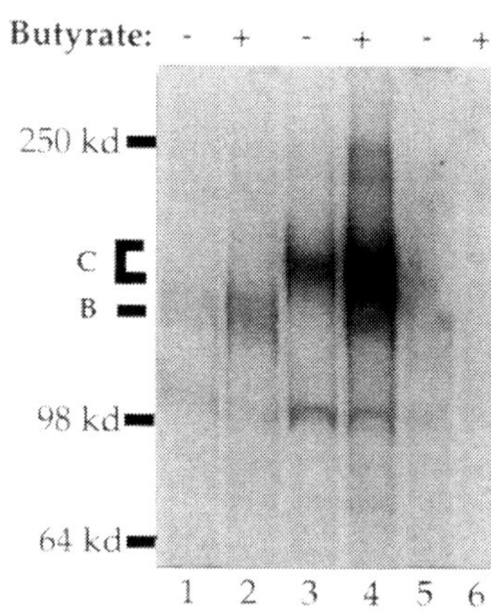

Figure 3. In vitro phosphorylation of CFTR. Lysates were prepared from 293 cells infected with Ad2CMVCFTR ΔF508 ΔE3-2.9 at MOI = 10 (lanes 1 and 2) or with Ad2 CFTR 16 at MOI = 10 (lanes 3 and 4) or mock-infected (lanes 5 and 6). Cells were treated with either 5 mM sodium butyrate or 1× PBS for 72 h before lysis. The catalytic subunit of protein kinase A and [γ-^{32}P]ATP were used to phosphorylate immunoprecipitates obtained using MAb 24-1. kd = kilodalton

fruitful, thus allowing all vectors to be released. With the current $pAd_{vantage}$ vector presented in this manuscript, most expression cassettes can be released. Site-directed mutagenesis can remedy gene cassettes that contain a *Sna*BI site. We are also working on a second-generation $pAd_{vantage}$ vector that will allow for transgene insertion not only in the E1 region but in the E3 and E4 regions as well. We are screening other commercial intron-encoded endonucleases that will serve as effective enzymes for the manipulation of full-length adenovirus genome in *E. coli*.

ACKNOWLEDGMENTS

I would like to thank Gary White and Cathy Sookdeo for providing key plasmids that aided in the construction of the $pAd_{vantage}$ cloning vectors. I would also like to thank Paula Boutin for naming the system and Richard Gregory and Sam Wadsworth for helpful suggestions.

REFERENCES

1. **Chartier, C., E. Degryse, M. Gantzer, A. Dieterle, A. Pavirani and M. Mehtali.** 1996. Efficient generation of recombinant adenovirus vectors by homologous recombination in *Escherichia coli*. J. Virol. *70*:4805-4810.
2. **Cheng, S.H., L.S.L. Fang, J. Zabner, J. Marshall, S. Piraino, S.C. Schiavi, D.M. Jefferson, M.J. Welsh and A.E. Smith.** 1995. Functional activation of the cystic fibrosis trafficking mutant ΔF508-CFTR by overexpression. Am. J. Physiol. *268*:L615-L624.
3. **Cheng, S.H., R.J. Gregory, J. Marshall, S. Paul, D. Souza, G.A. White, C.R. O'Riordan and A.E. Smith.** 1990. Defective intracellular transport and processing of CFTR is the molecular basis of most cystic fibrosis. Cell *63*:827-834.
4. **Graham, F.L., J. Smiley, W.C. Russell and R. Nairn.** 1977. Characteristics of a human cell line transformed by DNA from human adenovirus 5. J. Gen. Virol. *36*:59-72.
5. **Graham, F.L.** 1987. Growth of 293 cells in suspension culture. J. Gen. Virol. *68*:937-940.
6. **He, T.C., S. Zhou, L.T. da Costa, J. Yu, K.W. Kinzler and B. Vogelstein.** 1998. A simplified system for generating recombinant adenoviruses. Proc. Natl. Acad. Sci. USA *95*:2509-2514.
7. **Marshall, P., T.B. Davis and C. Lemieux.** 1994. The I-CeuI endonuclease: purification and potential role in the evolution of *Chlamydomonas* group I introns. Eur. J. Biochem. *220*:855-859.
8. **Marshall, P. and C. Lemieux.** 1992. The I-Ceu I endonuclease recognizes a sequence of 19 base pairs and preferentially cleaves the coding strand of the *Chlamydomonas moewusii* chloroplast large subunit rRNA gene. Nucleic Acids Res. *20*:6401-6407.
9. **Perlman, P.S. and R.A. Butow.** 1989. Mobile introns and intron-encoded proteins. Science *246*:1106-1109.
10. **Stow, N.D.** 1981. Cloning of a DNA fragment from the left-hand terminus of the adenovirus type 2 genome and its use in site-directed mutagenesis. J. Virol. *37*:171-180.
11. **Turmel, M., B. Lemieux and C. Lemieux.** 1988. The chloroplast genome of the green alda *Chlamydomonas moewusii*: localization of protein-coding genes and transcriptionally active regions. Mol. Gen. Genet. *214*:412-419.

Update to: Novel Cloning Method for Recombinant Adenovirus Construction in *Escherichia coli*

David W. Souza and Donna Armentano

Genzyme Corporation, Framingham, MA, USA

Since this report was published, a few modifications and/or additions to the pAdQUICK system (formally known as the pAdvantage system) have been made. First, the ampicillian gene in the shuttle vector pSv2 ICEU I was replaced with the zeocine gene (pSV2 ICEU I ZEO) in an attempt to reduce any potential shuttle vector background arising in the subcloning of the transgene from the shuttle vector into the viral plasmid pAdQUICK (formally pAdvantage). Second, pSV2 ICEU I-II Zeo was created to accommodate transgene cassettes harboring other promoters and poly(A) sites. This shuttle vector contains a polylinker with twelve restriction sites that is flanked by ICEU I sites. Lastly, another viral vector was created to address concerns of retaining E4 in our vectors. pAdQUICK III is E1 and E4 deleted. This plasmid contains a unique I CEU I site for shuttle vector subcloning. To generate virus with this vector, *Sna*BI or I-*Ppo* I restriction digest will liberate Ad2 sequences from pBr322 backbone.

CHAPTER 39

The Admid System: Generation of Recombinant Adenoviruses by Tn7-Mediated Transposition in *E. coli*

Cynthia A. Richards, Chadwick E. Brown, John P. Cogswell and Michael P. Weiner

Glaxo Wellcome, Research Triangle Park, NC, USA

(Reprinted from BioTechniques 29:146-154, 2000)

ABSTRACT

A new system has been developed for generating recombinant adenoviruses by Tn7-mediated transposition in E. coli. *Low copy number* E. coli *plasmids containing a full-length adenoviral genome with* lacZatt*Tn7 replacing E1 have been constructed. The adenovirus plasmid or admid, as well as high copy number progenitors, were stably maintained in* E. coli *strain DH10B. Several transfer vectors containing a mammalian expression cassette flanked by Tn7R and Tn7L were used as donors to transpose the mini-Tn7 into the E1 region of the adenoviral genome. Transposed recombinant admids are readily identified by their β-galactosidase phenotype. Transfection of admid DNA into producer cells resulted in the efficient production of infectious adenovirus. This easy-to-use, efficient system generates pure, clonal stocks of recombinant adenovirus without successive rounds of plaque purification.*

INTRODUCTION

Adenoviruses (Ads) are 36-kb, linear, dsDNA viruses that have been widely used for gene therapy, functional genetics, vaccines and protein production (16,19,20,24,26). Most Ad vectors are based on the Ad type 5 (Ad5) viral backbone in which an expression cassette replaces the E1 and/or E3 region. Such E1$^-$, E3$^-$ viruses are attractive vectors because they: *(i)* can accept about 8 kb exogenous DNA, *(ii)* can be produced at very high titer in cells expressing E1 and *(iii)* are replication defective. Other attractive features of Ad vectors include high-efficiency infection of a wide range of both dividing and quiescent cell types, high-level gene expression and historical association with mild, self-limiting pathologies in humans (4,16,19,20,24,26). However, for gene therapy applications, recent clinical results suggest the toxicity of high-level recombinant Ad infection has been underestimated (21). The increasing numbers of genes discovered by the human genome project and the expanding field of gene therapy are driving the need for more efficient systems to generate recombinant Ad vectors to facilitate rapid, high-throughput functional and therapeutic analysis.

Because of their complex genome, manipulation of Ad vectors to produce recombinant viruses has been difficult. Traditional methods of making recombinant Ads use homologous recombination between two transfected DNAs in Ad-producing cell lines (4). The homologous recombination method is inefficient, time-consuming and subject to contamination with replication-competent virus. Some newer methods of generating recombinant Ad vectors in yeast and *E. coli* have been reported (7,8,14,16). These methods shift the homologous recombination step from eukaryotic packaging cells into either yeast or *E. coli*. Another recent method (25) uses rare-cutting, intron-encoded endonucleases to insert transgenes into a plasmid bearing a full-length Ad genome. These newer methods allow the true cloning of the recombinant viral genome and remove the need for clonal selection with repeated rounds of plaque purification. A method using the Cre-loxP system to generate recombinant Ad by in vitro recombination has also been reported (1). This method does not allow cloning in *E. coli*, but the recombination results in only one molecule that can produce clonal Ads.

A method that has been used to construct recombinant baculoviruses in *E. coli* by Tn7-mediated, site-specific transposition (22) has proven to be very efficient for generating recombinant baculoviruses. We have developed a Tn7-based transposition system for generating recombinant Ads in *E. coli*. A low copy number adenovirus plasmid (admid) containing a full-length Ad genome with *lacZatt*Tn7 replacing E1 has been constructed. The admid system is easy, rapid and efficient, accommodates DNA inserts up to 5.8 kb in the current E1$^-$, E3$^-$ deleted vectors and generates truly clonal viruses. This system is also easily adapted to allow subcloning into a single universal transfer vector that can be used to transpose genes into any of several different expression or viral vector systems, thus facilitating cost-effective subcloning into Ad, baculoviral and *E. coli* vectors (M.W., personal communication).

Cloning and Expression Vectors for Gene Function Analysis
Edited by Quinn Lu and Michael P. Weiner

MATERIALS AND METHODS

Bacterial Strains, Plasmids, Cell Lines and Viral DNA

DH10B and DH10Bac competent cells and pFASTBAC1 were obtained from Life Technologies (Rockville, MD, USA). The plasmid pMON7124 (3) was isolated from DH10Bac cells. BJ5183 (13) was provided by D. Hanahan and pGL3-basic was purchased from Promega (Madison, WI, USA). The tetracycline-regulatable promoter (TetOp) and tetracycline-controlled transactivator (tTA) were obtained from H. Bujard (10). P. Fasani-Ghersa (Serono, Geneva, Switzerland) constructed and provided AdtTA. Other reagents include pFOS1 (18), IN340 viral DNA (15), HEK 293 cells (11) and IMR-90 cells (23).

General Methods

Competent cells were prepared by the $CaCl_2$ method and stored frozen (2). Plasmid minipreps were prepared using Wizard® minipreps (Promega); plasmids larger than 20 kb were eluted from the minicolumns with distilled water heated to 80°C. For large-scale isolation, 500–1000-mL cultures were grown in LB medium, and the admids were isolated using plasmid maxi kits following the procedure recommended for low copy number plasmids (Qiagen, Valencia, CA, USA). LB agar was supplemented with 100 µg/mL ampicillin (Ap), 15 µg/mL tetracycline (Tc), 20 µg/mL chloramphenicol (Cm), 300 µg/mL Bluo-gal (Life Technologies) and 40 µg/mL isopropylthio-β-galactoside (IPTG) as indicated. PCR was performed with AmpliTaq® (PE Biosystems, Foster City, CA, USA) using standard techniques (2).

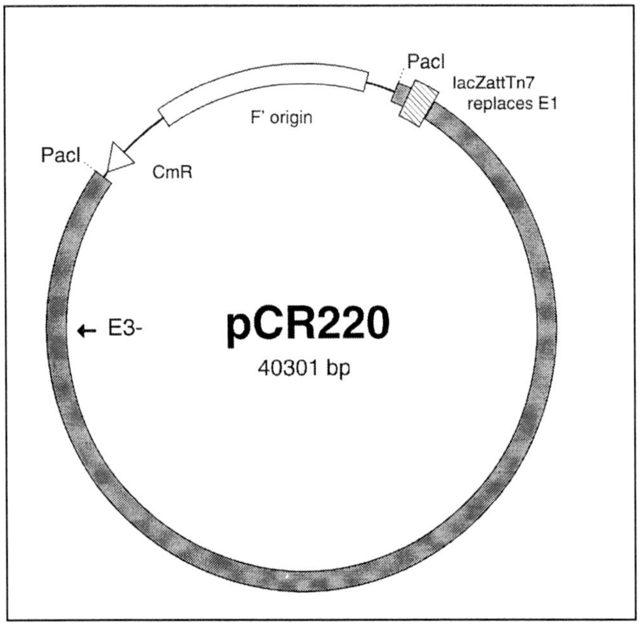

Figure 1. Diagram of admid vector pCR220. The shaded area represents Ad5 sequences and the hatched box indicates the *lacZatt*Tn7, which replaces approximately 3 kb from the Ad5 E1 region. Indicated are the F′ origin of replication (open box) and Cm^R marker (open arrow). An earlier vector, pCR213, has the same Ad genome with a 2.9-kb *Pac*I (left end of Ad sequences), *Swa*I (right end of Ad sequences) fragment containing the ColE1 origin and β-lactamase gene replacing the 6.5-kb *Pac*I fragment of pCR220.

Construction of Admid Homing Vectors and Transfer Vectors

The admid vector was constructed in a multistep procedure from IN340 viral DNA, bMON14272 and pFOS1. The final admid contains an $E1^-$, $E3^-$ Ad genome, the low copy number F′ origin and Cm-resistance gene (Figure 1). The *lacZatt*Tn7 (generated by PCR of DH10Bac DNA with primers priCR1 5′-GGTCGAGCGTCTTCGAAGCGC-3′ and priCR2 5′-CCGTCTTCGAACCAATCAGCAAACC-3′) replaces about 3 kb of the E1A/E1B region of Ad5 (bases 356–3329 deleted; GenBank® Accession No. M73260). The final construct, pCR220, is maintained as a low copy number plasmid in *E. coli*, and produces a β-gal^+ phenotype on media containing Bluo-gal and IPTG. Restriction of pCR220 with *Pac*I produces two linear DNA fragments. One fragment is essentially an $E1^-$, $E3^-$ recombinant Ad genome with *lacZatt*Tn7 replacing E1. The other fragment carries the DNA needed for replication and selection in *E. coli*.

Although functionally deleted for E3, the E3 mutation in pCR220 is derived from *dl*309, in which the deleted Ad5 E3 DNA was replaced with heterologous DNA, which results in little increase in the cloning capacity (5). To accommodate larger insert sizes, it was desirable to remake the E3 mutation as a physical deletion of DNA. This was accomplished using standard DNA cloning techniques to replace the *dl*309-derived E3 mutation with an E3 mutation, which was made by deleting the 1.9-kb *Xba*I fragment from wild-type Ad5 to create pCR249. The Ad genomes of pCR249 and pCR220 are 32.0 kb and 33.8 kb, respectively, allowing insertions up to 5.8 kb in pCR249 and 4.0 kb in pCR220 without exceeding the Ad packaging capacity (6).

Plasmid transfer vectors containing mini-Tn7 elements (Figure 2) were constructed using standard cloning techniques. Further details on the constructions of homing and transfer vectors are available from the authors.

Construction of *E. coli* Strains DH10B (Admid220) and DH10B (Admid249)

DH10B competent cells were transformed with extrachromosomal DNA isolated from DH10Bac and selected on Tc. Tc^R, Ap^S kanamycin-sensitive colonies containing pMON-7124 were isolated. Competent DH10B (pMON7124) cells were transformed with either pCR220 or pCR249. The transformants were plated on LB agar + Tc + Cm + Bluo-gal + IPTG and were designated DH10B (Admid220) and DH10B (Admid249), respectively.

Transposition of Mini-Tn7 Elements from Transfer Vectors into Admids

Transposition was performed by transforming 80 µL competent DH10B (Admid220) with 10 ng uncut transfer vector. The competent cells were mixed with the transfer vector and incubated for 20 min on ice. After a 45-s heat shock at 42°C and a subsequent 2-min incubation on ice, 900 µL SOC media (Life Technologies) were added, and the cells were grown for 4 h at 37°C with aeration. The cultures were then plated on LB agar + Cm + Tc + Ap + Bluo-gal + IPTG and grown for 24 h at 37°C. The β-gal^- transformants were verified as transposition products by PCR (one primer binding to the Ad sequences of the admid, the other primer binding to an sequence on the mini-Tn7 of the transfer vector). Some of the transposed constructs were further verified using restriction enzyme digests after segregating the low copy number recombinant admid from the high copy number pMON7124 helper plasmid and the transfer vector by $CaCl_2$ transformation.

Production and Purification of Ad

One day after plating about 2×10^6 HEK 293 cells per 60-mm dish, the cells were transfected using 2 µg plasmid DNA and 40 µg LIPOFECTAMINE™ (Life Technologies). The DNA/LIPOFECTAMINE complex was incubated with the cells for 6 h in serum-free media before diluting the transfection mixture with an equal volume of media containing 4% bovine calf serum. The cells were harvested after six days, usually before any cytopathic effect was observed. The cells were frozen and thawed three times, the cell debris pelleted and the virus-containing supernatant was titered and amplified.

The virus was titered by infecting confluent plates of HEK 293 cells with varying virus dilutions for 1 h. Eighteen hours later, the cells were fixed with 90% methanol, and the number of infected cells was determined by fluorescent microscopy [primary antibody: mouse Ab to Ad (Biodesign International, Kennebunk, ME, USA); secondary antibody: FITC-labeled goat anti-mouse antibody (ICN, Aurora, OH, USA)]. Amplified virus was extracted with Freon and purified by banding in CsCl gradients (12). The purified virus was dialyzed against 10 mM Tris HCl (pH 8.0), 10 mM $MgCl_2$, 150 mM NaCl, 10% glycerol and stored at -25°C.

Construction of Conventional Control Viruses

The control viruses, Adnull and Ad225, were constructed by the traditional method of homologous recombination in 293 cells (4). The recombinant viruses were identified by PCR analysis of the primary amplification of individual plaques, purified again and then amplified. The viruses were produced and purified as described for the viruses produced using the admid system.

Infection of Cell Lines and Luciferase Assays

IMR-90 cells were plated at 5×10^4 cells/well in 6-well dishes. Two days later, after undergoing two population doublings, triplicate wells were infected with Ad mixtures at the indicated MOI. Each infection contained equal amounts of luciferase virus and either Adnull or AdtTA virus. Two days later, the cells were harvested and assayed for luciferase activity using the luciferase assay system with cell culture lysis reagent as recommended by the supplier (Promega). Luciferase activity was quantitated using a ML3000 luminometer (Dynatech Labs, Chantilly, VA, USA). Where indicated, the cell culture medium was supplemented with Tc starting one day before viral infection.

RESULTS AND DISCUSSION

Construction of Plasmid Vectors

The development of the admid system required the construction of two plasmid components: *(i)* a low copy number admid homing vector containing an Ad genome with a *lacZatt*Tn7 site (Figure 1), and *(ii)* an admid transfer vector with a mini-Tn7 containing an expression cassette appropriate for expression in Ad vectors (Figure 2). The target site for Tn7 insertion, *att*Tn7, is located in the *lac*Z gene. Transposition into *att*Tn7 disrupts the *lac*Z reading frame, resulting in a β-gal⁻ phenotype. The final construct is maintained as a low copy number plasmid in *E. coli* to facilitate screening for the β-gal⁻ phenotype.

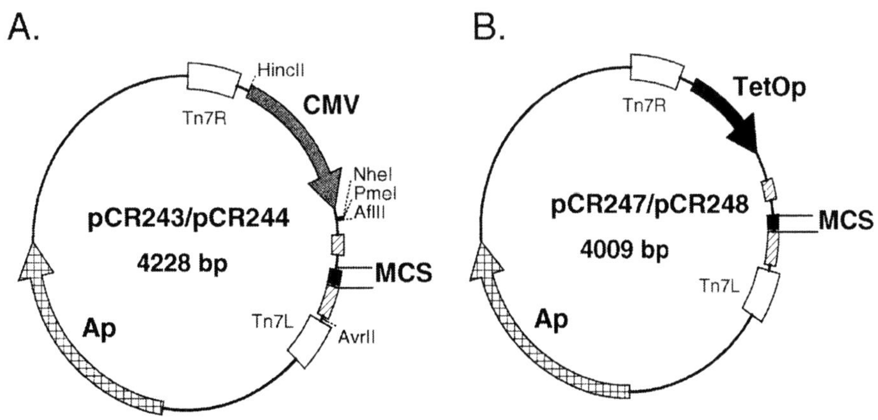

Figure 2. Diagram of transfer vectors. The ColE1 origin, Ap^R marker, Tn7R and Tn7L, are derived from the 3.1-kb *Sac*II, *Bcl*I fragment of pFASTBAC1. (A) The 1.0-kb CMV expression cassette has a multiple cloning site (MCS) between the simian virus 40 (SV40) splice and polyadenylation signals. (B) Vectors with the Tc-regulatable promoter were made by replacing the CMV promoter with the 474-bp *Xho*I, *Xba*I fragment from pUHG10-3 (9). Transfer vectors containing the luciferase reporter gene, pCR219 and pCR232, were made by inserting the 1.7-kb *Hin*dIII, *Xba*I fragment of pGL3-basic into the MCS of the CMV and TetOp vectors, respectively. Legend: Open boxes, left and right ends of Tn7; hatched filled arrow, Ap^R gene; small and large boxes with left slant, the splice site and polyadenylation signal from SV40, respectively; shaded arrow, CMV promoter; solid arrow, Tc-regulated promoter; filled box MCS-pCR243 and pCR247:*Pst*I/*Kpn*I/*Not*I/*Hin*dIII/*Nru*I/*Bam*HI; pCR244 and pCR248: *Pst*I/*Kpn*I/*Bam*HI/*Nru*I/*Hin*dIII/*Not*I. (The *Kpn*I site is not unique in the TetOp vectors because of the presence of a *Kpn*I site in the TetOp promoter.)

Table 1. Yield of Recombinant Ad from 293 Transfection

DNA	Treatment	Location of restriction enzyme site[a]	FFU[b]
pCR220::Tn219	*Pac*I	left, right	5.7×10^7
pCR220::Tn219/ pMON7124/pCR219	*Pac*I	left, right	7.7×10^5
pCR220::Tn232	*Pac*I	left, right	3.4×10^7
pCR220::Tn232/ pMON7124/pCR232	*Pac*I	left, right	8.5×10^5
pCR213	none	none	7.6×10^4
pCR213	*Pac*I	left	6.8×10^4
pCR213	*Swa*I	right	2.7×10^7
pCR213	*Pac*I,*Swa*I	left, right	1.1×10^7

[a]relative to the ends of the Ad genome
[b]60-mm dishes of 293 cells were transfected with 2 μg DNA using 40 μg LIPOFECTAMINE; five days later, monolayers were scraped from the dishes in 4 mL media, the suspension was frozen and thawed three times and titered on 293 monolayers.

The mini-Tn7 transposition cassette within the admid transfer vectors was designed to be as small as possible while providing efficient transposition, screening for transposed admids and transgene expression in mammalian cells. The "bacmid" system described by Luckow et al. (22) incorporates a gentamycin-resistance marker within the mini-Tn7 to aid in the selection of transposed constructs. We have found the antibiotic marker within the Tn7 transposition cassette to be unnecessary; similar transposition efficiencies were obtained whether or not an antibiotic marker within the mini-Tn7 cassette was used to select for the transposed constructs (data not shown). Several mini-Tn7 admid transfer vectors have been constructed (Figure 2). Convenient restriction sites are present in which to insert sequences under the control of the cytomegalovirus (CMV) or Tc-regulatable promoter. These sites may also be used to replace the CMV promoter with any other desired regulatory sequence.

Transposition and Confirmation of Transposition by PCR

The host for transposition was an *E. coli* strain DH10B that contained the plasmid pMON7124 [encoding the *trans*-acting Tn7 transposase genes and Tc-resistance (3)] and an admid (containing the *lacZatt*Tn7 gene). Transposition occurs following the introduction of a transfer vector containing a mini-Tn7 element into the host strain. The transposition and screening for recombinant admids containing a mini-Tn7 expression cassette occurs in a single step that is efficient and easily scored as a β-gal⁻ phenotype on media containing Bluo-gal and IPTG. The percentage of β-gal⁻ colonies obtained appears to vary based on an uncharacterized feature of the competent DH10B admid cells used for the transposition. For example, the nine transpositions performed with the first batch of frozen competent cells were carefully monitored. The transposition efficiency averaged 23%, range 9%–40% β-gal⁻ colonies. Although more recent batches of frozen competent cells have been less efficient, our experience using this system to make over 40 recombinant Ads has been that every transposition experiment has yielded β-gal⁻ colonies. Furthermore, all the β-gal⁻ colonies we have tested represent true transposition products (PCR product using one primer binding to sequences present on the admid and the second primer binding to the transfer vector).

Transfection of HEK 293 Cells and Initial Viral Titers

After clonal isolation of the β-gal⁻ isolates, admid DNA was isolated and restricted with restriction enzyme *Pac*I. Transfection of HEK 293 cells with admid DNA led to efficient production of infectious Ad (Table 1). Higher amounts of virus were produced if the cells were transfected with only the transposed recombinant admid DNA rather than a mixture of DNAs containing admid, pMON7124, and transfer vector. This likely reflects the increased molar amount of admid DNA present in 2 μg pure admid DNA versus the lesser molar amount of admid DNA in 2 μg mixed DNA (a low copy number admid and two high copy number plasmids).

Results obtained with pCR213, a predecessor of pCR220 that contains a *Pac*I site at the left end and a *Swa*I site at the right end of the Ad genome (see Figure 1 legend), demonstrated that the virus was produced even if the admid DNA was not restricted before transfecting 293 cells. However, approximately 100-fold more virus was obtained from admid DNA that was restricted at either the right or both ends of the adenoviral genome compared to undigested admid DNA or DNA restricted only at the left end of their Ad genome (Table 1). Therefore, in most cases, it will be beneficial to digest the admid with *Pac*I before transfection. However, in those rare cases when the expression cassette has a recognition site for *Pac*I, the digestion can be omitted if an additional round of amplification is performed before large-scale virus production.

Titers and Expression from Admid Vectors

We have also compared the yield of infectious virus and expression capabilities of E1⁻, E3⁻ Ads generated by conventional and admid methods. Both the traditional method of homologous recombination in 293 cells and the admid method were used to construct viruses expressing luciferase. The CMV-luciferase expression cassette is identical between Ad225 and Ad220::Tn219. The only differences between Ad225 and Ad220::Tn219 are the addition of *lac*Z and Tn7 sequences flanking both sides of the CMV-luciferase expression cassette in Ad220::Tn219 (Figure 3A). When equivalent numbers of 293 cells were infected and the virus was purified, similar amounts of virus were obtained [Ad225, 9.2×10^{10} focus-forming units (FFU); Ad220::Tn219, 6.5×10^{10}; Ad249::Tn219, 6.6×10^{10} FFU]. Neither the titer nor the final yield of virus was altered by the *lac*Z or Tn7 sequences of Ad220::Tn219 or the alternative E3 deletion of pCR249.

To test the expression levels obtained from each virus, IMR-90 cells were infected with recombinant Ad, and the

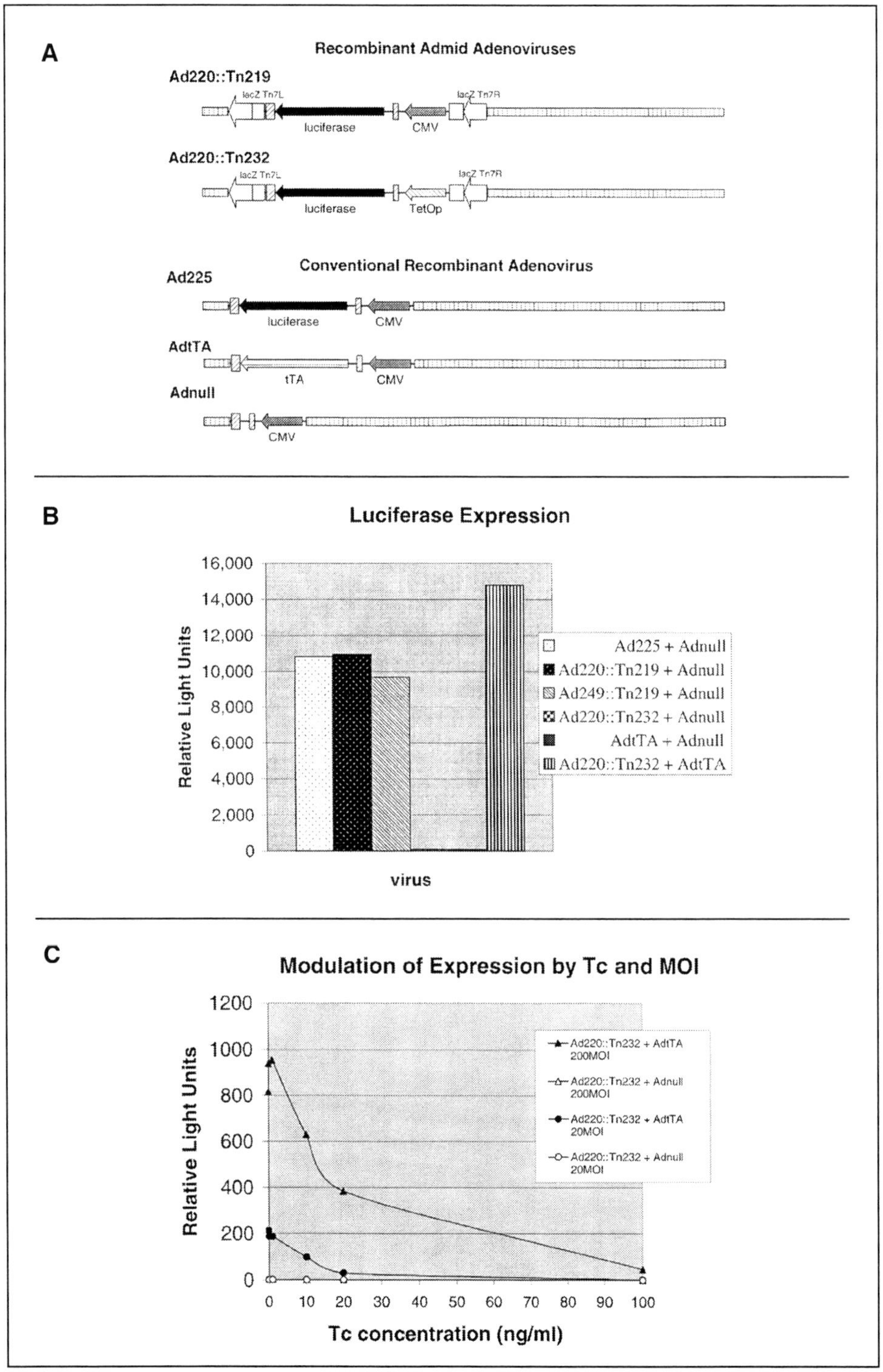

Figure 3. Luciferase expression from recombinant Ads. (A) diagram of recombinant Ads. Viruses that were made by the admid system have *lacZ* and Tn7 sequences (open symbols) flanking the luciferase expression cassette. Hatched boxes: Ad5 sequences; open arrow: *lacZatt*Tn7; open boxes: Tn7L and Tn7R; filled arrow: luciferase coding sequence; shaded arrow: CMV immediate early promoter; hatched arrow: Tc-regulated promoter; boxes with left slant fill: SV40 splice (small box) and SV40 polyadenylation signal (large box). (B). Luciferase expression 48 h after infecting IMR-90 cells at an MOI of 100 (50 MOI luciferase virus + 50 MOI either Adnull or AdtTA virus). (C) Tc-regulated expression 48 h after infecting IMR-90 cells at an MOI of 200 or 20 (half of virus contained luciferase gene and half of virus either Adnull or AdtTA). When indicated, Tc was added one day before viral infection.

luciferase activity was quantitated (Figure 3B). A comparison of the luciferase activity from cells infected with either Ad225 or Ad220::Tn219 shows the same luciferase expression from viruses made by either method. This indicates that the flanking *lacZ* and Tn7 sequences do not affect the expression of the CMV promoter. Essentially equal expression was also obtained from Ad249::Tn219 indicating the alternative E3 deletion also did not affect luciferase expression.

Ads for Efficient Functional Analysis

The ability to modulate the level of expression of an introduced gene over a broad range should facilitate efficient functional analysis experiments by allowing such studies to be carried out at very high, very low or physiological levels of gene expression with one set of viral reagents. Figure 3C demonstrates the broad range of expression that can be obtained using the Tc-regulated gene expression system (10) in Ad vectors. Little to no expression is detected in cells infected with the tetOp-luciferase virus (Ad220::Tn232) in the absence of the AdtTA transactivator virus (open symbols). However, when cells are co-infected with both viruses, a large amount of luciferase activity is observed (filled symbols). Also, adding Tc to the cell culture media allowed for the fine-tuning of the absolute level of luciferase expression.

The admid system could also be adapted to carry out efficient protein expression and gene therapy studies. By moving the promoter from the transfer vector onto the admid in such a way that after transposition the transposed gene is regulated by a promoter flanking the *lacZattTn7*, it should be possible to make one transfer vector that could be transposed into a series of admids that contain different promoters. Making similar adaptations to bacmids would allow the same "universal" transfer vector to be used to make both Ads and baculoviruses. It should also be possible to generate a wide range of viral vectors with attTn7 sites (e.g., herpes, AAV and retroviral vectors). If the appropriate homing vectors (that is, admids, bacmids, etc.) are made, a "universal" transfer vector could further increase the speed and lower the costs of expressing large amounts of functional protein for purification, for in vitro and in vivo functional analysis or to optimize delivery for gene therapy. Such increases in efficiency will be required to analyze the many genes being discovered by the human genome project and for rapid, cost-effective comparisons among viral delivery systems.

ACKNOWLEDGMENTS

We are grateful to D. Hanahan for providing *E. coli* BJ5183, P. Fasani-Ghersa for AdtTA and S. Neill, J. Bisi, L. Babiss and J. Hanvey for many helpful discussions.

REFERENCES

1. **Aoki, K., C. Barker, X. Danthinne, M.J. Imperiale and G.J. Nabel.** 1999. Efficient generation of recombinant adenoviral vectors by Cre-lox recombination in vitro. Mol. Med. *5*:224-231.
2. **Ausubel, F.M., R. Brent, R.E. Kingston, D.D. Moore, J.G. Seidman, J.A. Smith and K. Struhl (Eds.).** 1995. Current Protocols in Molecular Biology. John Wiley & Sons, New York.
3. **Barry, G.F.** 1988. A broad-host-range shuttle system for gene insertion into the chromosomes of gram-negative bacteria. Gene *71*:75-84.
4. **Berkner, K.L.** 1988. Development of adenovirus vectors for the expression of heterologous genes. BioTechniques *6*:616-629.
5. **Bett, A.J., V. Krougliak and F.G. Graham.** 1995. DNA sequence of the deletion/insertion in early region 3 of Ad5 *dl*309. Virus Res. *39*:75-82.
6. **Bett, A.J., L. Prevec and F.L. Graham.** 1993. Packaging capacity and stability of human adenovirus type 5 vectors. J. Virol. *67*:5911-5921.
7. **Chartier, C., E. Degryse, A. Dieterle, A. Pavirani and M. Mehtali.** 1996. Efficient generation of recombinant adeovirus vectors by homologous recombination in *Escherichia coli*. J. Virol. *70*:4805-4810.
8. **Crouzet, J., L. Naudin, C. Orsini, E. Vigne, L. Ferrero, A. Le Roux, P. Benoit, M. Latta et al.** 1997. Recombinational construction in *Escherichia coli* of infectious adenoviral genomes. Proc. Natl. Acad. Sci. USA *94*:1414-1419.
9. **Ghersa, P., R. Hooft van Huijsduijnen, J. Whelan, Y. Cambet, R. Pescini and J.F. DeLamarter.** 1994. Inhibition of E-selectin gene transcription through a cAMP-dependant protein kinase pathway. J. Biol. Chem. *269*:29129-29137.
10. **Gossen, M. and H. Bujard.** 1992. Tight control of gene expression in mammalian celss by tetracycline-responsive promoters. Proc. Natl. Acad. Sci. USA *89*:5547-5551.
11. **Graham, F.L., J. Smiley, W.C. Russell and R. Nairn.** 1977. Characteristics of a human cell line transformed by DNA from human adenovirus type 5. J. Gen. Virol. *36*:59-72.
12. **Greber, U.F., M. Willets, P. Webster and A. Helenius.** 1993. Stepwise dismantling of adenovirus 2 during entry into cells. Cell *75*:477-486.
13. **Hanahan, D.** 1983. Studies on transformation of *Escherichia coli* with plasmids. J. Mol. Biol. *166*:557-580.
14. **He, T.-C., S. Zhou, L.T. da Costa, J. Yu., K.W. Kinzler and B. Vogelstein.** 1998. A simplified system for generating recombinant adenoviruses. Proc. Natl. Acad. Sci. USA *95*:2509-2514.
15. **Hearing, P. and T. Shenk.** 1983. The adenovirus type 5 E1A transcriptional control region contains a duplicated enhancer element. Cell *33*:695-703.
16. **Imler, J.-L.** 1995. Adenovirus vectors as recombinant viral vaccines. Vaccine *13*:1143-1151.
17. **Ketner, G., F. Spencer, S. Tugendreich, C. Connelly and P. Hieter.** 1994. Efficient manipulation of the human adenovirus genome as an infectious yeast artifical chromosome clone. Proc. Natl. Acad. Sci. USA *91*:6186-6190.
18. **Kim, U.-J., H. Shizuya, P.J. deJong, B. Birren and M.L. Simon.** 1992. Stable propagation of cosmid sized human DNA inserts in an F factor based vector. Nucleic Acids Res. *20*:1083-1085.
19. **Kozarsky, K.F. and J.M. Wilson.** 1993. Gene therapy: adenovirus vectors. Curr. Opin. Genet. Dev. *3*:499-503.
20. **Lamarche, N., B. Massie, M. Richer, H. Paradis and Y. Langelier.** 1990. High level expression in 293 cells of the herpes simplex virus type 2 ribonucleotide reductase subunit 2 using an adenovirus vector. J. Gen. Virol. *71*:1785-1792.
21. **Lehrman, S.** 1999. Virus treatment questioned after gene therapy death. Nature *401*:517-518.
22. **Luckow, V.A., S.C. Lee, G.F. Barry and P.O. Olins.** 1993. Efficient generation of infectious recombinant baculoviruses by site-specific transposon-mediated insertion of foreign genes into a baculovirus genome propagated in *Escherichia coli*. J. Virol. *67*:4566-4579.
23. **Nichols, W.W., D.G. Murphy, V.J. Cristofalo, L.H. Toji, A.E. Greene and S.A. Dwight.** 1977. Characterization of a new human diploid cell strain, IMR-90. Science *196*:60-63.
24. **Noel, R.J., P.A. Antinozzi, J.D. McGarry and C.B. Newgard.** 1997. Engineering of glycerol-stimulated insulin secreation in islet beta cells. Differential metabolic fates of glucose and glycerol provide insight into mechanisms of stimulus-secretion coupling. J. Biol. Chem. *272*:18621-18627.
25. **Souza, D.W. and D. Armentano.** 1999. Novel cloning method for recombinant adenovirus construction in *Escherichia coli*. BioTechniques *26*:502-508.
26. **Zhao, H., L. Ivic, J.M. Otaki, M. Hashimoto, K. Mikoshiba and S. Firestein.** 1998. Functional expression of a mammalian odorant receptor. Science *279*:237-242.

Update to: The Admid System: Generation of Recombinant Adenoviruses by Tn7-Mediated Transposition in *E. coli*

Sue H. Kadwell, John E. Bisi, Chadwick E. Brown, John P. Cogswell, Anne-Marie Colapietro, Michael A. Luther, Suzanne D. Neill, Cynthia A. Richards and Michael P. Weiner

Glaxo Wellcome, Research Triangle Park, NC, USA

SECOND GENERATION ADMIDS AND THE DESIGN AND SYNTHESIS OF A UNIVERSAL EXPRESSION SYSTEM

Second Generation Admids

Although first generation E1, E3 deleted adenoviral vectors produced in 293 cells have been very useful, it is possible for a single homologous recombination event between the adenovirus E1 region present in 293 cells and the sequences flanking the E1 deletion in the first generation adenoviral vectors to generate replication competent E1$^+$ adenovirus. Additionally, we have found that some tumor cell lines (e.g., A549, HeLa, and HCT116 cells) are capable of complementing the E1 deletion of first generation vectors, allowing viral DNA replication and late gene expression to occur. Furthermore, the maximum insert size in first generation admids is approximately 6 kb. We have found that E1, E3, E4 deleted second

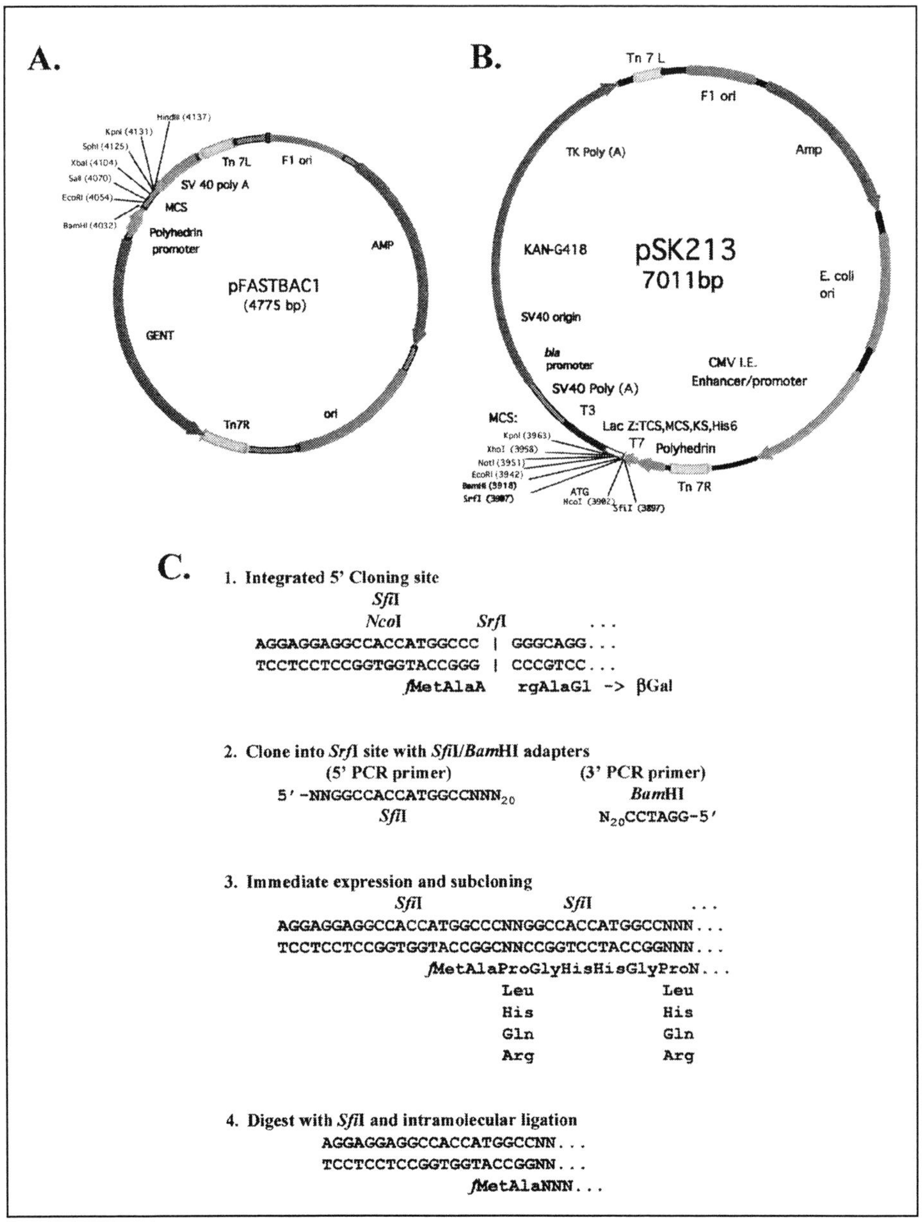

Figure 1. Plasmids and multiple cloning sites. (A) pFastbac1 showing features (13) and (B) pSK213. A series of synthetic oligonucleotides were cloned into pFastBac1 3′ of the baculovirus polyhedrin promoter (P_{bac}), between the *Bam*HI and *Hin*dIII sites. These oligonucleotides incorporated the following relevant regions downstream of the P_{bac} (from 5′ to 3′): T7 promoter (P_{T7}), the *Eco*RI isocaudamer *Esp3*I, *E. coli* ribosome binding site (*E.c.*$_{RBS}$), *Sfi*I, Kozak site, ATG start site, *Nco*I, *Srf*I, *Bam*HI, thrombin cleavage site (TCS), multiple cloning site (MCS), Kemptide, His6, *Hin*dIII. PCR was used to generate a fragment of the *lacZα* gene from plasmid pBK-CMV (Stratagene). The PCR primers used to generate the *lacZα* gene were designed to incorporate a 3′ T3 promoter (P_{T3}) and 3′ T7 transcription termination sequence (T7tt). The generated fragment includes the region between the P_{T7} and 3′ of the SV40 poly(A) site. This PCR fragment was cloned into the *Hin*dIII site of the modified pFastBac1. Site-directed mutagenesis was used to delete a 3′ duplicated poly(A) site in the vector. The sequence between the *Eco*RV and 5′ side of the P_{bac} (encoding the Gt^r gene of pFastBac1) was replaced by a PCR-generated fragment encoding the Kn^r gene of pBK-CMV. The CMV immediate-early enhancer/promoter and a chimeric intron ($P_{CMV.I/E}$) derived from the 5′ donor site from the first intron of the human β-globin gene and the branch and 3′-acceptor site from the intron of an immunoglobulin gene heavy chain variable region were amplified from pCI-Neo and cloned upstream of the Tn7L site as indicated. (C) Universal cloning site allowing precise fusion to a *Sfi*I site after *Srf*I-PCR cloning. See text for details.

generation adenoviral vectors circumvent these limitations of the first generation adenoviral vectors.

Unlike E3, which is completely dispensable for adenovirus growth in culture, E4 is essential for adenovirus replication. Since unregulated expression of the E4 gene products is toxic to cells (15), we constructed cell lines, 293 tTA/E4 and 293 rTA/E4, in which expression of the E4 gene is regulated by the TetOp promoter (5). We also constructed an E4 deleted admid, pCR293, by deleting bases 32918–35604 (based on GenBank® Accession No. M73260). The E4 deletion allows inserts of approximately 9 kb to be cloned into second generation admids.

Virus production curves demonstrated that E1, E3, E4 deleted viruses require cell lines expressing both E1 and E4 to replicate. There is no appreciable production of the E4⁻ viruses in 293 cells. In addition, tumor cell line complementation of viral DNA replication does not occur with E1, E3, E4 deleted viruses. As previously reported, the E4 deleted viruses are produced at reduced efficiencies, yielding titers approximately 10-fold lower than those obtained with first generation E4⁺ viruses (6). We believe the second generation admids to be of value when either the gene of interest is larger than 6 kb or the cells of interest complement viral DNA replication of the first generation adenoviral vectors.

Design of a Universal Expression System

Traditional methods for expression of heterologous proteins involve the subcloning the gene of interest into DNA vectors specific for the desired expression host. This requires either restriction digestion or PCR amplification of the gene, subcloning into an appropriate expression vector, and DNA sequence verification. To reduce the amount of subcloning needed we developed a single "homing" vector system capable of protein expression in three host systems commonly used for heterologous gene expression: *Escherichia coli*, baculovirus, and mammalian tissue culture. The vector system was designed with the following criteria: *(i)* a phenotypic color selection for positive cloning; *(ii)* a multiple cloning site for both restriction enzyme and PCR cloning; *(iii)* an affinity tag and C-terminal fusion; *(iv)* antibiotic selection for both eukaryotic and prokaryotic organisms; and *(v)* proper splice sites and enhancers for mammalian tissue culture expression.

pSK213 vector: general design features and construction. Using the above-mentioned design criteria, the baculoviral expression aspect of our vector system was based on a previously described baculovirus expression platform (12,13). This platform has several components needed to generate recombinant baculovirus in *E. coli* by a Tn7-mediated

```
      (PCMV)          CMV IE              BglII          Tn7 R
3432 TGTGGGCGGA CAATAAAGTC TTAAACTGAA CAAAATAGAT CTAAACTTTG ACAATAAAGT CTTAAACTAG

3502 ACAGAATAGT TGTAAACTGA AATCAGTCCA GTTATGCTGT GAAAAAGCAT ACTGGACTTT TGTTATGGCT

3572 AAAGCAAACT CTTCATTTTC TGAAGTGCAA ATTGCCCGTC GTATTAAAGA GGGGCGTGGC CAAGGGCATG
                                                 SacII           BglII
3642 GTAAAGACTA TATTCGCGGC GTTGTGACAA TTTACCGAAC AACTCCGCGG ATTTAAATAG ATCTTGGAGA

                                PPH
3712 TAATTAAATT GATAACCATC TCGCAAATAA ATAAGTATTT TACTGTTTTC GTAACAGTTT

3772 TGTAATAAAA AAACCTATAA ATATTCCGGA TTATTCATAC CGTCCCACCA TCGGGCGCGG

        SalI                        Plac               SalI    PT7
3832 ATCGTCGACT TGACAATTAA TCATCGGCTC GTATAAT T TGTGG AA GTCGACTAAT ACGACTCACT

                                             RBS   Kozak  fMet
          XbaI         SacI                     SfiI   NcoI   SrfI /SmaI BamHI
3897 ATCGGGATCT AGAAATATCT GAGCTCGTCT CGAATTAGGA GGAGGCCACC ATGGCCC GGGCAGGGGATCC
                                                      >MetAlaA rgAlaGlyAspPr
                                                               >GlyArgGlySer

                     EcoRI   SacII  NotI       KpnI
3967 ATGTACCCGC GTGGCAACGA ATTCTCCGCG GCCGCTCGAG GTACCTTACG TCGTGCTTCT CTGGGCAGAT
     >oCysThrArg ValAlaThrA snSerProAr gProLeuGlu ValProTyrV alValLeuLe uTrpAlaAsp
     >MetTyrProA rgGlyAsnGl uPheSerAla AlaAlaArgG lyThrLeuAr gArgAlaSer LeuGlyArgS
          TCS            MCS                               Kemptide
                      EcoRV
4037 CCAAACGTCG TGGAGCGCTC GATATCCACC ACCATCATCA CCACTAATAA TCTTGGCGCG CCAAGGGTTA
     >ProAsnValV alGluArgSe rIleSerThr ThrIleIleT hrThrAsnAs nLeuGlyAla ProArgValA
     >erLysArgAr gGlyAlaLeu AspIleHisH isHisHisHi sHis•••••••       βgal  (α)
          PT3                      His6

4107 ATTGCGCGCT TATGACCATG ATTACGGATT CACTGGCCGT CGTTTTACAA CGTCGTGACT GGGAAAACCC
     >snCysAlaLe uMetThrMet IleThrAspS erLeuAlaVa lValLeuGln ArgArgAspT rpGluAsnPr

                                                  PvuII
4177 TGGCGTTACC CAACTTAATC GCCTTGCAGC ACATCCCCCT TTCGCCAGCT GGCGTAATAG CGAAGAGGCC
     >oGlyValThr GlnLeuAsnA rgLeuAlaAl aHisProPro PheAlaSerT rpArgAsnSe rGluGluAla

4247 CGCACCGATC GCCCTTCCCA ACAGTTGCGC AGCCTGAATG GCGAATGGAG ATCCAATTTT TAAGTGTATA
     >ArgThrAspA rgProSerGl nGlnLeuArg SerLeuAsnG lyGluTrpAr gSerAsnPhe •••
```

Figure 2. Controlling element, DNA and protein sequence surrounding the start of recombinant protein translation. The start site for protein translation (ATG) is at nucleotide 3947. Controlling elements include the CMV intron/enhancer region (CMV IE), Tn7 right attachment site (Tn7R), polyhedron promoter (P_{PH}), lac promoter (P_{lac}), bacteriophage T7 promoter (P_{T7}), bacteriophage T3 promoter (P_{T3}) and (complete sequence not shown) CMV promoter (P_{CMV}), *E. coli* ribosome binding site (RBS), thrombin cleavage site (TCS), multiple cloning site (MCS), polyhistidine coding sequence (His6), and the first and third translated reading frames corresponding to β-gal and the affinity tag fusion frames, respectively.

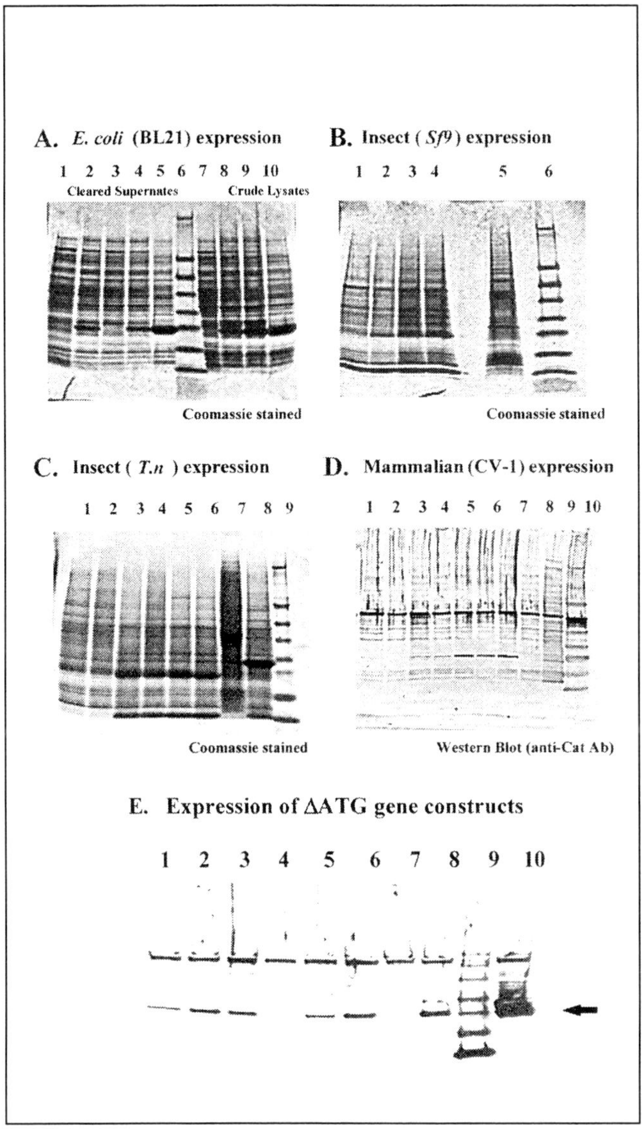

Figure 3. Expression of gene constructs in *E. coli*, insect, and mammalian tissue culture. (A) Expression in *E. coli*. Lanes 1: pSK212, 2: pSK215(tac+ATG), 3: pSK213, 4: pSK215, 5: pRSETcat, 6: molecular weight (mw) markers, 7: pSK212, 8: pSK212Δtac, 9: pSK215, 10: pRSETcat. Lanes 1–5, cleared supernatants, lanes 7–10 crude lysates. (B) Expression in *Sf9* cells. Lanes 1–4 are *Sf9* insect cells infected with 1: bacmidSK212, 2: bacmidSK215, 3: bacmidSK215, 4: bacmidFastbacCat; lanes 5: *Sf9* cells alone, 6: mw markers. Bacmids were generated as described in text. (C) Expression in *T.ni* cells [with culture collection post-infection (PI) time points as indicated]. Lanes 1–9 are *T.ni* insect cells infected with 1, 3, 5: bacmid pSK215, 2, 4, 6: bacmid pFastBacCat, 7: uninfected *T.ni* cells, 8: *T.ni* cells infected with wt baculovirus, 48-h PI, and 9: SeeBlue® (Novex/Invitrogen, Carlsbad, CA, USA) prestained mw standards. Bacmids were generated as described in text. (D) Expression in CV-1 cells. Lanes (changes from pSK215 indicated in parenthesis) 1: pSK215(tac+ATG), 2: pSK213, 3: pSK215, 4: pSK215(ΔP_{tac}), 5: pSK215(ΔTn7R), 6: pSK215(Δ5′intron-*Xba*I), 7: pcDNA3Cat, 8: pCI-Neo, 9: cells alone, 10: mw markers. (E) Expression of ΔATG gene constructs in mammalian tissue culture. Site-directed mutagenesis was used to change ATG DNA sequences to TTG between the CMV promoter and the first ATG of the Cm gene. The ATG of the β-gal gene was changed to GTG, which is able to initiate protein translation in *E. coli*. Lanes (changes from pSK215 indicated in parenthesis) 1: pSK215, 2: pSK215(ΔTn7R), 3: pSK215(ΔIPpoI-*Xba*I), 4: pSK213, vector only, 5: pSK215(Δchimeric intron-Tn7R), 6: pSK215(ΔATGs in Tn7R), 7: mock transfection, 8: pcDNA3Cat, 9: markers, 10: pCINeoCat. Arrow indicates Cm gene product, the upper band seen in all lanes is an artifact of the antibody preparation; it also appears in the mock transfection (lane 7).

transposition event in a much shorter time period than that required by standard homologous recombination transfection and plaque purification methods.

In the functional bacmid system, the donor plasmid is transformed into a strain of *E. coli* previously transformed for both the target and helper plasmids (13). In this *E. coli* host strain, transformed for all three plasmids, successful Tn7 transposition results in the disruption of the *lacZ*::attTn7 gene. The *lacZ*::attTn7 disruption can be differentiated on plates containing either Bluo-gal (Life Technologies, Rockville, MD, USA) or XGal. Transfection of insect cells, such as either *Spodoptera frugiperda* or *Trichoplusia ni*, with composite bacmid DNA generates recombinant, infectious baculovirus capable of producing high-level expression of encoded protein from the polyhedrin promoter (see Figure 3).

Because baculovirus is infectious for some mammalian cell lines (1,8), we decided to place our mammalian expression components (specifically, the CMV promoter and 5′ intron and enhancer regions) outside of the Tn7 transposition cassette (Tn7R; Figure 1). This necessitates driving expression through the Tn7R region. Additionally, it was found necessary to mutate several ATG codons to decrease translation initiating upstream from the desired *f*Met codon, which would inhibit production of the desired gene product. The first Met codon contains a Kozak concensus sequence (10,11)

Cloning and blue-white color selection. We placed a synthetic *tac* promoter 5′ to the T7 promoter. The resultant vector containing tandem *E. coli* and T7 promoters has a blue/white color selection phenotype that is visually comparable to pCRScript (3) on LB + Xgal plating medium. At the ATG (*f*Met) start site, we made allowance for both blunt-ended PCR cloning via a *Srf*I restriction enzyme site (5′-GCCC|CCCG-3′), and restriction enzyme cloning via *Sfi*I (5′-GGCCNNNN|NGGCC-3′) and *Nco*I (5′-CCATGG-3′) sites. The design of the *f*Met start site makes it possible to use the pSK213 vector for blunt-ended PCR cloning and initial gene expression with a 6 amino acid linker, and subsequent intramolecular ligation to generate a precise fusion. The *Sfi*I site has been designed to overlap with both the *Srf*I and *Nco*I sites (5′-GGCCACCATGGCCCGGGC-3′). Expression in *E. coli*, insect, and mammalian cells of this clone would result in a 6 amino acid N-terminal fusion [*f*Met and Ala from the initial ATG codon preceding the *Srf*I site, then either Leu, Pro, His, Gln or Arg (depending on the choice of NN at the 5′ end of the PCR primer), followed by Gly, His, His, Gly, and then either Leu, Pro, His, Gln or Arg, (depending on the choice of NN); it is assumed that the *f*Met preceding the Ala will be removed by formyl-methionine peptidase]. Digestion with restriction enzyme *Sfi*I at the two tandem sites and subsequent intramolecular ligation will result in both the loss of the 7 amino acid linker and a precise fusion. The pSK213 vector has been used for both blunt-ended *Srf*I PCR cloning and restriction enzyme cloning.

C-Terminal fusion peptide. Fusion tags can play a critical role in protein purification. Encoding a tag at the C terminus allowed us to produce a single vector that could be manipulated through subcloning to express a protein with or without fusion to a tag. To encode both a C-terminal affinity tag and β-gal phenotypic color selection further necessitated that the tag be encoded in a *lacZ*-alternative (second) reading frame. This was accomplished by recoding (4) the amino-terminus of the lacZα gene product for functionality in two different (+1 and +2) reading frames. To accomplish that, we *(i)* designed the protein sequence, *(ii)* encoded it using optimal *E. coli* codons, *(iii)* removed all termination codons by replacing specific nucleotides in the wobble positions in the three reading frames, and *(iv)* added two tandem termination codons (TAA) at the end of the C-terminal fusion that were themselves out-of-frame with respect to the *lacZ* gene itself. The affinity tag we chose was the polyhistidine (His6) sequence because of its small size, utility, and widespread use (7, 14). We also chose a thrombin cleavage site for tag removal, and a Kemptide sequence (Leu-Arg-Arg-Ala-Ser-Leu-Gly, 2, 9) for in vitro phosphorylation.

ACKNOWLEDGMENTS

Admid reagents are currently available from QBIOgene (**www.qbi.com**).

REFERENCES

1. **Boyce, F. and N.L. Bucher.** 1996. Baculovirus-mediated gene transfer into mammalian cells. Proc. Natl. Acad. Sci. USA *93*:2348-2352.
2. **Casnellie, J.E.** 1991. Assay of protein kinases using peptides with basic residues for phosphocellulose binding. Methods Enzymol. *200*:115.
3. **Costa, G.L. and M.P. Weiner.** 1994. Protocols for cloning and analysis of PCR-generated DNA fragments. PCR Methods Appl. *3*:S95-S106.
4. **Gesteland, R.F. and J.F. Atkins.** 1996. RECODING: Dynamic reprogramming of translation. Annu. Rev. Biochem. *65*:741-768.
5. **Gossen, M. and H. Bujard.** 1992. Tight control of gene expression in mammalian cells by tetracycline-responsive promoters. Proc. Natl. Acad. Sci. USA *89*:5547-5551.
6. **He, T.-C., S. Zhou, L.T. Da Costa, J. Yu, K.W. Kinzler and B. Vogelstein.** 1998. A simplified system for generating recombinant adenoviruses. Proc. Natl. Acad. Sci. USA *95*:2509-2514.
7. **Hochuli, E., W. Bannwarth, H. Dobeli, R. Gentz and D. Stuber.** 1988. Genetic approaches to facilitate purification of recombinant proteins with a novel metal chelate adsorbant. Biotechnology (NY) *6*:1321-1325.
8. **Hoffman, C., V. Sandig, G. Jennings, M. Rudolph, P. Schlag and M. Strauss.** 1995. Efficient gene transfer into human hepatocytes by baculovirus vectors. Proc. Natl. Acad. Sci. USA *92*:10099-10103.
9. **Kemp, B.E., D.J. Graves, E. Benjamini and E.G. Krebs.** 1977. Role of multiple basic residues in determining the substrate specificity of cyclic AMP-dependent protein kinase. J. Biol. Chem. *252*:4888.
10. **Kozak, M.** 1986. An analysis of 5′ noncoding sequences from 699 vertebrate messenger RNAs. Nucleic Acids Res. *15*:8125-8148.
11. **Kozak, M.** 1995. Adherence to the first-AUG rule when a second AUG codon follows closely upon the first. Proc. Natl. Acad. Sci. USA *92*:2662-2666.
12. **Leusch, M.S., S.C. Lee and P.O. Olins.** 1995. A novel host-vector system for direct selection of recombinant baculoviuses (bacmids) in *Escherichia coli*. Gene *160*:191-194.
13. **Luckow, V.A., S.C. Lee, G.F. Barry and P.O. Olins.** 1993. Efficient generation of infectious recombinant baculoviruses by site-specific, transposon-mediated insertion of foreign genes into a baculovirus genome propagated in *Escherichia coli*. J. Virol. *67*:4566-4579.
14. **Smith, M.C., T.C. Furman, T.D. Ingolia and C. Pidgeon.** 1988. Chelating peptide-immobilized metal ion affinity chromatography. J. Biol. Chem. *263*:7211-7215.
15. **Wang, Q. and M. H. Finer.** 1996. Second-generation adenovirus vectors. Nat. Med. *2*:714-716.

CHAPTER 40

Efficient Gene Transfer to Human Endothelial Cells Using DNA Complexed to Adenovirus Particles

Cora-Jean S. Edgell, David T. Curiel[1], Ping-Chuan Hu and Henry S. Marr

University of North Carolina, Chapel Hill, NC; [1]University of Alabama, Birmingham, AL, USA

(Reprinted from BioTechniques 25:264-272, 1998)

ABSTRACT

Human endothelial cells have been found to be relatively refractory to various methods of DNA transfection currently in common use. By using a transfection method involving DNA complexed with adenovirus particles, we have shown that 20% of a population of cultured endothelial cells can be transfected, and that high levels of transient expression are achieved using replication-deficient virus particles. Both early-passage human umbilical vein endothelial cells and the continuous differentiated line of human endothelium-derived EA.hy926 cells are responsive to this method of transfection. Efficient DNA transfection of endothelial cells is important for studies of endothelium-specific promoters and is a potentially useful route for transgenic therapy.

INTRODUCTION

Human endothelial cells are reportedly difficult to transfect (17,21). For this reason, the mechanisms of endothelium-specific gene expression have been difficult to study (11). The level of transfected reporter gene expression in human umbilical vein endothelial cells (HUVECs) has been reported to be less than 1% of that in HeLa cells (26). In a recent study, 5 different transfection methods were optimized for early-passage HUVECs, including electroporation, lipofection, transferrinfection, DEAE-dextran transfection and calcium phosphate transfection (23). The best HUVEC transfection rate achieved was 0.68% of the population. In the same study, the continuous human endothelium-derived EA.hy926 cell line (7), which maintains an extensively differentiated phenotype (20), was also found to be refractory to transfection, despite its vigorous growth properties. After optimization of 5 transfection methods for EA.hy926 cells, Teifel et al. (23) found that the best transfection efficiency achieved was 2%.

Because endothelial cells serve many important physiological transport functions involving specialized intracellular pathways, incoming macromolecules that do not trigger a specific pathway can be efficiently diverted to lysosomes for degradation. This could explain why endothelial cells are relatively refractory to most methods of transfection. However, DNA incorporated into viruses can avoid lysosomal degradation because viral coat proteins trigger the disruption of endosomes before they combine with lysosomes (22). Retroviruses and adenoviruses have been used to transfer DNA sequences that have been incorporated into the viral genome into many cell types including endothelial cells (8,13,27). However, this approach limits the size of the DNA that can be transferred because it must be packaged within the virus. Furthermore, nucleic acid constructions, viral packaging, amplification in a producer cell type (that might be sensitive to the transgene), purification and titration must be accomplished before transfer to the recipient cell of interest. In addition, it is less feasible to analyze DNA sequences for promoter functions after incorporation into viral genomes because of potential *cis* interactions with viral regulatory elements.

These major limitations of virus-mediated transfection can be avoided. It has been demonstrated that DNA that is not incorporated into the viral genome can nevertheless be transfected into cultured cells relatively efficiently if adenovirus particles are present during exposure of the cells to the DNA (4,9,19). DNA segments of up to 48 kb can be transfected into cells in this way (3). Reporter signals have been shown to increase manyfold when the virus particles are actually complexed with the DNA being transfected. Complexing the negatively charged DNA to adenovirus using polylysine that has been enzymatically conjugated to the virus has been effective in transfection of mouse hepatocytes (24). Adenovirus has also been shown to be effective in conjunction with DEAE-dextran in enhancing transfection of CHO, rat-1 fibroblasts and primary rat osteoblasts (9). Positively charged DEAE-dextran can link negatively charged DNA and virus particles and promote their attachment to negatively charged cell surfaces as well as promote endocytosis. Adenovirus added to plasmid DNA in the presence of LIPOFECTAMINE™ (Life Technologies, Gaithersburg, MD, USA) has been shown to cause the DNA to be incorporated into large electrophoretically stable complexes and to enhance bovine aortic endothelial cell transfection by orders of magnitude compared to DNA with just LIPOFECTAMINE (19). Virus-specific antibody

Cloning and Expression Vectors for Gene Function Analysis
Edited by Quinn Lu and Michael P. Weiner

conjugated to positively charged polylysine has also been used to complex the adenovirus to DNA and shown to be effective in enhancing the transfection of CHO cells (5).

This report demonstrates that differentiated human endothelial cells can be efficiently transfected by complexing the transfecting DNA with adenovirus particles. The relative contributions of the virus component and of the antibody-polylysine complexing reagent to the reporter signal obtained are determined. A linear relationship between DNA concentrations and the transfection signal is demonstrated in the clonally pure EA.hy926 endothelial cells. Early-passage HUVECs and cells of the continuous EA.hy926 line are transfected at similar frequencies. Efficient transfection was accomplished using this method even though human endothelial cells are relatively refractory to other methods of transfection.

MATERIALS AND METHODS

Cells

HUVECs were isolated as described by Jaffe et al. (12) and Gimbrone (10) and cultured on gelatin-coated tissue culture plasticware in EGM®-UV with bovine brain extract (BBE) (Clonetics, San Diego, CA, USA) supplemented with 12% fetal bovine serum (FBS). The cells were positive for von Willebrand antigen (an endothelial cell marker) as determined after the first subpassage using a previously described immunofluorescence assay (7). The HUVECs were used in these experiments prior to 10 population doublings in culture. The A549/8 cell line (1) is a relatively undifferentiated hypoxanthine guanine phosphoribosyltransferase-deficient derivative of the human lung carcinoma cell line A549 (14). The EA.hy926 cell line (7) is a continuous differentiated endothelial cell line derived by hybridization of an HUVEC with an A549/8 cell. EA.hy926 cells were cultured in Dulbecco's modified Eagle medium (DMEM) with high glucose, supplemented with 10% FBS, plus HAT (100 µM hypoxanthine, 0.4 µM aminopterin, 16 µM thymidine) and antibiotics. A549/8 cells were cultured under the same conditions except that the medium was supplemented with 100 µM 6-thioguanine instead of HAT.

Reporter DNAs

The luciferase reporter construct pCLuc4 contains the *Photinus pyralis* luciferase cDNA downstream of the cytomegalovirus (CMV) early promoter/enhancer (18). The β-galactosidase reporter construct pCMVβ contains the *Escherichia coli lacZ* gene downstream of the CMV immediate early enhancerpromoter (15). The plasmid pCLuc4 DNA was purified by cesium chloride banding, and the pCMVβ DNA was purified by passage through a pZ523™ column (5 Prime → 3 Prime, Boulder, CO, USA).

Adenovirus

Two types of adenovirus were used in this work: *(i)* P202-Ad5, which is replication-competent, and *(ii)* P259-Ad5, which is derived from P202-Ad5 by deleting the E1A/E1B gene region to make it replication-incompetent (5). Because E1A and E1B are the first genes expressed by adenovirus, on which all other viral gene expression is dependent, no effect of the viral genome is expected in transfectants involving P259-Ad5. Both these viruses include a useful heterologous epitope in the surface region of the viral hexon protein, which has no effect on infectivity (5).

Antibody

The mouse monoclonal antibody MP301 (5) is specific for a heterologous epitope expressed at the surface of the adenovirus strains P202-Ad5 and P259-Ad5. The antibody was purified from ascites fluid by protein A affinity.

Transfection Complex Formation

In advance, poly(L-lysine) with an average polymer length of 300 residues was conjugated at equal molar ratios to MP301 antibody and separately to transferrin as previously described (2,5,25) and stored at -20°C. Transfection complexes were prepared immediately before use in HEPES-buffered saline (HBS) (20 mM HEPES, pH 7.3 with 150 mM NaCl). For each cm^2 of culture to be transfected, 1.25×10^9 adenovirus particles were combined with 100 ng antibody-poly(L-lysine) in 12.5 µL HBS and incubated for 30 min at room temperature. Then, 300 ng of plasmid DNA in 6.25 µL HBS were added and incubated for 30 min at room temperature. Finally, 300 ng of transferrin-poly(L-lysine) in 6.25 µL HBS were added and incubated for another 30 min at room temperature. For one experiment, 2% (vol/vol) FBS was added to the transfection complex preparation just before it was applied to cell cultures.

Transfection Protocol

Replicate cultures were plated 2 days before transfection experiments at densities that would be subconfluent at the time of transfection. Just before adding the transfection complex, cultures were rinsed in culture medium with 10 mM HEPES (pH 7.3) and 2% FBS. For each cm^2 of culture, 25 µL of transfection complex were added, and the cultures were left at room temperature for 1 h. Then an equal volume of culture medium containing 2% FBS and 10 mM HEPES (pH 7.3) was added, and the culture incubated for another hour at room temperature, after which the usual volume of complete culture medium was added, and the culture was returned to the 37°C CO_2 incubator.

Reporter Gene Assays

The assay used for luciferase is based on that of de Wet et al. (6). One day after transfection, cultures were rinsed well with phosphate-buffered saline and then lysed in luciferase extraction buffer (1% Triton® X-100, 25 mM glycylglycine, pH 7.8, 15 mM $MgSO_4$, 4 mM EGTA, 1 mM dithiothreitol [DTT]) using 50 µL/cm^2 of culture. The extracts were cleared of cellular debris by brief centrifugation and stored at -20°C until assayed. For each assay, up to 100 µL of cytoplasmic extract were added to 400 µL luciferase assay buffer (25 mM glycylglycine, pH 7.8, 15 mM $MgSO_4$, 4 mM EGTA, 15 mM $KHPO_4$, 1 mM DTT, 2 mM ATP). Then 100 µL of substrate solution (0.2 mM D-luciferin, 25 mM glycylglycine, pH 7.8, 15 mM $MgSO_4$, 2 mM DTT) were injected into the assay tube, and the light units generated over a 30-s interval were recorded by a Lumat LB 9501 luminometer (EG&G Berthold, Nashua, NH, USA). A histochemical assay was used for β-galactosidase in fixed cells as described in detail by MacGregor et al. (16).

RESULTS AND CONCLUSIONS

A luciferase reporter gene signal of almost three orders of magnitude above background was obtained when early-passage HUVECs were transfected with DNA complexed with replication-defective adenovirus particles. Figure 1 shows the luciferase signals obtained from cytoplasmic extracts representing 2 cm^2 of endothelial cell culture transfected with the standard amount of pCLuc4, 300 ng/cm^2. An even larger signal is produced by cells of the continuous differentiated EA.hy926 line. The relatively undifferentiated A549/8 cells produce a still greater transfection signal. The variation among three replicate cultures was small for EA.hy926 and A549/8 cells, which represent clonally pure cell lines, but variation among replicate HUVEC cultures was greater as indicated in the figure. Total protein concentrations in the HUVEC lysates from replicate cultures differed from the means by 5% or less and did not correlate with differences in the reporter signals. It is noted that trypsinization to the point of monodispersion is contra-indicated for successful passage of HUVECs, so plate-to-plate variation in culture morphology or "patchiness" due to different distributions of cell aggregates might be expected, and these differences could conceivably affect the efficiency of transfection and/or expression of the transfected DNA.

The reporter signal generated was found to be linearly related to the amount of DNA transfected. Figure 2 shows the reporter signals generated by EA.hy926 cultures transfected with adenovirus complexed with varying amounts of reporter DNA. The reporter signal was proportional to the amount of DNA over multiple orders of magnitude. The signal variation among the EA.hy926 replicate cultures was small at all levels of DNA transfected. In HUVEC transfection experiments (data not shown), the reporter signals were also found to vary with the amount of DNA transfected. However, the signal variation among replicate HUVEC cultures was substantial at all levels of DNA transfected, and the signal-to-DNA profile differed from one HUVEC experiment to another even when consecutive early passages of the same line of HUVECs were used.

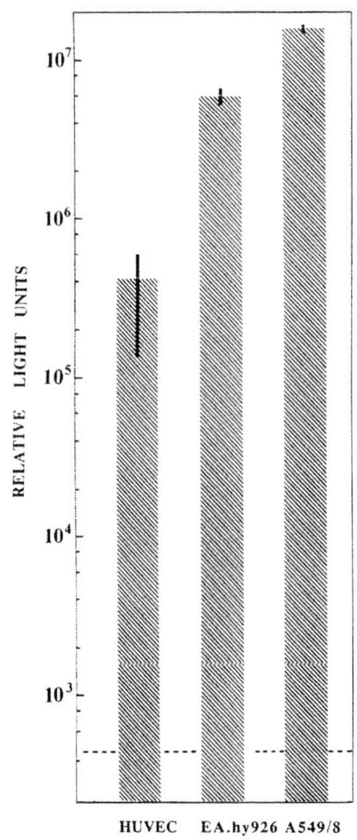

Figure 1. Transfection reporter signals obtained with cultures of differentiated human endothelium-derived and related cells. Cultures of HUVECs, EA.hy926 cells and A549/8 cells were transfected with 300 ng/cm^2 pCLuc4 DNA complexed with replication-defective adenovirus P259A and transferrin-polylysine as described. The cells were extracted after 24 h, and samples representing 2 cm^2 of culture were assayed for luciferase activity. For each cell type, the mean number of light units produced by three replicate cultures and the range are displayed. The dashed line indicates the assay background with no cellular extract.

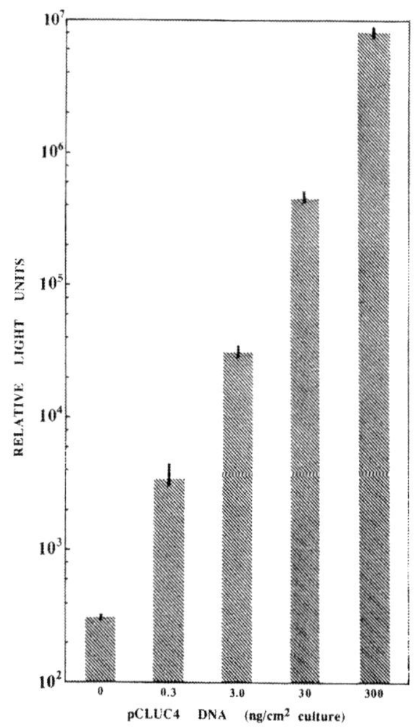

Figure 2. Linear relationship between the amount of DNA transfected and the reporter signal from EA.hy926 cells. Subconfluent cultures were transfected with various amounts of pCLuc4 DNA complexed with the usual amount of adenovirus P202 and transferrin-polylysine as described. The cells were extracted after 24 h and samples representing 2 cm^2 of culture were assayed for luciferase activity. The mean number of light units produced by three replicate cultures and the range are displayed.

The transient expression signals achieved were similar whether replication-competent or incompetent adenoviruses were used in forming the transfection complexes. This is indicated by data for 300 ng/cm^2 shown in Figures 1 and 2. The replication-competent P202 adenovirus was used for the experiment in Figure 2, and the incompetent E1A/E1B-deleted P259A virus was used for the experiment in Figure 1.

The adenovirus component of the transfection complex was the major determinant of gene transfer efficiency into EA.hy926 endothelial cells. The relative contributions of various components of the transfection complex to the efficiency of the transfection were examined by leaving out individual components when preparing the complexes as specified in Materials and Methods. In this experiment, the transfection complexes were applied to EA.hy926 cells at 4°C (instead of room temperature) to delay endocytosis and recycling of receptors. After 3 h at 4°C, the cultures were carefully rinsed using HEPES-buffered DMEM with 2% FBS to remove all of the transfection components not bound to the cell, complete growth medium was added, and the cultures were gradually returned to 37°C. After 24 h, cell extracts were prepared, and luciferase activity was assayed as usual. The reporter signals generated under these transfection conditions are shown in Figure 3. The complete transfection complex with adenovirus P202 resulted in a large signal (although not as large as that obtained when transfection is conducted at room temperature, which would allow recycling of receptors). When the DNA was deleted, the signal was at background levels as expected. When only the virus was deleted, the signal was also at background levels, indicating that this component is very important for transfection efficiency. When the antibody-polylysine component was deleted, only 6% of the full signal was obtained, suggesting that the physical linkage of the DNA to the virus by the antibody bridges contributes substantially to transfection efficiency. When the transferrin-polylysine component was deleted, only 20% of the full signal was obtained. This might indicate that transferrin binding sites and/or polylysine binding sites contributed to the cellular uptake of DNA, but the adenovirus component was required to obtain significant expression of the DNA-encoded reporter. When 2% FBS was present along with the transfection complex, there was no significant difference in the signal obtained, suggesting that serum components (such as transferrin) block few, if any, of the cellular receptors involved in binding the transfection complexes.

A significant fraction of endothelial cells can be transfected by adenovirus-complexed DNA. To determine what fraction of the cells are transfected by adenovirus-DNA complexes, the β-galactosidase reporter gene construct pCMV-β was used to transfect HUVECs and EA.hy926 cells. The cells were transfected using DNA complexed with adenovirus P202 as described above, fixed with paraformaldehyde 24 h after transfection and histochemically stained for β-galactosidase to assess individual cells for reporter gene activity. Figure 4 shows transfected HUVECs stained for β-galactosidase. Some of the cells stained intensely, and other cells were negative. Neither cell size nor local cell density seemed to correlate with β-galactosidase staining. About 3% of the HUVECs and 4% of EA.hy926 cells were found to be β-galactosidase-positive using the same transfection reagents. Thus, approximately the

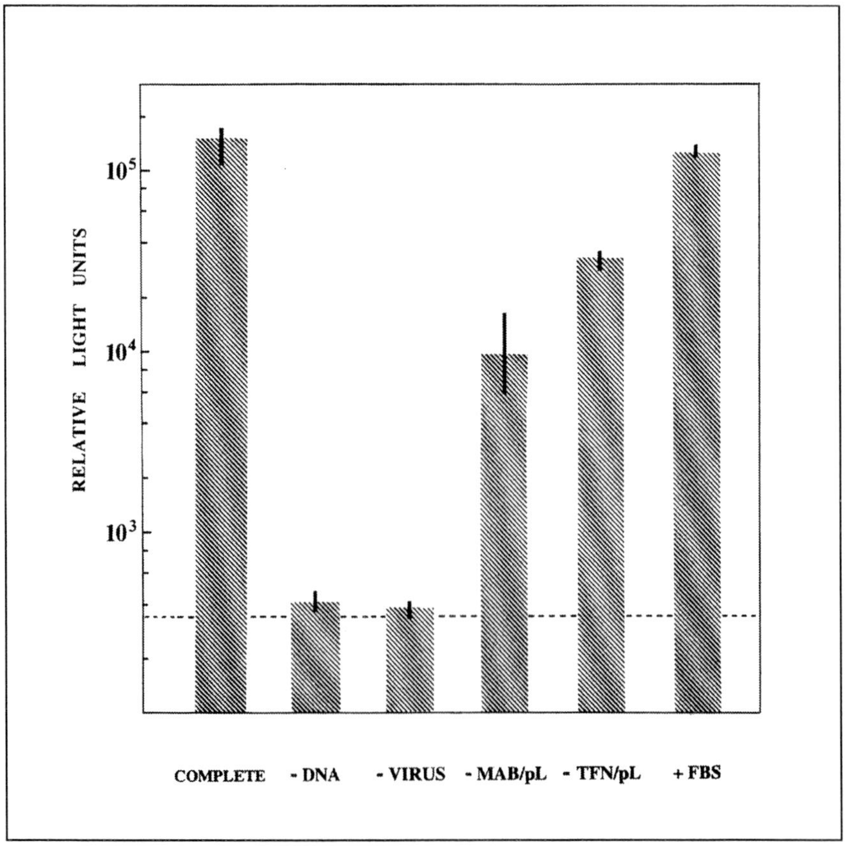

Figure 3. The dependence of transfection reporter signals obtained on various components of the transfection complex. A complete transfection complex involving pCLuc4 DNA and P202 adenovirus was prepared as described in Materials and Methods. Incomplete complexes were prepared identically except that one lacked DNA (- DNA), another lacked virus (- VIRUS), another lacked the monoclonal antibody polylysine conjugate (- Mab/pL) and another lacked the transferrin-polylysine conjugate (- TFN/pL). A complete complex preparation with FBS in addition to the usual components (+ FBS) was also tested. Replicate cultures of subconfluent EA.hy926 cells were exposed to the various transfection complexes at 4°C for 3 h, the unbound transfection complexes were then washed away and cultures were returned to 37°C with growth medium for 24 h. The cells were then extracted, and samples representing 2 cm^2 of culture were assayed for luciferase activity. The mean number of light units produced by three replicate cultures and the range are displayed. The dashed line indicates the assay background with no cellular extract.

Table 1. Frequency of Transfected Endothelial Cells Depends on Virus Particle Concentration

Cells	DNA ng/cm^2	P202 Virus P/cm^2	β-gal-positive after 24 h
HUVEC	30	1.2×10^9	3.3%
HUVEC	30	6.0×10^9	20%
EA.hy926	60	1.2×10^9	4.1%

Table 2. Comparisons of Adenovirus Complexes and LipofectAMINE Transfection Efficiencies

		EA.hy926 (β-gal-positive cells)	A549/8
Adenovirus (P202 at 1.2×10^9 P/cm^2)	24 h	11%	65%
	48 h	38%	57%
LipofectAMINE	24 h	0.1%	1.5%
	48 h	0.2%	1.2%

same proportion of these two cell types can be transfected by adenovirus-attached DNA, despite the fact that some HUVECs produce considerably lower levels of the extractable luciferase reporter compared to EA.hy926 cells (Figure 1). When the amount of virus was increased 5-fold in the β-galactosidase transfection complexes, the fraction of HUVECs that expressed the reporter gene also increased about 5-fold to 20% (Table 1), but considerable cytopathology was noted with the higher levels of this replication-competent virus. These data indicate that the virus concentration in the transfection complex is a limiting factor in transfection efficiency.

Adenovirus-complexed DNA transfection was more efficient than LipofectAMINE transfection for two cell types tested. Transfection of EA.hy926 endothelial cells with adenovirus (P202, at 1.2 P/cm^2) complexes was 100-fold more efficient than transfection with LipofectAMINE used according to the manufacturer's instructions. Replicate cultures of EA.hy926 and of A549/8 cells were transfected side by side using the same concentrations of DNA and exposure times. Table 2 shows that the transfection efficiency was much greater for both cell types using adenovirus complexes.

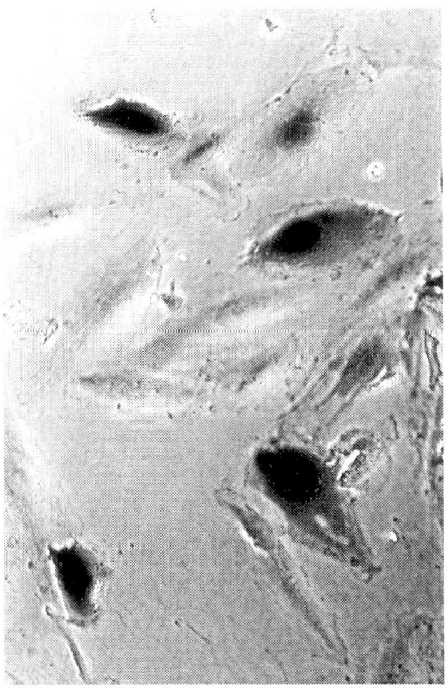

Figure 4. HUVECs transfected with pCMVβ DNA complexed with adenovirus P202 and histochemically stained for β-galactosidase activity after 24 h.

DISCUSSION

Our data indicate that transfection of human endothelium-derived cells is strongly facilitated by complexing the DNA with adenovirus particles. Using this method, a high level of reporter gene expression is achieved, and a much larger fraction of the exposed cells can be shown to express the transgene than that reported with 5 other transfection methods that have been optimized for endothelial cells (23). By complexing the transfecting DNA with adenovirus-antibody-polylysine plus transferrin-polylysine, more than 3% of an HUVEC population were found to express the β-galactosidase reporter (Table 1), and luciferase reporter signals three orders of magnitude above background were obtained (Figure 1). Using this method, even higher transient expression signals were obtained with the well-differentiated continuous human EA.hy926 endothelial cell line that has also been found to be resistant to transfection by other commonly used methods (23). The reporter response was linearly related to the amount of DNA transfected over 5 orders of magnitude (Figure 2). This strong linear response would be very convenient for analysis of endothelium-specific gene expression.

Data in Table 1 and Figure 3 indicate that the adenovirus component contributes prominently to transfection efficiency. Virus particles can escort the complexed reporter gene by cellular receptors into endosomes, which are then disrupted in response to viral coat proteins before combining with lysosomes (22), allowing the DNA to enter the cytoplasm and then the nucleus. We have used polylysine conjugated to a virus-specific monoclonal antibody to link the transfecting DNA to the virus, but other methods of associating them might be more convenient (9,19).

The adenoviral genome is not likely to have had a significant effect on the reporter gene expression in our experiments for several reasons. The viral DNA is not directly linked to the reporter DNA, ruling out any *cis* effect. An adenovirus strain deleted in the E1A/E1B genes (which must be expressed before the other viral genes) was approximately as effective in mediating transfection of EA.hy926 cells as was an adenovirus strain with all genes intact, as shown in Figures 1 and 2. Thus, *trans* effects of adenovirus gene products on expression of the reporter gene are not indicated. Functions of the adenovirus coat proteins in the transfection complex are likely to be responsible for the major effect of viral particles on the efficiency of the transfection method.

Using adenovirus-complexed DNA for transfection, the frequencies of transfected cells were similar for HUVEC and EA.hy926, but the two cell types produced significantly different levels of luciferase reporter signals in their lysates. This difference could indicate that the number of reporter genes that reach the nucleus differs for the two cell types and/or that the capacity to transcribe and translate the reporter gene differs for the two cell types. Because EA.hy926 cells replicate faster than HUVEC cells under common culture conditions, the former might have a greater overall capacity for transcription/translation than the latter.

There is considerable variation in the response of replicate HUVEC cultures to the same transfection complex but little variation in replicate EA.hy926 cultures and replicate A549/-8 cultures. The substantial variability in response of replicate HUVEC cultures might be related to the distribution of cell aggregates during plating of these cells. A greater fraction of the cells in some plates might have been in confluent configurations, even though protein assays indicate that the same total amount of cellular material was delivered to each plate. Cell density is known to affect transfection efficiency for many cell types.

Efficient transfection methods for endothelial cells are important for analyses of tissue-specific promoters by transient expression experiments. There is a great deal to be learned about the regulation of genes that are differentially expressed in endothelial cells, including genes involved in hemostasis, blood pressure regulation, the inflammatory response, wound healing and angiogenesis. The endothelium has also been considered a good potential target for gene therapy because secreted gene products can be introduced directly into the circulation. Transfection efficiency will be important for such applications whether transfection is accomplished in vivo or in culture. In vivo transfection of endothelial cells would have to occur before the transfecting complexes were removed from the circulation by various physiological mechanisms, and transfection in culture would involve a very limited number of homologous endothelial cells. Others have demonstrated that a gene incorporated into the adenovirus genome can be efficiently introduced into endothelial cells in culture and into umbilical vein endothelial cells in situ (13). However, there are several advantages to transfection by adenovirus-DNA complexes compared to infection by an adenovirus with an incorporated transgene. Transfection by adenovirus-polylysine-DNA complexes allows very large DNA sequences to be transferred. Potential *cis* effects of the viral genome on the transferred gene are avoided. Genetic effects of adenovirus on the recipient cells (or organism) can be minimized by using virus with mutationally and/or physically (UV) inactivated viral genomes. There are additional important practical advantages to adenovirus-DNA complex transfection over adenovirus infection for analyses of endothelial cell-specific promoters. DNA promoter/reporter constructs of any design can be used. Co-transfection with control constructs, with competitive regulatory sequences or with genes for *trans*-acting factors can be readily accommodated. Thus, efficient transfection of endothelial cells using adenovirus-DNA complexes represents significant advantages for tissue-specific promoter analyses as well as potential for gene therapy applications.

ACKNOWLEDGMENTS

This work was supported in part by National Institutes of Health Grants Nos. HL45100 and HL55452.

REFERENCES

1. **Carney, D.N., C.J. Edgell, A.F. Gazdar and J.D. Minna.** 1979. Suppression of malignancy in human lung cancer (A549/8) x mouse fibroblast (3T3-4E) somatic cell hybrids. J. Natl. Cancer Inst. *62*:411-415.
2. **Cotten, M., E. Wagner and M.L. Birnstiel.** 1993. Receptor mediated transport of DNA into eucaryotic cells. Methods Enzymol. *217*:618-644.
3. **Cotten, M., E. Wagner, K. Zatloukal, S. Phillips, D.T. Curiel and M.L. Birnstiel.** 1992. High-efficiency receptor-mediated delivery of small and large (48 kilobase) gene constructs using the endosome-disruption activity of defective or chemically inactivated adenovirus particles. Proc. Natl. Acad. Sci. USA *89*:6094-6098.
4. **Curiel, D.T., S. Agarwal, E. Wagner and M. Cotten.** 1991. Adenovirus enhancement of transferrin-polylysine-mediated gene delivery. Proc. Natl. Acad. Sci. USA *88*:8850-8854.
5. **Curiel, D.T, E. Wagner, M. Cotten, M.L. Birnstiel, S. Agarwal, C.M. Li, S. Loechel and P.C. Hu.** 1992. High-efficiency gene transfer mediated by adenovirus coupled to DNA-polylysine complexes. Hum. Gene Ther. *3*:147-154.
6. **de Wet, J.R., K.V. Wood, M. DeLuca, D.R. Helinski and S. Subramani.** 1987. Firefly luciferase gene: structure and expression in mammalian cells. Mol. Cell. Biol. *7*:725-737.
7. **Edgell, C.J.S., C.C. McDonald and J.B. Graham.** 1983. Permanent cell line expressing human factor VIII-related antigen established by hybridization. Proc. Natl. Acad. Sci. USA *80*:3734-3737.
8. **Eglitis, M.A. and W.F. Anderson.** 1988. Retroviral vectors for introduction of genes into mammalian cells. BioTechniques *6*:608-614.
9. **Forsayeth, J.R. and P.D. Garcia.** 1994. Adenovirus-mediated transfection of cultured cells. BioTechniques *17*:354-359.
10. **Gimbrone, M.A.** 1976. Culture of vascular endothelium. Prog. Hemost. Thromb. *3*:1-28.
11. **Ginsburg, D. and E.J.W. Bowie.** 1992. Molecular genetics of von Willebrand disease. Blood *79*:2507-2519.
12. **Jaffe, E.A., R.L. Nachman, C.G. Becker and C.R. Minick.** 1973. Culture of human endothelial cells derived from umbilical veins. J. Clin. Invest. *52*:2745-2756.
13. **Lemarchand, P., H.A. Jaffe, C. Danel, M.C. Cid, H.K. Kleinman, L.D. Stratford-Perricaudet, M. Perricaudet, A. Pavirani, J.P. Lecocq and R.G. Crystal.** 1992. Adenovirus-mediated transfer of a recombinant human α_1-antitrypsin cDNA to human endothelial cells. Proc. Natl. Acad. Sci. USA *89*:6482-6486.
14. **Lieber, M., B. Smith, A. Szakal, W. Nelson-Rees and G. Todaro.** 1976. A continuous tumor-cell line from a human lung carcinoma with properties of type II alveolar epithelial cells. Int. J. Cancer *17*:62-70.
15. **MacGregor, G.R. and C.T. Caskey.** 1989. Construction of plasmids that express *E. coli* β-galactosidase in mammalian cells. Nucleic Acids Res. *17*:2365.
16. **MacGregor, G.R., G.P. Nolan, S. Fiering, M. Roederer and L.A. Herzenberg.** 1989. Use of *Escherichia coli lacZ* (β-galactosidase) as a reporter gene. Methods Mol. Biol. *7*:217-236.
17. **Nathwani, A.C., K.M. Gale, K.D. Pemberton, D.C. Crossman, E.G.D. Tuddenham and J.H. McVey.** 1994. Efficient gene transfer into human umbilical vein endothelial cells allows functional analysis of the human tissue factor gene promoter. Br. J. Haematol. *88*:122-128.
18. **Plank, C., K. Zatloukal, M. Cotten, K. Mechtler and E. Wagner.** 1992. Gene transfer into hepatocytes using asialoglycoprotein receptor mediated endocytosis of DNA complexed with an artificial tetra-antennary galactose ligand. Bioconjug. Chem. *3*:533-539.
19. **Raja-Walia, R., J. Webber, J. Naftilan, G.D. Chapman and A.J. Naftilan.** 1995. Enhancement of liposome-mediated gene transfer into vascular tissue by replication deficient adenovirus. Gene Ther. 2:521-530.
20. **Rieber, A.J., H.S. Marr, M.B. Comer and C.J.S. Edgell.** 1993. Extent of differentiated gene expression in the human endothelium-derived EA.hy926 cell line. Thromb. Haemost. *69*:476-480.
21. **Schwachtgen, J.-L., V. Ferreira, D. Meyer and D. Kerbiriou-Nabias.** 1994. Optimization of the transfection of human endothelial cells by elec-

troporation. BioTechniques *17*:882-887.
22. **Seth, P., D. Fitzgerald, M. Willingham and I. Pastan.** 1986. Pathway of adenovirus entry into cells: some new observation on an old problem, p. 191-195. *In* R.L. Crowell and K. Lonberg-Holm (Eds.), Virus Attachment and Entry into Cells. Am. Soc. Microbiol., Washington DC.
23. **Teifel, M., L.T. Heine, S. Milbredt and P. Friedl.** 1997. Optimization of transfection of human endothelial cells. Endothelium *5*:21-35.
24. **Wagner, E., K. Zatloukal, M. Cotten, H. Kirlappos, K. Mechtler, D.T. Curiel and M.L. Birnstiel.** 1992. Coupling of adenovirus to transferrin-polylysine/DNA complexes greatly enhances receptor-mediated gene delivery and expression of transfected genes. Proc. Natl. Acad. Sci. USA *89*:6099-6103.
25. **Wagner, E., M. Zenke, M. Cotten, H. Beug and M.L. Birnstiel.** 1990. Transferrin-polycation conjugates as carriers for DNA uptake into cells. Proc. Natl. Acad. Sci. USA *87*:3410-3414.
26. **Wilson, D.B., D.M. Dorfman and S.H. Orkin.** 1992. A nonerythroid GATA-binding protein is required for function of the human preproendothelin-1 promoter in endothelial cells. Mol. Cell. Biol. *10*:4854-4862.
27. **Wilson, J.M., L.K. Birinyi, R.N. Salomon, P. Libby, A.D. Callow and R.C. Mulligan.** 1989. Implantation of vascular grafts lined with genetically modified endothelial cells. Science *244*:1344-1346.

Update to: Efficient Gene Transfer to Human Endothelial Cells Using DNA Complexed to Adenovirus Particles

Cora-Jean S. Edgell

University of North Carolina, Chapel Hill, NC, USA

In this report, we demonstrated that genes can be transferred into human endothelial cells relatively efficiently by linking the DNA to adenovirus particles. This method takes advantage of the viral particle as a transfection agent without having to construct a recombinant viral genome. The linked adenovirus particles can be replication-deficient. They are thought to facilitate entry of the complexed DNA into the cell by receptor-mediated endocytosis. The viral coat proteins may then also mediate release of the complex into the cytoplasm prior to endosome-lysosome fusion, so that the associated DNA is less likely to be degraded intracellularly. Although the method we used to link the reporter gene to the viral particles was complex, simpler methods cited in the article (see References 9 and 19 in original paper) are likely to be effective. A more recent report (1) documents the effectiveness of another simple method of linking the DNA of interest to replication-defective adenovirus particles for transfection.

REFERENCES

1. **Diebold, S.S., H. Lehrmann, M. Kursa, E. Wagner, M. Cotten and M. Zenke.** 1999. Efficient gene delivery into human dendritic cells by adenovirus polyethylenimine and mannose polyethylenimine transfection. Hum. Gene Ther. *10*:775-786.

Section VI

In Vitro Protein Expression Technologies

CHAPTER 41

Cell-Free Synthesis and Affinity Isolation of Proteins on a Nanomole Scale

A.P. Alimov, A.Yu. Khmelnitsky, P.N. Simonenko, A.S. Spirin and A.B. Chetverin

Institute of Protein Research, Russian Academy of Sciences, Pushchino, Moscow Region, Russia

(Reprinted from BioTechniques 28:338-344, 2000)

ABSTRACT

The performance of conventional cell-free gene expression systems based on the Escherichia coli *S30 extract can be significantly improved by using expression vectors that encode viral structural elements known to enhance translation in vivo and to protect mRNA from ribonuclease action. The expression vectors reported here are designed to produce a functionally active protein carrying the Strep-tag oligopeptide at its C terminus. They can be used in translation, transcription-translation or replication-translation reactions. Depending on its type, the reaction yields up to 40 μg per mL, or about 1 nmol of a standard protein. The presence of Strep-tag allows the synthesized protein to be easily isolated on a streptavidin-agarose column under mild conditions and the entire procedure to be completed within one working day. The results show that standard low-cost, cell-free systems can serve for rapid preparation of purified proteins in amounts that can satisfy a number of needs of a research laboratory.*

INTRODUCTION

Cell-free translation is a powerful tool for studying gene expression. It would be desirable to use cell-free translation systems as a source of proteins for biochemical and molecular biology studies. The expensive continuous flow cell-free (CFCF) reactors (26,32,33) or dialysis cells (15,16) can produce milligrams of a protein per 1 mL of reaction volume. However, many research laboratories need only smaller amounts. The goal of this study was to improve the performance of a conventional microcentrifuge tube reaction so that a substantial amount of a protein could be rapidly synthesized and isolated at a low cost by using the standard components readily available on the market.

A higher-performance, cell-free system can be achieved by manipulating the system composition (12–14,17). However, such improvements often lead to an increase in the system cost that is comparable with the increase in the protein yield. On the other hand, our previous observations indicated that the performance of a standard cell-free system could be greatly improved solely by manipulating the untranslated regions of mRNA templates. For example, embedding an mRNA into the genomic RNA of Qβ phage in place of its coat protein cistron enhanced the mRNA translation by up to two orders of magnitude (11). However, Qβ RNA is too large (>4.2 kb) and includes other concurrently translated viral cistrons. We decided to approach the goal from another end, by constructing a vector possessing only short essential virus-specific structural elements.

The vectors reported here (Figure 1) are suitable for gene expression in the bacterial cell-free systems and encode the following structural elements of *Escherichia coli* phages: (*i*) the promoter of phage T7 RNA polymerase, one of the most potent and selective RNA polymerases (34); (*ii*) the 133 nucleotide long cDNA sequence of an RQ RNA (RQ135-1 RNA) (21), a Qβ phage satellite RNA (5) that can be amplified by Qβ replicase (the RNA-directed RNA polymerase of phage Qβ) (6) with a strong secondary structure that resembles the terminal structures of the genomic Qβ RNA (1,4,6). Insertion of an mRNA into an RQ RNA (an RNA that is replicable by Qβ replicase) (21) enhances the expression of that mRNA by making it amplifiable by Qβ replicase in the presence of a coupled translation system (20,25) and protecting it from ribonucleases (35); (*iii*) the Epsilon (E) sequence, a powerful translational enhancer of phage T7 coat protein gene 10, which is also present in the initiator region of the coat protein cistron of Qβ RNA (23). These viral elements were expected to enhance the expression of a harbored gene in a translation, transcription-translation, or replication-translation, cell-free system based on the crude (S30) *E. coli* extract.

Also, the vectors contain a sequence coding for Strep-tag (ST), an oligopeptide selected for possessing a moderate affinity to streptavidin [K_d ca. 10^{-5} M (28,29)]. Proteins fused to ST at the C terminus can be affinity purified on a streptavidin-agarose column under mild conditions (30).

Cloning and Expression Vectors for Gene Function Analysis
Edited by Quinn Lu and Michael P. Weiner

We show here that using these vectors in the standard cell-free expression systems provides for the synthesis of up to 40 µg of an active ST-fused protein per 1 mL of reaction mixture and that the synthesized protein can be efficiently isolated on an affinity column.

MATERIALS AND METHODS

Plasmids

Plasmid pASK-IBA1 (Institut für Bioanalytik, Göttingen, Germany) was used as a source of the ST-coding sequence, which was PCR-amplified with primers 1 (5′-TTCTCG-AGGTAACATATGGGCCATGGTCTCAGCGCTTGGC-3′, introducing the upstream *Xho*I and *Nde*I sites, underlined) and 2 (5′-CCCTCGAGTTAACCACCGAACTGCGGGTGA-3′, introducing the downstream *Xho*I site). Plasmid pRQST was obtained by ligating the *Xho*I-digested PCR product at site *Xho*I of a pUC18-derived plasmid pRQXhoI (Reference 20 and Figure 1) with its site *Nde*I at position 183 of the original pUC18 vector eliminated. Plasmid pRQSTE was prepared by PCR amplification of the ST-coding sequence using pRQST as a template and primers 2 and 3 (5′-TTCTCGAGGTTA-ACTTTACCAAGAGGTAACATATGGGCCATGGTCTCA-3′, introducing the E sequence) (23) and ligating the *Xho*I-digested PCR product at site *Xho*I of pRQXhoI.

Plasmid pSP65DHFR$_{0.7}$ (20) was used as a source of *E. coli* dihydrofolate reductase (DHFR) gene that was PCR-amplified with primers 4 (5′ GGCATATGATCAGTCTG-ATTGCGGCGT-3′, introducing site *Nde*I at the beginning of the coding sequence) and 5 (5′-GACCATGGCCCCGCCGC-TCCAGAATCTCA-3′ and introducing site *Nco*I at the end of the coding sequence). The *Nde*I/*Nco*I-digested PCR product was ligated between these sites into plasmids pRQST and pRQSTE, resulting in plasmids pRQST-DHFR and pRQSTE-DHFR, respectively (Figure 1). Plasmid pRQSTEP-DHFR was obtained by PCR amplifying the STE-DHFR insert of pRQSTE-DHFR with primers 6 [5′-CTTCTCGAGCTTAA-CTTTAGTAGGAGGTATCATATG-3′, slightly changing the 5′ untranslated region (UTR)] and 7 (5′-TCCCTCGAGG-GCAGTTCGGTTTAACCACCGAACTGCGGGT-3′, introducing the anti-ST sequence, underlined) and ligating the *Xho*I-digested PCR product into site *Xho*I of pRQXhoI.

Plasmid pBAD-GFP (Maxygen, Redwood City, CA, USA) was used as a source of *Aequorea victoria* green fluorescent protein (GFP) gene (Cycle 3 mutant), which was PCR-amplified with primers 8 (5′-GATATACATATGGCTAGCAAA-GGAGAAGAAC-3′. This primer fully matched the template and contained site *Nde*I) and 9 (5′-GACCATGGCC-TTTGTAGAGCTCATCCATGCCATG-3′, introducing the downstream *Nco*I site). Since the GFP gene itself contains sites *Nde*I and *Nco*I, the PCR product was first cloned, after removing the overhanging A with T4 DNA polymerase and

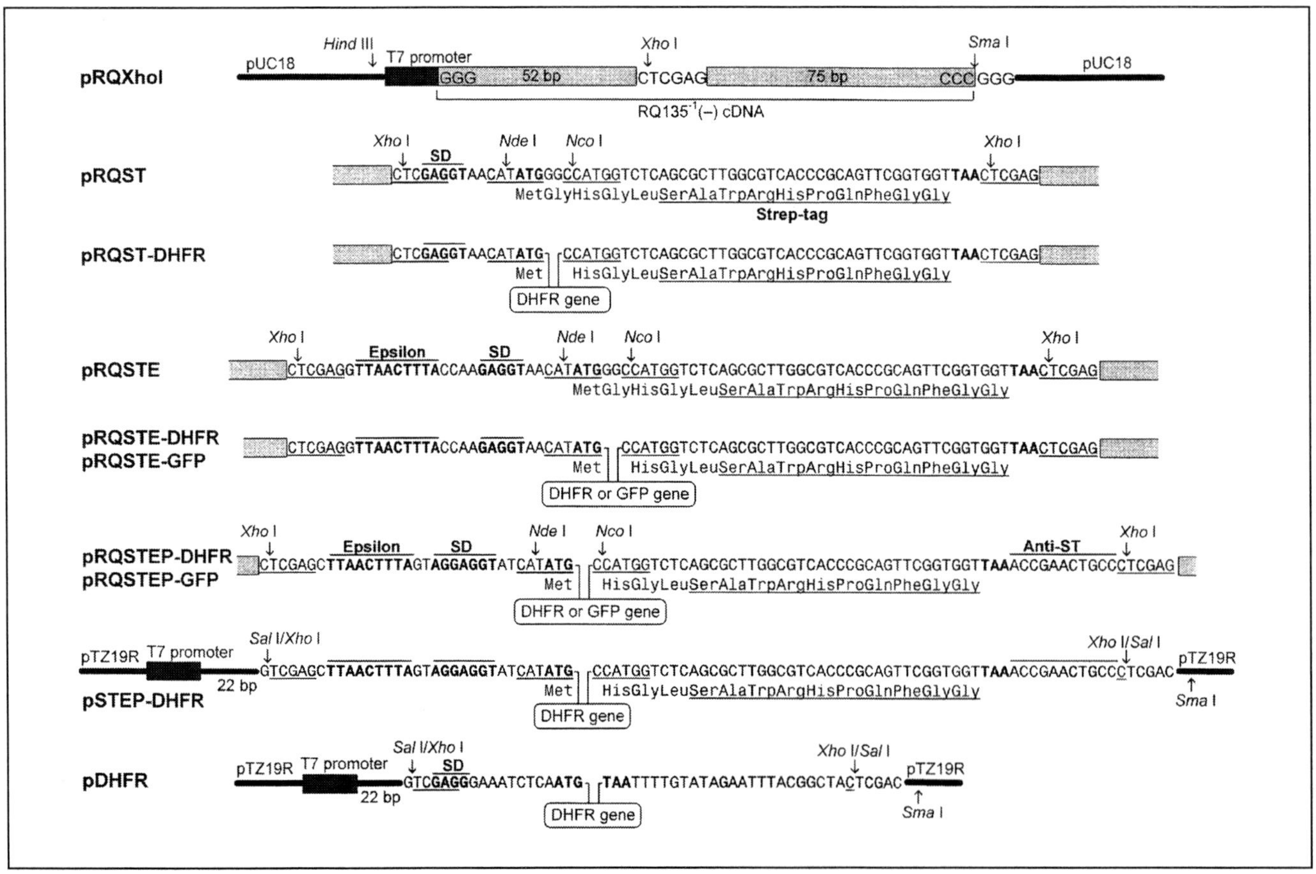

Figure 1. Schematic presentation of the expression vectors. Important restriction sites are underlined; E and Shine-Delgarno sequences, as well as the initiator and terminator codons are shown in bold; gray bars indicate RQ elements. In the plasmid names, "RQ" stands for RQ sequence, "ST" for Strep-tag, E for the Epsilon sequence and "P" for plus.

phosphorylating, at site *Pvu*II of pGEM2 (Promega, Madison, WI, USA). The *Nde*I/*Nco*I fragment of an appropriate size was then obtained by a limited digestion and agarose electrophoresis, and was then ligated between these sites into plasmids pRQSTE-DHFR and pRQSTEP-DHFR, resulting in plasmids pRQSTE-GFP and pRQSTEP-GFP, respectively (Figure 1).

Plasmids pSTEP-DHFR lacking the RQ135 sequences and pDHFR lacking both the RQ135 and STEP sequences (Figure 1) were prepared by excising the *Xho*I fragments from pRQSTEP-DHFR and pT7RQ135^{-1}(-)DHFR(+) (20), respectively, and recloning them at site *Sal*I of pTZ19R (Amersham Pharmacia Biotech, Piscataway, NJ, USA) downstream from the T7 promoter.

Cell-Free Reactions

*Sma*I-digested plasmids were transcribed with T7 RNA polymerase (8). Cell-free translations were carried out in microcentrifuge tubes at 37°C in a medium containing 30% (vol/vol) S30 extract prepared (37) from *E. coli* A19 cells (ca. 2 nmol/mL final ribosome concentration), 100 mmol/L HEPES-KOH, pH 7.6, 1 mmol/L EDTA, 2 mmol/L dithiothreitol, 4% glycerol, 0.03% NaN_3, 18 mmol/L $Mg(OAc)_2$, 120 mmol/L KOAc, 2 mmol/L ATP, 1.6 mmol/L GTP, 0.2 mmol/L each of amino acids (leucine omitted), 0.1 mmol/L [^{14}C]leucine at a specific activity of 38 mCi/mmol, 0.01 mg/mL folinic acid, 0.5 mg/mL total *E. coli* tRNA, 800 pmol/mL mRNA, 160 U/mL RNase inhibitor (Roche Molecular Biochemicals, Mannheim, Germany) and 40 mmol/L Li,K-acetylphosphate (served as a phosphate donor for endogenous kinases that regenerate NTPs) (24). A transcription-translation cocktail contained additionally 1 mmol/L CTP, 1 mmol/L UTP, 3000 U/mL T7 RNA polymerase and 40 µg/mL (ca. 20 pmol/mL) of a covalently closed circular plasmid instead of mRNA. The amount of synthesized protein (expressed in pmol/mL or µg/mL) was assessed by measuring the trichloroacetic acid-insoluble radioactivity in 5 µL aliquots at the indicated time points. The measurements took into account the number of Leu residues (11 in DHFR and 19 in GFP, plus 1 in the ST extension) and the molecular mass of the protein (18 kDa for DHFR and 26.98 kDa for GFP, plus 1.49 kDa for the ST extension).

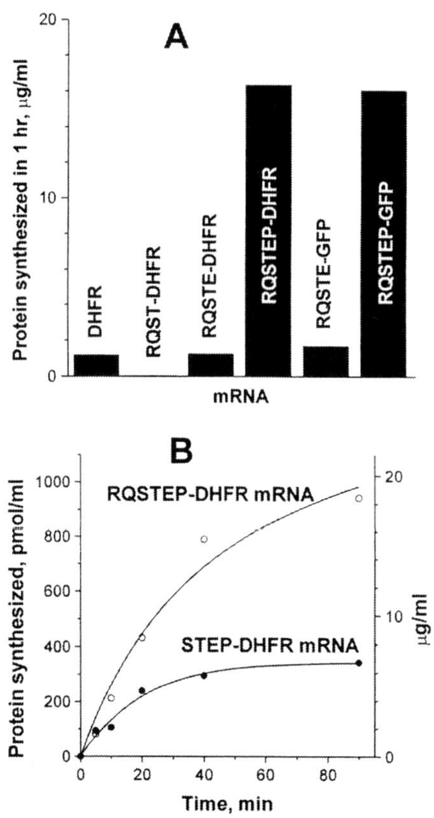

Figure 2. Protein synthesis in the translation system. (A) Performance of mRNAs differing by structural elements in the untranslated regions (cf. Figure 1). (B) Time course of DHFR synthesis on mRNAs possessing and lacking the RQ elements.

Affinity Isolation of Proteins

Strep-tag-fused proteins were isolated from 0.5 mL of a cell-free system using the Strep-tag® kit (Institute für Bioanalytik) according to the manufacturer's protocol. Briefly, a sample was applied to a 0.5 mL streptavidin-agarose column, followed by 3 mL of the washing buffer (100 mmol/L Tris-HCl, pH 8.0, 1 mmol/L EDTA, 0.02% NaN_3). The bound protein was then eluted by the same buffer containing 3 mmol/L diaminobiotin and collected in 0.5 mL fractions.

RESULTS AND DISCUSSION

Vector Design

As a starting vector, we used plasmid pRQXhoI, which already contained RQ135^{-1} cDNA downstream from the T7 promoter and whose RQ sequence was mutated to introduce a unique *Xho*I site convenient for insertion of foreign sequences (20). The plasmid was modified to introduce a Shine-Dalgarno sequence (31), the initiator codon, the ST-coding sequence (pRQST series) and the E sequence (pRQSTE series) and then used for the cloning of DHFR and GFP genes (Figure 1). The pRQSTEP plasmids also contained in the 3′ UTR a short oligonucleotide complementary to the ST-coding sequence (anti-ST sequence). The vector construction allows any gene, after it has been PCR-amplified with a pair of appropriate primers, to be inserted between sites *Nde*I and *Nco*I in frame with the downstream ST-coding sequence. The resulting protein is fused at its C terminus to the ST peptide using tetrapeptide Gly-His-Gly-Leu (Figure 1).

Vector Performance in the Translation System

The *Sma*I-runoff transcripts of the plasmids obtained were tested in a cell-free translation system (Figure 2). Earlier, we found that insertion of the DHFR gene into the RQ sequence resulted in a threefold enhancement of its expression due to a higher mRNA stability in the S30 translation system (35). Surprisingly, introduction of ST-coding sequence to the RQ-DHFR mRNA resulted in a null translation (Figure 2A,

RQST-DHFR mRNA). Further addition of the E sequence revived the template (Figure 2A, RQSTE-DHFR mRNA), but its translational activity was not higher than that of the authentic DHFR mRNA (Figure 2A) that lacked the RQ, ST and E sequences (Figure 1, pDHFR).

Figure 3 shows that the last portion of the ST-coding sequence is capable of forming base pairs with the initiator region of mRNAs containing the Shine-Dalgarno and E sequences. This could explain the observed inhibition of translation. A similar hiding of the initiator region within RNA secondary structure results in a complete suppression of translation of the A cistron of RNA coliphages (4).

To test this hypothesis, we introduced, just downstream from the terminator codon, UAA, an 11-nucleotide long segment complementary to the interfering part of the ST-coding sequence (anti-ST sequence). We expected that the anti-ST sequence would distract the ST-coding sequence from the E and Shine-Dalgarno sequences (Figure 3) and would thereby unblock the initiation of translation. We also introduced into the translation leader sequence a few nucleotide substitutions to additionally weaken the RNA secondary structure in the initiator region. The resulting vector was designated pRQSTEP (Figure 1). Figure 2A demonstrates that these minor changes in the vector structure resulted in a dramatic enhancement of translation. The enhancement was similar for DHFR and GFP templates (cf. RQSTE and RQSTEP mRNAs), as was the amount of synthesized protein. The translational potential of the pRQSTEP vector is comparable to that of Qβ RNA at its coat protein cistron (11), which is one of the most potent templates known; yet, it is a much simpler vector.

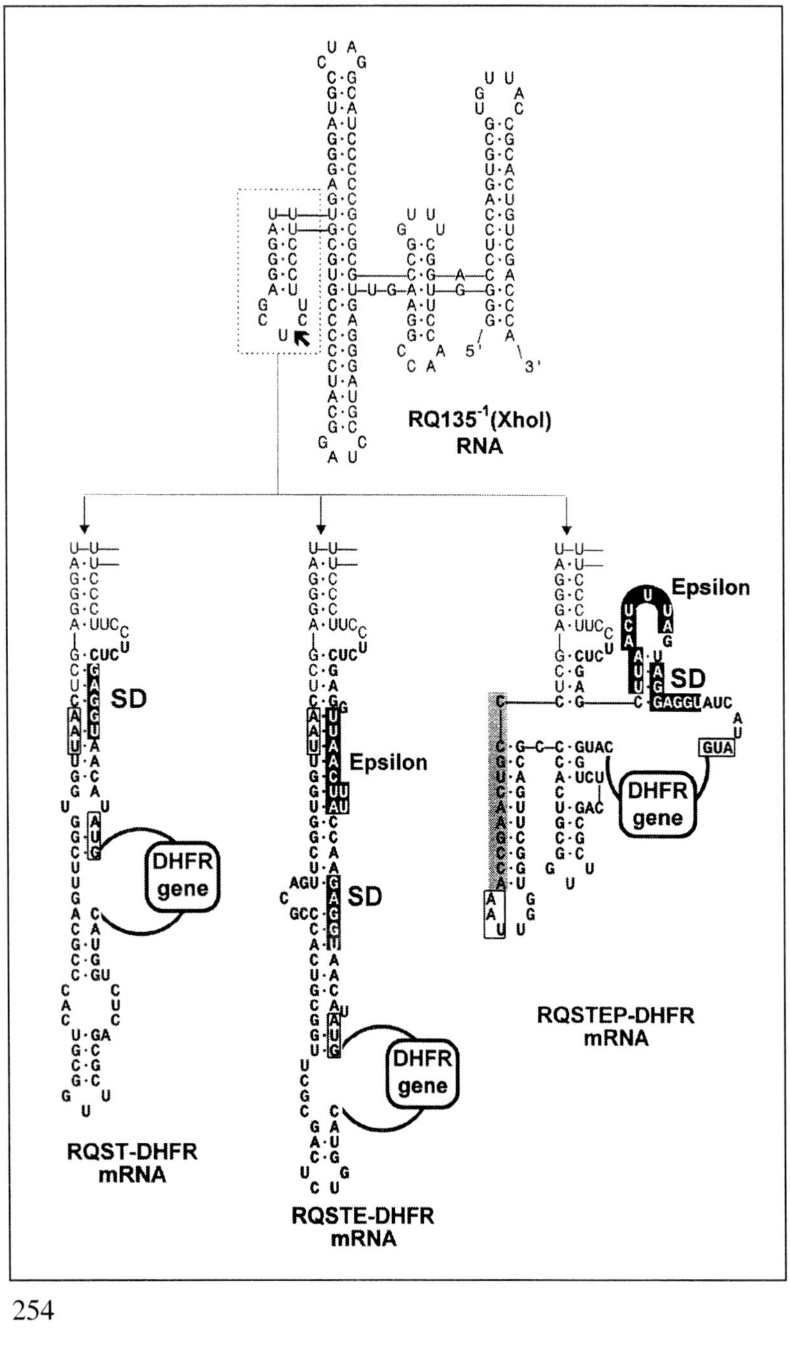

Figure 3. Possible secondary structure of the RQ moiety (upper panel) and of mRNA inserts within the RQST, RQSTE and RQSTEP vectors (lower panel). Arrow indicates the site in the RQ RNA sequence where mRNAs are inserted. The inserted sequences are shown in bold; E and Shine-Dalgarno sequences are shown on a black background; initiator (AUG) and terminator (UAA) codons are boxed; the anti-ST sequence is shown on a gray background. Secondary structures were predicted online at (http://www.ibc.wustl.edu/~zuker/) by program MFOLD 3.0 based on free energy minimization (9).

Figure 2B shows that the RQ moiety of the vector is responsible for about two-thirds of the translational activity of the RQSTEP-DHFR mRNA. With RQ elements, the protein synthesis lasted for a longer time, presumably due to the protection of mRNA against endogenous ribonucleases (35), and produced approximately 20 µg (1 nmol) of ST-DHFR per 1 mL of translation mixture by 90 min of incubation.

Vector Performance in the Coupled Transcription-Translation System

An even larger amount of a protein (ca. 40 µg/mL) can be synthesized in a coupled transcription-translation system, in which T7 polymerase continuously transcribes plasmids of the pRQSTEP series (Figure 4, A and B). In the coupled system, the RNA stability is no longer a limiting factor: the same translation kinetics are observed with pRQSTEP and pSTEP plasmids (Figure 4A), and there is still a large amount of undegraded RNA at 30 min (Figure 4C) when protein synthesis ceases (Figure 4, A and B). The initial rate of protein synthesis in the coupled system is about 4× higher than in the translation system, but the final protein yield is only 2× as large because of the earlier plateau (Figure 4A). In the coupled system, the full-length RNA begins to disappear at the same time point when the protein plateau is reached (cf. Figure 4, B and C), which indicates that translation and transcription shut down simultaneously, probably due to the depletion of common substrates, NTPs. Indeed, in a CFCF reactor that is continuously fed with substrates, the RNA and protein syntheses remain steady for many hours (data not shown).

Isolation of Proteins Synthesized in a Cell-Free System

Proteins synthesized from pRQSTEP or pSTEP vectors can be isolated because of their affinity to streptavidin. Figure 5 shows an example of such an isolation. In this experiment, ST-DHFR synthesized on RQSTEP-DHFR mRNA had a specific activity of 53 U/mg, which is close to that of DHFR isolated from *E. coli* cells, 55 U/mg (2). This result demonstrates

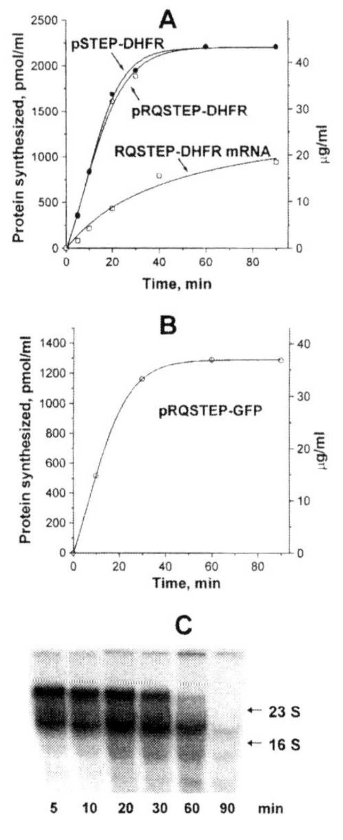

Figure 4. Protein synthesis in the coupled transcription-translation system. (A) Comparison of the kinetics of DHFR synthesis on a plasmid (with or without the RQ moiety) with that on an mRNA. (B) Time course of GFP synthesis. (C) RNA synthesized during the reaction displayed in Panel B. To monitor the RNA, a part of the reaction mixture was supplied with [^{32}P]ATP, and phenol-extracted samples were analyzed by electrophoresis through a denaturing 5% polyacrylamide gel (27) and autoradiography. Arrows indicate the position of ribosomal RNA bands visible after staining the gel with toluidine blue. Since no transcriptional terminator was introduced behind the RQSTEP cassette (Figure 1), the size of synthesized RNAs was determined by terminators present in the pUC18 vector.

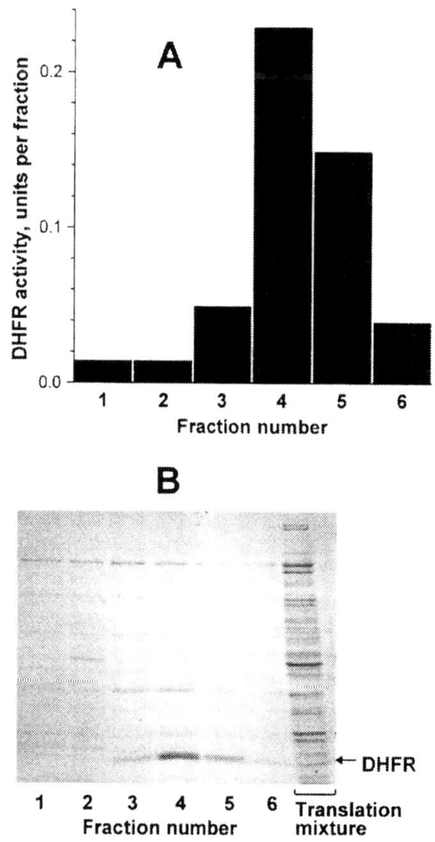

Figure 5. Affinity isolation of DHFR synthesized in the cell-free translation system. (A) Direct chromatography on a streptavidin-agarose column of 0.5 mL of the translation mixture containing 2.4 mg of total protein (19) and 15.5 µg (ca. 0.6%) of ST-DHFR as determined by [^{14}C]Leu incorporation. The increase of enzymatic activity (3) above that present in the *E. coli* extract (0.06 U) was 0.82 U. About 35% of the starting DHFR activity was found in the flow-through, and 61% (0.5 U) in fractions 3–6. (B) Coomassie blue G-250 staining of the samples subjected to electrophoresis through a 10%–20% gradient polyacrylamide gel in the presence of SDS (18). Arrow indicates the position of the labeled band as revealed by autoradiography.

that the in vitro synthesized protein is fully active and that the ST peptide does not affect the activity of DHFR. Figure 5 shows that a 100-fold purification of ST-DHFR was achieved by single-step affinity chromatography on a streptavidin-agarose column, yielding 19 μg (0.95 nmol) of a relatively pure enzyme per 1 mL of translation mixture. Similar results were obtained with ST-GFP. Its activity was tested by its green fluorescence under a 366 nm light (data not shown). If needed, the isolated proteins can be further purified by HPLC or by gel electrophoresis.

CONCLUSION

This paper demonstrates that a significant enhancement of protein synthesis can be achieved in a standard cell-free system by solely manipulating the UTRs of an expression vector by introducing the viral structural elements known to enhance translation in vivo and to protect mRNA from ribonuclease action. The expression vectors reported here are universal in that they can be used, at a researcher's discretion, in a translation reaction or in a coupled transcription-translation reaction. The vectors can also be used in a coupled replication-translation reaction. It has been shown that embedding an mRNA into an RQ RNA sequence makes the former amplifiable by Qβ replicase in the presence of a coupled translation system, resulting in a higher protein yield (20). Depending on the choice, 1 mL of a reaction yields 20–40 μg, or about 1 nmol of a standard protein. This amount meets many needs of a research laboratory, such as exploring the enzymatic, ligand-binding or immunological properties of a protein or its ability to form complexes with other proteins or with nucleic acids.

In this study, we did not attempt to manipulate the system composition, but the data obtained indicate that a higher protein yield may be achieved by increasing the concentration of certain substrates (e.g., NTPs). The volume yield can also be increased by using condensed cell-free extracts (15,17,22). If desired, the reported vectors can be used in a CFCF or dialysis-type protein synthesis reactors, in which one to two orders of magnitude larger amounts of a protein can be synthesized in the same volume using a continuous supply of substrates to the reaction mixture (15,16,26,32,33).

The reported vectors allow any gene to be inserted between the initiator codon and the ST-coding sequence, so that its translation results in a functionally active protein carrying ST at the C terminus. The results demonstrate that such proteins can be easily isolated from the products of cell-free translation on a streptavidin-agarose column at a good yield and in a relatively pure state. Recently, a modified ST and a mutated streptavidin were reported that may further improve the purity of affinity isolated proteins (36). Of course, other fusion tags (7,10) can also be used to recover the synthesized proteins.

Only one working day is required to carry out the entire procedure including the cell-free synthesis of a tagged protein and its affinity isolation. This may significantly advance any field in which rapid analysis of a gene expression product is important, such as screening the cDNA libraries, screening gene variants in protein engineering and functional identification of open reading frames whose sequences increasingly accumulate because of a progress in the genome projects.

AKNOWLEDGMENTS

We thank Dr. Herbert Stadler (Institute für Bioanalytik, Göttingen, Germany) for the gift of pASK-IBA1 and of a Strep-tag kit, and Oksana Kalita for technical assistance. This work was partially supported by Roche Molecular Biochemicals.

REFERENCES

1. **Adhin, M.R., J. Alblas and J. van Duin.** 1990. Secondary structures at the 3′ terminal region of RNA coliphages: comparison with tRNA. Biochim. Biophys. Acta *1050*:110-118.
2. **Baccanari, D.P., D. Averett and C. Briggs.** 1977. *Escherichia coli* dihydrofolate reductase: isolation and characterization of two isozymes. Biochemistry *16*:3566-3772.
3. **Baccanari, D.P. and S.S. Joyner.** 1981. Dihydrofolate reductase hysteresis and its effect of inhibitor binding analyses. Biochemistry *20*:1710-1716.
4. **Beekwilder, J., R. Nieuwenhuizen, R. Poot and J. van Duin.** 1996. Secondary structure model for the first three domains of Qβ RNA. Control of A-protein synthesis. J. Mol. Biol. *256*:8-19.
5. **Chetverin, A.B., H.V. Chetverina and A.V. Munishkin.** 1991. On the nature of spontaneous RNA synthesis by Qβ replicase. J. Mol. Biol. *222*:3-9.
6. **Chetverin, A.B. and A.S. Spirin.** 1995. Replicable RNA vectors: prospects for cell-free gene amplification, expression and cloning. Prog. Nucleic Acid Res. Mol. Biol. 51:225-270.
7. **Ford, C.F., I. Suominen and C.E. Glatz.** 1991. Fusion tails for the recovery and purification of recombinant proteins. Protein Expr. Purif. *2*:95-107.
8. **Gurevich, V.V., I.D. Pokrovskaya, T.A. Obukhova and S.A. Zozulya.** 1991. Preparative in vitro mRNA synthesis using SP6 and T7 RNA polymerases. Anal. Biochem. *195*:207-213.
9. **Jaeger, J.A., D.H. Turner and M. Zuker.** 1990. Predicting optimal and suboptimal secondary structure for RNA. Methods Enzymol. *183*:281-306.
10. **Jones, C., A. Patel, S. Griffin, J. Martin, P. Young, K. O'Donnell, C. Silverman, T. Porter and I. Chaiken.** 1995. Current trends in molecular recognition and bioseparation. J. Chromatogr. A *707*:3-22.
11. **Katanaev, V.L., O.V. Kurnasov and A.S. Spirin.** 1995. Viral Qβ RNA as a high expression vector for mRNA translation in a cell-free system. FEBS Lett. *359*:89-92.
12. **Kawarasaki, Y., T. Kawai, H. Nakano and T. Yamane.** 1995. A long-lived batch reaction system of cell-free protein synthesis. Anal. Biochem. *226*:320-324.
13. **Kawarasaki, Y., H. Nakano and T. Yamane.** 1998. Phosphatase-immunodepleted cell-free protein synthesis system. J. Biotechnol. *61*:199-208.
14. **Kigawa, T., Y. Muto and S. Yokoyama.** 1995. Cell-free synthesis and amino acid-selective stable isotope labeling of proteins for NMR analysis. J. Biomol. NMR *6*:129-134.
15. **Kigawa, T., T. Yabuki, Y. Yoshida, M. Tsutsui, Y. Ito, T. Shibata and S. Yokoyama.** 1999. Cell-free production and stable-isotope labeling of milligram quantities of proteins. FEBS Lett. *442*:15-19.
16. **Kim, D.M. and C.Y. Choi.** 1996. A semicontinuous prokaryotic coupled transcription/ translation system using a dialysis membrane. Biotechnol. Prog. *12*:645-649.
17. **Kim, D.M., T. Kigawa, C.Y. Choi and S. Yokoyama.** 1996. A highly efficient cell-free protein synthesis system from *Escherichia coli*. Eur. J. Biochem. *239*:881-886.
18. **LaemmLi, D.K.** 1970. Cleavage of structural proteins during the assembly of the head of bacteriophage T4. Nature *227*:680-685.
19. **Lowry, O.H., N.J. Rosebrough, A.L. Farr and R.J. Randall.** 1951. Protein measurement with the Folin phenol reagent. J. Biol. Chem. *193*:265-275.
20. **Morozov, I. Yu., V.I. Ugarov, A.B. Chetverin and A.S. Spirin.** 1993. Synergism in replication and translation of messenger RNA in a cell-free system. Proc. Natl. Acad. Sci. USA *90*:9325-9329.
21. **Munishkin, A.V., L.A. Voronin, V.I. Ugarov, L.A. Bondareva, H.V. Chetverina and A.B. Chetverin.** 1991. Efficient templates for Qβ replicase are formed by recombination from heterologous sequences. J. Mol. Biol. *221*:463-472.
22. **Nakano, H., T. Tanaka, Y. Kawarasaki and T. Yamane.** 1994. An increased rate of cell-free protein synthesis by condensing wheat-germ

extract with ultrafiltration membranes. Biosci. Biotechnol. Biochem. *58*:631-634.

23.**Olins, P.O. and S.H. Rangwala.** 1989. A novel sequence element derived from bacteriophage T7 mRNA acts as an enhancer of translation of the *lacZ* gene in *Escherichia coli*. J. Biol. Chem. *264*:16973-16976.

24.**Ryabova, L.A., L.M. Vinokurov, E.A. Shekhovtsova, Yu.B. Alakhov and A.S. Spirin.** 1995. Acetyl phosphate as an energy source for bacterial cell-free translation systems. Anal. Biochem. *226*:184-186.

25.**Ryabova, L., E. Volianik, O. Kurnasov, A. Spirin, Y. Wu and F.R. Kramer.** 1994. Coupled replication-translation of amplifiable messenger RNA. A cell-free protein synthesis that mimics viral infection. J. Biol. Chem. *269*:1501-1505.

26.**Ryabova, L.A., I.Yu. Morozov and A.S. Spirin.** 1998. Continuous-flow cell-free translation, transcription-translation, and replication-translation systems. Methods Mol. Biol. *77*:179-193.

27.**Sambrook, J., E.F. Fritsch and T. Maniatis.** 1989. Molecular Cloning: A Laboratory Manual, 2nd ed. CSH Laboratory Press, Cold Spring Harbor, NY.

28.**Schmidt, T.G.M., J. Koepke, R. Frank and A. Skerra.** 1996. Molecular interaction between the Strep-tag affinity peptide and its cognate target streptavidin. J. Mol. Biol. *255*:753-766.

29.**Schmidt, T.G.M. and A. Skerra.** 1993. The random peptide library-assisted engineering of a C-terminal affinity peptide, useful for the detection and purification of a functional Ig Fv fragment. Protein Eng. *6*:109-122.

30.**Schmidt, T.G.M. and A. Skerra.** 1994. One-step affinity purification of bacterially produced proteins by means of the "Strep-tag" and immobilized recombinant core streptavidin. J. Chromatogr. A *676*:337-345.

31.**Shine, J. and L. Dalgarno.** 1974. The 3′-terminal sequence of Escherichia coli 16S ribosomal RNA: complementarity to nonsense triplets and ribosome binding sites. Proc. Natl. Acad. Sci. USA *71*:1342-1346.

32.**Spirin, A.S.** 1992. Cell-free protein synthesis bioreactors, p. 31-43. *In* P. Todd, S.K. Sikdar and M. Bier (Eds.), Frontiers in Bioprocessing II. American Chemical Society, Washington, DC.

33.**Spirin, A.S., V.I. Baranov, L.A. Ryabova, S.Yu. Ovodov and Yu.B. Alakhov.** 1988. A continuous cell-free translation system capable of producing polypeptides in high yield. Science *242*:1162-1164.

34.**Studier, F.W. and B.A. Moffatt.** 1986. Use of bacteriophage T7 RNA polymerase to direct selective high-level expression of cloned genes. J. Mol. Biol. *189*:113-130.

35.**Ugarov, V.I., I. Yu. Morozov, G.Y. Jung, A.B. Chetverin and A.S. Spirin.** 1994. Expression and stability of recombinant RQ-mRNAs in cell-free translation systems. FEBS Lett. *341*:131-134.

36.**Voss, S. and A. Skerra.** 1997. Mutagenesis of a flexible loop in streptavidin leads to higher affinity for the Strep-tag II peptide and improved performance in recombinant protein purification. Protein Eng. *10*:975-982.

37.**Zubay, G.** 1973. *In vitro* synthesis of protein in microbial systems. Annu. Rev. Genet. *7*:267-287.

CHAPTER 42

E. coli-Based In Vitro Transcription/Translation: In Vivo-Specific Synthesis Rates and High Yields in a Batch System

Ranjan Patnaik and James R. Swartz

Genentech, South San Francisco, CA, USA

(Reprinted from BioTechniques 24:862-868, 1998)

ABSTRACT

A highly efficient Escherichia coli*-based, batch in vitro protein synthesis system using circular plasmid DNA is described. Compared to a presently available commercial kit, this improved system produced several hundredfold greater yields of the rDNA human protein thrombopoietin (ca. 450 µg/mL). The system is capable of obtaining specific synthesis rates similar to those in vivo, approximately a 1000-fold increase compared to the original methods previously described. It compares favorably in rates and yields to the recently published semicontinuous methods but with the convenience of a true batch system.*

INTRODUCTION

E. coli-based transcription/translation systems have been used for several decades and recently have been applied to: *(i)* expression of toxic rDNA proteins, *(ii)* radiolabeling of newly synthesized gene products, *(iii)* incorporation of unnatural amino acids and *(iv)* production of small quantities of proteins quickly and economically (5,7,12,19). Summaries of various other applications as well as accounts of batch system development have been reported (5,10,18,23,24). Most notable is the recent report of Kim et al. (11). By using condensed cell extracts and optimized concentrations of polyethylene glycol and phosphoenol pyruvate and by adding more amino acids, they were able to synthesize 0.4 mg/mL of chloramphenical acetyltransferase (CAT) in a batch reaction. Continuous, cell-free protein synthesis systems also have been described by several groups and offer extended synthesis periods but are more complex to operate and also have higher reagent costs (1,9,22).

More recently, an intermediate approach has been developed by Kim and Choi (10). Using a semicontinuous approach, they conducted the synthesis reaction in a chamber bounded by a dialysis membrane to allow interchange with a larger reservoir (25× volume) of low-molecular-weight reagents. During a 14-h reaction, 1.2 mg/mL of CAT accumulated in the smaller reaction chamber.

In an attempt to produce a simple and productive batch system, we have independently improved the methods of Pratt (19) and Zubay (27) to increase initial, specific synthesis rates as well as the duration of synthesis. We report a batch system that nearly reproduces in vivo-specific rates of protein synthesis and maintains them for at least 50 min. This procedure produced more than 0.4 mg/mL of aggregated human thrombopoietin polypeptide in a 1-hour reaction. The system offers simplicity and efficient use of reaction reagents and avoids the need to condense the cell extract and add exogenous T7 RNA polymerase. Curiously, the improved procedure preferentially synthesized β-lactamase from its native promoter rather than rDNA human growth hormone (hGH) from the *tac*II promoter even though more dilute reaction mixtures preferentially synthesized hGH. Although this latter phenomenon is not currently understood, we believe that our improved batch procedure offers a simple, cost-effective and highly productive approach for in vitro protein synthesis.

MATERIALS AND METHODS

Preparation of Cell Extract

Cells from a 10-L fermentation of an *E. coli* K-12 triple-protease strain 27C7 (4) (*degP*, *ompT*, *ptr-3*) were harvested in mid-log phase and were immediately washed three times with S30 buffer (10 mM Tris-acetate, pH 8.2, 14 mM magnesium acetate, 60 mM potassium acetate and 1 mM dithiothreitol [DTT]) (19) containing mercaptoethanol before freezing at -80°C. Frozen cells (200 g) were thawed at room temperature, resuspended thoroughly in 200 mL of S30 buffer using a tissue homogenizer and disrupted by passing twice through an M-110Y Microfluidizer® (Microfluidics, Newton, MA, USA)

Cloning and Expression Vectors for Gene Function Analysis
Edited by Quinn Lu and Michael P. Weiner

at 10 000 psi. Immediately, 10 µL of 1 M DTT were added per 10 mL of lysate collected. The crude lysate was then processed according to published protocols (19) except that a final spin at 135 000× *g* for 12 min was included in the protocol. This high-speed centrifugation removes small residual membrane particles from the cleared lysate (17). The lysate was separated into aliquots and frozen in liquid nitrogen for future use. The protein concentration of the lysate was measured by the Bradford assay (Bio-Rad, Hercules, CA, USA) with bovine serum albumin (BSA) as the standard. This cell extract is termed S135.

Preparation of Plasmid DNA Templates

Plasmid pMP331 (a kind gift from D. Yansura, Genentech) is a pBR322 derivative encoding for recombinant thrombopoietin (rTPO) driven from the *trp* promoter (6). Plasmid pRPHGH2 is a pUC19 derivative encoding for pre-recombinant human growth hormone (pre-rhGH) driven from a strong *tac*II promoter (6). In addition, both plasmids encode for pre-β-lactamase driven from its natural promoter. Plasmids were purified from 250 mL of an overnight LB culture by first lysing the cells at 4°C in 25% sucrose, 50 mM Tris-HCl (pH 8.0) with 1.4 mg/mL hen egg white lysozyme, 55 mM EDTA and 1% Triton® X-100. After centrifugation at 39 000× *g* for 30 min, 2 vol of 0.2 M NaOH, 1% sodium dodecyl sulfate (SDS) were added. After 5 min, 1 vol 2 M potassium acetate, 2 M acetic acid was added. The solution was clarified by centrifugation at 6800× *g* for 15 min and then precipitated at -20°C for 1–2 h after adding 2 vol of cold ethanol. The pellet was resuspended in 5 mL TE buffer (10 mM Tris-HCl, 1 mM EDTA, pH 8.0), adjusted to 2 M ammonium acetate and incubated on ice for 20 min. After centrifugation, the supernatant was again ethanol-precipitated and the pellet resuspended in 0.4 mL TE buffer for loading onto a Sepharose® CL 4B Gel Filtration Column (Amersham Pharmacia Biotech, Piscataway, NJ, USA). The column was equilibrated and eluted with TE buffer. Fractions were monitored for absorbance $(A)_{260}/A_{280}$, and relevant fractions were examined by running on a 1% agarose gel. Fractions containing the plasmid and devoid of any contaminating RNA were pooled and concentrated for further use. This procedure of plasmid purification avoids RNase, phenol/chloroform and high concentrations of salts that might be detrimental to the performance of the in vitro system.

Composition of In Vitro Transcription/Translation Systems

The original in vitro system is essentially the same as that published by Lesley et al. (15) and derives from the landmark work of Zubay (27) and Pratt (19). One exception is that 12 µL (rather than 15 µL) of S30 extract were added to give a final protein concentration of 7.9 mg/mL in the 50-µL reaction. The composition of the improved in vitro system described here is listed in Table 1 and compared to the original system. The composition in Table 1 does not, however, include the contribution of salts in the S135 or S30 cell extracts.

A typical 30-µL reaction contained 10 µL S135 cell extract, 3 µL ^{35}S-methionine (Amersham Pharmacia Biotech), 7.5 µL low-molecular-weight mixture, approximately 5 µL plasmid DNA and the remainder was nuclease-free water. Transcription in this system is dependent on endogenous *E. coli* RNA polymerase. All chemicals were purchased from Sigma Chemical (St. Louis, MO, USA) except tRNA (Boehringer Mannheim, Indianapolis, IN, USA) and phosphoenol pyruvate (PEP; Fluka Chemical, Ronkonkoma, NY, USA).

Monitoring ^{35}S-Methionine Incorporation

In vitro transcription/translation was initiated by adding plasmid DNA to the reaction mixture at 37°C. Small aliquots (3–5 µL) were withdrawn at indicated time points and spotted on Whatman GF/C Filter Paper (Clifton, NJ, USA). These were dried for 30 min at room temperature, soaked in cold methanol containing 10% trichloroacetic acid (TCA) and 0.1% cold methionine for 1 h and soaked again for 30 min in cold methanol with 5% TCA and 0.1% methionine. Finally, the filter papers were washed for 5 min in acetone and dried for 2 h at room temperature. Radioactivity was determined using 5 mL of scintillation cocktail. Counts per minute (cpm) were measured twice and the results averaged. The cpm of unwashed samples from the reaction were measured to determine the specific activity of ^{35}S-methionine in the reaction mixture.

Calculation of Specific Protein Synthesis Rate

To calculate the specific synthesis rate (µg new protein/mg total cell protein/h), the specific radioactivity of the methionine (hot and cold) used in the reaction was calculated by dividing the measured radioactivity of unwashed samples by the known concentration of methionine. The cpm of TCA-precipitated and washed samples were then used to determine the amount of methionine incorporated into protein. The measured cpm of samples from complete reaction mixtures were corrected for background by subtracting the corresponding time point cpm of reactions without plasmid DNA. The corrected cpm were used to estimate protein synthesis rates, and these were divided by the protein content of the cell extracts to determine specific expression rates.

SDS Polyacrylamide Gel Electrophoresis (PAGE)

Samples were withdrawn at the end of an in vitro reaction, mixed with equal volumes of 2× tricine buffer (NOVEX, San Diego, CA, USA), boiled for 5 min and loaded onto a 10%–20% tricine gel (NOVEX). ^{14}C standard markers (Amersham Pharmacia Biotech) were run in one of the wells. Gels were fixed for 30 min in 500 mL of 34.6 g/L sulphosalicylic acid and 115 g/L TCA, treated for 30 min each in gel-drying solution (NOVEX) and 1 M sodium salicylate, 0.1 *N* NaOH before drying for 2 h in a gel dryer. Dried gels were exposed to X-OMAT Film (Scientific Imaging Systems [Eastman Kodak], Rochester, NY, USA) at -70°C for 1–12 h to detect ^{35}S-labeled translation products.

RESULTS AND DISCUSSION

Following an extensive investigation of various parameters in our laboratory by Amina Sheikh, the initial specific rate of protein production for the original in vitro system of Pratt (19) and Zubay (27) was increased substantially to approxi-

Table 1. Comparison of Improved System and Original System (15)

Reagent	Improved System	Original System
Tris-acetate	56.4 mM (pH 8.2)	35 mM (pH 8.0)
DTT	1.7 mM	2 mM
Sodium ATP	2.9 mM	2 mM
Sodium CTP, GTP and UTP	2.0 mM	0.5 mM
PEG-8000	5%	3.5%
Sodium PEP	27 mM	20 mM
E. coli tRNA	0.17 mg/mL	1 mg/mL
Ammonium acetate	36.0 mM	30 mM
Potassium acetate	72.0 mM	None
Potassium glutamate	None	190 mM
Magnesium acetate	7.3 mM	8.0 mM
Folinic acid	52 µg/mL	20 µg/mL
Folic acid	2.7 µg/mL	None
Pyridoxine-HCl	2.7 µg/mL	None
β-NADP[a]	2.7 µg/mL	None
FAD[b]	2.7 µg/mL	None
PABA[c]	2.7 µg/mL	None
cAMP	None	1 mM
Methionine	2 mM	0.3 µM
19 amino acids	0.56 mM	0.5 mM
IPTG[d]	0.1 mM	0.8 mm
Cell extract (final protein concentration)	12 mg/mL (S135)	7.9 mg/mL (S30)
DNA	0.26 µg/µL	0.12 µg/µL

[a]Nicotinamide-adenine dinucleotide phosphate
[b]Flavin-adenine dinucleotide
[c]*p*-Aminobenzoic acid
[d]Isopropyl β-D-thiogalactopyranoside

mate in vivo rates of recombinant protein production. Details of this experimentation will be published elsewhere, but the main differences are shown in Table 1. The biggest advantage was gained from increasing the methionine concentration (42-fold improvement in rate) but adjusting the concentration of the DNA (8-fold improvement) and polyethylene glycol (PEG)-8000 (1.5-fold improvement), and low-molecular-weight translation initiation factors (1.5-fold improvement) also contributed.

This initial investigation was conducted using a T7 promoter for expression of pre-rhGH. When promoters recognized by native *E. coli* RNA polymerase were used, the duration of methionine incorporation was limited to approximately 30 min. We observed substantial precipitation in the reaction mixture after 15 min. Further efforts revealed that Ca^{++} (9.7 mM) was the source of this precipitation. Removal of calcium extended the period of protein synthesis to at least 50 min. It is of interest that the in vivo concentration of free calcium in *E. coli* is negligible (8,16). Many existing in vitro systems contain calcium in millimolar concentrations, and apparently that poses no problem (1,9,13,19,27). In fact, it might even be beneficial for relatively dilute systems using the T7 or SP6 RNA polymerases. However, as we seek to use the native RNA polymerase as well as higher concentrations of key components, calcium becomes detrimental.

To test the improved method with a different promoter and product, we used plasmid pMP331 encoding the heterologous protein rTPO, driven from a *trp* promoter. The protein yield was compared to that of a commercially available in vitro protein synthesis system. Figure 1 shows the time course of methionine incorporation for the improved system as measured by TCA-precipitable cpm. Methionine incorporation is linear for at least 50 min. The high background of methionine incorporation without plasmid addition is probably due to background transcription/translation from chromosomal DNA in the cell extract. This background was subtracted in determining rTPO expression rates and yields.

With the *E. coli* S30 Extract System for Circular DNA (Promega, Madison, WI, USA), methionine incorporation into rTPO is essentially complete after 10 min (data not shown). This short burst of incorporation was also observed with the positive control DNA (pBEST*luc*™) supplied with the kit. Specific synthesis rates and protein yields for the two systems are shown in Table 2. For calculation purposes, it is assumed that either rTPO or luciferase is the major protein product. This assumption is supported by the fluorograms shown in Figure 2. Faster rates and prolonged synthesis result in nearly a 500-fold increase in protein accumulation with the

Table 2. Performance of the Improved System vs. a Commercial Kit for Synthesis of Human Thrombopoietin Using pMP331

In Vitro System	Protein in Cell Extract (mg/mL)	Specific Protein Synthesis Rate (μg/mg protein/h)		Final Product Concentration (μg/mL)	
		pMP331	pBEST*luc*	pMP331	pBEST*luc*
Commercial kit	2.5	1.8	1.6	0.99	0.68
Improved system	12.0	47.3	ND	466	ND

ND: not determined. pBEST*luc* was supplied by the kit manufacturer as a positive control.

improved system. It is gratifying that conditions identified with the T7 promoter/pre-rhGH system provide excellent results with the *trp* promoter/rTPO plasmid. It is also relevant that both the rTPO and luciferase plasmids produced similar results with the commercial kit.

Figure 2, A and B, shows the fluorograms of samples from rTPO expression reactions run on 10%–20% tricine polyacrylamide gels. These results indicate that the major species in both in vitro systems is aggregated rTPO that migrates at the top of the gel at around 200 kDa. The identity of this band was confirmed by Western blotting. A minor amount of pre-β-lactamase is also made in both systems and migrates at around 31 kDa. Monomeric rTPO (37.5 kDa) accumulates in insignificant amounts. rTPO aggregates (200 kDa) are also the major species when rTPO is expressed in vivo in *E. coli* from the same plasmid construct (personal communication from D. Reilly, Genentech). Estimated in vivo rTPO accumulation rates are approximately 17 μg/mg cell protein/h. Thus, our improved system not only produces a polypeptide similar to that observed in vivo but also is capable of exceeding estimated in vivo-specific accumulation rates.

To compare our improved system to an in vitro system from the literature (15), we expressed pre-rhGH from a *tac*II promoter using plasmid pRPHGH2. Time courses of methionine incorporation as monitored by TCA-precipitable cpm are shown in Figure 3, A and B. Both the original and the improved systems showed linear rates of protein synthesis for at least 50 min. However, the ordinates have different scales because the quantities of methionine incorporated were significantly different between the two systems.

Figure 4, A and B, shows fluorograms of gels showing the products synthesized in the two in vitro reaction systems. To our surprise, the improved system synthesized predominantly pre-β-lactamase (31 kDa) while the original system synthesized pre-rhGH (22 kDa) as the major species. The identity of the pre-rhGH was confirmed by Western blotting (not shown).

Quantitative results are shown in Table 3. The specific rate of protein production in the original system is 0.009 μg pre-rhGH/mg cell protein/h, while the improved system yields a rate of 9.34 μg pre-β-lactamase/mg cell protein/h. The estimated in vivo-specific rate for rhGH is 12–19 μg/mg protein/h (3). Although it was encouraging that 94 μg/mL of pre-β-lactamase were made with a promoter that is relatively weak in vivo, it was disappointing that rhGH expression was so low from a promoter that is much stronger (*tac*II) (Figure 4B). The significant accumulation of pre-β-lactamase suggests that the in vitro transcription and translation reactions are working well. It is not clear at this stage whether the low pre-rhGH accumulation is attributable to weak recognition of the *tac*II promoter by the native *E. coli* RNA polymerase in our in vitro environment or is related to the stability of rhGH messenger RNA or, perhaps, the stability of the polypeptide.

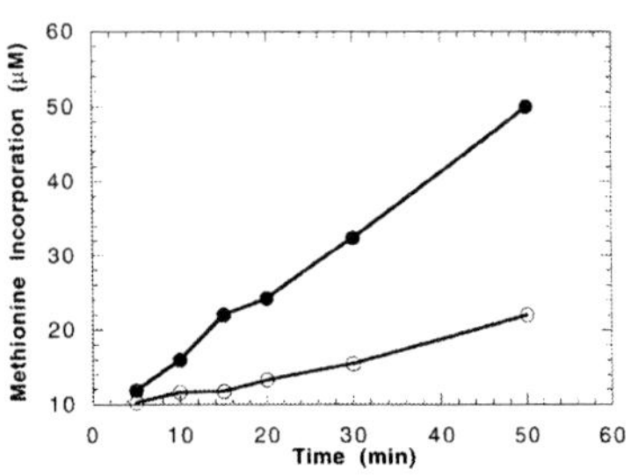

Figure 1. Time course of methionine incorporation while expressing TPO from a *trp* promoter with the improved system. Open circles denote methionine incorporation without the addition of plasmid DNA, and closed circles indicate results from reactions with the addition of pMP331.

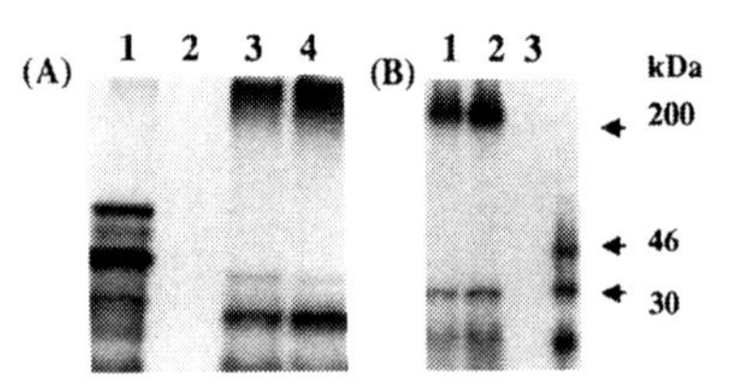

Figure 2. Fluorograms of gels showing products synthesized from pMP331 with the commercial kit (A) and the improved system (B). 5 μL of sample were withdrawn after a 50-min reaction (Figure 1) and boiled for 5 min with equal volumes of 2× tricine buffer before loading on to 10%–20% tricine gels. Gels were dried and processed for fluorography as described in Materials and Methods. The relative intensity of bands on these fluorograms does not correlate with quantitative estimates of proteins synthesized because the specific activity of ^{35}S-methionine used is different in each system. (A) Lane 1 shows results with the positive control plasmid pBEST*luc* supplied with the commercial kit. Lane 2 is a control without plasmid, and lanes 3 and 4 are two independent reactions with plasmid pMP331. (B) Lanes 1 and 2 represent the results of two independent reactions with plasmid pMP331. Lane 3 shows results without plasmid addition.

Table 3. Performance of the Improved System vs. the Original System (15) for Protein Synthesis from pRPHGH2 Encoding for Pre-β-Lactamase from its Native Promoter and for Pre-rhGH from the *tac*II Promoter

In Vitro System	Protein in Cell Extract (mg/mL)	Specific Protein Synthesis Rate (µg/mg protein/h)	Final Product Concentration (µg/mL)	
			pre-rhGH	pre-β-lactamase
Original system	7.9	0.009	0.04	negligible
Improved system	12.0	9.38	low	94

Until this is known, we would recommend the use of the *trp* rather than the *tac*II promoter.

In summary, we have developed an *E. coli*-based, in vitro protein synthesis system that reproduces in vivo-specific rates of heterologous protein synthesis. Specific rates of protein synthesis are 1000-fold higher than with the literature system used for comparison. The volumetric yields are also quite high, exceeding those of a commercially available batch system by more than two orders of magnitude. It is worth noting that the in vivo-specific rate of total protein synthesis for *E. coli* growing with a doubling time of 30 min is approximately 1.7 mg/mg soluble protein/h, considerably higher than the rate we have achieved in vitro and also much higher than the in vivo-specific expression rates for most heterologous proteins.

As a batch, coupled transcription/translation system with intrinsic ATP/GTP regeneration, our improved system offers simple operation. Also, the consumption of reagents such as amino acids and nucleotides is significantly reduced relative to published continuous in vitro systems (1,23) and is also lower than for the published semicontinuous system (10). We believe that our improved system can be scaled easily to larger volumes and offers significant advantages for the production of research proteins. In addition, approximate duplication of in vivo-specific rates of protein production is the first step towards developing an *E. coli*-based coupled transcription/translation system that mimics in vivo physiology. Other biochemical attributes of the cytoplasm, such as macromolecular crowding (25,26), ionic composition (2,14,20,21) and state of the plasmid DNA etc., also need to be considered in the search for true cytoplasmic replication.

ACKNOWLEDGMENTS

We thank Dot Reilly for sharing her data on in vivo expression of rTPO in *E. coli*. Antibodies for rhGH and rTPO were

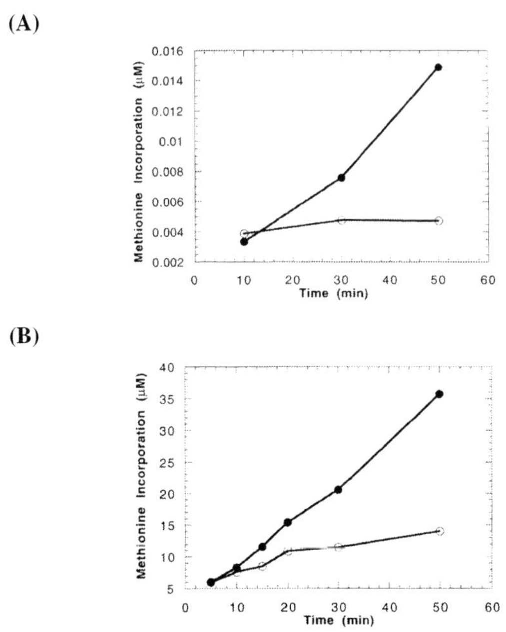

Figure 3. Time course of ^{35}S-methionine incorporation with pRPHGH2 using the original method (A) and improved system (B). Plot symbols are as described for Figure 1.

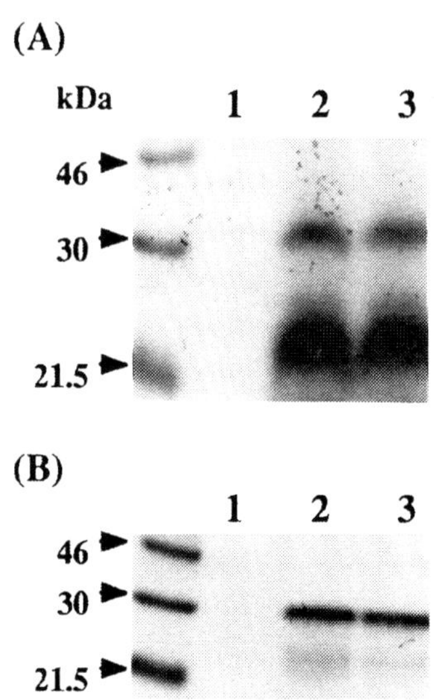

Figure 4. Fluorograms of gels showing different products synthesized from pRPHGH2 in the original (A) and improved systems (B). Lane 1 is a control with no plasmid and lanes 2 and 3 represent two independent reactions with plasmid pRPHGH2. Samples were processed as described for Figure 2. Again, relative intensity of bands on these fluorograms does not correlate with quantitative estimates of proteins synthesized because the specific activity of ^{35}S-methionine used is different in each system.

obtained from Jennifer Gaunce and Sylvia Wong, respectively; we acknowledge their help. Technical assistance provided by Susan Leung on fermentation procedures is highly appreciated. Lastly, we thank Brad Snedecor, Nancy McFarland and John Joly for stimulating discussions.

REFERENCES

1.**Baranov, V.I., I.Y. Morozov, S.A. Ortlepp and A.S. Spirin.** 1989. Gene expression in a cell-free system on the preparative scale. Gene *84*:463-466.

2.**Cayley, S., B.A. Lewis, H.J. Guttman and M.T. Record, Jr.** 1991. Characterization of the cytoplasm of *Escherichia coli* K-12 as a function of external osmolarity. J. Mol. Biol. *222*:281-300.

3.**Chang, C.N., M. Rey, B. Bochner, H. Heyneker and G. Gray.** 1987. High-level secretion of human growth hormone by *Escherichia coli*. Gene *55*:189-196.

4.**Chang, J.Y., N.C. McFarland and J.R. Swartz,** inventors. Genentech, assignee. Method for Refolding Insoluble, Misfolded Insulin-Like Growth Factor-I into an Active Conformation. US patent 5,288,931. 1994 February 22.

5.**Davis, J., D. Thompson and G.S. Beckler.** 1996. *In vitro* transcription/translation. Large scale dialysis reactions using *E. coli* S30 extract systems. Promega Notes *56*:14-21.

6.**de Boer, H.A., L.J. Comstock and M. Vasser.** 1983. The tac promoter: a functional hybrid derived from the trp and lac promoters. Proc. Natl. Acad. Sci. USA *80*:21-25.

7.**Ellman, J., D. Mendel, S. Anthony-Cahill, C.J. Noren and P.G. Schultz.** 1991. Biosynthetic method for introducing unnatural amino acids site-specifically into proteins, p. 301-336. *In* J.J. Langone (Ed.), Methods in Enzymology, Vol. 202. Academic Press, New York.

8.**Gangola, P. and B.P. Rosen.** 1987. Maintenance of intracellular calcium in Escherichia coli. J. Biol. Chem. *262*:12570-12574.

9.**Kigawa, T. and S. Yokoyama.** 1991. A continuous cell-free protein synthesis system for coupled transcription-translation. J. Biochem. (Tokyo) *110*:166-168.

10.**Kim, D.-M. and C.-Y. Choi.** 1996. A semicontinuous prokaryotic coupled transcription/translation system using a dialysis membrane. Biotechnol. Prog. *12*:645-649.

11.**Kim, D.-M., T. Kigawa, C.-Y. Choi and S. Yokoyama.** 1996. A highly efficient cell-free protein synthesis system from *Escherichia coli*. Eur. J. Biochem. *239*:881-886.

12.**Kudlicki, W., G. Kramer and B. Hardesty.** 1992. High efficiency cell-free synthesis of proteins: refinement of the coupled transcription/translation system. Anal. Biochem. *206*:389-393.

13.**Kudlicki, W., M. Mouat, J.P. Walterscheid, G. Kramer and B. Hardesty.** 1994. Development of a chaperone-deficient system by fractionation of a prokaryotic coupled transcription/translation system. Anal. Biochem. *217*:12-19.

14.**Leirmo, S., C. Harrison, D.S. Cayley, R.R. Burgess and M.T. Record, Jr.** 1987. Replacement of potassium chloride by potassium glutamate dramatically enhances protein-DNA interactions *in vitro*. Biochemistry *26*:2095-2101.

15.**Lesley, S.A., M.A.D. Brow and R.R. Burgess.** 1991. Use of *in vitro* protein synthesis from polymerase chain reaction-generated templates to study interaction of *Escherichia coli* transcription factors with core RNA polymerase and for epitope mapping of monoclonal antibodies. J. Biol. Chem. *266*:2632-2638.

16.**Liguri, G., C. Cecchi, A. Pieri, G. Raugei, M. Vecchi, A. Modesti, P. Nassi and G. Ramponi.** 1994. Expression of human acylphosphatase in *Escherichia coli* affects intracellular calcium levels. Biochem. Mol. Biol. Int. *34*:109-117.

17.**Muller, M. and G. Blobel.** 1984. *In vitro* translocation of bacterial proteins across the plasma membrane of *Escherichia coli*. Proc. Natl. Acad. Sci. USA *81*:7421-7425.

18.**Nishimura, N., Y. Kitaoka and M. Niwano.** 1995. Cell-free system derived from heat-shocked *Escherichia coli*. Synthesis of enzyme protein possessing higher specific activity. J. Fermentation Bioeng. *79*:131-135.

19.**Pratt, J.M.** 1984. Coupled transcription-translation in prokaryotic cell-free systems, p. 179-209. *In* B.D. Hames and S.J. Higgins (Eds.), Transcription and Translation, A Practical Approach. IRL Press, Oxford.

20.**Richey, B., D.S. Cayley, M.C. Mossing, C. Kolka, C.F. Anderson, T.C. Farrar and M.T. Record, Jr.** 1987. Variability of the intracellular ionic environment of *Escherichia coli*. J. Biol. Chem. *262*:7157-7164.

21.**Roe, J.H. and M.T. Record, Jr.** 1985. Regulation of the kinetics of the interaction of *Escherichia coli* RNA polymerase with the lambda P_R promoter by salt concentration. Biochemistry *24*:4721-4726.

22.**Spirin, S.A., V.I. Baranov, L.A. Ryabova, S.Y. Ovodov and Y.B. Alakhov.** 1988. A continuous cell-free translation system capable of producing polypeptides in high yield. Science *242*:1162-1164.

23.**Stiege, W. and V.A. Erdmann.** 1995. The potentials of the *in vitro* protein biosynthesis system. J. Biotechnol. *41*:81-90.

24.**Yamane, T., Y. Kawarasaki and H. Nakano.** 1995. *In vitro* protein biosynthesis using ribosome and foreign mRNA. An approach to construct a protein biosynthesizer. Ann. NY Acad. Sci. *750*:146-157.

25.**Zimmerman, S.B. and A.P. Minton.** 1993. Macromolecular crowding: biochemical, biophysical, and physiological consequences. Annu. Rev. Biophys. Biomol. Struct. *22*:27-65.

26.**Zimmerman, S.B. and S.O. Trach.** 1991. Estimation of macromolecule concentrations and excluded volume effects for the cytoplasm of *Escherichia coli*. J. Mol. Biol. *222*:599-620.

27.**Zubay, G.** 1973. *In vitro* synthesis of protein in microbial system. Annu. Rev. Genet. *7*:267-287.

Update to: *E. coli*-Based In Vitro Transcription/Translation: In Vivo-Specific Synthesis Rates and High Yields in a Batch System

James R. Swartz

Stanford University, Stanford, CA, USA

Although it has only been about two and a half years since this work was published, the project has experienced many changes as well as significant progress. My move to academia in the fall of 1998 allowed a more concentrated focus on cell-free protein synthesis. Just as importantly, the experiences described in this paper along with subsequent perplexing results convinced us that an empirical approach would not deliver ultimate success. We could not treat the cell extract and synthesis reagents as a "black box". Certainly, the system is very complex. The cell extract contains hundreds of enzymes. It was this complexity that intimidated us into the empirical approach. Ironically, the complexity is precisely the characteristic that limited the success of that empiricism.

Although we needed a more rational approach, we decided to continue our focus on batch reactions. This format is simple and is easily multiplexed, for example, in microplates. However, to simplify extract preparation and scale-up, we decided to go back to an S30 cell extract rather than the S135 extract described in the paper.

First, we began by examining key stoichiometric and kinetic characteristics. Noticing that a reaction system without a DNA template lost activity anyway, Dr. Dong-Myung

Kim (Roche Molecular Biochemicals) conducted experiments to see why. He found that both the energy source, phosphoenol pyruvate (PEP), and the amino acids were rapidly degraded even without protein synthesis (2). Although additional PEP and amino acids could restore activity, PEP hydrolysis accumulates inorganic phosphate. This phosphate inhibits further protein synthesis, apparently by complexing with Mg^{2+}. Further analysis showed that PEP is rapidly degraded even without nucleotides present, apparently by nonspecific phosphatases. We also detected the degradation of three amino acids: arginine, tryptophan and cysteine. Repeated additions of PEP, arginine, tryptophan, cysteine and magnesium prolonged protein synthesis, increasing yields to over 400 µg/mL, but we wanted a better system.

A projected cost analysis suggested that in addition to its instability, PEP is very expensive. In searching for an alternative, we noticed that the pyruvate oxidase isolated from such bacteria as *Pediococcus* sp. and *Lactobacillus* sp. produces acetyl phosphate. Ryabova et al. (5) had shown that acetyl phosphate could regenerate ATP for protein synthesis. Since pyruvate is much less expensive than PEP, we tested the combination of pyruvate kinase and acetate kinase and showed that pyruvate could indeed regenerate ATP as long as oxygen was supplied (1). Furthermore, even though the reaction was slower, yields increased. Since the phosphate was recycled and did not accumulate, a fed batch protocol with hourly additions could be extended for 6 hours, resulting in 700 µg/mL of chloramphenicol acetyl transferase (CAT). We were also able to scale-up this reaction by delivering oxygen by diffusion through a silicone tubing. In initial testing, a 5-mL reaction produced over 2 mg of CAT. Approximately 70% was soluble and fully active.

When we had studied amino acid degradation, we also noticed that alanine and aspartate/asparagine were accumulating. Known metabolic pathways suggested that pyruvate was being converted to PEP at a cost of 2 ATPs per PEP by the enzyme PEP synthetase (Pps). Since Narindrasorasak and Bridger (4) had reported that oxalic acid inhibits Pps, we found that the addition of 2.7 mM oxalic acid not only reduced aspartate and asparagine synthesis, but it also increased protein yields (3).

Meanwhile, in our search for better ATP regeneration methods, we realized that we could possibly use the system complexity to our advantage. We found that we could activate the glycolytic pathway by adding NAD and coenzyme A (CoA). We generated ATP from either pyruvate or glucose 6-phosphate using endogenous enzymes. Both substrates support substantial protein synthesis (D.-M. Kim and J.R. Swartz, manuscript submitted).

Since the immediate product of ATP regeneration by PEP is pyruvate, we found that the protein yield from PEP could be significantly improved by adding NAD and CoA. Finally, by combining this technique with oxalate addition and 2 mM initial amino acids, we have developed a simple batch reaction producing approximately 350 µg/mL in 1 h and a fed batch procedure producing 750 µg/mL of CAT in 3 h. Exogenous enzymes and oxygen transfer are not required, and the fed batch protocol requires only 2 additions.

These new studies have produced cell-free synthesis techniques that are both more productive and more robust, but we feel that we are just beginning to tap the full potential. In future work, we will modify the cell extracts to remove deleterious enzymes such as Pps and amino acid degrading enzymes. We are also developing fermentation protocols and extract preparation procedures to consistently provide highly active extracts. These advances will then form the foundation for further improvements in yields and productivities. Even more importantly, we are working on techniques for the simultaneous synthesis and folding of complex proteins and for expression from PCR products. I believe that this future work and work from other laboratories promises an important future for cell-free protein synthesis.

REFERENCES

1. **Kim, D.-M. and J.R. Swartz.** 1999. Prolonging cell-free protein synthesis with a novel ATP regeneration system. Biotechnol. Bioeng. *66*:180-188.
2. **Kim, D.-M. and J.R. Swartz.** 2000. Prolonging cell-free protein synthesis by selective reagent additions. Biotechnol. Prog. *16*:385-390.
3. **Kim, D.-M. and J.R. Swartz.** 2000. Oxalate improves protein synthesis by enhancing ATP supply in a cell-free system derived from *Escherichia coli*. Biotechnol. Lett. *22*:1537-1542.
4. **Narindrasorasak, S. and W.A. Bridger.** 1978. Probes of the structure of phosphoenol pyruvate synthetase: effects of a transition state analog on enzyme conformation. Can. J. Biochem. *56*:816-819.
5. **Ryabova, L.A., L.M. Vinokurov, E.A. Shekhovtsova, Y.B. Alakhov and A.S. Spirin.** 1995. Acetyl phosphate as an energy source for bacterial cell-free translation systems. Anal. Biochem. *226*:184-186.

CHAPTER 43

Construction of a Vector for Coupled In Vitro Transcription/Translation

Oren Ostersetzer and Zach Adam

The Hebrew University of Jerusalem, Rehovot, Israel

(Reprinted from BioTechniques 27:428-434, 1999)

In vitro expression of cloned genes is an important tool in molecular biology. For the expression of those genes, a variety of plasmids have been constructed. Subcloning cDNAs into those plasmid vectors, under the control of phage promoters such as SP6, T3 and T7, have allowed in vitro transcription by the respective RNA polymerase (2,8,9). For the translation step, the transcribed RNAs are incubated with in vitro translation systems, such as rabbit reticulocyte or wheat germ extract (1,6). Recently, commercially available coupled systems have been introduced, allowing the execution of transcription and translation reactions simultaneously.

The efficiency of the aforementioned translation systems is often variable. In addition to this variability, efficiency is correlated with the identity of the cloned gene itself. Some cDNAs are transcribed and translated very efficiently, whereas others serve as very poor templates. To enable transcription/translation of these poorly expressed cDNAs, we constructed a new vector, incorporating the 3′ and 5′ untranslated regions (UTRs) of a cDNA that is routinely highly expressed in in vitro systems. This vector was found to be efficient in directing the translation of cDNAs that are otherwise difficult to translate.

The cDNA encoding the Rieske Fe-S protein (RISP) from pea (7) (.Accession No. X63605), cloned into pSP64, was generously donated by Dr. J. Gray of Cambridge University, UK. Vectors, restriction enzymes and TNT® Coupled Wheat Germ Extract were obtained from Promega (Madison, WI, USA). Coupled transcription/translation reactions were performed using the TNT kit according to the manufacturer's instructions, using [^{35}S]Met as a radiolabel. For polymerase chain reaction (PCR), two oligonucleotides were synthesized: 5′-GGGGATATCCATGGTGGAAGCTCCTG-3′ (oligonucleotide 1) and 5′-GGGGATATCAGATCTTAGATTTCTTA-AACT-3′ (oligonucleotide 2). Oligonucleotide 1 was complementary to the 5′ UTR sequence and contained additional *Eco*RV and *Nco*I restriction sites in its 5′ end. Oligonucleotide 2 contained *Eco*RV and *Bgl*II sites, followed by a sequence complementary to the 3′ UTR sequence. These oligonucleotides were used for the construction of the pRUTR vector, as detailed in Figure 1.

Cloning of different cDNAs into pRUTR involved the following: (*i*) digestion of pSP72-DegP with *Xho*I and *Bam*HI and insertion of the fragment into the *Sal*I and *Bam*HI sites of

Table 1. Quantification of In Vitro Translation Products

	Met Incorporation	
Plasmid-Gene	**TCA Counts (× 10^3 cpm/µL)**	**TCA Counts/Met (× 10^3 cpm/µL)**
pSP64-RISP	210	42.0
pSP72-RISP	91	18.2
pGEM4-OE33	65	21.7
pSP72-SSU	94	15.7
pSP73-PC	158	31.6

pSP72-RISP was generated by digestion of pSP64-RISP with *Eco*RI and *Ssp*I, and the fragment containing the coding region was ligated into pSP72 digested with *Eco*RI and *Eco*RV. This resulted in the loss of two-thirds of the 3′ UTR of the RISP cDNA. The different cDNAs were transcribed and translated using the TNT kit with [^{35}S]Met as a radiolabel. At the end of the reaction, Met incorporation was assayed by TCA precipitation (6). RSIP, Rieske Fe-S protein; OE33, the 33-kDa subunit of the chloroplast oxygen-evolving complex; SSU, the small subunit of Rubisco; PC, the chloroplast electron carrier plastocyanin.

Cloning and Expression Vectors for Gene Function Analysis
Edited by Quinn Lu and Michael P. Weiner

pRUTR; (*ii*) digestion of pSP72-ΔTTD with *Sac*I and *Eco*RV and ligation into the respective site in pRUTR; (*iii*) digestion of pGEM®-T-FtsH-s (Promega) with *Nco*I and *Bgl*II and insertion into the respective sites in pRUTR; and (*iv*) digestion of pSP72-mRISP with *Sac*I and *Bgl*II and ligation into the respective sits in pRUTR.

The cDNA encoding the RISP is routinely transcribed and translated in vitro very efficiently. It translated 1.3–2.7 times better than other well-translated templates (OE33, SSU and PC; see Table 1). Both total trichloroacetic acid (TCA) counts, measuring the level of radiolabeled Met incorporated into protein and the TCA counts normalized to the number of Met residues in each translation product were higher for pSP64-RISP than for any other gene tested. However, subcloning of RISP into pSP72, which involved removal of most of its 3′ UTR, resulted in reduced translation, suggesting that the UTRs of RISP cDNA contribute to its high levels of translation.

If the RISP UTRs indeed promote high levels of translation, cloning cDNAs that are usually difficult to translate between these UTRs should increase their translatability. We thus constructed pRUTR, a vector containing a multiple cloning site (MCS) nested between the 5′ and 3′ UTRs of RISP, as described in Figure 1. Several chloroplast cDNAs that are usually translated with low yields were subcloned into pRUTR: (*i*) the thylakoid lumen protease DegP (4), (*ii*) OE33, a component of the oxygen-evolving complex, with a modified transit peptide (3), (*iii*) the soluble portion of the thylakoid-membrane protease FtsH (5) and (*iv*) the mature portion of RISP itself (7). Their translatability was then compared in their original and the pRUTR vectors. As shown in Figure 2, the four different cDNAs, when subcloned into pRUTR, translated significantly better than in their original vectors. DegP, which did not translate at all when cloned in pSP72, gave a clear, although faint band. The low intensity of this band can probably be attributed to the fact that this protein contains only a single Met residue. The three other templates, ΔTTD, FtsH-s and mRISP, translated in pRUTR 6–15 times better than in their original vectors, as can be seen in the quantification of the respective bands. These results supported our expectations that the UTRs of RISP cDNA contribute to its high levels of expression in vitro. Moreover, the constructed vector pRUTR appears to be suitable for the translation of cDNAs that are usually difficult to translate.

Three out of the four cDNAs that we tested (ΔTTD, FtsH-s and mRISP) resulted from manipulations of full-length cDNAs, which involved removal of UTRs. Thus, it appears that the removal of UTRs might result in a loss of translatability, not only of RISP, but also of other cDNAs. The new vector presented in this work might restore translatability to these cDNAs and increase that of other full-length cDNAs that are difficult to express in vitro for one reason or another.

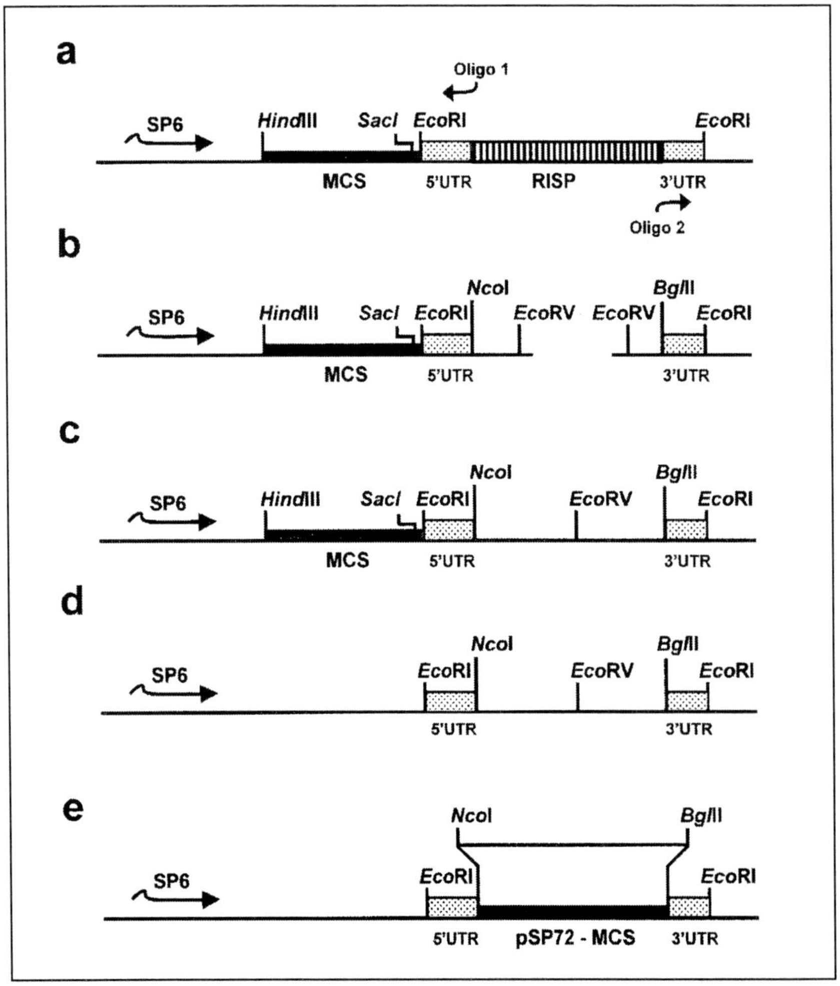

Figure 1. Construction of pRUTR. (a) Schematic presentation of the cDNA encoding RISP, flanked by its UTRs, subcloned in pSP64. Oligonucleotides 1 and 2 were used as primers in an "inverse PCR"-like reaction. (b) The PCR product started with the 5′ UTR of the cDNA, followed by the entire pSP64 vector and ended with the 3′ UTR of the cDNA. (c) Both ends were digested with *Eco*RV and ligated with T4 DNA ligase (Promega) to form a circular vector. (d) Removal of the original MCS, located upstream of the 5′ UTR, by digestion with *Hin*dIII and *Sac*I, was followed by blunt-ending and ligation. (e) The *Pvu*II-*Bgl*II fragment of the MCS of pSP72 was inserted between the *Eco*RV and *Bgl*II sites to form pRUTR. The relative location of the SP6 promoter is indicated.

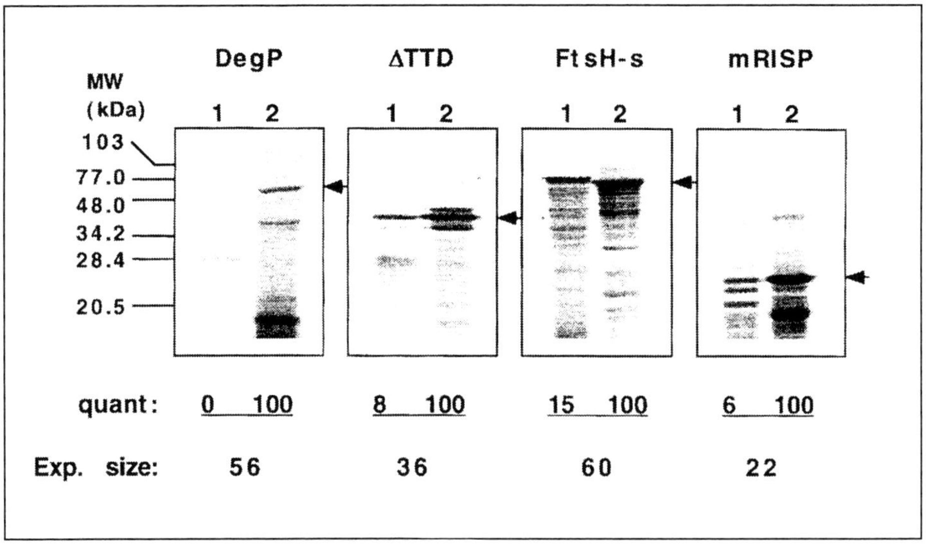

Figure 2. Translation of cDNAs subcloned into pRUTR and other vectors. cDNAs, subcloned into either their original vectors (lane 1) or pRUTR (lane 2), were transcribed and translated using the TNT kit. The translation products were resolved by sodium dodecyl sulfate polyacrylamide gel electrophoresis (SDS-PAGE) and exposed to a PhosphorImager® screen (Fuji, Kanagawa, Japan). The bands corresponding to the major translation products are indicated by arrows, and the relative quantities of each pair are indicated below the images. The expected sizes of the proteins (in kDa) are indicated below. DegP, the precursor protein of DegP protease (4); ΔTTD, the cDNA encoding OE33 in which the transit peptide was replaced (3); FtsH-s, the soluble portion of FtsH, protease (5); mRISP, the mature portion of RISP.

Because the UTRs incorporated into pRUTR and the cDNAs tested in this work are nuclear encoded, the described vector will be applicable to all plant nuclear-encoded cDNAs, and probably to other eukaryotic cDNAs as well.

ACKNOWLEDGMENT

We wish to thank Dr. J. Gray of Cambridge University, UK, for his generous donation of the RISP cDNA.

REFERENCES

1. **Erickson, A.H. and G. Blobel.** 1983. Cell-free translation of messenger RNA in a wheat germ system. Methods Enzymol. *96*:38-50.
2. **Gurevich, V.V., I.D. Pokrovskaya, T.A. Obukhova and S.A. Zozulia.** 1991. Preparative in vitro mRNA synthesis using SP6 and T7 RNA polymerases. Anal. Biochem. *195*:207-213.
3. **Halperin, T. and Z. Adam.** 1996. Degradation of mistargeted OEE33 in the chloroplast stroma. Plant Mol. Biol. *30*:925-933.
4. **Itzhaki, H., L. Naveh, M. Lindahl, M. Cook and Z. Adam.** 1998. Identification and characterization of DegP, a serine protease associated with the luminal side of the thylakoid membrane. J. Biol. Chem. *273*:7094-7098.
5. **Lindahl, M., S. Tabak, L. Cseke, E. Pichersky, B. Andersson and Z. Adam.** 1996. Identification, characterization, and molecular cloning of a homologue of the bacterial FtsH protease in chloroplasts of higher plants. J. Biol. Chem. *271*:29329-29334.
6. **Morch, M., G. Drugeon, W. Zagorski and A. Haenni.** 1986. The synthesis of high molecular weight proteins in the wheat germ translation system. Methods Enzymol. *118*:154-164.
7. **Salter, A., B.J. Newman, J.A. Napier and J.C. Gray.** 1992. Import of the precursor of the chloroplast Rieske iron-sulphur protein by pea chloroplasts. Plant Mol. Biol. *20*:569-574.
8. **Studier, F.W. and B.A. Moffat.** 1986. Use of bacteriophage T7 RNA polymerase to direct selective high-level expression of cloned genes. J. Mol. Biol. *189*:113-130.
9. **Studier, F.W., A.H. Rosenberg, J.J. Dunn and J.W. Dubendorff.** 1990. Use of T7 RNA polymerase to direct expression of cloned genes. Methods Enzymol. *185*:60-89.

Update to: Construction of a Vector for Coupled In Vitro Transcription/ Translation

Oren Ostersetzer and Zach Adam

The Hebrew University of Jerusalem, Rehovot, Israel

Unique restriction sites: 5′ *Nco*I, *Hin*dIII, *Pst*I, *Sal*I, *Acc*I, *Xba*I, *Bam*HI, *Sma*I, *Kpn*I, *Sac*I, *Cla*I, *Eco*RV, *Bgl*II 3′. All these sites are convenient for subcloning. *Sph*I and *Eco*RI sites are located after the *Hin*dIII and *Sac*I sites, respectively. However, such sites appear elsewhere in the vector and, thus, are inconvenient for subcloning.

Antibiotic resistance, Amp; promoter for transcription, SP6.

INDEX

C

D

E

F

G

H

M

N

O

P

R

T

U

V

W

Y